MACRO TO NANO SPECTROSCOPY

Edited by **Jamal Uddin**

Macro to Nano Spectroscopy
Edited by Jamal Uddin

Published by InTech

Janeza Trdine 9, 51000 Rijeka, Croatia

Publishing Process Manager Marina Jozipovic

Technical Editor Teodora Smiljanic

Cover Designer InTech Design Team

Additional hard copies can be obtained from orders@intechopen.com

Macro to Nano Spectroscopy, Edited by Jamal Uddin
p. cm.
ISBN 978-953-51-0664-7

For EU product safety concerns:
IN TECH d.o.o., Prolaz Marije Krucifikse Kozulić 3, 51000 Rijeka, Croatia,
info@intechopen.com or visit our website at intechopen.com.

Meet the editor

Dr Jamal Uddin is an Associate Professor of Coppin State University, Department of Natural Sciences, Baltimore, Maryland. He is the founder and director of the Coppin Nanotechnology Center. He has been teaching chemistry, physical science, environmental science and nanotechnology since 2005 in the Department of Natural Science. He is also working as a Visiting Scientist, Center for Fluorescence Spectroscopy, University of Maryland, School of Medicine, Department of Biochemistry & Molecular Biology, Baltimore. Dr Uddin's research interest is in the areas of Solar energy, Laser Photochemistry, Nanotechnology, Quantum dots, Single molecule imaging spectroscopy, Fluorescence correlation spectroscopy and Metal enhance fluorescence. Some of his significant publications focused on Electron Transfer Chemistry in Donor-Acceptor complexes. He is serving as an Editor-in-Chief of Emerging Scientific Resource (ESR) for the Research Journal of Engineering and Applied Sciences (RJEAS). He is a member of the American Chemical Society (ACS) and corresponding secretary of the American Association of University Professors (AAUP) Coppin State University, Maryland chapter. Dr Uddin was the recipient of the 2012 Daily Record Innovator of the Year for Maryland individuals who have had a positive effect and tremendous impact in Maryland, USA.

Contents

Preface

The book " Macro to Nano Spectroscopy" has been written to fulfill a need for an up-to-date text on spectroscopy. It has vast of applications, including study of Macro to Nanomaterial, remote sensing in terrestrial and planetary atmospheres, fundamental laboratory spectroscopic studies, industrial process monitoring, and pollution regulatory studies. The importance of spectroscopy in the physical and chemical processes going on in planets, stars, comets and the interstellar medium has continued to grow as a result of the use of satellites and the building of radio telescopes for the microwave and millimeter wave regions.

This book has wide variety of topics in spectroscopy including spectrophotometric apparatus and techniques. Topics covered absorption, emissions, scattering, causes of non-linearity, monochromators, detectors, photocells, photomultipliers, differential spectrophotometry, spectrophotometric titration, single beam, dual and multi wave length spectrophotometry, and diode array spectrophotometers.

The emphasis of this book serve both theorist and experimentalist. The authors present the models and concepts needed by theorists to understand the spectroscopic language spoken by molecules as translated by experimentalists and the tools and terminology needed by experimentalists to communicate with both molecules and theorists.

We (INTECH Publisher and Editor) owe our gratitude to the experts who gave much of their time and expertise in determining the scientific merit of the articles submitted to this special book.

Coppin State University professors Dr. Moses Wekesa, Dr. Hany F. Sobhi and Dr. Mintesnot Jiru helped greatly with the review of some chapters of this book and I would like to thank them very much for their critical work.

Jamal Uddin
Associate Professor of Natural Sciences
Director of Nanotech Center
Coppin State University, Baltimore, Maryland,
USA

Section 1

Atomic Absorption Spectroscopy

1

Atomic Absorption Spectroscopy: Fundamentals and Applications in Medicine

José Manuel González-López[1], Elena María González-Romarís[2], Isabel Idoate-Cervantes[3] and Jesús Fernando Escanero[4]
[1]*Miguel Servet University Hospital, Clinical Biochemistry Service, Zaragoza*
[2]*Galician Health Service, Clinical Laboratory, Santiago de Compostela*
[3]*Navarra Hospital Complex, Clinical Laboratory, Pamplona*
[4]*University of Zaragoza, Faculty of Medicine, Department of Pharmacology and Physiology, Zaragoza*
Spain

1. Introduction

Spectroscopy measures and interprets phenomena of absorption, dispersion or emission of electromagnetic radiation that occur in atoms, molecules and other chemical species. Absorption or emission is related to the energy state changes of the interacting chemical species which characterise them, which is why spectroscopy may be used in qualitative and quantitative analysis.

The application of spectroscopy to chemical analysis means considering electromagnetic radiation as being made up of discrete particles or quanta called photons which move at the speed of light. The energy of the photon is related to the wavelength and the frequency by Plank's constant ($h = 6.62 \times 10^{-34}$ J second) and the speed of light in a vacuum ($c = 3 \times 10^8$ m/s) according to the following equation (Skoog et al., 2001):

$$E = h\nu = hc/\lambda$$

The interaction of radiation with matter is produced throughout the electromagnetic spectrum which ranges from cosmic rays with wavelengths of 10^{-9} nm to radio waves with lengths over 1000 Km. Between both extremes, and from the shortest upwards, can be found the following: gamma rays, X-rays, ultraviolet rays (far, mid and near), the visible portion of the spectrum, infrared rays and radio microwaves. All radiations are of the same nature and travel at the speed of light, being differentiated by the frequency, wavelength and the effects they produce on matter (Skoog et al., 2008).

The Bouguer-Lambert-Beer Law is fundamental in molecular absorption spectrophotometry. According to this law, absorbance is directly proportional to the trajectory of the radiation through the solution and to the concentration of the sample producing the absorption although there are limitations to its application. The Law is only applied to monochromatic radiation although it has been demonstrated experimentally that the deviations with polychromatic light are unappreciable (Skoog & West, 1984).

According to the Bouguer-Lambert-Beer Law:

$$A = abc$$

when b (trajectory of the radiation) is expressed in cm and c (concentration of the substance) in $g.L^{-1}$, the units of a (absorptivity) are $L.g^{-1}.cm^{-1}$, or

$$A = \varepsilon bc$$

when b is expressed in cm and c in $mol.L^{-1}$, a (absorptivity) is called molar absorptivity and it is represented by the symbol ε and its units are $L.mol^{-1}.cm^{-1}$

The absorption of light (A) = log Po/P (= Optical density or extinction)

where:

Po: Incident radiation
P: Transmitted radiation
Absorptivity, a, is A/bc (= Coefficient of extinction)
Molar absorptivity, ε, is A/bc (= Coefficient of molar extinction)

The Bouger-Lambert-Beer Law is fulfilled with limitations in molecular absorption spectrophotometry (Skoog et al., 2008).

In 1927, Werner Heisenberg proposed the principle of uncertainty, which has important and widespread implications for instrumental analysis. It is deduced from the principle of superposition, which establishes that, when two or more waves cross the same region of space, a displacement is produced equal to the sum of the displacements caused by the individual waves. This is applied to electromagnetic waves in which the displacements are the consequence of an electric field, as well as to various other types of waves in which atoms or molecules are displaced. The equation $\Delta t \times \Delta E = h$, expresses the uncertainty principle, signifying that, for finite periods, the measurement of the energy of a particle or system of particles (photons, electrons, neutrons, protons) will never be more precise than $h/\Delta t$, in which h is the Planck´s constant. For this reason, the energy of a particle may be known as a zero uncertainty only if it is observed for an infinite period (Skoog et al., 2001).

In 1953, the Australian Physicist Alan Walsh laid the foundations and demonstrated that atomic absorption spectrophotometry could be used as a procedure of analysis in the laboratory (Willard et al., 1991). The theoretical background on which most of the work in this field was based is due almost entirely to this author (Elwell & Gidley, 1966).

1.1 Fundamentals

The spectra of atomic absorption of an element are made up of a series of lines of resonance from the fundamental state to different excited states. The transition between the fundamental state and the first excited state is known as the first line of resonance, being that of greatest absorption, and is the one used for analysis.

The wavelength of the first line of resonance of all metals and some metaloids is greater tan 200 nm, while for most non-metals it is lower than 185 nm. The analysis for these cases requires modifications of the optical systems which increases the cost of atomic absorption instruments.

In atomic absorption spectrometry, no ordinary monochromator can give such a narrow band of radiation as the width of the peak of the line of atomic absorption. In these conditions the Beer Law is not followed and the sensitivity of the method is reduced. Walsh demonstrated that a hollow-cathode, made of the same material as the analyte, emits narrower lines than the corresponding lines of atomic absorption of the atoms of the analyte in flame, this being the base of the instruments of atomic absorption. The main disadvantage is the need for a different lamp source for each element to be analysed, but no alternative to this procedure improves the results obtained with individual lamps.

The energy source most frequently used in atomic absorption spectroscopy is the hollow-cathode lamp.

1.2 Types

The field of atomic absorption spectroscopy (AAS) includes: flame (FAAS) and electrothermal (EAAS or ETAAS) atomic absorption spectroscopy (Skoog et al., 2008). The base is the same in both cases: the energy put into the free atoms of the analyte makes its electrons change from their fundamental state to the excited state, the resulting absorbed radiation being detected. However the fundamental characteristic of the FAAS is the stage of atomization which is performed in the flame and which converts the analyte into free atoms, whereas in the EAAS the stage of atomization goes through successive phases of drying, calcination and carbonization, and it is not required to dissolve the sample in the convenient matrix as occurs with the FAAS (Skoog et al., 2008; Vercruysse, 1984).

1.2.1 Flame atomic absorption spectroscopy

Prior steps to the stage of atomization in flame are the treatment of the sample, dissolving it in a convenient matrix, and the stage of pneumatic nebulisation. In FAAS the stage of atomization is performed in flame. The temperature of the flame is determined by the fuel/oxidant coefficient. The optimum temperatures depend on the excitation and ionization potentials of the analyte.

The concentration of excited and non-excited atoms in the flame is determined by the fuel/oxidant coefficient and varies in the different regions of the flame (Willard et al., 1991).

1.2.2 Electrothermal atomic absorption spectroscopy

The electrothermal atomic absorption spectrophotometer has three part: the atomizing head, the power source and the controls for feeding in the inert gas. The atomizing head replaces the nebulising-burning part of the FAAS. The power source supplies the work current at the correct voltage of the atomizing head. The computer control ot the atomizing chamber ensures reproducibility in the heating conditions, establishing a suitable profile of temperatures in the heating scale from environmental temperature to that of atomization so that the successive stages of drying, calcination and carbonization the sample must go through are those required. The working temperature and the duration of each stage of the electrothermal process must be carefully selected taking into account the nature of the analyte and the composition of the matrix of the sample. The control unit which measures and controls the flow of an inert gas within the atomizing head is designed to avoid the destruction of the graphite at high temperatures due to oxidation with the air.

One variant of the graphite oven is the carbon bar atomizer.

The main advantages of electrothermal atomic absorption are:

a. high sensitivity (absolute quantity of analyte of 10^{-8} to 10^{-11} g);
b. small volumes of liquid samples (5–100 µL);
c. the possibility of analysing solid samples directly without pretreatment and,
d. low noise level of the oven (Willard et al., 1991).

2. Applications of atomic absorption spectroscopy in medicine

Atomic absorption spectroscopy is a sensitive means for the quantification of some 70 elements and is of use in the analysis of biological samples (Skoog & West, 1984). FAAS allows the detection of Ag, Al, Au, Cd, Cu, Hg, Pb, Te, Sb and Sn with great sensitivity (Taylor et al., 2002). For most elements, the EAAS has lower detection limits than the FAAS. The incorporation of the new technology in the Laboratory of Clinical Biochemistry opened the possibility of approaches which had been unthinkable until then. For many of them it meant a reinforcement in the central position they held in hospital research. For those which incorporated spectroscopy it meant the possibility of new diagnostic, therapeutic and toxic controls.

In this chapter, a series of research studies are presented as example of the above mentioned. Thus, with respect to the Sr, refer to section 2.1, the first paper deals with the discrimination factor between Ca/Sr in absorption intestinal mechanisms. Afterwards, the different behavior of these metals in the binding to serum proteins is studied. And finally, the possibility of a hormonal regulation mechanism of serum levels of this element is evaluated, given the similarity of its biological behavior with the Ca.

The quantification of element bound to protein derived towards the direct applications for the study of medical problems such as the distribution of Zn in the acute and chronic overload and de Zn and Cu in the serum proteins in myocardial infarction.

Later, in section 2.2, a specific problem resulting from industrial development among other causes, threatening a part of the population –Pb poisoning– was tackled, analysing the serum and urine concentration of Pb and the hem biomarkers. This example is particularly useful not only because of what the technology meant for the diagnosis and control of this disorder but also because it has allowed to observe how the levels of this element in our city and, in general, in the West has declined over the years.

Finally, in section 2.3, an actual research is included: the design of a new strategy or approach possibility in the knowledge of the physiopathology of different neurological conditions based on the concentration of certain trace elements in the CSF, as well as of other parameters such as the cellularity, the proteins concentration, etc.

2.1 Strontium

This first group of papers serves as a model to analyze how this new technique allows evolving from the specific research problems (intestinal absorption, transport of element bound to proteins, etc.) to applications in medical pathology, such as the binding of Zn and Cu to plasma proteins after myocardial infarction.

2.1.1 Introduction

The disintegration of uranium and plutonium atoms in atomic explosions provokes the appearance of a series of elements with maxima in atomic weight of around 90 and 140. The isotopes of heavier atomic weight (140) fall in the area of the explosion while those of lower atomic weight (90) enter the troposphere and stratosphere. The particles which enter the troposphere spread out forming a gigantic belt around the area and are later deposited in local rain. Those others which reach the higher zones - the stratosphere - can be disseminated in over wide areas.

Atmospheric and tropospheric precipitation follow more or less quickly but the contaminants in the stratosphere may take many years before falling into the troposphere and being deposited in zones of greater rainfall.

Among the elements thrown into the troposphere and stratosphere are found those of the first peak of atomic weight (about 90), with two of the artificial isotopes of Sr, ^{89}Sr and ^{90}Sr, with different half lives. In particular, the half life of ^{90}Sr is 28.79 years.

The Sr deposited by the rain together with the Sr present in nature is absorbed by plants through the roots and that which is deposited on the leaves may also be absorbed. From here it enters the human organism, either directly by consuming the plants or, indirectly, by eating the animals which have eaten them.

Once the Sr has entered the organism it is carried in the blood to the cartilage and bone, choice sites for bonding. Far lower rates are found in other tissues.

In face of this threat, the analyses of animal milk for human consumption as markers of radioactive contamination and strategies to prevent the uptake (intestinal absorption) or to facilitate the elimination of ^{90}Sr from the bone once it has bonded are priority research into Sr domain (Escanero, 1974).

2.1.2 Development

2.1.2.1 A curiosity in biological barriers: discrimination between strontium and calcium

It is assumed axiomatically that biological organisms use Sr less effectively than Ca, which means that they discriminate against Sr in favour of Ca. This may be expressed in another manner by the concept "Strontium-Calcium Observed Ratio" (OR), the value of which is lower than 1. Comar et al. (1956) introduced the term to denote the overall discrimination observed in the movement of the two elements from one phase to another under steady-state conditions. The term OR denotes the comparative rates of Sr and Ca in balance between a sample and its precursor and is defined as:

$$OR= (Sr/Ca)\ sample/(Sr/Ca)\ precursor.$$

More precisely, the OR can be defined as the product of a number of 'discrimination factors' (DF), each of which is a measure of the extent to which the physiological process to which it refers contributes to the overall discrimination.

In this line it is shown one of the first studies which aimed to ascertain at what intestinal level the Sr/Ca discrimination takes place. Vitamin D_3, 25-hydroxy-cholecalciferol (25-0H-CC) and 1,25-dihydroxy-cholecalciferol (1,25 $(OH)_2$-CC) were administered to rats. The apparent and real intestinal absorption were analysed at those moments when their activity was maximum. Different concentrations of Sr were used and it was concluded that the discrimination occurred in the passive rather than in the vitamin-dependent transport (Escanero et al., 1976).

2.1.2.2 Hormonal regulation of strontium

From a general point of view, this research with Sr was designed to find the hormonal regulatory mechanisms in order to be able to act on them, provoking or forcing its elimination. This approach was based on several facts: a) The chemical similarity between Ca and Sr and their common participation in certain physiological processes lead one to think that both elements share a hormonal regulatory mechanism; b) The contribution of Chausmer et al. (1965) who had demonstrated the existence of specific action of calcitonin (CT) on the distribution of Zn, contrary to that exercised in the distribution of Ca, reducing the levels of this element in different tissues -thymus, testicles- seemed guiding; and c) Comar's idea (1967) that the plasmatic levels of Ca could participate in the regulation of Sr metabolism was particularly attractive. Taking these facts into account and with the possibility of a shared hormonal regulation for different trace elements (Zn and Sr among others), with specific properties for each one, the first steps in this approach were addressed to finding out more about serum Sr.

In the first study (Escanero, 1974), Comar's idea was proved since, at the same time as the concentration of Sr in bovine blood increased, so did that of Ca in total proteins, the contrary phenomena occurring with inorganic phosphorus and Ca in albumin.

In a later study (Alda & Escanero, 1985), the association constant and the maximal binding capacity for Ca, Mg and Sr to human serum proteins taken as a whole were determined. For Sr, a maximal binding capacity of 0.128 mmol/g of proteins and the association constant (K_{prs}) of 49.9 ± 16 M^{-1} were obtained. These values were not far from those of Ca (0.19 mmol/g of proteins and 55.7 ± 18 M^{-1}).

Later, Córdova et al. (1990) studied the regulating hormones of Ca and the effects of the administration of parathormone (PTH) and CT in rats and after thyroparathyroidectomy (TPTX) were analyzed. The PTH excess and defect (TPTX treated with CT + T_4) showed plasmatic increases in Sr. However, CT excess provokes decreases while the defect (administration of PTH + T_4 to TPTX rats) causes increases. Consequently, CT may be the hormone that plays a regulating role in the plasmatic Sr concentrations.

In this line, a study (Escanero and Córdova, 1991) was conducted in order to known the effect of the administration of glucagon on the serum levels of the alkaline earth metals since the phosphocalcic response to glucagon was already known in different animal species and it had been reported that, in mammals, glucagon triggered the release of CT by the thyroid gland. The study was completed with the analysis of the changes induced in these metals by the administration of CT. The effect (reduction) was observed two hours after administration and in daily administration the effect peaks on the 3rd day and CT significantly reduces the serum levels of these metals up to the 3rd day of treatment, just when the glucagon effect is highest.

2.1.2.3 Continuing with serum proteins

While the alkaline metals hardly bond in proteins and the alkaline earth metals do so in a proportion of 50% or slightly less, the trace elements do so almost completely. Zinc is deposited in a proportion of more than 99% and the proteins involved with albumin and the alpha$_2$ macroglobulin (α_2MG). The α_2MG bound tightly to the metal is responsible for 30-40%. The aminoacids are only responsible for about 2% of bonded Zn (Giroux & Henkin, 1972). The serum levels of Zn vary in different pathological symptomatologies and in relation to the amount of exercise. Several studies have shown these variations are related to the percentage of the element bound to albumin and another one established this relation between total serum Zn and α_2MG-bound Zn, in athletes after exercise (Castellano et al., 1988). This last study was aimed at analysing the variations in the serum levels of Zn after acute and chronic overload of this element and verifying whether these variations may be correlated to changes in the percentage of the element bound to albumin, α_2MG or both.

After a single intragastric administration of 0.5 mL of a solution containing 1000 ppm of Zn, the levels of this metal increased significantly ($p<0.01$) in serum 30 min after the beginning of the experiment, reaching maximum values at one hour and returning to normal levels 24 h later. It should be noted, however, that 8 h after administration the increases were no significant ($p>0.05$). The concentration of Zn bonded to albumin varied in parallel to the total serum levels of Zn. The values of bonded Zn in the globulins α_2MG also varied but the values returned to basal concentrations 4 h after the beginning of the experiment.

The chronic overload was performed with different groups who underwent a daily intragastric dosis of 0.5 mL of the same solution used for the acute overload. Different group of animals were sacrificed on days 7, 14, and 30 after the beginning of the experiment, with the aim of collecting values at different times of the overload. The control group and one other were used for recuperation and were kept for ten days with no treatment after day 30. Chronic overload of Zn caused significant increases in the serum levels of Zn throughout all the experiments. The highest values were found on the 14th day. The amount of Zn bonded to albumin varied in parallel with the total serum Zn; however, the concentrations of Zn bonded to globulins (α_2MG) showed a significant decrease ($p<0.05$) on the seventh day, increasing significantly ($p<0.01$) on the 14th day and the 30th day and returning to normal values 10 days after Zn overload was interrupted.

The results of the acute overload suggest a correlation with the proportion of element bonded to the albumin and those of the chronic overload showed a rapid response of the albumin while the increases of element bonded to the α_2MG responded slowly and remained constant until the end of the experiment. However, at this time the total serum Zn and the Zn bonded to albumin decreased.

These results suggest that albumin may play a new physiological role by adjusting its binding capacity to the serum Zn levels.

In the same line the levels and distribution of serum Cu and Zn were studied in patients diagnosed with acute myocardial infarction from the day of admission to the Cardiovascular Intensive Care Unit until the 10th day following the attack (Gómez et al., 2000). The results obtained showed that Cu increased significantly after the 5th day after the myocardial infarction, while Zn decreased significantly ($p<0.01$) with relation to the control group after the 1st day, the lowest values being found on the 3rd day after the attack.

Later, the total serum Cu showed an excellent correlation with the Cu bonded to the albumin and to the globulins (ceruloplasmin), as well as with the concentration of both fractions of serum proteins. In contrast, the total serum Zn only presented this correlation with the Zn bonded to the albumin, but not with the Zn bonded to the globulin or the albumin concentration.

These findings suggest the existence of some kind of relationship between the two fractions of the element bonded to proteins, which is probably different for each metal.

A further step was taken when wanting to analyse the possible role of albumin in the uptake of Zn by erythrocytes (Gálvez et al., 2001). Zinc is incorporated to erythrocytes by several mechanisms: i) passive transport, ii) anionic exchanger, iii) amino acid transport and iv) especially in the efflux, the Zn^{2+}-Ca^{2+} exchanger. In accordance with the free-ligand hypothesis only the free fraction could be used for the erythrocyte uptake. The results showed a significantly higher uptake ($p<0.01$) of Zn in the absence than in the presence of albumin for equimolar concentrations in both cases. However, the uptake of Zn in an albumin-free medium with a similar free-Zn concentration to Zn ultrafiltrable (20%) to another with albumin, a significantly ($p<0.01$) greater Zn uptake was observed in the latter. The DIDS (4-4´-diidothiocyanatostilbene-2.2´-disulphonic acid), that inhibits an important fraction of the Zn bonded to the anion carrier, also triggered a greater inhibition in the uptake of Zn when the albumin was present. Consequently, it was suggested that albumin must be directly or indirectly involved in Zn capture, facilitating the processes of passive transport and anionic exchanger.

Other properties of the uptake of Zn by erythrocytes were published in previous reports (Galvez et al., 1992, 1996a, 1996b):

- high dependence on temperature (Zn uptake was almost eliminated at 4° C)
- dependence on the concentrations of external Na^+ and K^+ and
- the apparent dissociation constant for the fast uptake step (15 minutes) is 0.46 µM for a medium without albumin and 0.121 µM with human albumin. For physiological concentrations of Zn the value was 15,3 µM (unpublished data).

2.2 Study of the effects of lead in hem biosynthesis

These studies are shown as an example of integration of the analysis provided for Pb by FAAS and EAAS with the results of the biomarkers of hem obtained with other laboratory techniques.

2.2.1 Introduction

Lead produces interferences in the hem biosynthesis pathway (Cambell et al., 1977), inhibiting the enzymatic effects of ALA-Dehydratase (ALA-D, EC 4.2.1.24) in cytosol and coproporphyrinogen oxydase (EC 1.3.3.3) and ferrochelatase (EC 4.99.1.1), both in mitochondria (González-López, 1992).

In the exposed organism, Pb produces affection of the target organs and critical effects which are characteristic of the disease known as saturnism since the days of Ancient Rome. Lead poisoning is diagnosed in the clinic and is shown by the analysis of Pb in blood. However, the concentration of Pb in blood depends on the metabolic condition of the individual (pH of the internal medium, bone activity, etc.) as well as on interactions with other metals such as Ca, Fe, Zn, Cu and Mg, among others. In order to ascertain the intensity and degree of affection of the Pb intoxication, it is necessary the study the so-called biomarkers of Pb exposure and poisoning. The most frequently analysed are the enzyme ALA-D in erythrocytes, protoporphyrin IX in erythrocytes, 5-aminolevulinic acid (5-ALA) in urine and the coproporphyrins in urine (Meredith et al., 1979).

The extreme sensitivity of ALA-D to divalent Pb ions has resulted in the measurement of its activity, as an indirect measurement of Pb in human blood (Berlin et al., 1977). Of all the enzymes involved in the hem biosynthesis pathway, it is the one which has been most studied due to the inhibiting effect that Pb has on its activity and the practical importance of the measurement of the enzymatic activity of ALA-D is considered to be of interest as a bioanalytic marker of environmental exposure to Pb. This has been assisted by the development of a method which has been standardised at the proposal of the executive council of the European Union. Hemberg & Nikkanen (1972) have published an extensive report on the biological meaning of ALA-D inhibition and its use as an exposure test.

As well as because of the effect of Pb, the activity of ALA-D may be also reduced by the effect of ethanol in alcoholics and by carbon monoxide in smokers although, in both cases, the reduction of activity is slight. In this line, porphyria of Doss is a recessive autonomous hereditary disease produced by the alteration of the gen which codifies the synthesis of ALA-D located in allele q34 of chromosome 9. It is a rarely presented porphyria, characterised by a deficit of ALA-D. In the homozygote form, a great reduction of ALA-D activity is observed in erythrocytes (2% of the control mean value), while in the heterozygotes, the enzymatic activity of ALA-D is reduced to 50%, being asymptomatic but especially sensitive to the toxic effect of Pb even with scarcely increased levels of Pb. The improvement of environmental and working conditions as well as the use of unleaded petrol, has led to the reduction of the concentration of Pb in blood in the population, a phenomenon observed over the last twenty years (González-López, 1992).

Taking into account the precedent facts, this section will analyse the effects of Pb poisoning in the biomarkers of hem studied in order to evaluate the recovery in the post-treatment with $CaNa_2EDTA$ as chelating agent.

2.2.2 Development: Subjects and methods

Subjects: The study of the biomarkers of Pb poisoning was performed in 377 adults (30-65 years old) and in 36 healthy children aged between 6 and 14. The adults were distributed as follows: 325 healthy (control group) and 52 cases of Pb poisoning: 24 severe, 15 slight and 13 treated with chelating agents. The children were another control group.

For inclusion in the group of healthy people, control groups, were required to:

a. No symptoms of lead poisoning and other diseases,
b. No changes in routine biochemical parameters and
c. Normal values of biomarkers characteristic of lead poisoning.

The group of the 13 patients treated with chelating agents were integrated as patients with severe poisoning and they were treated with $CaNa_2$-EDTA at doses of 50-75 mg/Kg body weight per day for five days, not exceeding the amount of 500 mg. It may be given an additional set after two days of interruption. This treatment requires hospitalization and clinical management with special attention to controlling renal function due to its nephrotoxicity.

Methods: The parameters analyzed were: Pb in blood and urine and the biomarkers of hem characteristic of Pb poisoning, ALA-D and protoporphyrin IX in blood and 5-ALA and coproporphyrins in urine.

The methods used were as follows:

Lead was analyzed in heparinized whole blood and 24-hour urine collection in container without additives. The first results of Pb were performed in both blood and urine by FAAS, using a modification of the method of Hessel (1968) by extraction into n-butyl acetate of the complex formed by the Pb with dithiocarbamate ammonium pyrrolidine (Pb-APDC).

Subsequently, the latest determinations were obtained by EAAS with Zeeman correction spectrometry, given the recent introduction of this technique to the laboratory. Blood was used diluted to 0.2% nitric acid (Pearson & Slavin, 1993).

Biomarkers: ALA-D of erythrocytes (in total blood with heparine) was determined applying the method of the European Standards Committee (Sibar Diagnostici) (Berlin & Schaller, 1974; Schaller & Berlin, 1984).

Erythrocyte-free Protoporphyrin IX (in total blood with heparine) was performed applying the method of Piomelli (1977), (Sibar Diagnostici).

At present, both products are manufactured by Immuno Pharmacology Research (IPR) Diagnostics.

ALA/PBG in urine was determined by means of column chromatography. For analysis a modification of the method of Mauzerall & Granick (1956), manufactured by Bio-Rad, was applied.

Determination of porhyrins in urine was performed after the separation of uro- and coproporphyrins by means of ethyl acetate which extracts the coproporphyrins from the aqueous phase and later absorption of uroporphyrins from the aqueous phase with activated Al_2O_3. Finally, the uroporphyrins are extracted from the Al_2O_3 activated and the coproporphyrins from the ethyl acetate by means of HCl 1.5 N, performing a fluorimetric reading of the extracts obtained (Schwartz et al., 1960).

In cases of the increased excretion of porphyrins, it is important to analyze the porphyrin biosynthesis pathway, applying the following analytic methods:

Analysis of carboxylic acids of free porphyrins in urine, by means of HPLC/FD. The chemical structure of the porphyrins presents the natural property of being fluorescent compounds, which makes them detectable by spectrofluorimetry.

The application of high pressure liquid chromatography (HPLC), constitutes a valuable resource in research applied to the study of the porphyrin biosynthesis pathway.

The analysis of the carboxylic acids of free porphyrins does not require any prior treatment of the samples (urine or faeces) to be chromatographed. They are only passed through a 22 µm Millex-GS (Millipore) filter. If the quantity of porphyrins contained in the sample is very high, it will be diluted with the eluyent.

The method used allows us to obtain a type of chromatogram in which the separation can be observed and the identification performed of the carboxylic acids of free porphyrins, from the octacarboxylic porphyrins (8-COOH, uroporphyrins) followed by the heptacarboxylic porphyrins (7-COOH), hexacarboxylic (6-COOH) and pentacarboxylic (5-COOH) and the tetracarboxylic porphyrins (4-COOH, coproporphyrins). The dicarboxylic form (2-COOH, protoporfirin) is not excreted in urine (Meyer et al., 1980a; Meyer, 1985).

As standard the Porphyrin acids chromatographic marker kit is used (Porphyrin products, INC., CMK-IA).

Analysis of type I and II isomers of uroporphyrins and coproporphyrins in urine by means of HPLC/FD. According to the disposition the substitutes may adopt around the tetrapyrrol ring of the porphyrin molecule, there are only four possible types of uroporphyrinogens (I, II, III and IV), but in nature only the type I and III uroporphyrinogens exist and, consequently, decarboxylation will only produce coproporphyrinogens I and III.

The use of high pressure liquid chromatography (HPLC) with fluorimetric detection (FD) allows to separate and identify the isomer forms of type I and III uro- and coproporphyrins as metabolytes derived from the oxidation of the corresponding porphyrinogens, produced in the process of natural metaloporphyrin biosynthesis. With this aim, the application method described by Jacob et al. (1985) was used.

With this method, a chromotagram is obtained in which the separation and identification can be observed of the isomers of uroporphyrins I and III and coproporphyrins I and III which are eluated and detected in this order.

The following standards are used: Uroporphyrin fluorescence standard, Uroporphyrin I (Porphyrin products, INC. UFS-I), Uroporphyrin III octamethyl ester (Sigma), Coproporphyrin I (Sigma), Coproporphyrin fluorescence standard, Coproporphyrin III (Porphyrin products, INC. CFS-3).

Analysis of protoporphyrin-Zn of erythrocytes in blood.

The alteration of the enzymatic activity of ferrochelatase due to the inhibition of this enzyme by the effect of Pb produces an increase in the protoporphyrin IX concentration in erythrocytes. This increase of protoporphyrin IX produces and accumulation of Zn-protoporphyrin I due to the complexation formation of this porphyrin with Zn^{2+}.

The method described by Meyer et al. (1980) was used for the analysis of porphyrins in erythrocytes. They developed a procedure for the separation of porphyrins from erythrocytes in blood with HPLC/FD in reverse phase by formation of the ionic pair (Meyer et al., 1982).

The standards used were Coproporphyrin fluorescente Standard (Porphyrin products, INC. CFS-3, Logan, UTA, USA), Protoporphyrin fluorescent Standard (Porphyrin products, INC. PFS-9, Logan, UTA, USA) and Mesoporphyrin IX (Porphyrin products, INC. M 566-9, Logan, UTA, USA).

2.2.3 Development: Results

The first group of values (results not published) were obtained in an early study and show the concentrations of Pb in blood and urine, as well as the values of various biomarkers of the porphyrin biosynthesis pathway. Moreover, it was also included the results of a second study from some years later, presenting the normality values of Pb in blood as an update, observing the difference of Pb concentration with respect to the earlier study.

I.A Control group. The results of the control group for Pb in blood and in 24 hours urine are shown in Table I. In Table 2 are presented the values of ALA-D and protoporphyrin IX in erythrocytes for the same population and in Table III those of the hemoglobin concentration and protoporphirin IX/g of haemoglobin coefficient. In table 4 are indicated the values of 5-ALA and porphobilinogen (PBG) in 24 hours urine and in the last (table 5) are presented the results found in urine of 24h of uroporphyrins and coproporphyrins.

	Pb in blood μg/dL x ± SD	**Pb in 24 hours urine** μg/dL x ± SD
Men (n = 104)	17.72 ± 6.01	45.10 ± 39.85
Women (n = 61)	14.00 ± 3.86	35.78 ± 27.26
Adults (n = 165)	16.54 ± 5.49	40.96 ± 35.01
Children (n = 36)	14.58 ± 2.79	-

Table 1. Values (x ± SD) of Pb in blood and in 24 hours urine

	ALA-D U. of CEE/mL erythrocytes* x ± SD	**Protoporphyrin IX** μg/dL of blood x ± SD
Men (n = 104)	44.70 ± 13.94	28.92 ± 11.50
Women (n = 61)	50.41 ± 17.27	26.80 ± 9.66
Adults (n = 165)	46.81 ± 15.45	28.13 ± 10.87
Children (n = 36)	64.42 ± 13.39	28.53 ± 9.28

(*): Units of the European Standards Committee

Table 2. Values (x ± SD) of ALA-D and protoporphyrine IX in erythrocytes

	Hemoglobin g/dL of blood x ± SD	**Protoporphyrin IX** μg/g Hb x ± SD
Men (n = 104)	15.27 ± 1.54	1.92 ± 0.84
Women (n = 61)	13.39 ± 1.16	2.02 ± 0.78
Adults (n = 165)	14.57 ± 1.68	1.96 ± 0.82
Children (n = 36)	12.94 ± 0.71	2.19 ± 0.65

Table 3. Values of Hb in blood and protoporphyrin IX/g of Hb coefficient

	5-ALA mg/24 hours x ± SD	PBG mg/24 hours x ± SD
Men (n = 223)	2.96 ± 1.30	0.41 ± 0.42
Women (n = 102)	2.42 ± 1.17	0.35 ± 0.35
Adults (n = 325)	2.79 ± 1.28	0.39 ± 0.40
Children (n = 36)	2.84 ± 0.91	0.46 ± 0.43

Table 4. Normal values of 5-ALA and PBG in urine of 24 hours

	Uroporphyrins µg/24 hours x ± SD	Coproporphyrins µg/24 hours x ± SD
Men (n = 223)	11.09 ± 6.00	101.95 ± 55.46
Women (n = 102)	9.08 ± 4.95	64.34 ± 35.07
Adults (n = 325)	10.46 ± 5.76	90.15 ± 52.88
Children (n = 36)	4.18 ± 2.98	56.59 ± 34.03

Table 5. Normal values of uroporphyrins and coproporphyrins in 24 hours urine

The values of Pb in blood and urine (Table 1) for the different population groups studied are within the range of those reported by other authors of the time the study was conducted (Carton, 1985; Carton, 1988). Because of that environmental improvements and labor have reduced Pb concentrations in the environment, the serum concentration of Pb have been also reduced (Trasobares, 2010).

With regard to the values of the biomarkers of Pb poisoning in blood analyzed (Tables 2, 3, 4 and 5), all of them are within the range reported by other authors (Goldberg, 1972; Tomokuni, K. & Ogata, 1976; Campbell et al., 1977; Goldberg et al. 1978; Granick et al., 1978; Meredith et al. 1979; Sakai et al., 1982; Barbosa et al., 2005). Although the standard deviation values can be considered high for some parameters, this should not be attributed to the methodology used given the biological variability that is observed in the study population.

There have also been included the values for Hb in the blood and the ratio of protoporphyrin IX/g of Hb. This last value increases in the Pb poisoning (protoporphyrin increased and decreased hemoglobin) while in the protoporphyria the Hb did not decrease and consequently the ratio does not increase as much as in the Pb poisoning. Likewise the values of 5-ALA, PBG and porphyrins (uro-and copro-) are included in the study as they provide a more complete picture of potential changes in Pb poisoning.

I.B Pb poisoning. In table 6 and 7 are presented the statistical tests of comparison of means observed in large samples with independent data and their degree of significance, performed for each biomarker of Pb poisoning analysed in each group studied with respect to the control group. Specifically, in table 6 are presented the results in blood of Pb, ALA-D, protoporphyrine IX and Protoporphyrine IX/g Hb and in table 7 the results in urine (24 hours) of Pb, 5-ALA, PBG, uroporphyrins and coproporphyrins.

	BLOOD			
	Pb μg/dL x ± SD	**ALA-D U/mL x ± SD**	**Protoporphyrine IX μg/dL x ± SD**	**Protop. IX/Hb μg/g Hb x ± SD**
Control group (n = 165)	16.54 ± 5.49	46.81 ± 15.45	28.13 ± 10.87	1.96 ± 0.82
Severe pois.	101.04±58.03[b]	12.00 ± 5.56[b]	198.49 ± 89.51[b]	17.24 ± 8.74[b]
Slight pois.	41.60±15.86[b]	26.47 ± 20.50[b]	26.57 ± 92.15[b]	9.65 ± 7.35[b]
Post Treatment	40.92±18.74[b]	37.08 ± 14.20[a]	116.94 ± 106.21[b]	8.70 ± 8.26[b]

Concentration of Hb (g/ dL blood) in each group studied:
Control group: 14.57±1.68;
Severe poisoning: 11.46±1.98b
Slight poisoning: 13.59±1.46a;
Post treatment: 14.12±1.17

Degree of statistical significance: a ($p < 0.050$); b ($p < 0.001$);

Table 6. Values (x ± SD) of Pb, ALA-D, protoporphyrine IX and Protoporphyrine IX/ g Hb in blood of the indicated group

The above table shows that in both, severe and slight Pb poisoning, the increases in the concentration of blood Pb are associated with significant inhibition of ALA-D activity (Campbell et al., 1977; Goldberg et al., 1978; Sakai et al., 1982) and with significant increases of protoporphyrin IX concentrations (Goldberg et al., 1978; Meredith et al., 1979). After treatment the values still remain significantly altered, indicating that the patients need a new series of treatment, because of the guidelines therapy is performed during five days with breaks in which they carry out checks on biomarkers. The fact of treating patients with different series of treatment would explain the dispersion of the results.

	URINE				
	Pb μg/24h x ±SD	**5-ALA mg/24h x ±SD**	**PBG mg/24h x ±SD**	**Uroporphyrins μg/24h x ±SD**	**Coproporphyrins μg/24h x ±SD**
Control group (n = 325)	41±35	2.79±1.28	0.39±0.60	10.46±5.76	90.15±52.88
Severe pois.	361±189[b]	47.3±21.8[b]	2.21±1.84[a]	26.4±18.8[b]	1612.9±682.8[b]
Slight pois.	114±81[b]	4.89±1.96[b]	0.49±0.62	5.61±2.25	87.7±54.2
Post treatment	197±167[b]	5.43±3.52[b]	0.42±0.62	9.67±6.50	148.9±119.1[b]

Degree of statistical significance: [a] ($p<0.002$); [b] ($p<0.001$).

Table 7. Values (x ±SD) of Pb, 5-ALA, PBG, uroporphyrins and coproporphyrins in 24 hours urine

The identification and quantification of high levels of porphyrins in erythrocytes, mainly protoporphyrin IX and its chelated form, Zn-protoporphyrin, are essentials in the diagnosis of Pb poisoning and in erythropoietic porphyrias (Meyer et al., 1980b).

The results of the above table present significant differences in the Pb urinary elimination in Pb poisoning both severe as slight in relation to the control group. Moreover, increases in the elimination of all parameters (biomarkers) analyzed in severe poisoning, being the most important those of the 5-ALA and coproporphyrins. In contrast, in slight poisoning has only significantly increased the elimination of 5-ALA. After treatment of severe cases there is a significant decrease in excretion of 5-ALA and coproporphyrins, keeping the levels still increased. Of the above comments it can be seen that the most effective urinary biomarkers are the urinary elimination of 5-ALA and coproporphyrins (type isomeric III).

These results agree, in the literature review performed, with the studies published by numerous authors and refered in the Doctoral Thesis of González-López (1992) and others (Sakai, 2000; Gurer-Orhan et al., 2004). All of them showed the influence of Pb on the heme biosynthesis pathway and the effects produced in some of the enzymes which take part in the biosynthesis of porphyrins, phenomena demonstrated in clinical research, *in vitro* studies and in experiments on animals.

In the diagnosis and evolutionary control of Pb poisoning, the study of the biomarkers of the hem biosynthesis pathway is very efficient, even more so than that demonstrated by the concentration of Pb in blood and urine due to the susceptibility of these to hormonal influences as well as metabolic ones such as the pH of the internal medium and the activity of bone turnover. According to this, Pb in blood means the degree of uptake; Pb in urine, the degree of elimination; Pb in urine provoked by EDTA or some other chelating agent, the degree of accumulation; ALA-D, the degree of exposure and is directly related to Pb in blood; erythrocyte free protoporphyrin IX, especially Zn-protoporphyrin IX, the degree of intake and chronic evolution; and the coproporphyrins (coproporphyrin III) the severity of the poisoning.

II. In the second study was performed a review of the values of Pb in blood in the present. The values were obtained by EASS in a graphite furnace with correction of the Zeeman effect, in a sample of 156 individuals. The results obtained are shown in Table 8.

	Pb in blood µg/dL x ± SD
Men (n = 83)	3.51 ± 2.16
Women (n = 73)	2.29 ± 1.64
Adults (men + women) (n = 156)	2.94 ± 2.02

Table 8. Normal values (x ± SD) of healthy individuals of Pb concentration in blood

These results indicate a decrease of the concentration obtained in 1989 which was 13.17 ± 3.47 µg/dL, attributed to the improvement of environmental and working conditions as well as to the suppression of Pb as antiknock agent in petrol (Izquierdo-Álvarez et al., 1985; Trasobares, 2010).

In the review of results for 2008 (results not published), the values for ALA-D in the adult sample were 60.59 ± 16.49 (x ± SD), higher than those obtained in 1990: 46.81 ± 15.45 (x ± SD), which is logical if we take into account the fact that the activity of ALA-D in erythrocytes has a negative or inverse correlation with respect to the concentration of Pb in blood. In the 1990 study, Pearson's coefficient of linear correlation obtained was: $r = -0.568$ ($p<0.001$).

2.3 Study of some elements in cerebrospinal fluid: physiopathological evaluation

This study, still in progress, can serve as an example of using the atomic absorption spectrometry technique for the assistance, together with other techniques, in the clinical diagnosis of some diseases.

2.3.1 Introduction

The cerebrospinal fluid (CSF) fills the subarachnoid space between the arachnoid membranes and the pia mater called leptomeninges which protect the Central Nervous System -CNS- (encephalon and spinal cord). Seventy percent of the CSF is formed in secretory structures called choroid plexi and the remaining 30% is produced from the cerebral capillaries (Carpenter, 1985).

Although the composition of CSF is similar to a plasma filtrate, there are differences which indicate that the CSF formed is produced both by a process of filtration and by active secretion, an osmotic balance being observed between CSF and plasma. A similarity can be seen in the composition of CSF and the extracellular liquid of the nervous system, indicating an easy interchange between both compartments. The CSF and in the cerebral interstice are separated from the blood circulation by the hematoencephalic and hematocephalorachideal barriers which prevent the free passing of substances. This is why they are considered to be functional elements of protection of the nerve cells (Nolte, 1994). In comparison with plasma, the CSF contains a greater concentration of Na, Cl and Mg and a lower one of glucose, proteins, amino acids, uric acid, K, bicarbonate, Ca and phosphate (Guyton, 1990). These differences indicate that the CSF is produced by a mechanism of active secretion and varies according to the location of CSF extraction with regard to the structures it bathes.

Water passes from the stroma to the CSF following the concentration gradient produced by the ATPase-dependent carrier proteins, Cl, Ca and Mg (Nolte, 1994). Cellular metabolytes also enter the extracellular liquid from neurones and glial cells.

The CSF maintains an appropriate chemical environment for neurotransmission and removes metabolic products and substances which are harmful for the CNS.

2.3.2 Development

In this research, it has been studied the concentration of Ca, Mg, Zn, Cu, Fe and Mn in CSF in order to analyze their influence on the pathogeny of some neuropathies (González-Romarís et al., 2011).

The mineral chemical elements and the trace elements were analysed in the CSF extracted from 37 people (17 men and 20 women, between 27 and 73 years of age) who were considered to be healthy after performing a clinical and analytical study. They made up the control group.

In addition, analysis was made of the CSF of 136 individuals from the Services of Neurology, Neurosurgery and Emergency. The CSF was extracted by lumbar puncture in all cases.

The analysis of Ca, Mg, Zn, Cu, Fe and Mn was performed by flame atomic absorption using the corresponding hollow cathode for each metal. The wavelengths for the reading of the absorbance corresponding to each cation analysed were the indicated by the manufacturer.

The values (mean and standard deviation) obtained in the control group for each metal analysed were as follows: Ca (mg/dL), 4.95 ± 0.70; Mg (mg/dL), 2.74 ± 0.10; Zn (μg/dL), 17.40 ± 7.50; Cu (μg/dL), 15.70 ± 4.50; Fe (μg/dL), 13.10 ± 3.60; and Mn (μg/dL), 2.50 ± 0.70. These values agree with the findings published by other authors (Hazell, 1997; Kapaki et al., 1997; Levine et al., 1996).

With regard to the results obtained in the pathological CSF, significant increases were found ($p<0.05$) in the concentrations of Ca, Cu, Fe, Zn and Mn in the groups classified with cell and protein increase in CSF in comparison to the control group. It was also see that the significant increase of the Ca, Zn and Cu concentrations is greater in those groups which present a higher concentration of proteins, while the increase of Mn corresponds to the increase of cell count. With regard to magnesium, it was seen that the significant reduction of its concentration in relation to the control group corresponds equally both if the cell count or the protein concentration was increased.

Interest in analysing these metals in CSF is directed to explaining the pathogenesis of some dysfunctions of the CNS. Clinical and experimental studies reviewed in the literature confirm the influence that these metals have on the pathogeny of some CNS dysfunctions and diverse neuropathies.

In this way the Ca and Mg ions play an important role in the action of glutamate, which is one of the most important neurotransmitter of vertebrates in the brain. The receptor of glutamate N-methyl-D-aspartate (NMDA) can only be activated in certain conditions of depolarisation of the membrane (Johnson & Ascher, 1988). The Mg ion blocks the channel, not being permeable to the Ca ions. When the receptor of glutamate are activated, the receptor reduces its affinity for Mg and the channel becomes permeable, permitting the entry of Ca ions to the neurone, a phenomenon which has been related to memory and the learning process (Hammond & Tritsch, 1990; Thomson, 1986). A reduction has been found in the glutamate and Mg concentration in the CSF in schizophrenic patients (Levine et al., 1996).

Some studies have found an increase of Cu in serum and CSF and an increase of Mn in the spinal cord in Amyotrophic Lateral Sclerosis (ALS), which suggests that this metal plays a role in the pathogeny of this disease (Kapaki et al., 1997).

The clinical association of Pb poisoning and ALS with an increase of Pb in blood and bone has been reported (Kamel et al., 2006). It has been suggested that patients with polymorphism in the gene of the ALA-D enzyme might be more at risk of presenting ALS in exposure to Pb (Kamel et al., 2003). Other studies have found no association between exposure to metals and ALS (McGuire et al., 1997; Bergomi et al., 2002).

It has been suggested that, regarding Mn, the binding of this metal in the basal ganglions of the brain may contribute in the pathogeny of the symptomatology of hepatic encephalopathy (Kuliseusky & Puyols, 1992; Weissenborn, 1995; Noremberg, 1998).

Moreover, it has been demonstrated that Mn reduces the uptake of glutamate in cultivated astrocytes. The great capacity of astrocytes to accumulate Mn suggests that its uptake by these cells may play an important role in the development of Alzheimer's type II astrocytosis (Hazell, 1997; Aschner & Gannon, 1992).

In conclusion, this research line, still in progress, can be highly promising to clarify the pathogenesis of some brain conditions.

3. Conclusions

After incorporating the atomic absorption spectrophotometry (spectroscopy) to the hospital laboratories has been observed that the medical research has improved in these laboratories. In this chapter the highlighted technique has been presented through a few research examples with different metals.

The conclusions from the research with the metal studies have been the following:

a. Strontium:
   - The intestinal absorption discrimination between Ca and Sr takes place in the passive transport mechanism and not in the active one or of vitamin D dependent mechanism.
   - With respect to the hormonal regulation of the plasmatic Sr, the CT is the only hormone that caused consistent changes in the concentrations of this element.
   - The addition of Sr in vitro to equimolar concentrations with Ca to bovine serum forces the binding of this last element with the total serum proteins. An opposite phenomenon takes place when the experience is conducted with albumin bovine as a single protein.

b. Zinc and Copper:
   - The Zn bound to albumin varies more consistently than Zn bound to globulins after acute and chronic overload. This fact allows suggesting that the albumin could act as a buffer.
   - In the myocardial infarction, Cu in serum increases significantly after the 5th day after the heart attack, while the Zn in serum decreases from the 1st day; being the lowest values of Zn found on the 3rd day after the attack. The total Cu in serum showed an excellent correlation with the Cu bonded to the albumin and to the globulins (ceruloplasmin), while, the total Zn in serum only presents a positive correlation with the Zn bonded to the albumin.

c. Lead:
   - In the diagnosis and the control of the Pb poisoning, the study of the biomarkers of the hem biosynthesis pathway is very efficient, even more so than its concentration of Pb in blood and urine.
   - According to this, the Pb concentration in blood means the degree of Pb intake; the Pb concentration in urine, the degree of its elimination; the Pb concentration in urine after the administration of EDTA or some other chelating agent (Dimercaprol -BAL, British Anti-Lewisite- and penicilamine), the degree of its accumulation; the ALA-D activity, the degree of exposure and these ones are directly related to the levels of Pb in blood; the Zn-protoporphyrin IX concentration, the degree of intake and chronic poisoning evolution; and finally the coproporphyrins (coproporphyrin III) gives an indication of the severity of the poisoning.

- In recent years the concentration of Pb in blood has decreased significantly in the aragonese population.

d. Metals in cerebrospinal fluid (Ca, Mg, Zn, Cu, Fe and Mn):
   - The concentration of different metals jointly with the rates of cellular and protein concentration has been proved to be a useful tool for the understanding of the pathogenesis of some brain conditions.

4. References

Alda, JO. & Escanero, JF. Transport of calcium, magnesium and strontium by human serum proteins. *Rev. Esp. Fisiol.* 1985; 41: 145-150.

Aschner, M. & Gannon, M. Manganese uptake and efflux in cultured rat astrocytes. *J. Neurochem.* 1992; 58: 730-735.

Barbosa, F.; Tanus-Santos, JE.; Gerlach, RF. & Parsons, PJ. A Critical Review of Biomarkers Used for Monitoring Human Exposure to Lead: Advantages, Limitations, and Future Needs. *Environ. Health Perspect.* 2005; 113(12): 1669–1674.

Bergomi, M.; Vinceti, M.; Nacci, G.; Pietrini, V. & Bratter, P. Environmental exposure to trace elements and risk of amyotrophic lateral sclerosis: a population-based control study. *Environ. Res.* 2002; 89: 116-23.

Berlin, A. & Schaller, KH. European Standardized Method for the determination of delta-amino-levulinic acid dehydratase activity in blood. *Z. Klin. Chem. Klin. Biochem.* 1974; 12: 389-390.

Berlin, A.; Schaller, KH.; Grimes, H.; Langevin, M. & Trotter J. Environmental exposure to lead: analytical and epidemiological investigations using the European Standardized Method for blood delta-amino-levulinic acid dehydratase activity determination. *Int. Arch. Occup. Environ. Health* 1977; 39: 135-141.

Campbell, BC.; Brodie, MJ.; Thompson, GG.; Meredith, PA.; Moore, R. & Goldberg, A. Alterations in the activity of the enzymes of haem biosynthesis in lead poisoning and acute hepatic porphyria. *Clin. Sci. Mol. Med.* 1977; 53: 335-340.

Carpenter, MB. Neuroanatomía Humana de Strong y Edwin (5ª Edición, 2ª reimpresión). Buenos Aires: El Ateneo, 1985, 1-19.

Cartón, JA. Saturnismo: epidemiología y diagnóstico. *Med. Clin. (Barc)* 1985; 84: 492-499.

Cartón, JA. Saturnismo. *Med. Clin. (Barc)* 1988; 91: 538-540.

Castellano, MªC.; Soteras, F.; Córdova, A.; Elósegui, LMª. & Escanero, JF. Zinc distribution between protein serum ligands in rats: acute and chronic overload of zinc. *Med. Sci. Res.* 1988; 16: 1229-1230.

Córdova, A.; Soteras, V.; del Villar, V.; Elósegui, LMª. & Escanero, JF. Efecto de la tiroparatiroidectomía, la parathormona y la calcitonina sobre el estroncio plasmático en rata. *Rev. Esp. Fisiol.* 1990; 46(2): 139- 146.

Chausmer, AB.; Weiss, P. & Wallach, S. Effect of thyrocalcitonin on calcium exchange in rat tissues. *Endocrinology* 1965; 77: 1151-1154.

Comar, CL. In: "Strontium metabolism", Leniham, JMA; Loutit, JF; Martin, JH. Eds. New York: Academic Press, 1967; 17-31.

Comar, CL.; Wasserman, RH.; & Nold, MM. Strontium-Calcium Discrimination Factors in the Rat. *Proc. Soc. Exp. Biol., N.Y.*, 1956; 92(4): 859-863.

Escanero, JF. Inferencia del estroncio en el metabolismo del calcio. Tesis Doctoral. Departamento de Fisiología. Facultad de Medicina. Universidad de Zaragoza, 1974.

Escanero, JF.; Carre, M. & Miravet, L. Effets des différents métabolites de la vit. D_3 et de la concentration calcique sur l´absorption intestinale de strontium. *C. R. Soc. Biol.* 1976; 170: 47-53.

Escanero, JF. & Córdova, A. Effects of glucagon on serum calcium, magnesium and strontium levels in rats. *Miner. Electrolyte Metab.* 1991; 17: 190-193.

Elwell, WT. & Gidley, JAF. Atomic-Absorption Spectrophotometry (2nd edition). International Series of Monographs in Analytical Chemistry (vol. 6). Oxford : Pergamon Press, 1966.

Gálvez, M.; Elósegui, LMª.; Guerra, M.; Moreno, JA. & Escanero, JF. Zinc Exchange between erythrocytes and médium with and without albumin at different temperaturas. In: "Metal Ions in Biology and Medicine", Anastassoupoulos, J; Collery, P; Theophanides, T; Etienne, JC, eds. Paris: John Libbey Eurotext, 1992, 2: 89-90.

Gálvez, M.; Moreno, JA.; Elósegui, LMª. & Escanero, JF. Zinc uptake by human erythrocytes with and without serum albumin in the medium. *Biol. Trace Elem. Res.* 2001; 84: 45-56.

Gálvez, M.; Moreno, JA.; Elósegui, LMª. & Escanero, JF. Zinc uptake by human erythrocytes. 1. Effect of Na and K in medium at different and a medium at different temperatures. In: "Metal Ions in Biology and Medicine", Anastassoupoulos, J; Collery, P; Theophanides, T; Etienne, eds. Paris: John Libbey Eurotext, 1996a, 4: 218-221.

Gálvez, M.; Moreno, JA.; Elósegui, LMª. & Escanero, JF. Zinc uptake by human erythrocytes. 2. Effetc of the temperature on Zn-uptake sensitive to the stilbenes. In: "Metal Ions in Biology and Medicine", Anastassoupoulos, J; Collery, P; Thephanides, T; Etienne, JC, eds. Paris: John Libbey Eurotext,1996b, 4: 222-224.

Giroux, EL. & Henkin, RI. Competition for zinc among serum albumin and amino acids. *Biochim. Biophys. Acta* 1972; 273: 64-72.

Goldberg, A. Lead poisoning and haem biosynthesis. *Br. J. Haematol,* 1972, 23: 521-524.

Goldberg, A., Meredith, PA., Miller, S., Moore, MR. & Thomson, GG. Hepatic drug metabolism and haem biosynthesis in lead poisoned rats. *Br. J. Pharmacol.* 1978, 62: 529-536.

Gómez, E.; de Diego, C.; Orden, I.; Elósegui, LMª.; Borque, L. & Escanero, JF. Longitudinal study of serum cooper and zinc levels and their distribution in blood proteins alter acute myocardial infarction. *J. Trace Elements Med. Biol.* 2000; 14: 65-70.

González López, JM. Influencias del plomo en el metabolismo de las porfirinas. Tesis Doctoral. Resúmenes De Tesis Doctorales, Curso 1989-1990. Universidad de Zaragoza (España), Comisión de Doctorado, 1992, 65. I.S.B.N.: 84-7733-314-9.

González-Romarís, EMª.; Idoate-Cervantes, I.; González-López, JM. & Escanero-Marcén, JF. Concentration of calcium and magnesium and trace elements (zinc, cooper, iron and manganeso) in cerebrospinal fluid: A try of a pathophysiological classification. *J. Trace Elements Med. Biol.* 2011; 25 Supl.: S45-S49.

Granick, JL., Sassa, S. & Kappas, A. Some Biochemical and clinical aspects of lead intoxication. *Advan. Clin. Chem.* 1978, 20: 287-339.

Gurer-Orhan, H.; Sabır, HU. & Özgüneş, H. Correlation between clinical indicators of lead poisoning and oxidative stress parameters in controls and lead-exposed workers. *Toxicology,* 2004; 195(2-3): 147-154.

Guyton, AC. Anatomía y fisiología del sistema nervioso. Neurociencia básica, 2ª reimpresión. Buenos Aires: Editorial Médica Panamericana S.A., 1990, 75-84, 21-36 y 129-149. Edición original: Guyton, AC. Basic Neuroscience. Anatomy and Physiology. Philadelphia: W.B. Saunders, 1987.

Hammond, C. & Tritsch, D. Neurobiologie cellulaire. Paris: Doin Éditeurs, 1990, 439-462.

Hazell, AS. Manganese decreses glutamate uptake in cultured astrocytes. *Neurochem. Res.* 1997; 22: 1443-1447.
Hernberg, S. & Nikkanen, J. Effect of lead on delta-amino-levulinic acid dehydrase - A selective review - . *Pravoc Lék.* 1972; 24: 2-3.
Hessel, DW. A simple and rapid quantitative determination of lead in blood. *At. Absorpt. Newsletter* 1968; 7: 55.
Izquierdo-Álvarez, S.; Calvo-Ruata, MªL.; González-López, JM.; García de Jalón-Comet, A. & Escanero-Marcén, JF. The Need to Update Reference Values for Lead in Zaragoza, Spain. *Biol. Trace Elem. Res.* 2008; 123: 277-280.
Jacob, K.; Sommer, W.; Meyer, HD. & Vogt, W. Ion-pair high-performance liquid chromatographic separation of porphyrin isomers. *J. Chromatogr.* 1985; 349: 283-293.
Johnson, JW. & Ascher, P. The NMDA receptor and its channel modulation by magnesium and by glicine. In: Lodge, D, ed. Excitatory amino acids in health and disease. New York: John Wiley and Sons 1988: 143-64.
Kamel, F.; Umbach, DM.; Hu, H.; Munsat, TL.; Shefner, JM.; Taylor, JA. et al. Lead exposure as a risk factor For amyotrophiv lateral sclerosis. *Neurodegener. Dis.* 2006; 2: 195-201.
Kamel, F.; Umbach, DM.; Lehman, TA.; Park, LP.; Munsat, TL.; Shefner, JM. et al. Amyotrophic lateral sclerosis, lead, and genetic susceptibility: polymorphism in in test delta-aminolevulinic acid dehydratase and vitamin D receptor genes. *Environ. Health Perspect.* 2003; 111: 1335-9.
Kapaki, E.; Zournas, C.; Kanias, G.; Zambelis, T.; Kakami, A. & Papageorgiou, C. Essential trace elements Alterations in amyotrophic lateral sclerosis. *J. Neurol. Sci.* 1997; 147: 171-5.
Kuliseusky, J. & Puyols, J. Pallidal hyperintensity on magnetic resonance imaging in cirrotic patiens: Clinical correlation. *Hepatology* 1992; 16: 1382.
Levine, J.; Rapoport, A.; Mashiah, M. & Dolev, E. Serum and cerebrospinal levels of calcium and magnesium in acute versus remitted schizophrenic patiens. *Neuropsychobiology* 1996; 33(4): 169-72.
McGuire, V.; Logstreth, WT. Jr; Nelson, LM.; Koepsell, TD.; Checkoway, H.; Morgan, MS. et al. Occupational exposure and amyotrophic lateral esclerosis: a population-based control study. *Am. J. Epidemiol.* 1997; 145: 1076-88.
Mauzerall, D. & Granik, S. The occurrence and determination of δ-aminolevulinic acid and porphobilinogen in urine. *J. Biol. Chem.* 1956; 219: 435-446.
Meredith, PA.; Moore, MR. & Goldberg, A. Erythrocyte δ-aminolevulinic acid dehydratase activity and blood Protoporphyrin concentrations as indices of lead exposure and altered haem biosynthesis. *Clin. Sci.* 1979; 56: 61-69.
Meyer, HD. Porphyrins. In: Henschen, A; Hupe, KP; Lottspeich, F; Voelter, W, eds. High Performance Liquid Chromatography in Biochemistry. Weinheim (FRG): VCH Verlagsgessellschaft GmbH 1985; 445-479.
Meyer, HD.; Jacob, K. & Vogt, W. Rapid and Simple Direct Determination of Porphyrins in Urine by Ion-Pair Reversed-Phase High Performance Liquid Chromatography. *Journal of HRC & CC* 1980a; 85-86.
Meyer, HD.; Jacob, K. & Vogt, W. Ion-Pair-Reversed-Phase High-Performance Liquid Chromatographic Determination of Porphyrins from Red Blood Cells. *Chromatographia* 1982; 16: 190-191.
Meyer, HD.; Jacob, K.; Vogt, W. & Knedel, K. Diagnosis of porphyries by ion-pair high-performance liquid Chromatography. *J. Chomatogr.* 1980b; 199: 339-343.

Nolte, J. El Cerebro Humano. Introducción a la anatomía funcional. Primera Edición española. Madrid: Mosby/Doyma libros, 1994, pp. 33-75. Edición original: Nolte, J. The Human Brain. Third edition. Mosby - Year Book, Inc, MCMXCIII.

Norenberg, MD. Astroglial dysfunction in hepatic encephalopathy. *Metab. Brain Dis.* 1998; 13: 319-35.

Pearson, PJ. & Slavin, W. A rapid Zeeman graphite furnace atomic absorption spectrophotometric method for the determination of lead in blood. *Spectrochem. Acta* 1993; 48: 925-939.

Piomelli, S. Free erythrocyte porphyrins in the detection of undue absorption of Pb and Fe deficiency. *Clin. Chem.* 1977; 23(2): 264-269.

Sakai, T. Biomarkers of Lead Exposure. *Ind. Health* 2000; 38: 127–142.

Sakai, T., Yanagihara, S., Kunugi, Y. & Ushio, K. relationships between distribution of lead in erythrocytes in vivo and in vitro and inhibition of ALA-D. *Br. J. Ind. Med.* 1982, 39: 382-387.

Schaller, KH. & Berlin, A. Δ-Aminolaevulinate Dehydratase. In: Methods of Enzymatic Analysis (Volume IV, Third Edition). Bergmeyer, HU, Editor-in-Chief; Bergmeyer, J; Graβl, M; eds. Weinheim: Verlag Chemie GmbH, 1984; 363-368.

Schwartz, S.; Berg, MH.; Bossenmaier, I. & Dinsmore, H. Determination of porphyrins in biological materials. In: Methods of Biochemical Analysis, Glick, D., ed. New York: Interscience, 1960, Vol. 8, 221-293.

Skoog, DA. & West, DM. Análisis Instrumental (2ª Edición). México, D.F.: Nueva Editorial Interamericana, S.A. de C.V., 1984, 158-177 y 317-369.

Skoog, DA.; Holler, FJ. & Nieman, TA. Principios de Análisis Instrumental (5ª Edición). Madrid: McGraw-Hill/Interamericana de España, S.A.U., 2001, pp. 122-150 y 219-244.

Skoog, DA.; West, DM.; Holler, FJ. & Crouch, SR: Fundamentos de Química Analítica (8ª Edición, 2ª reimpresión). Madrid: Paraninfo, 2008, 719-723 y 870-880.

Taylor, A.; Branco, S.; Halls, D.; Patriarca, M.; & White, M. Atomic spectrometry update. Clinical and biological materials, foods and beverages. *J. Anal. Atom. Spectr.* 2002; 17: 414- 455.

Thomson, AM. A magnesium-sensitive post-synaptic potential in rat cerebral cortex resembles neuronal responses to N-methylaspartate. *J. Physiol.* 1986; 370: 531-49.

Tomokuni, K. & Ogata, M. Relationship between lead concentration in blood and biological response for porphyrin metabolism in Workers occupationally exposed to lead. *Arch. Toxicol.* 1976, 35: 239-246.

Trasobares, EM. Plomo y mercurio en sangre en una población laboral hospitalaria y su relación con factores de exposición. Tesis Doctoral. Madrid: Universidad Complutense, 2010. ISBN: 978-84-693-6339-3.

Vercruysse, A., ed. Techniques and Instrumentation in Analytical Chemistry, Volume 4. Evaluation of Analytical Methods in Biological Systems, Part B: Hazardous Metals in Human Toxicology. Amsterdam, The Netherlands: Elsevier Science Publishers, B.V. 1984.

Weissenborn, K. Pallidal lesion in patiens with liver cirrosis: Clinical and MRI evaluation. *Metab. Brain Dis.* 1995; 10: 219-231.

Willard, HH.; Merritt, LL.; Dean, JA. & Settle, FA. Jr. Métodos Instrumentales de Análisis. México, D.F.: Grupo Editorial Iberoamérica, 1991, 95-100 y 219-252.

2

Analysis of Environmental Pollutants by Atomic Absorption Spectrophotometry

Cynthia Ibeto[1], Chukwuma Okoye[2],
Akuzuo Ofoefule[1] and Eunice Uzodinma[1]
[1]Biomass Unit, National Center for Energy Research & Development,
University of Nigeria, Nsukka, Enugu State,
[2]Department of Pure and Industrial Chemistry, Faculty of Physical Sciences,
University of Nigeria, Nsukka, Enugu State,
Nigeria

1. Introduction

Environmental pollution as a result of man's increasing activities such as burning of fossil fuels and automobile exhaust emission has increased considerably in the past century due mainly to significant increases in economic activities and industrialization. Burning of fossil fuels and petroleum industry activities have been identified as primary sources of atmospheric metallic burden leading to environmental pollution. Several studies have shown that heavy metals such as lead, cadmium, nickel, manganese and chromium amongst others are responsible for certain diseases (Hughes, 1996). In general, heavy metals are systemic toxins with specific neurotoxic, nephrotoxic, fetotoxic and teratogenic effects. Heavy metals can directly influence behavior by impairing mental and neurological function, influencing neurotransmitter production and utilization, and altering numerous metabolic body processes. Systems in which toxic metal elements can induce impairment and dysfunction include the blood and cardiovascular, eliminative pathways (colon, liver, kidneys, skin), endocrine (hormonal), energy production pathways, enzymatic, gastrointestinal, immune, nervous (central and peripheral), reproductive and urinary that have lethal effects on man and animals. These diseases include abdominal pain, chronic bronchitis, kidney disease, pulmonary edema (accumulation of fluid in the lungs), cancer of the lung and nasal sinus ulcers, convulsions, liver damage and even death (Hughes, 1996).

Heavy metals get into the environment: water, soil, air and land through activities like intense agriculture, power generation, industrial discharges, seepage of municipal landfills, septic tank effluents e.t.c. Many authors have reported high levels of heavy metal ions in the soil, rivers and groundwater in different areas of Nigeria (Ibeto & Okoye, 2010a). To save the environment from further deterioration and also maintain sound public health, a strategy can be effectively utilized which is the use of organic materials such as municipal solid waste, agricultural waste and industrial waste to produce biogas. Biogas is a suitable alternative fuel which burns with similar properties to natural gas. Unlike natural gas, it is clean and has no undesirable effects on the environment. It is a mixture of gases consisting of around 60 to 70% of methane produced by the process of anaerobic digestion in a digester. The effluent of this process is a residue rich in the essential inorganic elements needed for healthy plant growth known as bio fertilizer, which when applied to the soil enriches it with no detrimental effects on the environment. Many authors have also reported the utilization of various wastes found in the environment, ranging from animal wastes, plant wastes to leaf litters and food wastes (Ofoefule et al., 2010; Uzodinma et al., 2011). It is also recommended that other alternative fuels such as bioethanol, which are becoming increasingly important not only because of the diminishing petroleum reserves, but also because of the environmental consequences of exhaust gases from petroleum fueled engines be made available for use in Nigeria. Good quality biodiesel fuel which is derived from triglycerides has attracted considerable attention during the past decade as a renewable, biodegradable and non-toxic fuel producing less particulate matter, hydrocarbons, aromatics, carbon-monoxide and soot emissions when burnt in the engines. Its production, marketing and use should therefore be highly encouraged as is the case in Europe, America and some other parts of the world.

Several spectroscopic methods have been used to monitor the levels of heavy metals in man, fossil fuels and environment. They include; flame atomic absorption spectrometry (AAS), atomic emission spectroscopy (AES), graphite furnace atomic absorption spectrometry (GFAAS), inductively coupled plasma-atomic emission spectroscopy (ICP/AES), inductively coupled plasma mass spectrometry (ICP/MS), x-ray fluorescence spectroscopy (XRFS), isotope dilution mass spectrometry (IDMS), electrothermal atomic absorption spectrometry (ETAAS) e.t.c. Also other spectroscopic methods have been used for analysis of the quality composition of the alternative fuels such as biodiesel. These include Nuclear magnetic resonance spectroscopy (NMR), Near infrared spectroscopy (NIR), inductively coupled plasma optical emission spectrometry (ICP-OES) e.t.c.

2. Sources of heavy metal pollution of the environment

2.1 Lead

Lead is a common industrial metal that has become widespread in air, water, soil and food. It is a naturally occurring metal that has been used in many industrial activities and therefore many occupations may involve exposure to it such as auto-mechanic, painting, printing, welding e.t.c putting the workers at risk of potential high exposure. In the atmosphere, lead exists primarily in the form of $PbSO_4$ and $PbCO_3$. Lead in paints and automobile exhausts are still recognized for its toxicity (Hughes, 1996). Episodes of poisoning from occasional causes such as imperfectly glazed ceramics (Matte et al., 1994), the use of medicines which may contain as much as 60% lead available from Asian healers and cosmetic preparations, may affect any age group and cases may present as acute emergencies (Bayly et al., 1995). The main source of adult exposure is food, air inhalation accounts for 30% and water of 10% (John et al., 1991).

Some individuals and families may be exposed to additional lead in their homes. This is particularly true of older homes that contain lead based paint. In an attempt to reduce the amount of exposure due to deteriorating leaded paint, the paint is commonly removed from homes by burning, scraping or sanding. These activities have been found to result to at least temporarily, in higher levels of exposure for families residing in those homes. Special population at risk of high exposure to tetra ethyl lead produced by reacting chloroethane with a sodium–lead alloy, includes workers at hazardous sites and those involved in the manufacturing and dispensing of tetraethyl lead (Gerbeding, 2005a). The production process is illustrated with the equation below.

$$4\ NaPb + 4\ CH_3CH_2Cl \rightarrow (CH_3CH_2)_4Pb + 4\ NaCl + 3\ Pb$$

Individuals living near sites where lead was produced or sites where lead was disposed and also hazardous waste sites where lead has been detected in some environmental media also may be at risk for exposure (Hazdat, 2005).

2.2 Cadmium

The principal form of cadmium in air is cadmium oxide, although some cadmium salts, such as cadmium chloride, can enter the air, especially during incineration. Environmental discharge of cadmium due to the use of petroleum products, combustion of fossil fuels (petroleum and coal) and municipal refuge contribute to airborne cadmium pollution (De Rosa et al., 2003) and possibly introduce high concentrations of this potential reproductive toxicant into the environment. This may be particularly true for Nigeria where refuse are burnt without control. In addition, humans may be unwittingly exposed to cadmium via contaminated food or paper (Wu et al., 1995) cosmetics and herbal folk remedies (Lockitch, 1993). All these factors put Nigerian population at high risk of cadmium toxicity (Okoye, 1994).

The greatest potential for above average exposure of the general population to cadmium is from smoking which may double the exposure of a typical individual. Smokers with additional exposure are at highest risk (Elinder, 1985). Soil distribution of urban waste and sludges is also responsible for significant increase in cadmium content of most food crops (WHO, 1996). Persons who have cadmium-containing plumbing, consume contaminated drinking water or ingest grains or vegetables grown in soils treated with municipal sludge or phosphate fertilizer may have increased cadmium exposure (Elinder, 1985). Persons who consume large quantities of sun flower kernels can be exposed to higher levels of cadmium. Reeves & Vanderpool (1997) identified specific groups of men who were likely to consume sunflower kernels. The groups included: basket ball and soft ball players, delivery and long distance divers and line workers in sunflower kernel processing plants.

2.3 Nickel

A person may be exposed to nickel by breathing air, drinking water, or smoking tobacco containing nickel. Skin contact with soil, bath or shower water, or metals containing nickel, as well as metals plated with nickel can also result in exposure. Coins contain nickel. Some jewellery are plated with nickel or made from nickel alloys (Gerbeding, 2005b). Exposure of an unborn child to nickel is through the transfer of nickel from the mother's blood to fetal blood. Likewise, nursing infants are exposed to nickel through the mother to breast milk. However, the concentration of nickel in breast milk is either similar or less than the concentration of nickel in infant formulas and cow's milk. Children may also be exposed to nickel by eating soil. Normally, the exact form of nickel one is exposed to is not known. It could be in form of nickel sulphate, nickel oxide, nickel silicate, iron-nickel oxides, nickel subsulfide or metallic nickel (Gerbeding, 2005b).

Patients may be exposed to nickel in artificial body parts made from nickel-containing alloys which are used in patients in joint prostheses, sutures, clips, and screws for fractured bones. Corrosion of these implants may lead to elevated nickel levels in the surrounding tissue and to the release of nickel into extracellular fluid. Serum albumin solutions used for intravenous infusion fluids have been reported to contain as much as 222 µg nickel/L, but are very rarely encountered. Dialysis fluid has been reported to contain as much as 0.82 µg nickel/L. Studies of nickel in serum pre- and post-dialysis show between 0 and 33% increases in nickel concentrations in patients (IARC, 1990).

2.4 Manganese

Populations living in the vicinity of ferromanganese or iron and steel manufacturing facilities, coal-fired power plants, or hazardous waste sites are exposed to elevated manganese particulate matter in air or water, although this exposure is likely to be much lower than in the workplace (Koplan, 2000a). Manganese is eliminated from the body primarily through the bile. Interruption of the manufacture or flow of bile can impair the body's ability to clear manganese. Several studies have shown that adults and children as well as experimental animals with cholestatic liver disorders have increased manganese levels in their blood and brain and are at risk from potentially increased exposure to manganese due to their decreased homeostatic control of the compound (Devenyi et al., 1994). In addition to oral diets, people on partial and total parenteral nutrition may be exposed to increased amounts of manganese. Forbes & Forbes (1997) found that of 32 patients receiving home parenteral nutrition due to digestive problems, 31 had elevated serum manganese levels (0.5–2.4 mg/L compared to normal range of 0.275–0.825 mg/L).

In comparison to other groups within the general population, persons living close to high density traffic areas, automotive workers, and taxi drivers may be exposed to higher concentrations of manganese arising from the combustion of methylcyclopentadienyl manganese tricarbonyl (MMT). MMT is actually a fuel additive developed in the 1950s to increase the octane level of gasoline and thus improve the antiknock properties of the fuel. Farmers, people employed as pesticide sprayers, home gardeners, and those involved in the manufacture and distribution of maneb and mancozeb may also be exposed to higher concentrations of these pesticides than the general public. People who ingest fruits and vegetables that have been treated with these pesticides and that contain higher-than-usual residues of the compounds (due to incomplete washing or over-application) may be exposed to increased concentrations of the pesticides. It is possible that medical workers may be exposed to higher concentrations of mangafodipir than the general population, although exposure routes other than intravenous are not expected to pose a significant risk (Koplan, 2000a). Manganese in the environment is in the form of their oxides or carbonates e.g MnO_2, $MnCO_3$ e.t.c.

2.5 Chromium

Blue prints, primer paints, household chemicals and cleaners, cements, diesel engines utilizing anti-corrosive agents, upholstery dyes, leather tanning processes, welding fumes, battery, rubber, dye, candles, printers and matches are occupational and environmental sources of chromium (Koplan, 2000b). In addition to individuals who are occupationally exposed to chromium, there are several groups within the general population that have potentially high exposures (higher than background levels) to chromium. These populations include individuals living in proximity to sites where chromium was produced or sites where chromium was disposed. Persons using chromium picolinate as a dietary supplement will also be exposed to higher levels of chromium than those not ingesting this product (Anderson, 1998). People may also be exposed to higher levels of chromium if they use tobacco products, since tobacco contains chromium. Workers in industries that use chromium are one segment of the population that is especially at high risk to chromium exposure. Occupational exposure from chromate production, stainless steel welding, chromium plating, and ferrochrome and chrome pigment production is especially significant since the exposure from these industries is to chromium (VI) (EPA 1984a).

Persons using contaminated water for showering and bathing activities may also be exposed via inhalation to potentially high levels of chromium(VI) in airborne aerosols. Elevated levels of chromium in blood, serum, urine, and other tissues and organs have also been observed in patients with cobalt-chromium knee and hip arthroplasts (Koplan, 2000b). Chromium in the environment can exist in many forms e.g chromium trioxide, potassium dichromate, sodium dichromate, potassium chromate, sodium chromate or ammonium dichromate e.t.c.

3. Environmental pollution

3.1 Fossil fuels combustion

The major sources of heavy metal pollution in urban areas of Africa are anthropogenic while contamination from natural sources predominates in rural areas. Anthropogenic sources of pollution include those associated with fossil fuel i.e. the non-renewable energy resources of coal, petroleum or natural gas (or any fuel derived from them) combustion, mining and metal processing (Nriagu, 1996). Fossil fuel consumption in Nigeria has risen ten-fold in the last two decades and consumption by urban households accounts for a large percentage, a trend which is expected to continue in the future. In a survey on urban household energy use patterns in Nigeria with respect to fuel preferences, sources and reliability of energy supply, it was found that kerosene, fuel wood, charcoal and electricity are the major fuels for urban use in Nigeria. Dependence on biomass fuels is rapidly giving way to the use of fossil fuels. Pollution problems associated with incidents of oil spills around automobile repair workshop resulting in metal contamination have been the subjects of many reports (Onianwa et al., 2001). Lead, cadmium, nickel, manganese and chromium are associated with automobile related pollution. They are often used as minor additives to gasoline and various auto-lubrication and are released during combustion and spillage (Lytle et al., 1995).

It is estimated that 8.5 million kg of nickel are emitted into the atmosphere from natural sources such as windblown dust and vegetation each year. Five times that quantity is estimated to come from anthropogenic sources (Nriagu & Pacyna, 1988) and the burning of residual and fuel oil is responsible for 62% of anthropogenic emissions. Chromium is released into the atmosphere mainly by anthropogenic stationary point sources including industrial, commercial and residential fuel combustion via the combustion of natural gas, oil, and coal. It has been estimated that emissions from the metal industry ranged from 35% to 86% of the total chromium and emissions from fuel combustion ranged from 11% to 65% of the total chromium. The main sources of manganese release to the air are industrial emissions, combustion of fossil fuels and re-entrainment of manganese-containing soils (EPA, 1987). High concentration of cadmium is released by human activities such as mining smelting operations and fossil fuel combustion. Coal, wood and oil combustion can all contribute cadmium to the atmosphere. It has been suggested that coal and oil used in classical thermal power plants are responsible for 50% of the total cadmium emitted to the atmosphere (Thornton, 1992). Anthropogenic sources of lead also include the mining and smelting of ore, manufacture of lead-containing products, combustion of coal and oil most notably leaded gasoline that may still be used in some countries including Nigeria. It is important to note that land is the ultimate repository for lead, and lead released to air and water ultimately is deposited in soil or sediment. For example, lead released to the air from leaded gasoline or in stack gas from smelters and power plants will settle on soil, sediment, foliage or other surfaces (Gerbeding, 2005a).

Atmospheric lead emissions in Nigeria have been estimated to be 2800 metric tonnes per year with most (90%) derived from automobile tail pipe (Nriagu et al., 1997). Lead in the form of tetra-ethyl lead $Pb(C_2H_5)_4$ is the most common additive to petrol to raise its octane number. Upon combustion in the petrol engine, the organic lead is oxidized to lead oxide according to the following reaction:

$$2Pb\,(C_2H_5)_4 + 27O_2 \longrightarrow 2PbO + 16CO_2 + 20H_2O$$

The lead oxide (PbO) formed, reacts with the halogen carriers (the co-additives) to form particles of lead halides- $PbCl_2$, PbBrCl, $PbBr_2$- which escape into the air through the vehicle exhaust pipes. By this, about 80% of lead in petrol escapes through the exhaust pipe as particles while 15-30% of this amount is air borne. Human beings, animal and vegetation are the ultimate recipients of the particulate (Ademoroti, 1996).

The lead level in Nigeria's super grade petrol is in the range 210-520 mg/L (Ademoroti, 1986). Automobile exhausts are also believed to account for more than 80% of the air pollution in some urban centres in Nigeria. The highest level of lead occurs in super grade gasoline with a concentration range of 600 to 800 mg/L (with a mean of 70µg/mL) and aviation gas with a concentration of 915µg/mL (Shy, 1990), which is much higher than permissible levels in some other countries. The comparable maximum levels in United States and Britain (UK) are 200µg/mL and 500µg/mL, respectively (Osibanjo & Ajaiyi, 1989). Automobiles in Nigeria may still be using leaded gasoline. Many cars are poorly maintained and characteristically emit blue plumes of bad odour and unburnt hydrocarbons (Baumbach et al., 1995), implying that a higher percentage of the lead in gasoline is emitted to the atmosphere.

Gasoline sold in most African countries contains 0.5–0.8g/L lead. In urban and rural areas and near mining centers, average lead concentrations are up to 0.5–3.0µg/m^3 in the atmosphere and >1000µg/g in dust and soils (Nriagu et al., 1996). In Nigeria, the level of lead in petrol is estimated at 0.7g/L. The national consumption of petrol in the country is estimated at 20 million litres per day with about 150 people per car. It is therefore predicted that at least 15 tonnes of lead is emitted into the environment through combustion of fossil fuel (Agbo, 1997). The annual motor gasoline consumption in 2000 was 56 litres per person. In 2005, Nigerian National Petroleum Corporation (NNPC) recorded domestic consumption of Premium Motor Spirit (Petrol) as 9,572,014,330 litres, while 2,361,480,530,000 litres of Automotive Gas Oil (Diesel) were equally recorded. Therefore, an average car in Nigeria uses over 1800 liters of petrol per year. The number of cars in Nigeria is assumed to be 3.27 million and fuel consumption for an average Nigerian car is 9.0 km L^{-1} (Ajao & Anurigwo, 2002). In 2006, 9.13 billion Litres of gasoline were consumed in Nigeria. Presently, Nigeria's daily fuel consumption stands at 30 million litres per day (Energy ministry, 2008). As reported by Agbo, 1997, the level of lead in petrol used in Nigeria is estimated at 0.7g/L. This probably is still the case as the use of leaded petrol may still be obtainable in Nigeria even though several countries have banned it's use (United Nations, 2006). It is therefore predicted that if it is so, at least 21,000kg of lead is emitted into the environment through combustion of petrol. Infact, Bayford & Co Ltd brought leaded petrol back to the UK market in 2000, primarily to service the needs of the classic car owners. As the only government approved supplier of leaded fuel in the UK, following the decision of the major oil companies to withdraw the product, Bayford remains committed to do all it can to make leaded fuel as widely available as possible and presently supplies over 36 petrol stations in the UK and still advertising for more supplies (Jonathan, 2008).

3.2 Indiscriminate disposal of waste

Anthropogenic sources of environmental pollution include those associated with industrial effluents, solid waste disposal and fertilizers (Nriagu, 1996). Heavy metals may enter soil and aquatic environments via sewage sludge application, mine waste, industrial waste disposal, atmospheric deposition and application of fertilizers and pesticides (Adaikpoh, et al., 2005). In Nigeria, recent reports indicate that the major contaminants found in drinking water especially from wells are heavy metals. These heavy metals find their way into the soil and groundwater through activities like intense agriculture, power generation, industrial discharges, seepage of municipal landfills, septic tank effluents, to mention a few. Infact, many authors have reported high levels of heavy metal ions in the soil, rivers and groundwater in different areas of Nigeria (Okuo et al., 2007). Indiscriminate disposal of toxic wastes therefore poses a great threat to human health.

4. Highlighted spectroscopic methods for heavy metals determination

4.1 Lead

Several analytical methods are available to analyze the level of lead in biological samples like blood. The most common methods employed are flame atomic absorption spectrometry (AAS). GFAAS and Anode stripping voltametry (ASV) are the methods of choice for the analysis of lead. In order to produce reliable results, background correction, such as Zeeman background correction that minimizes the impact of the absorbance of molecular species, must be applied. Limits of detection for lead using AAS are on the order of µg/mL (ppm) for flame AAS measurements, while flameless AAS measurements can detect blood lead levels at about 1ng/mL (Flegal & Smith, 1995). Inductively coupled plasma mass spectrometry (ICP-MS) is also a very powerful tool for trace analysis of lead and other heavy metals. ICP/MS not only can detect very low concentrations of lead but can also identify and quantify the lead isotopes present. Other specialized methods for lead analysis are X-ray fluorescence spectroscopy (XRFS), neutron activation analysis (NAA), differential pulse anode stripping voltametry, and isotope dilution mass spectrometry (IDMS). The most reliable method for the determination of lead at low concentrations is IDMS but due to the technical expertise required and high cost of the equipment, this method is not commonly used (Gerbeding, 2005a).

The primary methods of analyzing for lead in environmental samples are AAS, GFAAS, ASV, ICP/AES and XRFS. Less commonly employed techniques include ICP/MS, gas chromatography/photoionization detector (GC/PID), isotope dilution mass spectrometry (IDMS), electron probe X-ray microanalysis (EPXMA) and laser microprobe mass analysis (LAMMA). Chromatography (GC, HPLC) in conjunction with ICP/MS can also permit the separation and quantification of organometallic and inorganic forms of lead. Various methods have been used to analyze for particulate lead in air. The primary methods, AAS, GFAAS, and ICP/AES are sensitive to levels in the low $\mu g/m^3$ range (0.1–20 $\mu g/m^3$). Chelation/extraction can also be used to recover lead from aqueous matrices. GC/AAS has been used to determine organic lead, present as various alkyl lead species, in water. XRFS has been shown to permit speciation of inorganic and organic forms of lead in soil for source elucidation (Gerbeding, 2005a).

4.2 Cadmium

The most common analytical procedures for measuring cadmium concentrations in biological samples use the methods of atomic absorption spectroscopy (AAS) and atomic emission spectroscopy (AES). Methods of AAS commonly used for cadmium measurement are flame atomic absorption spectroscopy (FAAS) and graphite furnace (or electrothermal) atomic absorption spectroscopy (GFAAS or ETAAS). A method for the direct determination of cadmium in solid biological matrices by slurry sampling ETAAS has been described (Taylor et al., 2000).

Analysis for cadmium in environmental samples is usually accomplished by AAS or AES techniques, with samples prepared by digestion with nitric acid. Since cadmium in air is usually associated with particulate matter, standard methods involve collection of air samples on glass fiber or membrane filters, acid extraction of the filters, and analysis by AAS. Electrothermal inductively coupled plasma mass spectrometry (ETV-ICP-MS) has also been used to analyze size classified atmospheric particles for cadmium. The accuracy of the analysis of cadmium in acid digested atmospheric samples, measured by ACSV, was evaluated and compared with graphite furnace atomic absorption spectrometry (GFAAS) and inductively coupled plasma mass spectrometry (ICP-MS) (Koplan, 1999). Sediment and soil samples have been analyzed for cadmium using the methods of laser-excited atomic fluorescence spectroscopy in a graphite furnace (LEAFS), GFAAS and ETAAS preparation of the samples is generally accomplished by treatment with HCl and HNO_3.

Electrothermal vaporization isotope dilution inductively coupled plasma mass spectrometry (ETV-ID-ICP-MS) has been utilized for the analysis of cadmium in fish samples. Radiochemical neutron activation analysis (RNAA), differential pulse anodic stripping voltametry (ASV) and the calorimetric dithizone method may also be employed. The AAS techniques appear to be most sensitive, with cadmium recoveries ranging from 94 to 109% (Koplan, 1999).

4.3 Nickel

Analytical methods used in the determination of nickel in biological materials are the same as those used for environmental samples. Nickel is normally present at very low levels in biological samples. Atomic absorption spectrometry (AAS) and inductively coupled plasma-atomic emission spectroscopy (ICP-AES), with or without preconcentration or separation steps, are the most common methods. These methods have been adopted in standard procedures by EPA and the International Union of Pure and Applied Chemistry. Direct aspiration into a flame and atomization in an electrically heated graphite furnace or carbon rod are the two variants of atomic absorption. The latter is sometimes referred to as electrothermal AAS (ETAAS). Typical detection limits for ETAAS are <0.4 μg/L, while the limit for flame AAS and ICP-AES is 3.0 μg/L (Todorovska et al., 2002). Good precision was obtained with flame AAS after preconcentration and separation, electrothermal AAS, and ICP-AES. Inductively coupled plasma mass spectrometry (ICP-MS) techniques have been used to quantify nickel in urine with detection sensitivities down to approximately 1 μg/L. Voltammetric techniques are becoming increasingly important for nickel determinations since such techniques have extraordinary sensitivity as well as good precision and accuracy. Direct measurement of nickel in urine in the presence of other trace metals (e.g., cadmium, cobalt, and lead) was demonstrated using adsorption differential pulse cathodic stripping voltammetry at a detection limit of 0.027 μg/L (Gerbeding, 2005b).

The most common methods used to detect nickel in environmental samples are AAS, either flame or graphite furnace, ICP-AES, or ICP-MS. Nickel can also be analyzed in ambient and marine water using stabilized temperature graphite furnace atomic absorption (STGFAA) detection techniques as described in EPA methods 1639 and 200.12 respectively, which give limits of detection for nickel concentrations ranging between 0.65 and 1.8 μg/L and recoveries of >92%. Two other EPA standard test methods, 200.10 and 200.13, also use preconcentration techniques in conjunction with ICP-MS or graphite furnace AAS detection techniques, respectively, for analysis of nickel in marine water. One method uses activated charcoal to preconcentrate nickel in natural waters, followed by elution with 20% nitric acid and analysis by inductively coupled plasma-optical emission spectrometry (ICP-OES). This method achieved a detection limit of 82 ng/L (Gerbeding, 2005b).

4.4 Manganese

Flame atomic absorption analysis is the most straightforward and widely used method for determining manganese. In this method, a solution containing manganese is introduced into a flame, and the concentration of manganese is determined from the intensity of the colour at 279.5 nm. Furnace atomic absorption analysis is often used for very low analyte levels and inductively coupled plasma atomic emission analysis is frequently employed for multianalyte analyses that include manganese. Simple methods for the direct determination of Mn in whole blood by ETAAS have been described. Methods for measuring manganese therefore include spectrophotometry, mass spectrometry, neutron activation analysis and X-ray fluorimetry (Koplan, 2000a).

Atomic absorption spectrometry has been the most widely used analytical technique to determine manganese levels in a broad range of foods, as well as other environmental and biological samples. Tinggi et al., (1997) carried out a wet digestion technique using a 12:2 (v/v) nitric:sulfuric acid mixture for their determination, and for food samples with low levels of manganese, they found that the more sensitive graphite furnace atomic absorption analysis was required. Because manganese is often found at very low levels in many foods, its measurement requires methods with similarly low detection limits; these researchers identified detection limits of 0.15 mg/kg (ppm) and 1.10 µg/kg (ppb) for flame and graphite furnace atomic absorption spectrometry respectively (Tinggi et al. 1997).

A number of analytical methods for quantifying MMT in gasoline have been described including simple determination of total elemental manganese by atomic absorption and gas chromatography followed by flame-ionization detection (FID). In a certain method, in which MMT is detected in gasoline by gas chromatography coupled with flame photometric detection (FPD); the chemiluminescence of manganese is measured to determine MMT levels in a method that uses simple, inexpensive, and commercially available instrumentation (Koplan, 2000a).

4.5 Chromium

Prior to 1978, numerous erroneous results were reported for the chromium level in urine using electrothermal atomic absorption spectrometry (ETAAS) because of the inability of conventional atomic absorption spectrometry systems to correct for the high nonspecific background absorption. The use of GC-MS and ETAAS to determine ^{53}Cr and total Cr in biological fluids in order to investigate the distribution of Cr in lactating women following oral administration of a stable ^{53}Cr tracer have been reported. The authors detected ^{53}Cr in blood within 2 h of administration. They noted, however, that blood Cr changes in response to oral administration were variable and they considered that blood Cr was not tightly regulated. Similarly, the reported serum and plasma chromium concentrations of normal subjects have varied more than 5,000-fold since the early 1950s (Taylor et al., 2000).

The four most frequently used methods for determining low levels of chromium in biological samples are neutron activation analysis (NAA), mass spectrometry (MS), graphite spark atomic emission spectrometry (AES), and graphite furnace atomic absorption spectrometry (GFAAS). Of these four methods, only the GFAAS is readily available in conventional laboratories, and this method is capable of determining chromium levels in biological samples when an appropriate background correction method is used. The three commonly used methods that have the best sensitivity for chromium detection in air are GFAAS, instrumental neutron activation analysis (INAA), and graphite spark atomic emission spectrometry. Measurements of low levels of chromium concentrations in water have been made by specialized methods, such as inductively coupled plasma mass spectrometry (ICP-MS), capillary column gas chromatography of chelated chromium with electron capture detection (ECD), and electrothermal vaporization inductively coupled plasma mass spectrometry (Koplan, 2000b).

4.6 Biofuels

There are different spectrophotometric techniques for analysis of contaminants in biofuels. Simultaneous detection of the absorption spectrum and refractive index ratio with a spectrophotometer for monitoring contaminants in bioethanol has been carried out by Kontturi et al., 2011. Inductively Coupled Plasma Atomic Emission Spectrometry and optical emission spectral analysis with inductively coupled plasma (ICP-OES) have also been used to analyze biodiesel samples for trace metals (ASTM, 2007; ECS, 2006). An ICP-MS instrument fitted with an octopole reaction system (ORS) was used to directly measure the inorganic contents of several biofuel materials. Following sample preparation by simple dilution in kerosene, the biofuels were analysed directly. The ORS effectively removed matrix- and plasma-based spectral interferences to enable measurement of all important analytes, including sulfur, at levels below those possible by ICP-OES. A range of commonly produced biofuels was analysed, and spike recovery and long-term stability data was acquired. Also, suitably configured ICP-MS has been shown to be a fast and very sensitive technique for the elemental analysis of biofuels (Woods & Fryer, 2007).

A flow system designed with solenoid micro-pumps is proposed for fast and greener spectrophotometric determination of free glycerol in biodiesel. Glycerol was extracted from samples without using organic solvents. The determination involves glycerol oxidation by periodate, yielding formaldehyde followed by formation of the colored (3,5-diacetil-1,4-dihidrolutidine) product upon reaction with acetylacetone. The coefficient of variation, sampling rate and detection limit were estimated as 1.5% (20.0 mg L^{-1} glycerol, $n = 10$), 34 h^{-1}, and 1.0 mg L^{-1} (99.7% confidence level), respectively. A linear response was observed from 5 to 50 mg L^{-1}, with reagent consumption estimated as 345 µg of KIO_4 and 15 mg of acetylacetone per determination. The procedure was successfully applied to the analysis of biodiesel samples and the results agreed with the batch reference method at the 95% confidence level (Sidnei & Fábio, 2010).

5. Review of heavy metals in the environment using atomic absorption spectrophotometry

The negative effect on air quality will be unavoidable, if solid wastes are incinerated under uncontrolled conditions or left to biologically decompose in open areas, because waste gas will be given off to the atmosphere. Besides, heavy metals and hazardous organic pathogens are disseminated with organic wastes. Effluents from point sources change the characteristics of the receiving environment and its suitability for marinating its living communities and their ecological structure. Some metals when discharged into natural waters at increased concentration in sewage, industrial effluent or from mining and refining operations can have severe toxicological effects on aquatic environment and humans. Nigeria has a population of over 120 million. Degradation of water quality is most severe in the four states that contain 80 percent of the nations industries; Lagos, Rivers, Kano and Kaduna States, with the highest level of emission of 8000 tones of hazardous waste per year from Lagos State (Alamu, 2005).

5.1 Heavy metals in soils

In a study of soil samples of refuse dumps in Awka (Anambra State, Nigeria) the lead level (2467mg/kg) exceeded the limits set by the US Environmental Protection Agency. This study suggests that the refuse dumps in Awka may increase the level of environmental heavy metals in Nigeria (Nduka et al., 2006). Concentrations of cadmium, chromium, manganese, nickel and lead were determined in surface sediments of the Lagos Lagoon, Nigeria. The results revealed largely anthropogenic heavy metal enrichment and implicated urban and industrial waste and runoff water transporting metals from land - derived wastes as the sources of the enrichment. Okoye (1991) also reported that urban and industrial wastes discharged into the Lagos lagoon have had a significant impact on the ecosystem following the relative enrichment in the Lagoon fish with lead.

Several attempts have been made to assess the impact of the use of fossil fuels on the environment. Results obtained from the study on heavy metals (chromium, lead, cadmium, and nickel) concentrations and oil pollution in Warri area revealed that the concentrations of the heavy metals considered were higher in the oil-spilled sites relative to the control sites. Similarly, when compared with the European Community standards, the concentration is said to be quite significant. The results indicate the contribution of the oil industry to heavy metals contamination in the Niger-Delta area of Nigeria and that the operations of the oil industry in this study area have not been sufficiently accompanied by adequate environmental protection. To safeguard agricultural land in the area and hence human health, there is an urgent need for government to address the incidence of oil spills in this area (Essoka et al., 2006).

Concentrations of lead, cadmium, nickel, chromium and manganese were determined to assess the impact of automobiles on heavy metal contamination of roadside soil. The lead levels in polluted sites varied from 70 to 280.5$\mu g g^{-1}$ and it rapidly decreased with depth. Similarly, mean concentrations of cadmium, nickel, chromium, and manganese were significantly higher at polluted sites and followed a decreasing trend with increase in depth. Correlation coefficients between heavy metals and traffic density were positively significant except for nickel. Profile samples showed that lead, cadmium, manganese were largely concentrated in the top 5cm confirming airborne contamination (Ramakrishnaiah & Somashekar, 2002).

In a study of the effect of traffic density on heavy metal content of soil and vegetation along roadsides in Osun State Nigeria, the concentration of the heavy metals decreased with increasing soil depth and horizontal distance from the road. Metal contamination correlated positively with traffic volume. Concentrations of lead, cadmium and nickel along the low traffic density were lower than the high traffic density (Amusan et al., 2003). Reclamation of auto repair workshop areas for residential and agricultural purposes makes high the risk assessment of heavy metal contamination (Ayodele et al., 2007). The levels of lead, cadmium and nickel were determined in the roadside topsoil in Osogbo, Nigeria, with the view to determining the effect of traffic density and vehicular contribution to the soil heavy metal burden. The levels of the metals at the high density roads were significantly higher than the corresponding levels at the medium and low traffic density roads. The average levels of lead, cadmium, and nickel in all road locations at a distance of 5m from the roads were 68.74±34.82, 0.60±0.31 and 8.38±2.40mg/kg respectively. Lead and cadmium were of average levels of 92.07±21.25 and 0.76±0.35 mg/kg respectively at a distance of 5m from the road at high traffic density roads, while the levels of nickel averaged 9.65±2.61mg/kg respectively. There was a rapid decrease in the level of the metals with distance, with the metal levels at a distance of 50m from the road almost reaching the natural background levels of the metals at the control sites (Fakayode & Olu-Owolabi, 2003a). The levels of the metals were also determined at the four major motor parks and at the seven mechanic workshop settlements. The levels of the metals at the motor parks and mechanic workshops were far above the levels at the control sites. The levels of lead, cadmium and nickel at the motor parks were 519±73.0, 3.6±0.8, and 7.3±4.6 mg/kg respectively, with the levels of lead, cadmium and nickel at the mechanic workshops averaging 729.57±110.93, 4.59±1.01 and 30.21±9.40mg/kg respectively (Fakayode & Olu-Owolabi, 2003a).

5.2 Heavy metals in food

Heavy metals have been analyzed and found to be in considerable quantities in Food in Nigeria. In the assessment of heavy metal levels in fish species of Lagos Lagoon, lead levels in the fishes were beyond W.H.O. acceptable limit of 1 ppm with a concentration range of 10.81-152.42 ppm (Akan and Abiola, 2008). Also, 86% and 84% of the 50 beverages (canned and non-canned respectively) obtained in Nigeria failed to meet the US EPA criteria for acceptable lead and cadmium levels in consumer products. 79.3% of the non-canned beverages showed lead levels that exceeded the US EPA's maximum contaminant level (MCL) of 0.015 mg/dm^3, 100% of the canned beverages had lead levels that were greater than the MCL. The range of the lead in the canned beverages was 0.002–0.0073 and 0.001–0.092 mg/dm^3 for the non-canned beverages. The cadmium levels ranged from 0.003–0.081 mg/dm^3 for the canned and 0.006–0.071 mg/dm^3 for non-canned beverages. About 85.71% of the canned beverages had cadmium levels that exceeded the maximum contaminant level (MCL) of 0.005 mg/dm^3 set by US EPA while 82.7% non-canned beverages had cadmium levels exceeding the MCL (Maduabuchi et al., 2006). In addition, Fakayode and Olu-Owolabi (2003b), reported that concentrations of lead and cadmium 0.59 mg/kg and 0.07 mg/kg respectively in chicken eggs in Ibadan were comparatively greater than levels found in other countries e.g lead concentrations of 0.048 ppm and 0.489 ppm obtained in China and India respectively and cadmium concentrations of 0.01 ppm and 0.004 ppm obtained in Canada and finland respectively.

Some reported works have also shown that planted crops and vegetations along major roads where there was high traffic volume contained high levels of lead content due to automobile exhaust. For instance, cadmium levels (0.12 ± 0.03 - 0.28 ± 0.03ppm) and nickel levels (3.02 ± 0.14 - 6.50 ± 0.25ppm) of staple foods (yam, cassava, cocoyam and maize) from oil-producing areas of Rivers and Bayelsa States of Nigeria were higher than those of non-oil producing areas (Abakaliki). Because of this high trace metal level, the staple foods from oil-producing areas examined are likely to be the major source of exogenous contamination of these metals in the populace (Akaninwor et al., 2005).

The concentration of cadmium has been found to be higher in some Nigerian foods as compared to those of some other countries as shown in Table 1.

5.3 Heavy metals in water

Groundwater and soil samples from 16 locations near petrol stations (PS) and mechanic workshops (MW) around Calabar, Nigeria, were analyzed for heavy metals and hydrocarbons to determine their concentrations and assess the impact of the PS and MW on groundwater in the area. Results show that mean concentrations of cadmium, chromium, manganese, nickel, and lead in groundwater are higher than the maximum admissible concentration (Nganje et al., 2007).

Results from the evaluation of ground water quality characteristics near two waste sites in Ibadan and Lagos revealed that some of the ground-water quality constituents determined exceeded the World Health Organization (WHO) standards for drinking water irrespective of source of pollution. Some of the ground-water samples were poor in quality in terms of cadmium, chromium, lead and nickel recorded (Ikem et al., 2002). The levels of heavy metals (cadmium, chromium, nickel, and lead) were analysed in the River Ijana (Ekpa-

Warri, Nigeria). Generally, excessive levels of the parameters of pollution above W.H.O. standards recommended for surface waters were observed (Emoyan et al., 2005). The possible sources of these parameters of pollution are diverse: originating from anthropogenic/ natural and point sources. Coal contains diverse amounts of trace elements in their overall composition. Certain trace elements such as lead, cadmium and chromium if present in high amount could preclude the coal from being used in environmentally sensitive situations. Ekulu River is the largest body of inland waters in Enugu Urban, which is of considerable importance industrially, culturally, and in agriculture. Ekulu coal mine is located by the bank of the Ekulu river. The coal mine station discharges its effluents directly into River Ekulu. Enugu coal mine occurs in the area where River Ekulu takes its source. Metal concentrations were generally higher in the coal samples than in the sediments. The metals (manganese, chromium, cadmium, nickel, and lead) analysed for were present throughout the period monitored in both the sediment and coal samples with some variations. Mean concentrations of Mn (0.256-0.389mg/kg) and Cr (0.214-0.267mg/kg) were high relative to concentrations of Cd (0.036-0.043mg/kg), Ni (0.064-0.067mg/kg) and Pb (0.013-0.017mg/kg). The presence of toxic metals in the area is established, calling for the assessment of their impact on the health of human and aquatic lives around the area (Adaikpoh et al., 2005). Other industrial effluents also contribute to the level of the heavy metals such as lead in the environment as reported by (Ayodele et al., 1996).

Commodity	**Greece**	**Japan and China**	**Nigeria**	**European Countries**
Rice	0.006	0.070	0.060	0.010
Cereal—other	0.002	0.023	0.075	0.016
Roots and tubers	0.022	0.015	0.103	0.025
Soya bean	–	0.041	0.200	0.021
Pulses—other	0.004	0.019	0.140	0.019
Sugars and honey	–	0.003	0.015	0.004
Groundnuts—shelled	–	–	0.370	0.050
Oilseeds—other	–	0.021	0.100	0.119
Vegetable oils—other	0.002	0.001	0.127	0.002
Stimulants—other	–	0.017	0.160	0.006
Spices	–	0.005	0.191	0.055
Leafy vegetables	0.054	0.025	0.155	0.034
Vegetables—other	0.024	0.020	0.343	0.013
Fish and other seafood —other	0.034	0.035	0.207	0.014
Eggs	0.001	0.003	0.500	0.003
Fruits	0.009	0.006	0.067	0.004
Milks	0.001	0.004	0.006	0.001
Milk products	0.004	0.004	0.375	0.005
Poultry meat	0.013	0.005	0.110	0.002
Meats—other	0.027	0.006	0.083	0.006

Source: Moriyama et al., (2002)

Table 1. Average concentrations of cadmium in foods (mg/kg)

Several works have been done to access the impact of improper waste management on the environment. The elevated level of heavy metal in the Niger Delta aquatic environment as a result of industrial discharges from refining operations has been elaborated by Spiff & Horsfall, (2004). Therefore it can be said that there is unregulated discharge of untreated effluents into natural receptors by industries in Nigeria. Samples of industrial effluents from Sharada industrial area Kano Nigeria were assessed for heavy metals. The study showed that about 60% of the industries discharge effluents with heavy metal concentration higher than 0.30 mg/L. Lead and chromium ions were the most prevalent with values above the minimum tolerable limit. The presence of these metal ions could pose a serious public health hazard. It is therefore recommended that these effluents be adequately treated before discharge. Table 2 shows the nickel content in naturally occurring waters.

WATER TYPE	LOCATION	CONCENTRATION RANGE (μg/L)
River water	Poland Germany-Rhine USA	2-75 8.9-24 0-71
Lake water	Poland-Lakes of Wielkopolska National Park * Poland-Lakes of the Golanieckie stream	2-11 1-8
Underground water	Poland-Poznzn Poland-Pozan voivodship Poland-Szczecin	0.5-20 1-30 1-15
Drinking water	USA Poland-Pozcan	0.5-7 0-5

Source: Barałkiewicz and Siepak (1999)

Table 2. Content of nickel in naturally occurring waters

6. Reports of research works done on heavy metals analysis in Nigerian environment

6.1 Blood

6.1.1 Methodology

3 ml of blood were collected directly from the select population comprised of 60 children, 114 women (pregnant, nursing mothers, others) and 66 men. This was carried out by venous puncture by a qualified nurse under contamination controlled conditions using pyrogen-free sterile disposable syringes and placed into 5 ml capacity EDTA plastic bottles containing K_3EDTA as anticoagulant. Each sample (3 ml) was transferred into 100 ml conical flasks. The EDTA bottle was rinsed with a little nitric acid and transferred into 100ml conical flask. Perchloric acid and nitric acid which were of analytical grade was added in the ratio 1:3 as follows: 2 ml perchloric acid and 6 ml nitric acid. The conical flask was covered with an evaporating dish and the mixture digested at low temperature using a thermostated Bitinett hot plate until a clear solution was obtained. The digest was made up to 20 ml with deionized water in a 20 ml standard flask (Rahman et al., 2006). The sample solutions were then analyzed for lead, cadmium, nickel, manganese and chromium using a GBC atomic absorption spectrophotometer, model A6600 AVANTA PM.

6.1.2 Results

As shown in Table 3, lead was detected in 235 of the 240 samples (97.92 %), the concentration range was from 0.039-0.881 ppm. Cadmium was detected in 205(85.42 %) samples, the concentration range was from 0.007-0.293 ppm. Nickel was detected in 137(57.08 %) samples while in 103(42.92 %) of the samples, the concentration range was from 0.007-0.849 ppm. Manganese was detected in 203(84.58 %) samples, the concentration range was from 0.006-0.861 ppm. Chromium was detected in 113(47.08 %) samples, the concentration range was from 0.006-0.829 ppm. Comparing the concentrations obtained from this study with the WHO (1996) guideline for heavy metals in blood, all the detectable samples had concentrations higher than the permissible levels stipulated for all the heavy metals except for 5 that were within the range stipulated for manganese i.e 0.008–0.012 ppm and 24 that were within the stipulated range for lead i.e 0.05-0.15 ppm. Thus there is a clear indication of high concentrations of the heavy metals in the general population in Nigeria especially the Southeast (Ibeto & Okoye, 2009; 2010a; 2010b).

Group		**Heavy metal** **Lead**	**Cadmium**	**Nickel**	**Manganese**	**Chromium**
Men	N	66	65	40	63	26
	Mean ± sd	0.394 ± 0.126	0.093 ± 0.048	0.122 ± 0.079	0.119 ± 0.075	0.305 ± 0.228
	Range	0.07 - 0.76	0.01 - 0.21	0.01 - 0.33	0.01- 0.40	0.01- 0.83
Pregnant women/ Nursing mothers	N	56	48	31	52	35
	Mean ± sd	0.288 ± 0.198	0.099 ± 0.064	0.096 ± 0.061	0.088 ± 0.040	0.201 ± 0.150
	Range	0.04 - 0.72	0.01 - 0.28	0.01 0.21	0.01 - 0.21	0.01 - 0.57
Other women	N	54	50	20	53	28
	Mean ± sd	0.328 ± 0.121	0.080 ± 0.046	0.096 ± 0.067	0.121 ± 0.059	0.214 ± 0.167
	Range	0.12 - 0.67	0.01 - 0.29	0.01 - 0.25	0.02 - 0.31	0.01 - 0.67
Children	N	59	42	46	35	24
	Mean ± sd	0.488 ± 0.153	0.088 ± 0.056	0.411 ± 0.240	0.091 ± 0.147	0.267 ± 0.228
	Range	0.12 - 0.88	0.01 - 0.23	0.01 - 0.85	0.01 - 0.86	0.01 - 0.68
WHO guideline		0.05-0.15	0.0003-0.0012	0.001-0.005	0.008-0.012	<0.005

Source: Ibeto & Okoye, 2009; 2010a; 2010b

Table 3. Concentrations (ppm) of heavy metals in blood of different categories of the urban population in Enugu State Nigeria

However certain measures can be taken to reduce the effects of these heavy metals in the body. All of the currently available methods to obviate the toxic effects of the heavy metals are mainly by chelation. The chelating agents bind to the heavy metals, enhance its excretion by facilitating their transfer from soft tissues to where it can be excreted. Some of the standard chelating agents currently in use are meso-2,3- dimercaptosuccinic acid for cadmium, triethylenetetramine and cyclam (1,4,8,11-tetraazacyclotetradecane) for nickel, and ethylenediamine tetraacetic acid for lead, manganese and chromium. Also, through specific dietary supplementation, for example, sufficient iron or calcium stores, as opposed to a deficiency in these or other minerals, may reduce the heavy metals absorption, and thus reduce potential toxicity (Koplan, 2000a).

6.2 Fruit juices

6.2.1 Methodology

100ml of each of ten different brands of fruit juice was measured into a 200ml conical flask and heated till the volume reduced to 10ml. Perchloric acid and nitric acid was then added in a ratio of 1:2 with perchloric acid being 6ml and the nitric acid 12ml. The solution was then digested at low heat until a clear solution was obtained. It was then allowed to cool and made up to 25ml with distilled water using a standard flask. Heavy metals were then determined by atomic absorption spectrophotometry using Alpha 4 Serial no 4200 with air acetylene flame.

6.2.2 Results

As shown in Table 4, all the samples except one of guava brands contained lower concentration of copper than the 5ppm permissible limit set for the metal. All samples had concentrations of zinc well below the 5ppm maximum permissible level. The iron concentrations were below the limit of 15ppm in all the samples except for the pineapple brand, which showed a concentration of 50ppm. This could be due to many reasons such as the fact that the fruit juice brand was acidic and the fruit acids could pick up the metal from the equipment during processing or storage. As minerals are soil and species dependent, the fruit acids might also have picked up iron and other metals from the soil during growth. Iron could also be added for fortification.

Cadmium was more wide spread, occurring in seven brands with a range of 0.16 to 0.38ppm. Lead occurred in four brands with range 0.11 to 0.33 ppm. Only the foreign made apple juice brand with the lead content of 0.33ppm exceeded the maximum permissible level of 0.3ppm by FAO/W.H.O. The limit for cadmium was not stipulated but compared with the limit set for lead (since they are both non-nutritive elements), the foreign made guava brand and the pineapple brand may be considered to be high in cadmium (Okoye & Ibeto, 2009).

6.3 Soil

6.3.1 Methodology

Soil samples were collected from twenty different locations in three Local Government Areas in Enugu State. Soil samples were collected in duplicates at a dept of 15-20cm and transferred into a pre-washed polyethylene nylon bag to avoid contamination. Soil samples were dried at 105°C and sieved with 100mesh (152μm BS Screen 410). The samples were prepared for analysis by cold extraction. 1g of the dried soil sample was weighed into a labeled 100ml conical flask and 20ml of mixture of conc. HCl and conc. HNO_3 (1:1) were added and well shaken. The solution was kept overnight after which it was filtered through a whatman No 1 filter paper formerly leached by pouring cupious quantity of dilute HNO_3 on the filter paper while in the funnel. The clear solution obtained was made up to 50ml using a standard flask and transferred into a plastic bottle (Okoye, 2001). The sample solutions were analysed at various wavelengths for each metal using Buck Scientific Atomic Absorption Spectrophotometer 205.

SAMPLES	MEAN CONCENTRATIONS (mg/l) OF PARAMETERS				
	Cr	Mn	Ni	Cd	Pb
Lime	0.06	0.11	0.08	<0.002	<0.004
Mango	<0.002	0.67	<0.05	0.17	<0.004
Orange	<0.002	0.19	0.13	<0.002	0.11
Guava	<0.002	0.42	0.13	0.27	<0.004
Guava*	<0.002	0.67	0.03	0.37	<0.004
Black Currant	0.03	0.24	<0.05	0.16	0.13
Mixed fruit	<0.002	0.42	<0.05	<0.002	<0.004
Apple	<0.002	0.19	<0.05	0.26	0.33
Apple	<0.002	0.14	0.03	0.25	<0.004
Pine-apple	0.09	6.96	0.15	0.38	0.20

Source: Okoye & Ibeto, 2009

Table 4. Concentrations (ppm) of metals in fruit juice samples

6.3.2 Results

The ranges of concentrations were: Pb(30.3-235), Cr(9.0-15.5) and Cd(5.5-42.25) ppm in Igbo-Eze North. Pb (0.2-100), Cr(9.5-10.8) and Cd(0.51-44.8) ppm in Nsukka and Pb(14.8-165) and Cd(0.43-5.0) ppm in Udi. The order of abundance in the soil follow the order Pb>Cd> Cr. Compared with the work done in an automobile spare parts market the values for chromium and cadmium were relatively high. Compared with the Indian standard for heavy metals in soils, some of the samples exceeded the stipulated range of 3-6ppm for cadmium, indicating considerable cadmium contamination of some of the sampling points. However, the variations in the mean concentration of each metal in the three Local Government Areas were not significant (P>0.05) (Okoye & Ibeto, 2008).

6.4 Water

6.4.1 Methodology

Samples were collected from 17 different locations in Southeast Nigeria at various occasions covering the dry and wet seasons. In collecting samples from rivers, lakes and streams, the polyethylene sampling containers were dipped just below the surface to minimize the contamination of the water sample by surface films. For borehole samples, the mouth of the tap was cleaned with cotton wool and was left to run to waste for several minutes before collection while for spring water, samples were collected at different outlets of the spring. All the samples were collected with 2 L polyethylene cans which were leached with a 1:1 HCl and water and rinsed with distilled-de-ionized water (Okoye et al., 2010).

The samples were concentrated by evaporating 500 mL of water sample to about 100 mL followed by addition of 1 mL conc HCl and digesting until volume was about 15-20 mL. This was later made up to mark with distilled-de-ionized water in a 25 mL standard flask and later transferred into an acid-leached polyethylene bottle prior to analysis. Trace metals were determined with AAS (ALPHA Series 4200 CHEM TECH ANALYTICAL Ltd, UK) equipped with air-acetylene flame.

6.4.2 Results

The metal analysis gave values (mg/L) with ranges as follows: Pb (nd-13.5); Cd (nd-0.60); Ni (nd-0.075) and Cr (nd-0.10). Less than 40% had high levels of lead and cadmium which are indicative of the impact of indiscriminate discharge of untreated industrial effluents, domestic waste and inputs from other human activities on the pollution of the environment by trace metals. Concentrations of lead and cadmium in five locations were higher than the WHO limits of 0.01 mg/L and 0.003 mg/L respectively. Water containing high levels of lead and cadmium is not fit for drinking purposes. This study has created awareness concerning the risk of drinking from the identified water sources which have high concentrations of lead and cadmium (Okoye et al., 2010).

6.5 Chicken

6.5.1 Methodology

The samples of the liver, gizzards, muscles of chickens and also their feed were prepared by wet digestion. 10ml of nitric acid and 5ml of perchloric acid were added to 1g of each finely ground sample into different 100ml conical flasks covered with watch glasses for overnight predigestion. It was then heated on a hot plate until a clear solution was obtained. The contents were cooled and transferred to a 25ml standard flask and made up to mark with deionised water. These were then transferred to sample bottles until clear solutions were obtained. Each digested sample was transferred to prewashed sample bottles (Ibeto & Okoye, 2010a).

The sample solutions were then analyzed for the heavy metals: lead, cadmium, copper and zinc at required wavelength using a GBC atomic absorption spectrophotometer, model no A6600 AVANTA PM.

6.5.2 Results

The concentrations in ug/g of the heavy metals were in the range of 1.78 - 15.32, 9.7 - 147.07, 15.82 - 47.79 and 0.03 - 2.29 for cadmium, lead, copper and zinc respectively. Concentrations of cadmium were higher than the permissible limit of 0.5 ppm set by FAO/WHO and concentrations of lead were above the permissible limit of 1 ppm set by Australia New Zealand Food Authority. The high concentrations of the toxic metals obtained show a certain level of pollution of the environment. However, the low concentration of the essential metals in the feed shows there was no addition of nutritive supplements to the feed (Okoye et al., 2011).

7. Conclusion

Atomic absorption spectrophotometry was used to determine the heavy metal content of various samples from the environment and also human blood. The heavy metal content of the environmental samples indicated a certain level of heavy metal pollution in the Nigerian environment which can be attributed to fossil fuels combustion and indiscriminate disposal of wastes. This is also reflected in the level of heavy metals in the blood of the select population which on accumulation in the human system has led to low level of life expectancy globally. It is therefore recommended that utilization of alternative fuels be aggressively pursued and integrated into the energy mix of countries globally. These fuels include biogas, biodiesel and bioethanol and are becoming increasingly important not only because of the diminishing petroleum reserves but also because of the environmental consequences of exhaust gases from petroleum fuelled engines.

8. References

Adaikpoh, E. O., Nwajei, G.E. & Iogala, J. E. (2005). Heavy metals concentrations in coal and sediments from River Ekulu in Enugu, Coal City of Nigeria. J. Appl. Sci. Environ. Mgt. 9 (3) 5 – 8.

Ademoroti, C.M.A. (1986). Levels of heavy metals on bark and fruit of trees in Benin City, Nigeria. Environmental pollution. 11: 241-243.

Ademoroti, C.M.A. (1996). Environmental Chemistry and toxicology. March prints and Consultancy. Foludex Press Ltd. Ibadan. pp 177-195.

Agbo, S. (1997). Effects of lead poisoning in children. In: Proceeding at a workshop on vehicular emission and lead poisoning in Nigeria. Friends of the environment. pp 20-28.

Ajao, E.A. & Anurigwo, S. (2002). Land-based sources of pollution in the Niger Delta, Nigeria. AMBIO: Journal of the Human Environment. 31, (5) 442–445.

Akan, B.W. & Abiola, R.K. (2008): Assesment of trace metal levels in fish species of Lagos Lagoon. Conference Proceedings of Chemical Society of Nigeria. 31st Annual International Conference and Exhibition, Warri. 22nd-26th 2008. Delta State Nigeria. pp 394-399.

Akaninwor, J. O., Onyeike, E. N. & Ifemeje, J.C. (2005). Trace metal levels in raw and heat processed Nigerian staple foods from oil-producing areas of Rivers and Bayelsa States. Journal of Applied Sciences and Environmental Management. Vol. 10, No. 2, pp. 23-27.

Alamu, O. (2005). Watershed management to meet water quality standards and emerging TMDL (Total maximum daily load). Proceedings of the Third Conference 5-9 March 2005 (Atlanta, Georgia USA). American Society of Agricultural and Biological Engineers, St. Joseph, Michigan. www.asabe.org, 701P0105.

Amusan, A.A., Bada, S.B. & Salami, A.T. (2003). Effect of traffic density on heavy metal content of soil and vegetation along roadsides in Osun State, Nigeria. West African Journal of Applied Ecology. 4: 107-114.

Anderson, R.A. (1998). Effects of chromium on body composition and weight loss. Nutr. Rev. 56(9):266- 270.

ASTM (2007). International Standard Test Method for Determination of Additive Elements in Lubricating Oils by Inductively Coupled Plasma Atomic Emission Spectrometry. Annual Book of ASTM Standards, 2007, Vol. *05.03*, ASTM D4951-02.

Ayodele, J.T., Momoh, R.U. & Amm, M. (1996). Determination of heavy metals in Sharada Industrial effluents, in water quality monitoring and environmental status in Nigeria. Proceedings of the National Seminar on Water Quality Monitoring and Status in Nigeria, organized by Federal Environmental Protection Agency and National Water Resources Institute. October 16-18. pp 158-166.

Ayodele, R.I., Dawodu, M. & Akande, Y. (2007). Heavy metal contamination of topsoil and dispersion in the vicinities of reclaimed auto repair workshops in Iwo, Nigeria. Research Journal of Applied Sciences. 2(11): 1106-1115.

Barałkiewicz, D. & Siepak, J. (1999). Chromium, nickel and cobalt in environmental samples and existing legal norms. Polish Journal of Environmental Studies. Vol. 8, No. 4: 201-208.

Baumbach, G.U., Vogt, K.R.G., Hein, A.F., Oluwole, O.J., Ogunsola, H.B. & Akeredolu, F.A. (1995). Air pollution in large tropical city with high traffic density: results of measurements in Lagos, Nigeria. Sci. Total Environ. 169: 25-31.

Bayly, G.R., Braithwaite, R.A., Sheehan, T.M.T., Dyer, N.H., Grimley, C. & Ferner, R.E. (1995). Lead poisoning from Asian traditional remedies in the West Midlands-report of a series of five cases. Hum. Experiment Toxicol. 14: 24-28.

De Rosa, M., Zarrilli, S., Paesano, L., Carbone, U., Boggia, B., Petretta, M., Masto, A., Cimmino, F., Puca, G., Colao, A. & Lombardi, G. (2003). Traffic pollutants affect infertility in men. Human Reproduction. 18: 1055-1061.

Devenyi, A.G., Barron, T.F. & Mamourian, A.C. (1994). Dystonia, hyperintense basal ganglia, and whole blood manganese levels in Alagille's syndrome. Gastroenterology. 106:1068-1071.

EPA. 1984a. Health assessment document for chromium. Research Triangle Park, NC: Environmental Assessment and Criteria Office, U.S. Environmental Protection Agency. EPA 600/8-83-014F.

E.P.A. (1987). Toxic air pollutant/source crosswalk: A screening tool for locating possible sources emitting toxic air pollutants. Research Triangle Park, NC: U.S. Environmental Protection Agency, Office of Air Quality Planning and Standards. EPA-450/4-87-023a.

Elinder, C.G. (1985). Cadmium: uses, occurrence and intake. In: Friberg, L., Elinder, C.G., Kjellstrom, P. et al eds. Cadmium and health: A toxicological and epidermiological appraisal. Vol 1. Exposure, dose and metabolism. Effects and response. Boca Raton, FL. CRS. Press. pp 23-64.

Emoyan, O.O., Ogban, F.E. & Akarah, E. (2005). Evaluation of heavy metals loading in River Ijana in Ekpa-Warri, Nigeria. J. Applied Sciences and environmental management. Vol 10, No 2. pp 121-127.

Energy Ministry (2008). Business summary 21st to 28th 2008. Accessed from www.bpeng,org/NR/rdonlyres on 15/1/09.

Essoka, P.A., Ubogu, A.E. & Uzu, L. (2006). An overview of oil pollution and heavy metal concentration in Warri area, Nigeria. Management of environmental quality. Vol 10(2) 209-215.

European Committee for Standardization (ECS) (2006). Fat and oil derivatives -Fatty acid methyl ester (FAME) -Determination of Ca, K, Mg and Na content by optical emission spectral analysis with inductively coupled plasma (ICP-OES). EN 14538, 2006.

Fakayode, S. O. & Olu-Owolabi, B. I. (2003a). Heavy metal contamination of roadside topsoil in Osogbo, Nigeria: its relationship to traffic density and proximity to highways. Environmental Geology. Volume 44, (2) 150-157.

Fakayode, S. O. & Olu-Owolabi, B. I. (2003b). Trace metal content and estimated daily human intake from chicken eggs in Ibadan, Nigeria. Archives of Environmental Health. http:www.encyclopedia.com/beta/doc/IGI-111732614.

Flegal, A.R. & Smith, D.R. (1995). Measurements of environmental lead contamination and human exposure. Rev Environ Contam Toxicol. 143:1-45.

Forbes, G.M. & Forbes, A. (1997). Micronutrient status in patients receiving home parenteral nutrition. Nutrition. 13:941-944.

Gerbeding, J.L. (2005a). Toxicological profile for lead. Public health service, Agency for toxic substances and diseases. Atlanta Georgia. pp 3-5, 31, 113-130, 224-228 and 312-350.

Gerbeding, J.L. (2005b). Toxicological profile for nickel. Public health service. Agency for toxic substances and diseases. Atlanta Georgia. pp 27, 79, 134-144, 166-167.

Hazdat (2005). Hazdat data base. ASTDR's. Hazardous substance release and health effect data base. Atlanta. Agency for toxic substance and disease registry. www.astsdr.cdc.gov/hazdat-html. April 13, 2005.

Hughes, W.W. (1996). Essentials of environmental toxicology. The effects of environmental hazardous substances on human health. Loma, Lind Califonia. Tay and Francais Publishers. pp 3, 87-95.

IARC. (1990). IARC (International Agency for Research on Cancer) monographs on the evaluation of carcinogenic risks to humans. Chromium, nickel and welding. Lyon, France: International Agency for Research on Cancer. World Health Organization. Vol 49: 257-445.

Ibeto C. N. & Okoye C. O. B. (2009). Elevated Cadmium Levels in Blood of the Urban Population in Enugu State Nigeria. World Applied Sciences Journal 7 (10): 1255-1262, 2009. ISSN 1818-4952.

Ibeto, C.N. & Okoye, C.O.B. (2010a). High levels of heavy metals in blood of the urban population in Nigeria. Research Journal of Environmental Sciences. ISSN 1819-3412. 4 (4): 371-382.

Ibeto C. N. & Okoye C. O. B. (2010b). Elevated Levels of Lead in Blood of Different Groups in the Urban Population of Enugu State, Nigeria. International Journal of Human and Ecological Risk Assessment. Human and Ecological Risk Assessment: An International Journal, 16: 5, 1133 – 1144. DOI: 10.1080/10807039.2010.512257. http://dx.doi.org/10.1080/10807039.2010.512257.

Ikem, A., Osibanjo, O., Sridhar, M. K. C. & Sobande, A. (2002). Evaluation of groundwater quality characteristics near two waste sites in Ibadan and Lagos, Nigeria. Water, air and soil pollution. Vol 140, Nos 1-4. pp 307-333.

John, H., Cheryl, H., Richerd, S. & Christine, S. (1991). Toxics A-Z- A guide to everyday pollution hazards. University of Califonia, Press. Berkley. Angeles. Oxford. pp 47-104.

Jonathan, T. (2008). Leaded fuels update. Press information - Thursday 15th May 2008. Accessed from www.bayfordgroup.co.uk and www.leadedpetrol.co.uk. on the 24th of January 2008.

Kontturi, V., Hyvärinen, S., García, A., Carmona, R., Yu Murzin, D., Mikkola, J.P. & Peiponen, K.E. (2011). Simultaneous detection of the absorption spectrum and refractive index ratio with a spectrophotometer: monitoring contaminants in bioethanol. Measurement Science and Technology Vol. 22, No. 5, 055803 doi: 10.1088/0957-0233/22/5/055803.

Koplan, J.H. (2000a). Toxicological profile for manganese. Public health service. Agency for toxic substances and disease registry. Atlanta Georgia. pp 21-50, 175-207 and 295-400.

Koplan, J.H. (2000b). Toxicological profile for chromium. Public health service. Agency for toxic substances and disease registry. Atlanta Georgia. pp 1-9, 16-50, 122-157 and 301-315.

Koplan, J.P. (1999). Toxicological profile for cadmium. Public health service. Agency for Toxic Substance and Disease Registry (ATSDR). Atlanta Georgia. pp 126-140, 207 and 260- 270.

Lockitch, G. (1993). Prospective on lead toxicity. Clin Biochem. 26:371-81.

Lytle, C.M., Smith, B.N. & Mckinwu, C.Z. (1995). Manganese accumulation along the Utah roadways. A possible indication of motor exhaust pollution. Sci Total Environ. 162: 1056-109.

Maduabuchi, J.M.U., Nzegwu, C.N., Adigba, E.O., Aloke, R.U., Ezomike, C.N. Okocha, C.E., Obi, E. & Orisakwe, O.E. (2006): Lead and cadmium exposures from canned and non-canned beverages in Nigeria: A public health concern. Science of the Total Environment. Vol 366, Issues 2-3, pp 621-626.

Matte, T.D., Proops, D., Palazeulos, E., Graef, J. & Avila, H.A. (1994). Acute high dose lead exposure from beverage contaminated from traditional Mexican pottery. Lancet. 344: 1064-1065.

Moriyama, T., Taguchi, Y., Watanabe, H. & Joh, T. (2002). Changes in the cadmium content of wheat during the milling process (in Japanese). In: Report on Risk Evaluation of Cadmium in Food, Research on Environmental Health, Health Sciences Research Program, Ministry of Health, Labour and Welfare, pp. 153–160.

Nduka, J.K.C., Orisakwe, O.E., Ezenweke, L.O., Abiakam, C.A., Nwanguma, C.K. & Maduabuchi, U.J.M. (2006). Metal contamination and infiltration into the soil at refuse dump sites in Awka, Nigeria. Archives of Environmental and Occupational Health. Vol. 61, No 5, 197 - 204.

Nganje, T. N. Edet, A. E. & Ekwere, S. J. (2007). Concentrations of heavy metals and hydrocarbons in groundwater near petrol stations and mechanic workshops in Calabar metropolis, southeastern Nigeria. *Environmental Geosciences Vol.* 14; no. 1; p. 15-29; DOI:10.1306/eg.08230505005.

Nriagu, J.O. (1996): History of global metal pollution. Sci. 272:223-224.

Nriagu, J. O., Blankson, M. L. & Ocran, K. (1996). Childhood lead poisoning in Africa: a growing public health problem. Journal of Science of the Total Environment. 181(2, 15):93-100.

Nriagu, J.O. & Pacyna, J.M. (1988). Quantitative assessment of worldwide contamination of air, water and soils by trace metals. Nature. 333:134-139.

Ofoefule, A.U., Uzodinma, E.O. & Anyanwu C.N. (2010). Studies on the effect of Anaerobic digestion on the microbial flora of animal wastes 2: Digestion and modelling of process parameters. Trends in Appl. Sci. Res. 5(1) 39-47.

Okoye, C.O.B. (1991). Heavy metals and organisms in the Lagos lagoon. Inter. J. Environmental studies. Vol. 37, pp 285-292.

Okoye, C.O.B. (1994). Lead and other metals in dried fish from Nigerian markets. Bull. Environ. Contain. Toxicol. 52: 825 - 832.

Okoye, C.O.B. (2001). Trace metal concentrations in Nigerian fruits and vegetables. Intern. J. Environ. Studies. Vol. 58 pp 501-509.

Okoye C.O.B. & Ibeto C.N. (2008). Determination of Bioavailable Metals in Soils of Three Local Government Areas in Enugu State, Nigeria. Proceedings of the 31st Annual International Conference and Exhibition, Delta Chem, 2008. Petroleum Training Institute (PTI), Conference Centre Complex, Effurun-Warri, Delta State, Nigeria, pp 767-771.

Okoye C.O.B. & Ibeto C.N. (2009). Analysis of different brands of fruit juice with emphasis on their sugar and trace metal content. Bioresearch Journal. 7 (2): 493-495.

Okoye, C. O. B., Aneke A.U., Ibeto, C. N. & Ihedioha, J. N. (2011). Heavy Metals Analysis of Local and Exotic Poultry Meat. International Journal of Applied Environmental Sciences. ISSN 0973-6077 Vol. 6, No 1. pp. 49-55.

Okoye, C.O.B., Ugwu, J.N. & Ibeto, C.N. (2010). Characterization of rural water resources for potable water supply in some parts of South-eastern Nigeria. J. Chem. Soc. Nig. Vol. 35. No.1. pp 83-88.

Okuo, J.M., Okonji, E.I. & Omeyerere, F.R. (2007). Hydrophysico-chemical assessment of the Warri coastal aquifer, Southern Nigeria. J. Chem Soc. Nig. Vol 32, No 2. 53-64.

Onianwa, P. C.; Jaiyeola, O. M. & Egekenze, R. N. (2001). Heavy metals contamination of topsoils in the vicinities of auto-repair workshops, gas stations and motor parks in a Nigeria city. Toxicol. and Environ. Chem. 84(1-4), 33 -39.

Osibanjo, O. & Ajayi, S.O. (1989). Trace metal analysis of petroleum products by flame atomic absorption spectrophotometry. Nigeria Journal of Nutritional Health. 4: 33-40.

Rahman, S., Khalid, N. Zaidi, J.H., Ahmad, S. & Iqbal, M. Z. (2006): Non occupational lead exposure and hypertension in Pakistani adults. J.Zhepang University Science B. 9: 732-737.

Ramakrishnaiah, H. & Somashekar, R.K. (2002). Heavy metal contamination in roadside soil and their mobility in relations to pH and organic Carbon. Soil and sediment contamination: An International Journal, Vol.11, Issue 5, pp 643 - 654.

Reeves, P.G. & Vanderpool, R.A. (1997). Cadmium burden in men and women who report regular consumption of confectionery sunflower kernels containing a natural abundance of cadmium. Environ. Health. Perspect. 105 (10), 98-104.

Shy, C. M. (1990). Lead in petrol. The mistake of the 20th century. World health statistics, Quaterly. 43: 168-176.

Sidnei, G. S. & Fábio, P.R. (2010). A flow injection procedure based on solenoid micro-pumps for spectrophotometric determination of free glycerol in biodiesel. Talanta. Volume 83, Issue 2, 15 December 2010, Pages 559-564. doi:10.1016/j.talanta.2010.09.061.

Spiff, A. I. & Horsfall. M. Jnr. (2004). Trace metal concentrations in inter-tidal flate sediments of the upper new Calabar River in the Niger Delta area of Nigeria. Scientia African. Vol 3. 19-28.

Taylor, A., Branch, S. Halls, D.J., Owen, L.M.W. & White, M. (2000). Atomic Spectrometry update: Clinical and biological material, food and beverages. J. Anal. At. Spectrom. 15, 451-487.

Thornton, I. (1992). Sources and pathways of cadmium in the environment. IARC Sci Publ. 118:149-162.

Tinggi, U., Reilly, C., & Patterson, C. (1997). Determination of manganese and chromium in food by atomic absorption spectromety after wet digestion. Food Chem 60:123-128.

Todorovska, N., Karadjova, I. & Stafilov, T. (2002). ETAAS determination of nickel in serum and urine. Anal. Bioanal. Chem. 373(4-5):310-313.

United Nations (2006). Interim review of scientific information on lead. Overview of existing and future national actions, including legistation, relevant to lead. Appendix. Acessed from http://www.chem.unep.ch/pb_and_cd/SR/Files/Interim_reviews/UNEP-Lead-review-Interim-APPENDIX-Oct 2006.doc on December 20th 2008.

Uzodinma, E.O. Ofocfulc, A.U. & Enwere N.J. (2011). Optimization of biogas fuel production from blending maize bract with biogenic wastes. Amer. J. Food and Nutr. 1 (1): 1-6.

WHO (1996). Trace elements in human nutrition and health. International atomic energy agency. WHO Library Publication Data. Geneva. pp 194-215, 256-259.

Woods, G.D. & Fryer, F.I. (2007). Direct elemental analysis of biodiesel by inductively coupled plasma-mass spectrometry. Anal Bioanal Chem. 389(3):753-761.

Wu, T.N., Yang, G.Y., Shen, C.Y. & Liou, S.H. (1995). Lead contamination of candy: an example of crisis management in public health. *Lancet*. 346: 1437-1442.

3

Estimation of the Velocity of the Salivary Film at the Different Regions in the Mouth – Measurement of Potassium Chloride in the Agar Using Atomic Absorption Spectrophotometry

Shigeru Watanabe
Meikai University
Japan

1. Introduction

Saliva is secreted into the mouth at a rate of 0.3 to 0.4 ml per minute. Retained saliva in the mouth physiologically triggers swallowing to carry the saliva out of the mouth. Dawes (1983) have reported the volume of saliva in the mouth just before swallowing, the rate of swallowing, the volume swallowed per swallow, and the volume of saliva in the mouth just after swallowing. Clearance of materials from the mouth is facilitated by alternately-performed saliva secretion and swallowing, and thereby the oral environment is maintained relatively constant (Fig.1). The unstimulated salivary flow rate and saliva volume in a single swallowing have the most influence on the efficiency of clearance.

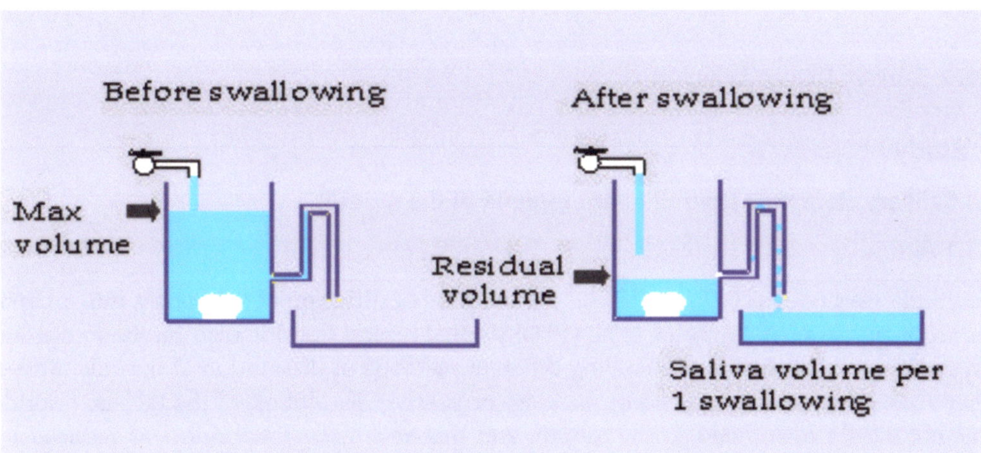

Fig. 1. Saliva volume in the mouth

Saliva is a crucial factor for protection of the oral environment. The rate of oral clearance of sugar and acid is inversely related to the onset and progression of dental caries, as shown particularly in persons with severe hyposalivation.

Saliva secreted into the mouth flows slowly as a thin film, over the tooth surfaces and mucosa and is cleared from the mouth by swallowing (Fig. 2). However, saliva does not flow equally throughout the mouth, and there are differences in the different areas. Measurement of the volume of saliva and velocity of the salivary film at different locations in the mouth are important for understanding the site-specificity of dental caries and periodontal disease.

Using agar as an artificial-plaque, we have conducted studies on the five following items by measuring the clearance of potassium chloride from the agar using an atomic absorption spectrophotometer.

1) Salivary clearance from different regions of the mouth. 2) Salivary clearance in children with complete primary dentitions. 3) Influence of the location of the parotid duct orifice on oral clearance. 4) Effect of salivary flow rate on fluoride retention in the mouth. 5) Estimation of the velocity of the salivary film at different locations in the mouth.

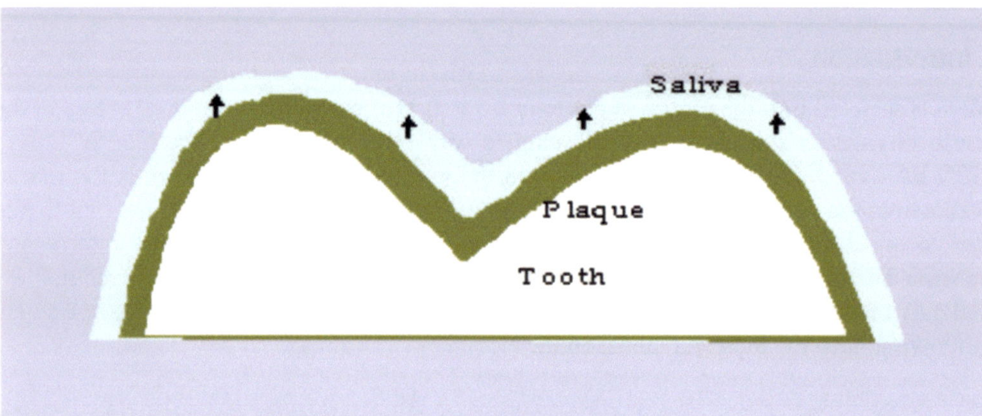

Fig. 2. Salivary film and plaque

2. Study

2.1 Salivary clearance from different regions of the mouth

2.1.1 Aim

Very little research has been carried out on the rates of diffusion of substances into or from dental plaque in vivo. Primosch et al. (1986) studied topical fluoride distribution in the oral cavity and rates of clearance following different methods of dissolution of fluoride tablets. They found that after the chewing, sucking, or passive dissolution of the tablets, fluoride was not evenly distributed in the mouth, and that retention of fluoride was reduced by increased salivary flow rate. Thus, it would seem likely that the rate of renewal of the film of saliva over plaque must influence diffusion rates into and from plaque.

The aim of this study was to determine the velocity of the salivary film by determining the rate of diffusion of potassium chloride from an artificial plaque at different sites in the mouth.

2.1.2 Materials and methods

- Determination of the rate of potassium chloride clearance:

A 1-mol/L solution of potassium chloride was mixed with sufficient agarose (Electrophoresis Purity Reagent; BioRad Laboratories, Richmond, CA) to give a 1.0% solution which was heated until the agarose dissolved. The acrylic chambers (Fig. 3) to hold the gel were rectangular (16 mm in length, 8 mm wide, and 1.5 mm thick) with a cylindrical central depression (6 mm diameter and 1.5 mm depth).

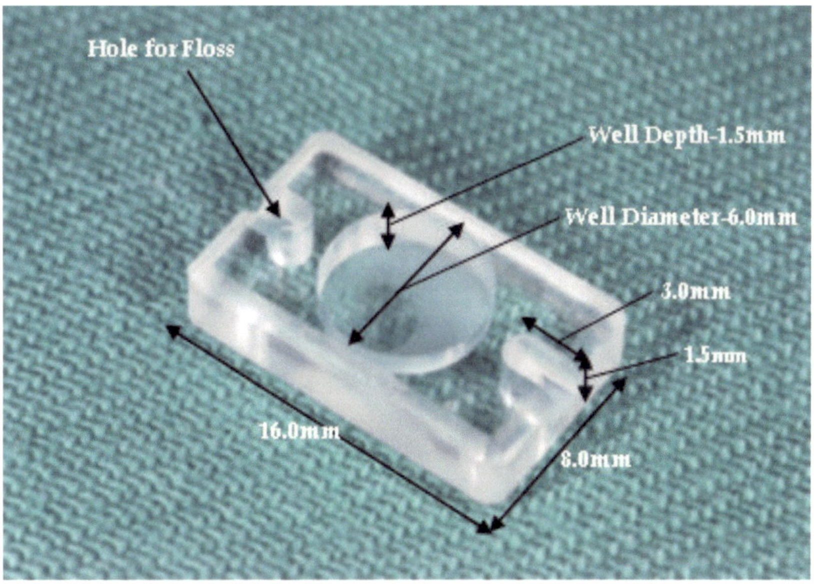

Fig. 3. Diagram of design of the diffusion chambers

The weight of the agarose held in the center well of each chamber was measured six times using an electronic balance (FX-3200; A&D, Tokyo, Japan), and chambers in which the mean weight of agarose was more than 2 SD from the mean were excluded.

Two chambers were initially covered with a layer of Parafilm (American Can, Greenwich, Conn., USA), were attached bilaterally by floss to the teeth, with the gel surface away from the teeth. The chambers were attached to the upper first molars for measurement of the posterior sites (UPB) and to both upper incisors for measurement of the anterior site (UAB) (Fig. 4).

After temperature and salivary flow equilibration, the Parafilm was removed at time 0. The first diffusion chamber was removed from the mouth after being exposed to saliva for a selected period of time and the gel was transferred to flasks containing 400 ml of (100 ppm) sodium chloride. Subsequently the second chamber was removed and the potassium chloride extracted by the same procedure. The fluid was agitated intermittently for 90 min, and the potassium concentration was assayed by atomic absorption spectrophotometry (Shimadzu AA-6105, Kyoto, Japan). The times were chosen so that between about 30 and 60% of the potassium chloride would have diffused from the agarose discs. The initial KCl concentration in the agarose discs, which had not been placed into the mouth, was also measured.

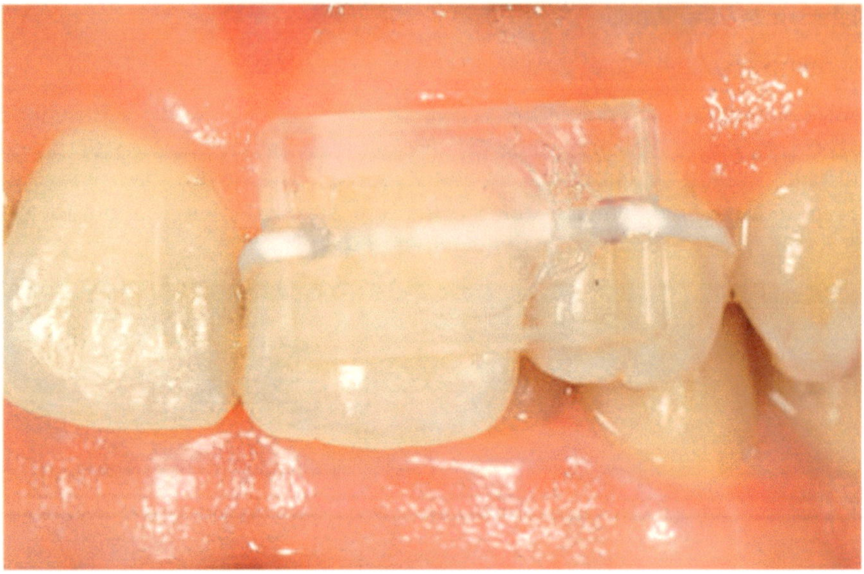

Fig. 4. Acrylic chamber attached to the upper central incised

- Calculation of the half-time (the time for half the KCl to diffuse from the gel) (Lecomte & Dawes, 1987) for clearance.

The rate of potassium chloride clearance from the gels into a large, stirred volume was determined. One involved suspending the filled chambers in one liter of 100 ppm NaCl, stirred by a magnetic stirrer, either at room temperature or at 37℃. The diffusion chambers were taken from the fluid at selected time intervals and the gels transferred quantitatively with a sewing needle to flasks containing 500 ml of 100 ppm sodium chloride. The fluid was agitated intermittently for 90 min, since preliminary studies showed that the remaining potassium chloride was extracted from the gel in this time interval. The potassium concentration was also measured in identically prepared agarose discs which had not been put into the 100 ppm NaCl, to give the initial concentration.

A least-squares straight line was fitted, by computer, to the potassium concentration plotted against the square root of time. This gives a very good approximation of the theoretical clearance curve until about 65% of the diffusant has been lost from the gel (see 2-1-5). From the results, the half-time was calculated.

2.1.3 Subjects and locations

The subjects were 6 adults with a mean age of 26 years. They had a complete dentition up to the second molar and no malocclusion.

Seven different sites in the mouth were chosen for measurements. These were the Lower anterior lingual (LALi) and buccal (LAB), lower posterior buccal (LPB) and lingual (LPLi), upper posterior lingual (UPLi) and buccal (UPB), and upper anterior buccal (UAB). The flow rate of unstimulated whole saliva was measured on each occasion for 5 min by being allowed to drip off the lower lip into a weighed container.

2.1.4 Result

The half-times in the mouth varied with locations and with salivary flow rate, as shown in Fig. 5 When the flow rate was unstimulated, the shortest halftimes occurred in the LALi site and the longest in the UAB site. In both groups, the difference was significant at $p<0.001$.

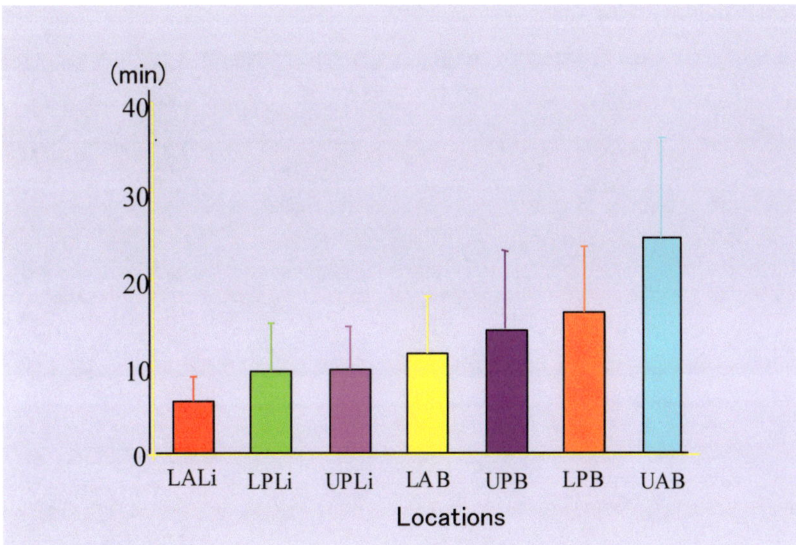

Fig. 5. Half-time when salivary flow rate is unstimulated

2.1.5 Discussion

In this study, we measured the concentration of residual potassium in agarose gel to determine the velocity of salivary flow for the 7 different sites. The reason why potassium chloride was used as the target substance (with an agarose gel used as artificial plaque) is that it is readily soluble in water, harmless, has a low molecular weight enabling it to diffuse easily, is present at a low concentration (around 20 mM) in saliva and can be measured relatively easily. Because the potassium concentration in the agarose gel used in this study was much higher than that in the saliva of the subjects, it was unlikely that the potassium concentration in the saliva affected that in the gel.

The relationship between time and the quantity of potassium diffused from the gel into the saliva was pre-determined in a pilot study. Clearance was evaluated by determining the half-time, the time at which the concentration at time 0 is reduced to half, from the relationship between 3 time points, including time 0, and potassium concentration, as well as by comparing the mean half-time between different sites. Since the correlation between time and the quantity of potassium eluted from the agarose gel was found to decrease in the early and late phases of the test (Lecomte P, Dawes C, 1987), the time to hold the holder in the mouth was determined so that the half-time would be almost at the mid-point of the test. For the measurement of potassium concentration, sodium chloride solution was used as the solvent to avoid errors in measurements due to ionization of potassium.

2.2 Salivary clearance in children with complete primary dentitions

2.2.1 Aim

Nursing bottle caries (Fig.6) is a specific form of rampant decay on the buccal surface of the upper anterior primary teeth. Some etiological factors, such as the types of microorganisms, tooth structure, and diet, have been reported, but there is little information about the influence of the salivary flow rate.

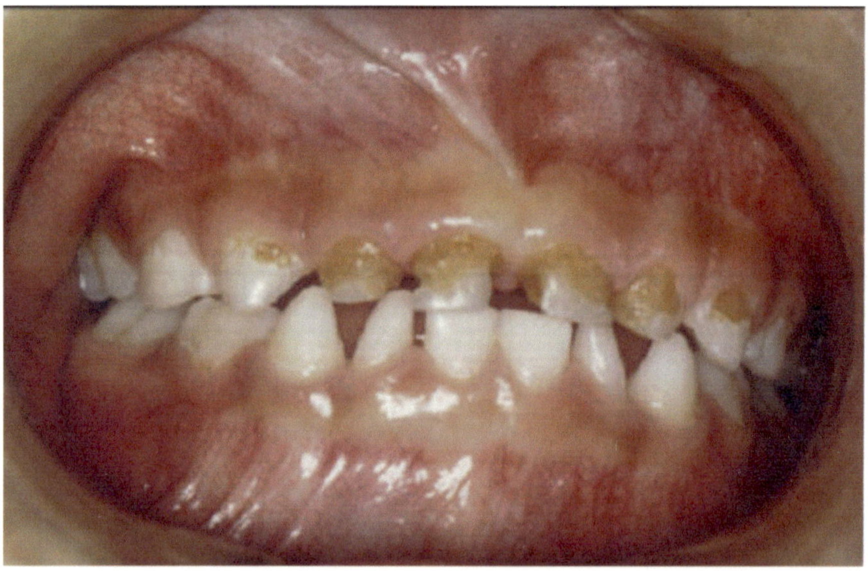

Fig. 6. Nursing bottle caries

Very little research has been carried out on the salivary flow rate or salivary clearance in children. Although the average thickness of the salivary film covering teeth and oral mucosa in children is essentially identical with values reported for adults, marked differences were found between children and adults for such parameters as unstimulated and stimulated whole-salivary flow rates, the volume of saliva in the mouth before and after swallowing, and the surface area of the mouth.

The aim of this study was to evaluate the rates of salivary clearance at different locations in the mouths of children and the effect of the spaces in the primary dentitions to determine whether prolonged clearance would occur in sites particularly susceptible to nursing bottle caries.

2.2.2 Materials and methods

The determination of the rate of potassium chloride clearance was done by the same methods as in study 1. A 1-mol/L solution of potassium chloride was mixed with sufficient agarose (Electrophoresis Purity Reagent; BioRad Laboratories, Richmond, CA) to give a 1.0% solution which was heated until the agarose dissolved. The acrylic chambers (Fig. 3) to hold the gel were rectangular (16 mm in length, 8 mm wide, and 1.5 mm thick) with a cylindrical central depression (6 mm diameter and 1.5 mm depth). The potassium concentration in agarose was analyzed by absorption spectrophotometry.

The subjects were 4 boys and 8 girls, 5 years of age, who were all in good health and with complete primary dentitions. Six subjects had primary spaces (located mesial to the maxillary and distal to the mandibular canines) and developmental spaces (present between the remaining teeth) (Fig. 7), and the other 6 subjects had no spaces in their arches (Fig. 8). The mean values of right and left primary spaces and total developmental spaces for 6 subjects who have spacing arch were 1.1 ± 0.4 mm, 1.2 ± 0.4, and 4.2 ± 1.9 mm for the upper dentition and 0.8 ± 0.3 mm, 0.7 ± 0.3, and 3.8 ± 1.4 mm for the lower dentition, respectively.

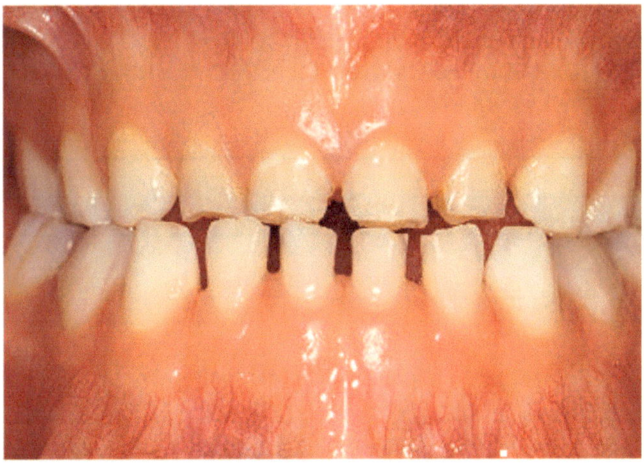

Fig. 7. Spacing arch at 5 years old.

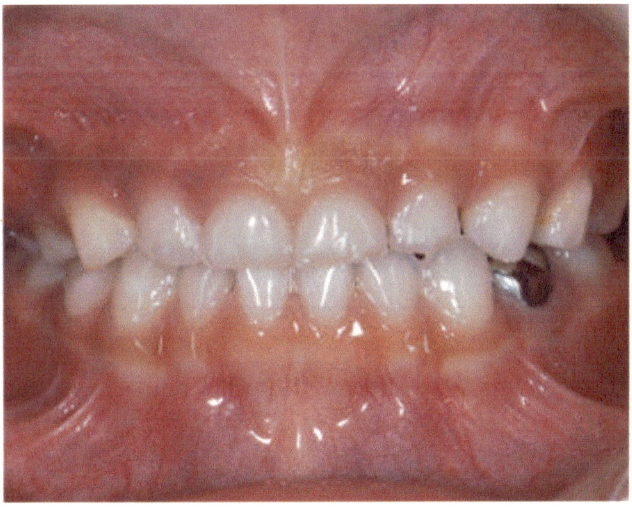

Fig. 8. No spacing arch at 5 years old.

Seven different sites in the mouth were chosen for measurements. These were the lower anterior lingual (LALi) and buccal (LAB), lower posterior buccal (LPB) and lingual (LPLi), upper posterior lingual (UPLi) and buccal (UPB), and upper anterior buccal (UAB). The flow rate of unstimulated whole saliva was also measured on each occasion for 5 min by being allowed to drip off the lower lip into a weighed container.

The rate of potassium chloride clearance from the gels into a large volume (1 liter) of 100 ppm NaCl fluid at 37℃ stirred by a magnetic stirrer was determined for the estimation of the half-time.

2.2.3 Result

The mean half-times for in vitro clearance into the large volume of stirred 100 ppm NaCl at 37°C was 3.9 ± 0.5 min.

The half-times in the mouth varied with location as shown in Table 1. The half-times of all sites for the spacing arch and of the LALi site for the no-spacing arch were reduced ($p<0.05$) as compared with the values when the flow rate was unstimulated.

( )SD

	Location						
	LALi	UPLi	LPLi	UPB	LAB	LPB	UAB
Spacing arch	5.9***	9.6*	9.4*	14.2	11.5	16.1	24.6
	(±2.8)	(±4.8)	(±5.4)	(±8.8)	(±6.4)	(±7.4)	(±11.4)
No-spacing	5.3 ***	9.1 **	10.1 *	13.8	13.8	17.4	25.9
arch	(±2.1)	(±4.4)	(±5.3)	(±6.8)	(±7.0)	(±8.1)	(±9.4)

Statistical analyses werw carried out between UAB site and the other sites in each group: *$p<0.05$, **$p<0.01$, ***$p<0.001$.
Mean unstimulated salivary flow rates were 0.47±0.2 ml/min.

Table 1. Half-times (mean ± SD) and salivary flow rates when salivary flow was unstimulated

When the saliva flow rate was stimulated (Table 2), the shortest halftimes occurred in the LALi site and the longest in the UAB site. The clearance from the LAB site in the spacing arch showed almost the same value as those from the LALi sites in both groups.

Halftimes(mean±SD)and salivary flow rates when salivary flow was stimulated

	Location		
	LALi	LAB	UAB
Spacing arch(n=6)			
Halftime, min	4.0±0.3**	4.5±0.8**	11.3±4.5
Flow rate, ml/min	3.4±1.5	4.1±2.1	3.4±1.7
No-Spacing arch(n=6)			
Halftime, min	4.0±0.2**	9.4±2.8*	15.4±4.2
Flow rate, ml/min	3.7±1.2	3.9±1.7	4.8±1.7

Halftime in large volume at 37°C=3.9 ± 0.5min.Statistical analyses were carried out detween UAB site and the other sites in each group :*$p<0.05$ **$p<0.01$

Table 2. Halftimes (mean ± SD) and salivary flow rates when salivary flow was stimulated

2.2.4 Discussion

The present study showed that the rate of clearance of substances from agarose gels into saliva varies markedly in different regions of the primary dentition. The location closest to the submandibular and sublingual ducts (LALi) showed the lowest half-time, whereas the UAB site had a clearance half-time 6.5 times longer than that for clearance into a large volume in vitro. As the opening of the parotid duct is situated on the rearward of the upper second primary molar in the children's mouth, there was a relatively long half-time in the UPB site. Since the mean salivary film thickness in 5-year-old children has been estimated to have almost the same value (0.06-0.09 mm) (Watanabe and Dawes, 1990) as in adults (Collins and Dawes, 1987), these results suggest that the velocity of the salivary film varies in different regions.

The relative order of the half-times at the different sites in the no-spacing arches was identical with that found in a study on adult subjects when saliva flow was unstimulated. Although in the spacing arch, the LAB site had a shorter clearance half-time than the UPB site. These results may be due to the fact that in the spacing arches, the tongue pushes out a portion of saliva from the lingual to the buccal side during swallowing, and this is in accordance with clinical findings that these sites are not susceptible to caries. The ideal arch in the primary dentition has spacing between the teeth (Pinkham et al, 1988), but Foster and Hamilton (1969) reported that only 33% had spacing between all the incisors in the upper and lower arches and that only 12% had spacing between all teeth in both arches in 100 British children aged 30-36 months.

Although it is known that nursing bottle caries depends on the feeding pattern in infancy, the present results suggest that the upper anterior buccal site in a no-spacing arch will be the most cariogenic site in a child's mouth because it has the lowest rate of salivary clearance.

2.3 Influence of the location of the parotid duct orifice on oral clearance

2.3.1 Aim

The rate of oral clearance was shown to vary markedly at different locations in the mouth. Oral clearance is slower for teeth in the maxilla than for those in the mandible and slower for the buccal surfaces of the teeth than for the lingual. Oral clearance on the labial surface of the upper anterior region is the slowest, while that for the lingual surface of the lower anterior region is the fastest. The lingual surface of the lower anterior region is near the openings of the ducts of the submandibular and sublingual glands, which probably accounts for the fastest rate of oral clearance being there. The effect of unstimulated parotid saliva on clearance around the maxillary first molar is not very striking, perhaps because the volume ratio of parotid saliva to total saliva is only about 15% at rest for each side. However, with stimulation, the proportion of parotid saliva increases, increasing clearance over the maxillary first molar, which is closest to the parotid duct. Few studies have examined positional relationships between the parotid duct orifice and the maxillary molars or individual differences in this positional relationship (Suzuki et al. 2009). The present study sought to ascertain the location of the parotid duct orifice in relation to the maxillary molars and whether oral clearance at locations 1 cm mesial and distal to the duct opening would be as rapid as that directly opposite the opening of the duct.

2.3.2 Materials and methods

2.3.2.1 Location of the parotid duct orifice

- Subjects

These were 35 consenting adults (20 men, 15 women) with a mean age of 27.1 years (range, 23-35 years). They had a complete dentition up to the second molar and no malocclusion. In each subject, plaster models were made after taking impressions of the upper and lower dentitions.

- Impressions of the right and left parotid duct orifice

Before taking an impression of the parotid duct orifice, a 2-mm hole was made at the centre of an adhesive therapeutic agent for aphthous stomatitis (Aftach; Teijin, Tokyo) and the agent was placed on the mucosa so that the hole matched the parotid duct orifice. Next, using a vinyl siloxane impression material (Stat BR; Car Japan, Tokyo), an impression of the buccal tooth surfaces and mucosa around the Aftach was taken with the teeth in centric occlusion to localize the duct opening in relation to the teeth.

- Reference plane setting

To take standard photos, a horizontal reference plane was set for each maxillary plaster model. This was a triangular plane defined by the occlusal plane at the maxillary midline and the distobuccal cusp of the left and right maxillary first molars.

- Taking standard photos

The standard plane was set horizontally and the plaster model was matched with the impression of the parotid duct orifice. In order to take standard photos from the same angle, the line connecting the disto- and mesio-buccal interdental papillae of the maxillary first molar was set orthogonal to the imaging direction.

- Location of the parotid duct orifice

After defining the reference plane on standard photos as the X axis and the line perpendicular to the X axis passing through the distal plane of the first molar as the Y axis, the location of the parotid duct orifice was measured in relation to the reference point.

In one subject the location of one parotid duct was determined six times in order to assess the reliability of the method.

2.3.2.2 Oral clearance on the buccal surface of the upper molar region

- Subjects

Subjects comprised 12 (8 men, 4 women mean age 28.3 years) of the original 35 subjects whose parotid duct orifice fell within 1 SD of the mean values for the X and Y coordinates obtained in Study 2-3-2-1.

- The rate of secretion by the parotid gland

The subjects had not eaten for at least one hour prior to the study and the studies were done in either the mid-morning or mid-afternoon. In the 12 subjects for whom oral clearance was measured, Lashley cups were attached over the left and right parotid duct orifices and with the agar holders in position, parotid saliva was collected on 5 separate occasions for a 5-min period without stimulant.

- Diffusing substance and agar holder

Oral clearance was assessed using the same methods of the study 2-1. 1% agar containing 1 mol/l potassium chloride was placed into cylinders (diameter, 4 mm; depth, 1 mm) held by an acrylic holder (width, 30 mm; height, 10 mm; thickness, 2 mm). The open surfaces of the cylinders were initially covered with microscope slides to allow the agar to set. In each agar holder, 3 cylinders were placed horizontally at 6-mm intervals (Fig. 9).

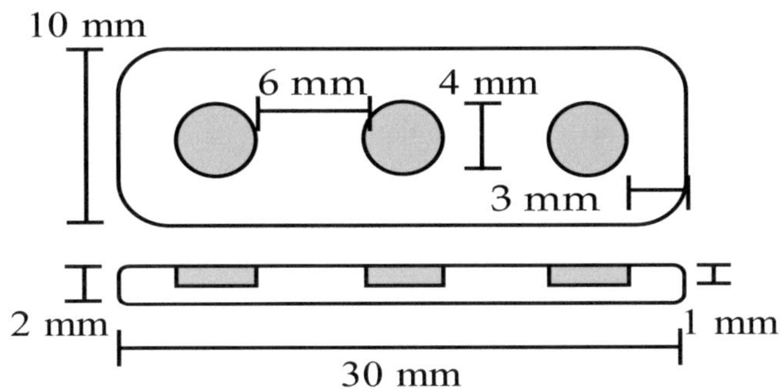

Fig. 9. Agar holder

The cylinders were attached to the teeth using Hydroplastic (TAK, Tokyo) so that the central cylinder would be on the buccal surface of the first molar, at the coordinates of the mean X and Y values obtained in Study 2-4-2-1. After salivary secretion stabilized, which took about 1 minute when parotid flow was measured with a Lashley cannula, the Parafilm was removed to initiate the experiment. On separate occasions, the holder was retained for 5, 10, 20 or 40 min without stimulant. At each time point, the concentration of residual potassium in the agar was measured for calculation of the half- time (the time for half of the potassium chloride to diffuse from the gel), as described by the study 1. Concentrations of potassium were measured by removing the agar cylinders from the holder, soaking each in 300 ml of 100 ppm sodium chloride solution for 90 min, and measuring the levels of eluted potassium by atomic absorption spectroscopy using an ANA-182 spectroscope (Tokyo Koden, Tokyo). The experiment was performed 3 times on both sides of each subject, and mean values were calculated. During the experiment, subjects were asked to refrain from touching the agar holder with their tongue or talking.

2.3.3 Result

Along the X axis, the location of the left and right parotid duct orifices varied within a range of −7.5 to +6.1 mm (Mean ± S.D.) from the reference point. Mean location (-0.36 ± 3.76) was just mesial to the reference point. Along the Y axis, the orifice was always located on the positive side of the reference point, ranging from +3.8 to +10.4 mm (mean value: 7.21 ± 2.15) (Fig. 10). This suggests that the parotid duct orifice is located above the reference plane near the contact surface between the maxillary first and second molars. Also, ranges of 13 mm in the mesiodistal direction and 6 mm in the perpendicular direction were noted, showing that there was a high degree of inter-individual variation. No significant left-right differences were identified. The intra-individual right-left differences were significantly less ($P< 0.001$) than the overall variability among subjects.

The unstimulated parotid saliva flow rates for left and right sides were 0.02 ± 0.02 and 0.02 ± 0.02 ml/min, respectively and no significant difference was found between results for the two sides. No significant differences in half-time could be detected between comparable left and right regions. Fig.11 shows the half-times for the right and left sides without stimulant. The half-time of the central cylinder was the shortest, followed by the mesial and then the distal cylinders, in that order, for both left and right sides. The half-time values among the 3 cylinders were all significantly different.

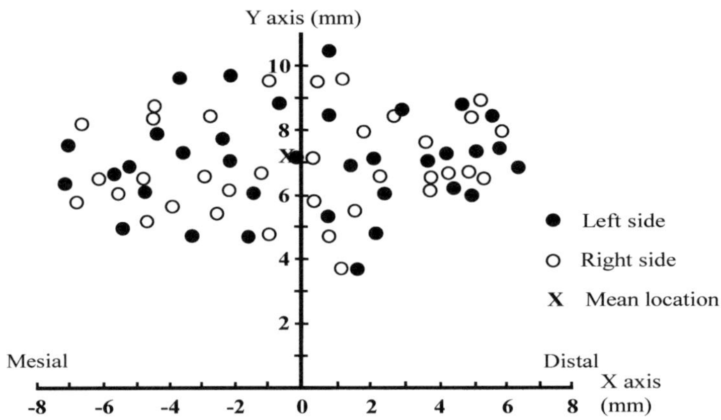

Fig. 10. The location of the parotid duct orifice. The symbols indicate the individual results for the 35 subjects. The x indicates the mean position of the duct orifice.

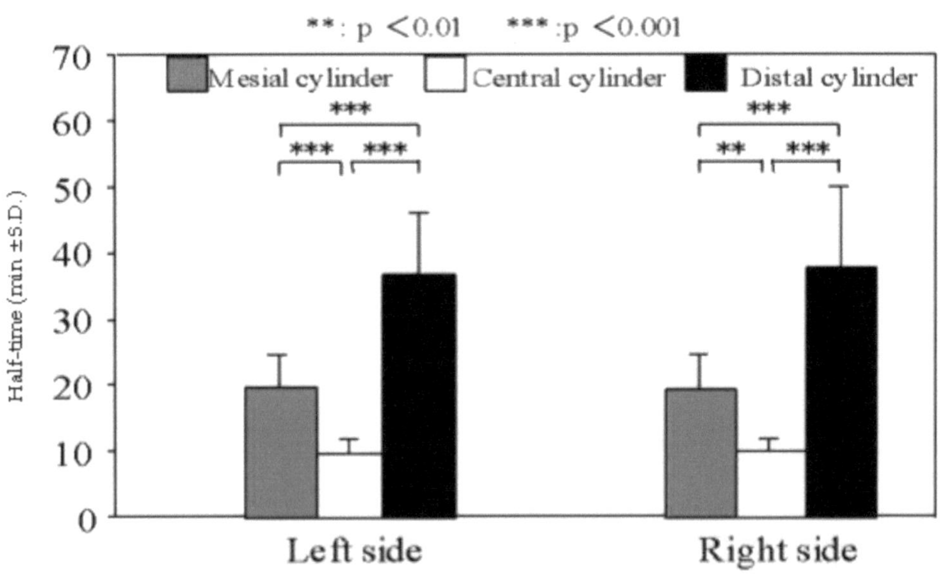

Fig. 11. Half-times when saliva flow was unstimulated

2.3.4 Discussion

The main finding from the second study was that clearance from a site directly opposite the opening of the parotid duct was significantly faster than from sites only one cm either mesially or distally. When salivary flow was unstimulated or stimulated, the clearance half-times mesial or distal to the duct opening were two or more times longer than those opposite the duct opening.

The present results are in conformity with those of Weatherell et al. (1986) who found that when a fluoride tablet was placed in the buccal vestibule, the fluoride concentration peaked in the fluid adjacent to the tablet but was much lower both mesially and distally. The previous reports and our results suggest that when parotid saliva exits the parotid duct, it primarily flows downwards and then, from the results of Weatherell et al. (1986), probably lingually over the occlusal surface of the teeth, rather than flowing mesially or distally in the buccal sulcus. If it flowed primarily in either of these two directions, one would have expected very little difference between the clearance rates from the mesial or distal agar cylinder and that from the cylinder positioned over the parotid duct opening. Sass and Dawes (1977) also reported that very little parotid saliva appeared to flow mesially when flow was either unstimulated or stimulated by the use of chewing gum.

In conclusion, the degree of individual variation in the location of the parotid duct orifice is great and its exact location will markedly affect oral clearance at different positions on the buccal surfaces of the upper molars.

2.4 Effect of salivary flow rate on fluoride retention in the mouth

2.4.1 Aim

Salivary clearance rates in different parts of the mouth are known to vary. The clearance half-times on the buccal surfaces of the upper anterior teeth were the longest of any site in the mouth. These show that the saliva secreted into the oral cavity is not perfectly mixed. Weatherell et al (1986) reports the difference by the fluoride distribution in the mouth after fluoride rinsing. Duckworth and Morgan (1991) and Heath et al. (2001) have also reported oral fluoride retention after use of fluoride rinse. These researches demonstrate the mechanism of the salivary clearance reported by Dawes (1983). According to Lear et al (1965), the salivary flow rate in the sleep is almost similar to the zero, but there are few reports the clearance of the fluoride in the sleep.

The purpose of this research was to measure the site-specificity of fluoride clearance when the subjects were awake and when they had been sleeping.

2.4.2 Materials and methods

40 mg of NaF and 5 ml distilled water were mixed with 0.15 g agarose which was heated until the agarose dissolved. Aliquots were pipetted into holders (diameter 4 mm, depth 1 mm) and these were bonded onto mouthguards produced from plaster casts of each subject (Fig.12).

The bonding sites were on the labial of maxillary incisors (UAB), the buccal of left maxillary molars (UPB) and the lingual of lower incisors (LAL). When the subjects were awake, the upper and lower mouthguards were fixed in the mouth and exposed to saliva for 15, 45 minutes. The agarose was taken out of the holder and put into 2 ml of distilled water mixed with 0.1 ml of the total ion strength adjustment buffer (TISABⅢ, Thermo Orion, IL, USA) for 90 minutes and the fluoride concentration was measured by atomic absorption spectrophotometry (Shimadzu AA-6105, Kyoto, Japan) as described in study 1. The fluoride concentration of the agarose held in the holder of each mouthguard was measured six times, and holders in which the mean concentration of agarose was more than 2 SD from the mean were excluded. To examine the retention of fluoride in the mouth during sleep, the mouthguards were placed before going to bed (0:00 a.m.) and removed at 6:30 a.m. and the fluoride concentration measured by a fluoride electrode (Thermo Fisher Scientific , MA, USA). The subjects, 6 adults who were all in good health and whose salivary flow rates exceed 0.3 ml/min were selected. Before the experiment, the subjects were explained the purpose and got their cooperation. In order to determine the effects of site specificity of salivary clearance, the data were analyzed by analysis of variance in randomized blocks and by Duncan`s New Multiple Range Test.

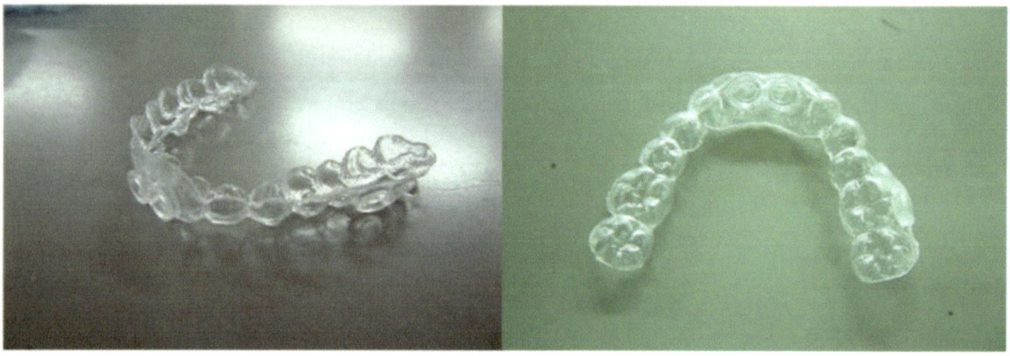

Fig. 12. Mouthguard with agarose holders. (Left: Upper, Right: Lower)

2.4.3 Results

Fig.13 showed the comparison of the mean half-times of each place, expressed as a standard in the value at LAL. The half-times were lowest in LAL and were highest in UAB. There were significant differences between the LAL and UAB ($p<0.01$), and between the LAL and UPB ($p<0.05$).

Table 3 showed the comparison of the mean volume of fluoride retention at 6:30 am. when the subjects had been sleeping. The fluoride concentrations were expressed as a percentage of that of the initial agarose which did not expose to saliva in the mouth. The values in LAL were also lowest, and UAB were highest. There were significant differences between the LAL and UAB ($p<0.05$) and between the LAL and UPB ($p<0.05$).

Most studies on fluoride clearance in the mouth have been carried out when the subjects were awake, and there is little information when they were sleeping. Ekstrand et al. (1986) and Featherstone et al.(1986) have suggested that fluoride, even at low concentrations, is necessary in the oral fluids to obtain maximum caries inhibition and have concluded that continuous or frequent elevation of the fluoride concentration in the oral fluids would be advantageous.

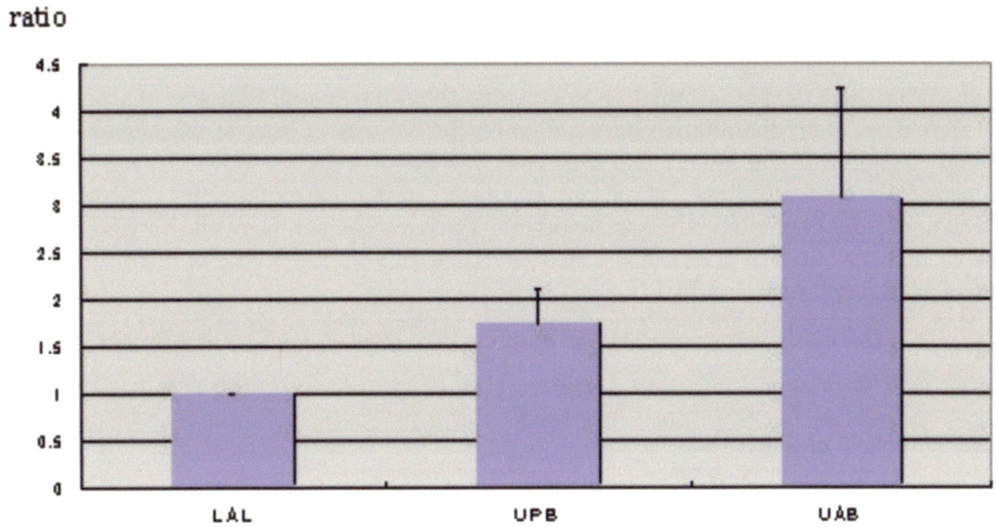

Fig. 13. The comparison of the mean half-times of each place, expressed in relation to the value for LAL. ($p<0.01$: LAL vs. UAB, $p<0.05$: LAL vs. UPB)

	LAL Mean S.D.	UPB Mean S.D.	UAB Mean S.D.
Subject A	1.2 ±0.3	1.5 ±0.2	3.4 ±2.2
Subject B	1.6 ±0.6	2.6 ±1.7	4.0 ±1.7
Subject C	0.9 ±0.2	0.6 ±0.2	2.6 ±1.8
Subject D	1.8 ±0.6	1.8 ±0.7	3.0 ±2.4
Subject E	1.7 ±0.8	4.1 ±1.2	3.3 ±2.6
Subject F	2.1 ±0.8	2.3 ±1.0	3.5 ±1.4
Mean	1.6 ±0.6*	2.2 ±1.2*	3.3 ±1.8

(*$p<0.5$: Significantly different from the mean volume of UAB)

Table 3. The mean volume (%) of fluoride retention at 6:30 a.m. when the subject had been sleeping

In this study it was shown that the fluoride concentration in the saliva was kept at high level for a long time during sleeping. In order to prevent dental caries at the buccal surfaces of the upper anterior teeth, it seems to be good to use a fluoride rinse before going to bed.

2.5 Estimation of the veloci1ty of the salivary film at different locations in the mouth

2.5.1 Aim

Although a great deal of information is available about the overall flow rate of whole saliva in man, there is no quantitative information on the velocity of flow of the salivary film in different regions of the mouth. Once secreted into the oral cavity, saliva forms a thin film, approximately 0.1 mm thick, which moves around inside the mouth until it is eventually swallowed. The higher the saliva secretion rate, the more frequently swallowing occurs, and the cleaner the mouth will be remain, However, this salivary film does not distribute evenly or reach all parts of the mouth.

The aim of this study was to estimate of the velocity of the salivary film at different locations in the mouth.

2.5.2 Materials and methods

- The equipment used in the salivary film velocity studies.

An extraoral device was used to adjust the flow rate of a 0.1-mm-thick film of artificial saliva over an agarose disk to determine the clearance half-time in the same manner as that performed intraorally (Dawes et al, 1989). Then, from the relationship between the intraoral and extraoral half-times, the salivary film velocities of the UAB and UPB sites were estimated. The half-time at UAB and UPB were evaluated by the method of study 2.1.

Fig. 14 shows the equipment used. The diameter of the well in the lower part of the device was 6 mm, the same as the width of the 0.1-mm-deep slot in the upper part. Thus, the fluid was directed over the surface of the gel. The well was 4 mm from the end of the device.

The well in the lower part of the device was filled with 1 mol/L KCl in 1 % agarose, as described for study 2.1 (Fig.3) and the upper and lower parts of the device were held together with three spring clamps.

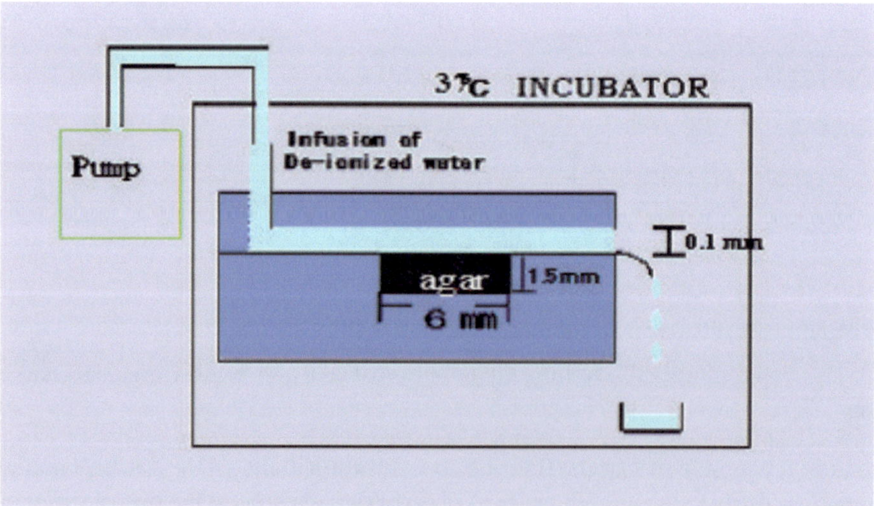

Fig. 14. An extraoral device for salivary flow rate study.

The device was maintained at 37 °C, and de-ionized water at the same temperature was infused with an infusion pump (Model 2000 IW, Harvard Apparatus Co., USA), the flow rate of which was adjustable over a wide range. The pump activated a 5-mL syringe connected to the device via polyethylene tubing.

For an experiment to be initiated, the flow rate of the pump was set to 1.07 ml/min. As soon as the water filled the tubing and completely covered the gel, a stop watch was started, and the flow rate of the pump was set to the desired value. After a pre-determined time, the pump was stopped, the two halves of the acrylic device were separated, and the agarose gel was removed with a needle and transferred to an appropriate volume of 100-ppm NaCl. The potassium concentration was determined by atomic absorption spectrophotometry.

For each flow rate, the experiment was repeated using five different gels for different durations to enable up to 70% of the KCl to be cleared from the gel. For each flow rate, a control gel that had not been exposed to water was used to determine the initial potassium concentration. The experiment was repeated three times for each flow rate. A least-squares straight line was fitted by computer to the potassium concentration plotted against the square root of time, and the half-time was calculated.

2.5.3 Results

There was a significant difference in the mean clearance half-time for the UAB between unstimulated (51.2 ± 19.3 min) and stimulated (40.1 ± 15.1 min) ($P < 0.05$) salivary flow rates. A significant difference was also found in the mean half-time for the UPB between unstimulated (19.2 ± 6.9 min) and stimulated (12.1 ± 5.2 min) ($P < 0.01$) salivary flow rates.

The results on the effect of velocity on the clearance half-time are shown in Table 4. With the flow rates set, the film velocity varied from 0.67 to 100 mm/min. The clearance half-times were inversely related to the velocity of fluid flow, and varied from 2.2 min to 58.3 min.

Fluid Flow Rate (ml/min)	Velocity of Fluid flow (mm/min)	Half-time (min) 6 mm in diam
0.06	100	2.2 ± 1.9
0.005	8.33	8.3 ± 1.2
0.003	5.00	15.1 ± 1.9
0.001	1.66	21.5 ± 2.8
0.0005	0.83	39.4 ± 4.2
0.0004	0.67	58.3 ± 6.1

Table 4. Effect of velocity of fluid flow on the mean half-time ± S.D. for clearance of KCL from an agarose gel, 1.5 mm in depth

Table 5 shows the in vivo clearance half-times for the UAB and the UPB as well as the estimated velocities of flow of the salivary film, as determined from the data in Table 1.

When salivary flow was unstimulated, the velocity of the salivary film of the UAB was estimated as 0.8 mm/min, whereas for the UPB it was estimated as 40.1 mm/min. When salivary flow was stimulated, the velocity of flow for the UAB was estimated as 2.3 mm/min, whereas for the UPB it was estimated as 12.1 mm/min.

Site	Salivary Flow	Clearance Half-time* (min)	Estimated Velocity of Salivary Film** (mm/min)
UAB	U	51.2 ± 19.3	0.8
	S	40.1 ± 15.1	2.3
UPB	U	19.2 ± 6.9	40.1
	S	12.1 ± 5.2	12.1

U = unstimulated (mean S.D = 0.42 ± 0.21 ml/min).
S = stimulated (mean S.D. = 4.5 ± 2.3 ml/min).
* = in vivo data.
** = from the data for the half-time in Table 1 of this study.

Table 5. Estimated velocity of the salivary film at UAB and UPB sites in the mouth

2.5.4 Discussion

Extraoral device for estimating salivary velocity at both sites:

Lagerlöf and Dawes (1984) measured oral salivary volume immediately before and after the onset of swallowing and reported the mean volumes to be 1.07 and 0.77 ml, respectively. Collins and Dawes (1987) and Watanabe and Dawes (1990) measured the surface area of the mouth, and based on the oral salivary volumes reported by Lagerlöf and Dawes (1984), they estimated the mean thickness of the salivary film in the mouth to be 0.1 mm. The extraoral device on which a 0.1-mm-thick salivary film flows on an agarose gel was designed on the basis of these reports to reproduce the situation in the mouth extraorally. Artificial saliva was allowed to flow onto an agarose gel in the same holder as that used in the mouth at different flow rates to determine the relationship between flow rate and clearance half-time, based on which salivary velocity at the two sites in the mouth was calculated. The velocity estimated with this device appears to be more useful for comparing salivary velocity between the different sites in the mouth than in determining actual salivary velocity. The mean half-time at stimulated salivary flow was 12.1 ± 5.2 min in the present study, which is substantially different from that obtained at unstimulated salivary flow. This may be attributable to the substantial difference in the secretion rate of saliva between when saliva is unstimulated and stimulated condition.

3. Conclusion

In clinical dentistry, it is generally accepted that the mandibular front teeth has low caries sensitivity, and the maxillary front teeth has high (e.g. nursing bottle caries) and the caries incidence of the buccal side of teeth in old person is higher than the lingual side. The results of this study have confirmed these clinical situations.

The author has also concluded that the velocity of salivary film is different by each site in the oral cavity and the slow velocity of the salivary film over the different surfaces of the teeth will retard clearance from diffusants of plaque such as acid. This suggests that the pH of each site in the oral cavity is slso different (Watanabe 2008),

4. Acknowledgment

The author thanks to emeritus proffesor C. Dawes (University of Manitoba) for helpful suggestions and comments.

5. References

Collins, L.M.C., & Dawes, C. (1987). The surface area of the adult human mouth and thickness of the salivary film covering the teeth and oral mucosa. *Journal of Dental Research,* Vol.66, No.8, pp. 1300 -1302, ISSN

Dawes, C. (1983). A mathematical model of salivary clearance of sugar from the oral cavity. *Caries Research,* Vol.17, No.4, pp. 321-334, ISSN 0008-6568

Dawes, C. & Watanabe, S., Biglow-Lecomte, P., & Dibdin, GH. (1989). Estimation of the velocity of the salivary film at some different locations in the mouth. *Joural of Dental Research,* Vol.68, No.11, pp.1479-1482, ISSN

Duckworth, RM., & Morgan, SN. (1991). Oral fluoride retention after use of fluoride dentifrices. *Caries Research,* Vol.25, No.2 , pp. 123-129, ISSN 0008-6568

Ekstrand, J. Lagerlöf, F. & Oliveby, A.(1986). Some aspects of the kinetics of fluoride in saliva; In: *Factors relating to demineralization and remineralisation of the teeth.* Leach, SA., & Edgar, WM. (Ed), ISBN 0-947946-73-X , pp 91-98, IRL Press, Oxford, UK

Foster, TD., & Hamilton, NC. (1969). Occlusion in the primary dentition. Study of children at 2.5 to 3 years of age. *British Dent Journal, Vol.*126, No.2, pp. 76-79, ISSN

Featherstone, JDB., Oreilly, MM., Shariati, M., & Brubler, S.(1986). Enhancement of remineralisation in vitro and in vivo, In *Factors relating to demineralization and remineralisation of the teeth.* Leach, SA., & Edgar, WM. (Ed), ISBN 0-947946-73-X , pp 23-34, IRL Press, Oxford, UK

Heath, K., Singh, V., Logan, R., & McIntyre. J.(2001). Analysis of fluoride levels retained intraorally or ingested following routine clinical applications of topical fluoride products. *Australian Dental Journal,* Vol.46, No. 1, pp.24-31, ISSN 00450421

Lagerlöf, F., & Dawes, C.(1984). The volume of saliva in the mouth before and after swallowing, *Journal of Dental Research.,* Vol.63, No.5 , pp. 618-621, ISSN 0022-0345

Lecomte, P., & Dawes, C.(1987).The influence of salivary flow rate on diffusion of potassium chloride from artificial plaque at different sites in the mouth. *Journal of Dental Research,* Vol.66, No.11, pp.1614 1618, ISSN

Pinkham, JR., Casamassimo, PS., Fields, HW., McTigur., DJ., & Nowak, AJ.(1988) *Infancy through adolescence.* Pediatric Dentistry, ISBN 0-7216-2106-6, Philadelphia, Saunders, USA

Primosh, RE., Weatherell, JA., & Strong, M.(1986). Distribution and retention of salivary fluoride from a sodium fluoride tablet following various intraoral dissolution methods. *Journal of Dental Research,* Vol.65, No.7, pp. 1001-1005, ISSN 0022-0345

Sass, R., & Dawes, C.(1997).The intra-oral distribution of unstimulated and chewing-gum-stimulated parotid saliva. *Archives Oral Biology,* Vol.42, No.7, pp. 469-474, ISSN 0003-9969

Suzuki, A., Watanabe, S., Ono, Y., Ohashi, H., Pai, C., Xing, X., & Wang, X. (2009). Influence of the location of the parotid duct orifice on oral clearance. *Archives of Oral Biology,* Vol.54, No.3, pp. 274-278, ISSN 0003-9969

Watanabe, S. & Dawes, C.(1992) Salivary flow rate and salivary film thickness in five-year-oldchildren. *Journal of Dental Research.,* Vol.69, No.5, pp. 1150-1153 0003-9969

Watanabe, S. (1992). Salivary clearance from different regions of the mouth in children. *Caries Research,* Vol.26, No.6, pp. 423-427 ISSN 0008-6568

Watanabe, S., Ogihara, T., Takahashi, S. 、Watanabe, K., Xuan, K., & Suzuki, S.(2010). Salivary clearance and pH in the different regions. IADR *General Session. http://iadr.confex.com/iadr/2010barce/webprogram/Paper132884.html,* (accessed 2012-05-08)

Weatherell, JA., Strong, M., Robinson, C., & Ralph, JP. (1986). Fluoride distribution in the mouth after fluoride rinsing. *Caries Research,* Vol.20, No.2, pp. 111-119, ISSN 0008-6568

4

An Assay for Determination of Hepatic Zinc by AAS – Comparison of Fresh and Deparaffinized Tissue

Raquel Borges Pinto[1], Pedro Eduardo Fröehlich[2], Ana Cláudia Reis Schneider[3], André Castagna Wortmann[3], Tiago Muller Weber[2] and Themis Reverbel da Silveira[1,*]

[1]Post Graduate Program in Medicine: Pediatrics,
[2]Post Graduate Program in Pharmaceutical Sciences, UFRGS, Porto Alegre,
[3]Post Graduate Program in Medical Sciences: Gastroenterology and Hepatology,
Universidade Federal do Rio Grande do Sul (UFRGS),
Hospital de Clínicas de Porto Alegre,
Brazil

1. Introduction

Atomic absorption spectroscopy (AAS) is a reliable method to determine metal concentrations. Zinc is a fundamental trace element because of its role in several essential biochemical functions. It is a component or co-factor of several enzymes, such as alcohol dehydrogenase and superoxide dismutase. It is of fundamental importance in cell division, genetic expression, and physiological processes, such as growth and development, immunity and wound healing; also, it plays a structural role in stabilizing biomembranes (Hambidge, 2000; Kruse-Jarres, 2001).

The importance of the determination of the hepatic concentration of certain metals is clearly established in the investigation of hereditary hemochromatosis and Wilson's disease (Pietrangelo, 2003, Roberts et al; 2003). As well, some interesting studies were published on the importance of zinc related to hepatic diseases. Decreases in plasma (Halifeoglu et al., 2004; Schneider et al., 2009; Pereira et al., 2011) or serum (Hamed et al., 2008, Matsuoka et al., 2009) zinc concentrations have been described in patients with chronic liver diseases. Authors that measured zinc in the liver parenchyma of adults and children with liver cirrhosis found low zinc levels (Milman et al., 1986; Göksu & Özsoylu, 1986; Sharda & Bhandari, 1986; Kollmeier et al., 1992; Adams et al., 1994). In patients with alcoholic cirrhosis, studies found abnormal zinc concentrations not only in the liver parenchyma (Rodriguez-Moreno et al., 1997), but also in subcellular fractions of the liver (Bode et al., 1988). In other diseases, such as biliary atresia (Bayliss et al., 1995; Sato et al., 2005), Indian childhood cirrhosis (Bhardwaj et al., 1980; Sharda & Bhandari, 1986), and chronic hepatitis B (Gür et al., 1998), also were found low liver zinc concentrations.

---

* Corresponding Author

Paraffin-embedded liver tissue usually stored in Pathology Laboratories may be used for analysis when fresh tissue is not available. Some hepatic diseases present a severe imbalance in the metabolism of metals, for example the excess of copper in Wilson's disease and iron in hemochromatosis. Recently, Wortmann and colleagues have validated an analytical method similar to ours for hepatic iron quantification, following the guidelines recommended by the International Conference on Harmonisation (ICH) of Technical Requirements for Registration of Pharmaceuticals for Human Use (Wortmann et al., 2007). Due to the fact that zinc is a protective metal to human health, the assessment of zinc status has a great importance in clinical investigation. Since there is only a few data available in the literature about methods to determine zinc concentration in hepatic tissue, the assay proposed can be helpful to this analysis. Therefore, the purpose of this study was to compare zinc concentrations in fresh and deparaffinized tissues and to determine whether the concentration of this metal in liver specimens is influenced by tissue processing in paraffin blocks.

2. Material and methods

This study was conducted after the validation of the graphite furnace AAS method to determine zinc concentration in bovine liver tissue (Fröehlich et al., 2006), in accordance with the guidelines established by ICH, FDA and ANVISA (ICH, 2005; FDA, 2001; ANVISA, 2003). We used a standard zinc solution (1 mg/ml) and standard bovine liver material from the National Institute of Standards and Technology (NIST, SRM 1577b) with known zinc concentrations (127 ± 16μg/g dry weight).

A steel scalpel was used to obtain 29 wedge biopsies from the same bovine liver obtained from a local market. Each specimen measured about 4 x 2 cm, and 2 were excluded from each group due to contamination and loss of material. Each of the 27 remaining specimens was divided in half, and two groups of samples were formed.

This study was approved by the Ethics in Research Committee of the Research and Graduate Studies, Hospital de Clínicas de Porto Alegre, Porto Alegre, Brazil.

2.1 Group 1

The samples in group 1 were placed in eppendorf tubes previously decontaminated with 10% nitric acid (HNO_3). Samples were lyophilized, using a lyophilizer (Micro moduli 97, Edwards®) for 72 hours. After lyophilization, 500 μL of concentrated HNO_3 (twice distilled, Zn concentration < 0.5 μg/g) was added to each sample, and sonication was applied for 1 hour. The samples were then placed in an incubator (303, Biomatic®) at 60° C for 1 hour to complete digestion of organic matter. From each of the 27 solutions prepared, 50 μl were poured into automatic sampler vials, and 950 μl of pure water (Milli-Q Plus, Millipore®) was added. Concentrations were then determined with a graphite furnace (HGA 800, Perkin-Elmer®) AAS (AAnalyst-3000, Perkin-Elmer®) and an automatic sampler (AS-72).

2.2 Group 2

Samples in the second group were fixed in formalin, embedded in paraffin, and later deparaffinized. The material was kept in paraffin blocks for about 7 days. The samples were deparaffinized by placing them in an incubator at 60° C for about 30 minutes to dissolve the paraffin block, and then in an average of two xylene baths (about 30 minutes each) alternating with 4 baths of alcohol (99%) and distilled water until the paraffin was totally removed. The samples were lyophilized for 72 h and concentrations were determined using AAS, following the same procedure described for samples in group 1. Mean dry weight of the 27 samples after lyophilization was 39.9 mg, and, after deparaffinization, 38 mg. Reagents were analyzed to assess contamination by zinc, which was negligible.

2.3 Statistical analysis

The software Microsoft Excel for Windows® and the Statistical Package for Social Sciences® 12.0 (SPSS) were used to create a database and to conduct statistical analysis. Measures of central tendency and dispersion were used to describe data, with means and standard deviations for quantitative variables. The Student t test for paired samples was used for comparisons between groups. The level of significance was established at $p<0.05$.

3. Results

3.1 Group 1 – Analysis of fresh liver tissue

Zinc concentration in group 1 was 173.6 ± 37.9 µg/g dry tissue (mean ± standard deviation). Table 1 shows the results for the 27 fresh bovine liver samples analyzed after lyophilization.

3.2 Group 2 – Liver tissue analyzed after deparaffinization

Zinc concentration in group 2 was 220.2 ± 127.0 µg/g dry tissue (mean ± standard deviation). Table 1 shows the results of 27 bovine liver samples that were embedded in paraffin, deparaffinized, lyophilized and then analyzed.

3.3 Comparison between analysis of fresh and deparaffinized samples

Liver zinc concentrations in fresh and deparaffinized samples were compared by Student t test for paired samples, and there were no statistically significant differences between these two groups ($p = 0.057$).

Samples	*Concentration fresh liver (µg/g dry tissue)*	*Concentration deparaffinized liver (µg/g dry tissue)*
1	213.3	246.6
2	178.5	148.0
3	169.5	163.3
4	241.2	153.8
5	154.3	201.8
6	154.0	201.4
7	179.4	171.6
8	169.2	193.7
9	170.8	185.1

Samples	*Concentration fresh liver (µg/g dry tissue)*	*Concentration deparaffinized liver (µg/g dry tissue)*
10	168.0	200.2
11	188.2	215.4
12	221.6	212.9
13	136.6	216.4
14	299.0	744.7
15	170.0	523.7
16	179.8	166.5
17	160.5	260.8
18	151.1	153.1
19	155.9	220.2
20	163.8	175.0
21	124.1	141.9
22	145.6	161.4
23	140.0	164.4
24	190.3	179.9
25	174.2	149.0
26	103.7	169.3
27	184.3	225.2
Mean	173.6	220.2
SD[a]	37.9	127.0
RSD[b]	21.9	57.5

[a] SD = standard deviation,[b] RSD = relative standard deviation (%).

Table 1. Liver zinc concentration in samples of fresh and deparaffinized bovine liver tissue.

4. Discussion

Zinc is an essential trace element to human health. It plays an important role in membrane stabilization and in cell protection against oxidative stress because it is part of structure of superoxide dismutase, the main enzyme in endogenous control of some types of free oxygen radicals. It also inhibits transition metals, such as copper and iron, from producing reactive types of oxygen (Powell, 2000). This metal is also essential for DNA and RNA polymerase, which has an important effect in hepatic regeneration (Sato et al., 2005).

The liver is one of the main organs in the metabolism of zinc. Disorders in zinc metabolism have been described in patients with chronic liver disease, and several studies found a decrease in plasma, serum or liver zinc concentrations (Loguercio et al., 2001; Halifeoglu et al., Schneider et al., 2009; Matsuoka et al., 2009, Milman et al., 1986; Göksu & Özsoylu, 1986; Sharda & Bhandari, 1986; Bode et al., 1988; Kollmeier et al., 1992). The decrease of zinc in liver disease seems to be associated with decreased intake, poor absorption associated with portal hypertension, and greater urinary excretion (Loguercio et al., 2001). Collagenase is a zinc-metalloenzyme and zinc is the most effective inhibitor for prolylhydroxylase, an enzyme which plays a key role in collagen synthesis. These two assumptions could explain the role of zinc in collagen deposition and reabsorption in liver disease, the role played by zinc in liver fibrosis and in the evolution of chronic hepatitis toward cirrhosis (Faa et al., 2008).

Some reports found an increase in liver zinc concentrations in chronic liver disease. An increase in copper and zinc liver concentrations was found in Canadian children with chronic cholestasis (Phillips et al., 1996). Another case report described the increase in zinc concentration in hepatic tissue of a child with hepatosplenomegaly and symptoms of zinc deficiency, and the authors speculated about the existence of a zinc metabolism disorder (Sampson et al., 1997). A study that investigated the concentration of metals in liver tissue of adults with hereditary hemochromatosis found an increase in zinc in the liver parenchyma. The authors suggested that the concurrent increase in iron and zinc might be explained by the greater intestinal absorption of these metals (Adams et al., 1991).

The test usually conducted to determine body zinc is the measurement of plasma zinc concentration. However, plasma zinc concentrations do not seem to reflect the concentration found in the liver parenchyma (Göksu & Özsoylu, 1986; Sato et al., 2005). This may be explained by the fact that there are very efficient homeostatic mechanisms to correct plasma or serum zinc deficiencies, which makes it difficult to diagnose marginal deficiency by using this method. Therefore, the investigation of zinc concentration in liver tissue is important.

Studies report a great variation in liver zinc concentrations, maybe due to the different techniques used (Table 2). Kollmeier et al. (1992) studied the distribution of zinc in adult liver parenchyma from necropsy material, and found a small variation in intraorgan metal concentrations. They reported that zinc concentrations in the liver do not seem to be associated with sex or age (Kollmeier et al., 1992). Another study, conducted with children by Coni et al. (1996), confirmed these findings. They measured the concentration of metal in necropsy material from infants that died of sudden infant death syndrome and from pediatric control subjects, and found that a small liver sample is representative of liver concentration in the whole liver.

Author and year	Subjects and technique	Zinc concentration in liver tissues (μg/g dry tissue)
Adams et al. (1991)	Healthy controls (n=21) Flame AAS[a]	326.3 ± 65.4
Kollmeier et al. (1992)	Unselected necropsies (n=58) Flameless AAS[a]	280.0 ± 178.0
Bush et al. (1995)	Unselected necropsies (n=30) ICP-ES[b] AAS[a]	191.0 ± 56.3
Treble et al. (1998)	Unselected necropsies (n=73) Graphite furnace AAS[a]	118.3 ± 44.4
Hatano et al. (2000)	Healthy controls (n=21) Particle-induced X-ray emission	281.0 ± 25.5

[a] AAS = atomic absorption spectrophotometry.
[b] ICP-ES = inductively coupled plasma emission spectroscopy.

Table 2. Zinc concentration in hepatic tissue according to the literature.

Fresh tissue is not always available for chemical analysis, but formalin-fixed tissue often is. In our study, zinc concentration measurements in fresh liver tissue and deparaffinized tissue have been shown to be concordant. We have found in fresh e deparaffinized tissues respectively, 173,6 ± 37,9 µg/g dry weight and 220,2 ± 127 µg/g dry weight. Two deparaffinized tissue samples showed higher values of zinc than the others samples, causing a higher standard deviation. There was no evidence of mineral contamination during the embedding process to account for these divergent values.

We found only one study in the literature, conducted at the Mayo Clinic by Bush et al. (1995) that compared zinc concentration in fresh and deparaffinized liver tissue. Their study investigated the concentration of metals in several organs using material obtained from autopsy of 30 presumably healthy individuals. Zinc concentration found in fresh liver tissue was 191 ± 56.3 µg/g dry weight, and, in formalin-fixed tissue, 204 ± 63.2 µg/g dry weight. They concluded that formalin fixation long-term storage has little effect on zinc concentrations in tissue and that zinc was homogeneously distributed in liver.

Due to the clinical importance of zinc in liver diseases, the use of paraffin-embedded specimens for analysis is extremely useful when fresh tissue is not available. Stored material for analysis may be available even years after the biopsy or autopsy sample was obtained.

4.1 Conclusion

More than the results themselves, the proposed protocol for paraffinization/deparaffinization as well as for sample preparation for zinc determination by atomic spectroscopy in paraffinized samples were adequately established. According to the results of this study, paraffin embedding and deparaffinization do not significantly affect the determination of zinc concentrations in liver tissue, and, therefore, stored material can be used for analysis.

5. Glossary

AAS - Atomic Absorption Spectrophotometry.
ANVISA - Agência Nacional de Vigilância Sanitária (Sanitary Surveillance National Agency).
FDA - Food and Drug Administration.
ICH - International Conference on Harmonisation of Technical Requirements for Registration of Pharmaceuticals for Human Use.

6. Acknowledgements

We are grateful to Coordenação de Aperfeiçoamento de Pessoal de Nível Superior (CAPES), Conselho Nacional de Desenvolvimento Científico e Tecnológico (CNPq) and Fundo de Incentivo a Pesquisas - Hospital de Clínicas de Porto Alegre (FIPE-HCPA).

7. References

Adams PC, Bradley C, Frei JV. Hepatic zinc in hemochromatosis. Clin Invest Med.1991;14:16-20.

Adams PC, Bradley C, Frei JV. Hepatic iron and zinc after portocaval shunting for nonalcoholic cirrhosis. Hepatology. 1994; 19(1):101-105.

ANVISA - Agência Nacional de Vigilância Sanitária. Guidelines for validation of analytical and bioanalytical methods. Diário Oficial da União, Brasília, 2003.

Bayliss EA, Hambidge KM, Sokol RJ, et al. Hepatic concentrations of zinc, copper and manganese in infants with extrahepatic biliary atresia. J Trace Elem Med Biol 1995; 9:40-43.

Beilby JP, Prins AW, Swanson NR. Determination of hepatic iron concentration in fresh and paraffin-embedded tissue. Clin Chem 1999; 45(4):573-574.

Bhardwaj S, Miglani N, Gupta BD, et al. Hepatic zinc levels in Indian childhood cirrhosis. Indian J Med Res 1980; 71:278-281.

Bode JC, Hanisch P, Henning H, et al. Hepatic zinc content in patients with various stages of alcoholic liver disease and in patients with chronic active and chronic hepatitis. Hepatology 1988; 8:1605-1609.

Bush VJ, Moyer TP, Batts KP, et al. Essential and toxic element concentrations in fresh and formalin-fixed human autopsy tissue. Clin Chem 1995; 41:284-294.

Coni P, Ravarino A, Farci AM, et al. Zinc content and distribution in the newborn liver. J Pediatr Gastroenterol Nutr 1996; 23:125-129.

Faa G, Nurchi VM, Ravamino A, et al. Zinc in gastrointestinal and liver disease. Coordination Chem Rev 2008; 252:1257-1269

Fröelich PE, Pinto RB, Wortmann AC, et al. Full validation of an electrothermal atomic absorption for zinc in hepatic tissue using a fast sample preparation procedure. Spectroscopy 2006; 20:81-87.

Göksu, N, Özsoylu S. Hepatic and serum levels of zinc, copper and magnesium in childhood cirrhosis J Pediatr Gastroenterol Nutr 1986; 5:459-462.

Gür G, Bayraktar Y, Ozer D, et al. Determination of hepatic zinc content in chronic liver disease due to hepatitis B virus. Hepatogastroenterology 1998; 45:472-476.

Hambidge M. Human zinc deficiency. J Nutr 2000; 130:1344-1349.

Hamed SA, Hamed EA, Farghaly MH, et al. Trace elements and flapping tremors in patients with liver cirrhosis. Is there a relationship? Saudi Med J. 2008; 29(3):345-351.

Halifeoglu I, Gur B, Aydin S, et al. Plasma trace elements, vitamin B12, folate, and homocysteine levels in cirrhotic patients compared to healthy controls. Biochemistry (Mosc) 2004; 69(6):693-696.

Hatano R, Ebara M, Fukuda H, et al. Accumulation of copper in the liver and hepatic injury in chronic hepatitis C. J Gastroenterol Hepatol 2000; 15:786-791.

ICH - Harmonized Tripartite Guideline, Test on Validation of Analytical Procedures - Q2(R1). In: International Conference on Harmonisation of Technical Requirements for Registration of Pharmaceuticals for Human Use, 2005.

Kollmeier H, Seemann J, Wittig P, et al. Zinc concentrations in human tissues. Liver zinc in carcinoma and severe liver disease. Pathol Res Pract 1992; 188:942-945.

Kruse-Jarres JD. Pathogenesis and symptoms of zinc deficiency. Am Clin Lab 2001; 20:17-22.

Loguercio C, De Girolamo V, Federico A, et al. Trace elements and chronic liver diseases. J Trace Elem Med Biol 2001; 11:158-161.

Matsuoka S, Matsumura H, Nakamura H, et al. Zinc supplementation improves the outcome of chronic hepatitis C and liver cirrhosis. J Clin Biochem Nutr 2009; 25:292-303.

Milman N, Laursen J, Podenphant, et al. Trace elements in normal and cirrhotic human liver tissue. I. Iron, copper, zinc, selenium, manganese, titanium and lead measured by X-ray fluorescence spectrometry. Liver 1986; 6:111-117.

Olynyk JK, O'Neill R, Britton RS, et al. Determination of hepatic iron concentration in fresh and paraffin-embedded tissue: diagnostic implications. Gastroenterology 1994; 106:674-677.

Pereira TC, Saron ML. Carvalho WA, et al. Research on zinc blood levels and nutritional status in adolescents with autoimmune hepatitis. Arq Gastroenterol 2011; 48(1):62-65.

Pietrangelo A. Haemochromatosis. Gut 2003; 52 Suppl 2:ii23-ii30

Phillips MJ, Ackerley CA, Superina RA, et al. Excess zinc associated with severe progressive cholestasis in Cree and Ojibwa-Cree children. Lancet 1996; 347:866-868.

Powell SR. The antioxidant properties of zinc. J Nutr 2000; 130:1447-54.

Roberts EA, Schilsky ML. A practice guideline on Wilson disease. Hepatology 2003; 37:1475-1492.

Rodriguez-Moreno F, González-Reimers E, Santolaria-Fernandez F, et al. Zinc, copper, manganese and iron in chronic alcoholic liver disease. Alcohol 1997; 14:39-44.

Sampson B, Kovar IZ, Rauscher A, et al. A case of hyperzincemia with functional zinc depletion: a new disorder? Pediatr Res 1997; 42:219-225.

Sato C, Koyama H, Satoh H, et al. Concentrations of copper and zinc in liver and serum samples in biliary atresia patients at different stages of traditional surgeries. Tohoku J Exp Med 2005; 207:271-277.

Schneider AC, Pinto RB, Fröelich PE et al. Low plasma zinc concentrations in patients with cirrhosis. J Pediatr (Rio J). 2009; 85(4):359-364.

Sharda B, Bhandari B. Studies of trace elements in childhood cirrhosis. Acta Pharmacol Toxicol (Copenh) 1986; 59 Suppl 7:206-10.

Treble RG, Thompson TS, Lynch HR. Determination of copper, manganese and zinc in human liver. Biometals 1998; 11:49-53.

U.S., Department of Health and Human Services. Food and Drug Administration. Guidance for industry: bioanalytical method validation. Available at: http://www.fda.gov/CDER/GUIDANCE/4252fnl.htm. Accessed: June 30, 2011.

Wortmann AC, Froehlich PE, Pinto RB, et al. Hepatic iron quantification by atomic absorption spectrophotometry: Full validation of an analytical method using a fast sample preparation. Spectroscopy 2007; 21:161-167.

Section 2

UV-VIS Spectroscopy

5

Synthesis and Characterization of CdSe Quantum Dots by UV-Vis Spectroscopy

Petero Kwizera[1,*], Alleyne Angela[1], Moses Wekesa[2,*], Md. Jamal Uddin[2,*] and M. Mobin Shaikh[3]

[1]*Department of Mathematics, Edward Waters College, Jacksonville, Florida,*
[2]*Department of Natural Sciences, Coppin State University, Baltimore, Maryland,*
[3]*Sophisticated Instrument Centre (SIC), School of Basic Science, Indian Institute of Technology Indore, Indore,*
[1,2]*USA*
[3]*India*

1. Introduction

CdSe nanocrystals are effective visual aid to demonstrate quantum mechanics, since their transition energies can be explained as a Particle in a Box, where a delocalized electron is the particle and the nanocrystal is the box. Kippeny and co-workers[1] have provided more background information and theoretical discussion. Additionally, Ellis et al.[2] have stated that modern science is becoming increasingly interdisciplinary. One example is material science, a broad, chemically oriented view of solids that results from the combined viewpoints of chemistry, physics, engineering, and for biotechnology, the biological sciences. Schulz[3] has suggested that nanotechnology is an exciting emerging field that involves the manipulation of the atoms and molecules at the nano scale. It is projected that important advances in engineering will come from understanding of the properties of matter constructed from building blocks whose size and shape is uniform and on the 1-100 nm scale. These consequences include technologies to be used in medicine[4], advances in computer technologies[5], defense[6] and everyday applications[3].

Several methods exist for synthesizing Cd-Se Quantum Dots. The Molecular Beam Epitax (MBE) is expensive and not readily accessible. Kippeny et al. have used dimethyl cadmium, which is expensive, explosive, and pyrophonic making the system difficult to control and reproduce. Peng and others[7-9] have pioneered the kinetic synthesis of Cd-Se nanocrystals from CdO and elemental Se. Boatman et al.[10] have prepared Cd-Se nanocrystals using a kinetic synthesis with a quenching technique where the temperature was 225°C. The visible absorption and emission spectra of individual samples collected at various time intervals during the experimental run were recorded and the maximum wavelength peak were determined. In this paper we report a modified technique of kinetic synthesis of Cd-Se nanocrystals that is safer, simple and can easily be carried out by students in the normal chemistry lab.

* Corresponding Authors

2. Experimental

UV-Vis spectrophotometer (Perkin Elmer Lambda 950) was used for spectroscopic measurements. The scan speed was 11.54 nm /min, integration time was 0.52 s and the data interval was 0.10 nm. The hotplate was Labcongo (115 V, 12 A) and the heating was set at level 3. All chemicals used were bought from Sigma-Aldrich and were of analytical grade. 60 mg of Se, 10 cm^3 of 1-octadecene and 0.8 cm^3 of trioctylphospine were mixed together in a round-bottomed flask. The solution was then continuously stirred with a magnetic stirrer on a hot-plate and warmed for a few minutes in a fume-hood. Separately, 26 mg of CdO was added to a 25 cm^3 round-bottomed flask and clamped in a heating mantle. 1.2 cm^3 of acid and 20 cm^3 of octadecene were added and mixed together. The solution was heated until CdO dissolved. The CdO solution was then sub-divided into 5 Erlenmeyer flasks each containing 4 cm^3 of the stock solution. 0.5 cm^3 of Se stock solution was then transferred into the CdO solution with pipette. The samples were heated for 50 s, 60 s, 70 s, 80 s and 120 s, respectively.

3. Results and discussion

The colloidal suspensions of Cd-Se quantum dots of increasing size from left to right are shown in Figure 1.

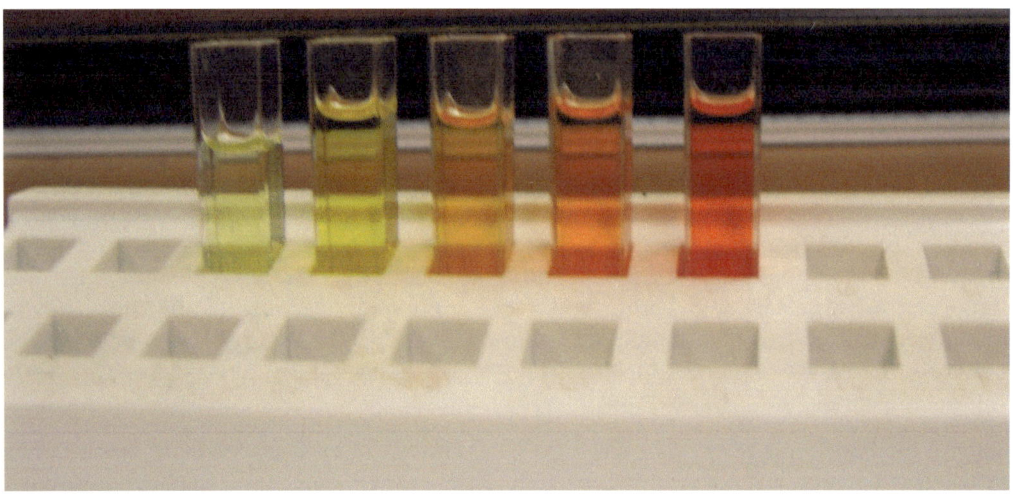

Fig. 1. Colloidal suspensions of Cd-Se quantum dots

The samples viewed in ambient light vary from green-yellow to orange-red. These changes in color which have been noted by other workers[1, 10] are attributed to the increasing size of the Cd-Se nano-crystals.

Figure 2 presents the calculated diameter of nanocrystal with time[1]. The diameter of nanocrystals increases with increasing time. As the nano-crystal size increases, the energy of the first excited state decreases.

The Cd-Se nano-crystals stay suspended in solution and cannot be filtered out. The oleic acid acts as a surfactant, binding to the exterior of the crystal lattice and allowing for the crystal to remain soluble in the octadecene[10]. The diameter of the nanocrystal was calculated using Kippeny[1] method and was found to be in the range found by other workers[10]. The Cd-Se crystal growth has been found to be temperature dependent. Transmission electron microscope (TEM) measurements of Cd-Se nanocrystals by others suggest that such wavelengths correspond to 2- 4 nm diameter crystals[10] with at most a few hundred atoms.

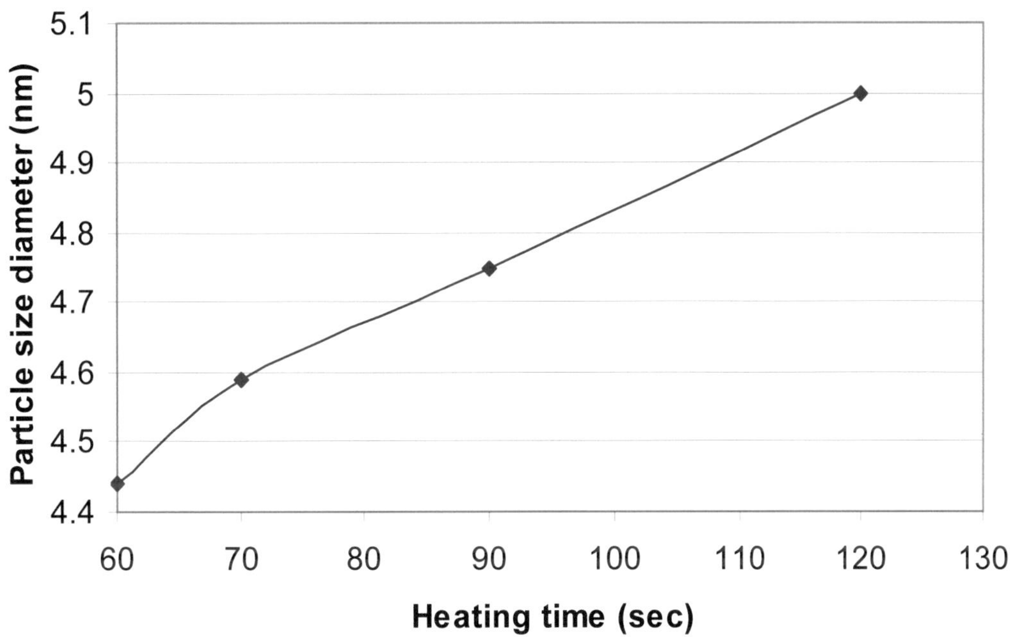

Fig. 2. The effect of particle size growth with time

Figure 3 presents ground state peak wavelengths as a function of reaction time. As the reaction progresses, the peak wavelength decreases. As nanocrystals grow, it has been suggested that their peak emission quickly approaches the band gap of bulk Cd-Se (730 nm).

The observable peak maximum shifts from violet to green with increasing crystal size. The absorption shows peak maxima with additional absorption at lower wavelengths due to the starting materials and oleic acid polymerization. Heating oleic acid and octadecene alone yields increasing visible absorption at increasing wavelengths over time as the effects of oleic acid polymerization become noticeable.

Figure 4 presents UV-Vis spectra of Cd-Se colloidal nanocrystals. The scan range was between 400 nm to 600 nm. The maximum peak shifted toward the longer wavelength. This observation is expected because as the crystal size increases, the energy absorbed or emitted decreases. The sample heated for only 50 sec did not show any peak.

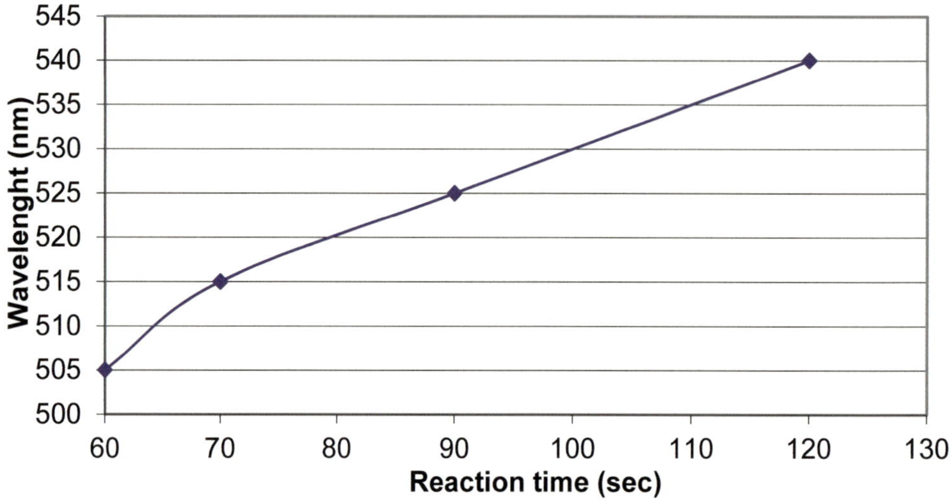

Fig. 3. The change of wavelength with reaction time

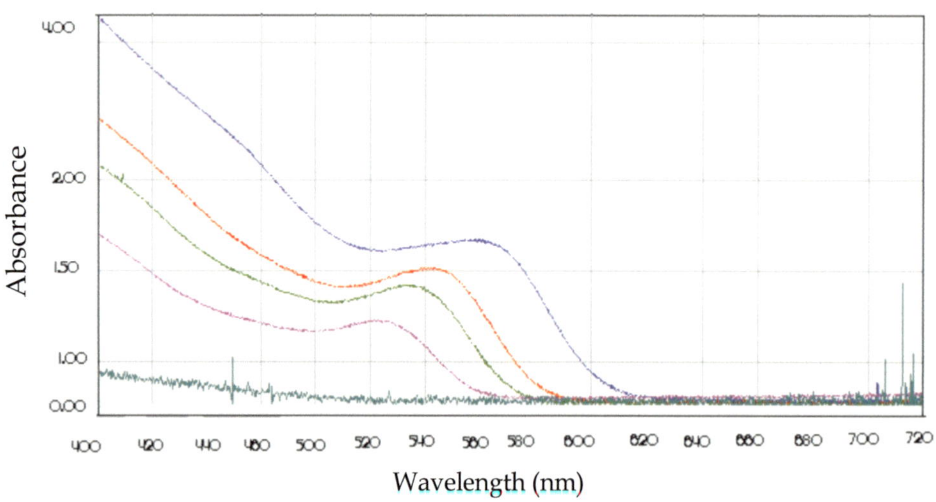

Fig. 4. UV-Vis spectra of Cd-Se colloidal suspension

4. Discussion of the optical measurements and results

As the nanocrystal size increases, the energy of the first excited state decreases qualitatively following particle in a box behavior[1]. The optical absorption results using Perkin Elmer Lambda 950 spectrometer are indicated in Figure 4.

a. Energy Shift and Nano- Crystal Size.

Using the L E. Brus [1,11,12] model we assume the following:

1. The nanocrystal is spherical with a radius R.
2. The interior of the nanocrystal consists of uniform medium and the excited electron and hole pair.
3. The potential energy outside the radius R is infinite—the radius R defines the confining boundary of the box.

The solution to the spherical Schrödinger equation leads to the energy of the exciton—electron hole pair as[1]:

$$E_{ex} = \frac{h^2}{8R^2}\left(\frac{1}{m_e}+\frac{1}{m_h}\right) - \frac{1.8e^2}{4\pi\varepsilon_{CdSe}\varepsilon_0 R} + \frac{e^2}{R}\left\langle \sum_{k=1}^{\infty}\alpha_k\left(\frac{S}{R}\right)^{2k}\right\rangle$$

The first term is the kinetic energy and the second term the Coulomb potential attractive energy; and the third term is the polarization energy.

b. Using the first term to calculate the exciton energy

At small R the predominant term is the first term (because of the inverse square R dependence since R<1 for a simple example $\frac{1}{(0.5)^2} > \frac{1}{0.5}$).

We can therefore use the first term to approximate R the radius of the nanoparticles as follows:

The energy needed to create the first peak - corresponding to the peak position in the spectra is $E_u = E_g + E_{ex}$ this energy corresponds to 500 nm from Figure 4.

The energy then converts to 2.48 eV (using the well known conversion formula $\frac{1.24}{1\mu m}$ for photon energy to eV.

The energy gap of bulk CdSe corresponds to 730 nm (0.73 μm) and is 1.70 eV [1,10]

This leads to the exciton energy of 0.78 eV

Using the formula:

$$E_{ex} = \frac{h^2}{8R^2}\left(\frac{1}{m_e}+\frac{1}{m_h}\right) - \frac{1.8e^2}{4\pi\varepsilon_{CdSe}\varepsilon_0 R} + \frac{e^2}{R}\left\langle \sum_{k=1}^{\infty}\alpha_k\left(\frac{S}{R}\right)^{2k}\right\rangle$$

and the first term alone as the approximation for small R

$$E_{ex} = \frac{h^2}{8R^2}\left(\frac{1}{m_e}+\frac{1}{m_h}\right) = 0.78eV$$

Using h as Planck's constant ; the electron effective mass m_e = 0.13 mass of a free electron and m_h equals 0.45 times the free electron mass

R can then be calculated to be the following:

$$R^2 = \frac{h^2}{8E_{ex}}\left(\frac{1}{m_e} + \frac{1}{m_h}\right)$$

$$R = \sqrt{\frac{(6.626)^2 x10^{-68} x9.91452}{9.1095x10^{-31} x8x0.78x1.602x10^{-19}}} = 2.18x10^{-9} m .$$

This leads to diameter of about 4 nm

5. Conclusion

We have demonstrated a more convenient synthesis method for colloidal CdSe quantum dots. This method does not involve quenching. This makes it easier for students to make the semiconductor nanoparticles. This synthesis method depends on different heating times for premixed CdO and Se solutions.

6. Acknowledgements

We would like to thank;

- Congresswoman Corrine Brown who was instrumental in procuring EWC grant to purchase the optical laboratory equipment.
- Army Research Laboratory and Dr N. Sundaralingam, Chair Department of Math and Sciences, Edward Waters College for their assistance.
- Dr. Elias Towe, ECE, MSE and Director-CNXT, Carnegie Mellon University for collaborative advice.
- Finally EWC who made the time available and provided the necessary resources to conduct this research possible.

7. References

[1] Tadd, K.; Laura, A. S.; Sandra, J. R. *J. of Chem. Edu*. 2002, 79, 9.
[2] Ellis, A. B.; Geselbracht, M. J.; Johnson, B.J.; Lisensky, G.C.; Robinson, W.R.; *American Chemical Society*, Washington, D.C., 1993.
[3] Schulz, W.G.; *Chem. Eng. News*, 2000, 78, 41.
[4] Rawls, R.L.; *Chem. Eng. News*, 2003, 81, 39.
[5] Halford, B.; *Chem. Eng. News,* 2004, 82, 5.
[6] Wilson, E.K.; *Chem. Eng. News*, 2003, 81, 29.
[7] Peng, Z.A.; Peng, X,; *J. Am. Chem. Soc*., 2001, 123, 183-184.
[8] Yu, W.W.; Peng, X.; *Angew. Chem. Int. Ed. Engl*., 2002, 41, 2368-2371.
[9] Peng, Z.A.; Peng, X.J.; *J. Am. Chem. Soc*., 2002, 124, 3343-3353.
[10] Elizabeth M. B.; George C. L.; *J. of Chem. Edu*., 2005, 82, 1697-1699.
[11] Brus, L.E.; *J. Chem. Phys*., 1983, 79, 5566-5571.
[12] Brus, L.E.; *J. Chem. Phys*., 1984, 80, 4403-4409.

6

The Use of Spectrophotometry UV-Vis for the Study of Porphyrins

Rita Giovannetti
University of Camerino, Chemistry Section of School of Environmental Sciences, Camerino Italy

1. Introduction

The porphyrins (Fig. 1) are an important class of naturally occurring macrocyclic compounds found in biological compounds that play a very important role in the metabolism of living organisms. They have a universal biological distribution and were involved in the oldest metabolic phenomena on earth. Some of the best examples are the iron-containing porphyrins found as heme (of haemoglobin) and the magnesium-containing reduced porphyrin (or chlorine) found in chlorophyll. Without porphyrins and their relative compounds, life as we know it would be impossible and therefore the knowledge of these systems and their excited states is essential in understanding a wide variety of biological processes, including oxygen binding, electron transfer, catalysis, and the initial photochemical step in photosynthesis.

The word porphyrin is derived from the Greek porphura meaning purple. They are in fact a large class of deeply coloured pigment, of natural or synthetic origin, having in common a substituted aromatic macrocycle ring and consists of four pyrrole rings linked by four methine bridges (Milgrom, 1997; D. Dolphin, 1978).

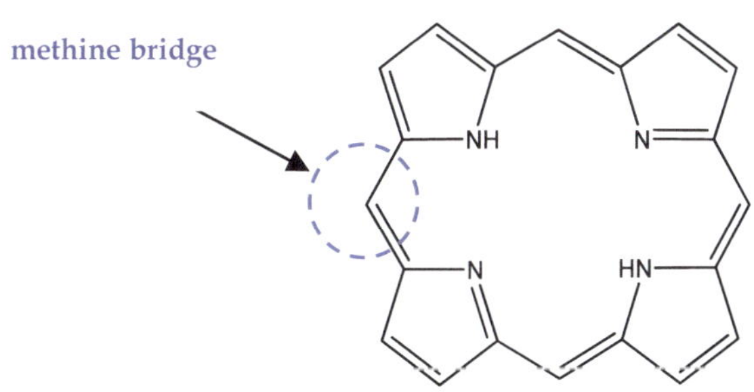

Fig. 1. The structure of porphyrin.

The porphyrins have attracted considerable attention because are ubiquitous in natural systems and have prospective applications in mimicking enzymes, catalytic reactions, photodynamic therapy, molecular electronic devices and conversion of solar energy. In particular, numerous porphyrins based artificial light-harvesting antennae, and donor acceptor dyads and triads have been prepared and tested to improve our understanding of the photochemical aspect of natural photosynthesis.

The porphyrins play important roles in the nature, due to their special absorption, emission, charge transfer and complexing properties as a result of their characteristic ring structure of conjugated double bonds (Rest et al.,1982).

As to their electronic absorption, they display extreme intense bands, the so-called Soret or B-bands in the 380–500 nm range with molar extinction coefficients of 10^5 M^{-1} cm^{-1}. Moreover, at longer wavelengths, in the 500–750-nm range, their spectra contain a set of weaker, but still considerably intense Q bands with molar extinction coefficients of 10^4 M^{-1} cm^{-1}. Thus, their absorption bands significantly overlap with the emission spectrum of the solar radiation reaching the biosphere, resulting in efficient tools for conversion of radiation to chemical energy. In such a conversion, the favourable emission and energy transfer properties of porphyrin derivatives are indispensable as in the case of chlorophylls, which contain magnesium ion in the core of the macrocycle. Also, metalloporphyrins can be utilized in artificial photosynthetic systems, modelling the most important function of the green plants (Harriman et al., 1996).

The studies of the wavelength shift of their adsorption band and the absorbance changes as function of pH, temperature, solvent change, reaction with metal ions and other parameters permits to obtained accurate information about equilibrium, complexation, kinetic and aggregation of porphyrins.

This review, resumes the best successes in the use of spectrophotometer UV-Vis for explained the chemical characteristics of this extraordinary group of natural occurring molecules and clarifies the potential of these molecules in many fields of application.

2. The chemical characteristics of porphyrins

The synthetic world of porphyrins is extremely rich and its history began in the middle of 1930s. An enormous number of synthetic procedures have been reported until now, and the reason can be easily understood analysing the porphyrin skeleton. In principle, there are many chemical strategies to synthesized porphyrins, involving different building blocks, like pyrroles, aldehydes, dipyrromethanes, dipyrromethenes, tripyrranes and linear tetrapyrroles.

The most famous monopyrrole polymerization route to obtain porphyrins involves the synthesis of tetraphenyl porphyrins, from reaction between pyrrole and benzaldehyde (Atwood et al., 1996). This procedure was first developed by Rothemund (Rothemund, 1935) and, after modification by Adler, Longo and colleagues (Adler et al., 1967), was finally optimized by Lindsey's group (Lindsey et al.,1987). In the Rothemund and Adler/Longo methodology the crude product contains between 5 and 10% of a byproduct, discovered later to be the *meso*-tetraphenylchlorin which is converted in the product under oxidative conditions (Fig. 2).

Rothemund in 1935 set up the synthesis of porphyrins in one step by reaction of benzaldehyde and pyrrole in pyridine in a sealed flask at 150 °C for 24 h but the yields were low, and the experimental conditions so severe that few benzaldehydes could be converted to the corresponding substituted porphyrin (Rothemund, 1936; Menotti, 1941). The reason in the low yield is that the main by-product of reaction was *meso*substituted chlorin and in understanding the nature of its formation, Calvin and coworkers (Calvin et al., 1946) discovered that the addition of metal salts to the reaction mixture, such as zinc acetate, increases the yield of porphyrin from 4-5% for the free-base derivative, and decreases the amount of chlorin compound. Others improvement were obtained by changing opportunely the reaction conditions and substituents in benzaldehyde molecule framework.

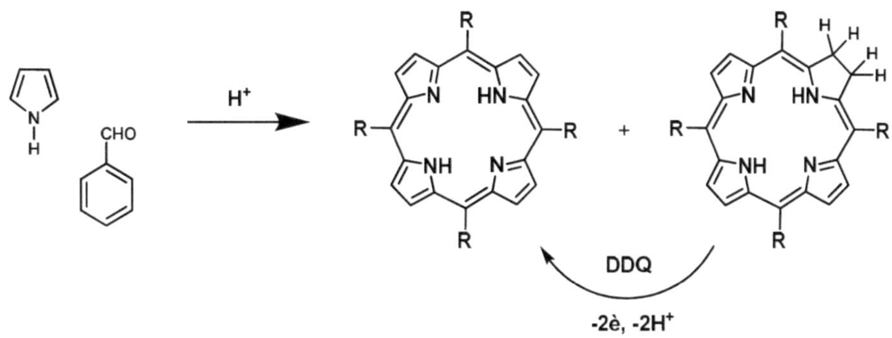

Fig. 2. Synthesis of 5,10,15,20-tetraphenyl porphyrin.

Adler, Longo and coworkers, in the 1960s (Adler et al. 1967), re-examined the synthesis of *meso*substituted porphyrins and developed an alternative approach (Fig. 3) with a method that involves an acid catalyzed pyrrole aldehyde condensation in glassware open to the atmosphere in the presence of air. The reactions were carried out at high temperature, in different solvents and concentrations range of reactants with a yields of 30-40%, and with chlorin contamination lower than that obtained with the Rothemund synthesis.

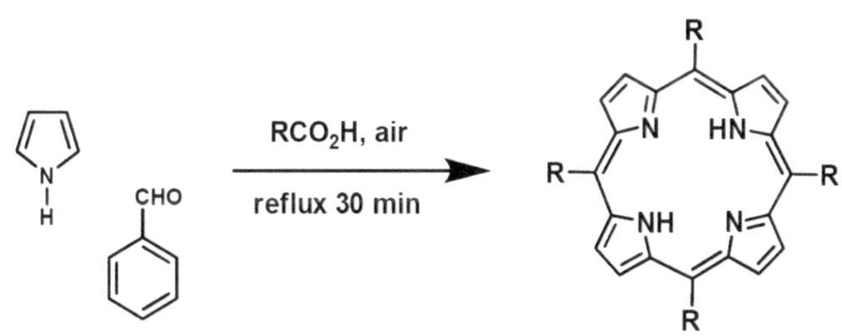

Fig. 3. Adler-Longo method for preparing *meso*-substituted porphyrins.

Over the period 1979-1986, Lindsey developed a new and innovative two-step room temperature method to synthesize porphyrins, motivated by the need for more gentle conditions for the condensation of aldehydes and pyrrole, in order to enlarge the number of the aldehydes utilizable and then the porphyrins available (Anderson et al., 1990; Acheson et al., 1976; Dailey, 1990; Porra, 1997; Mauzarall, 1960). The method has been a new strategy for the synthesis of porphyrins, using a sequential process of condensation and oxidation steps. The reactions were carried out under mild conditions in an attempt to achieve equilibrium during condensation, and to avoid side reactions in all steps of the porphyrin-forming process (Fig. 4)

The porphyrin macrocycle is a highly-conjugated molecule containing 22 π-electrons, but only 18 of them are delocalized according to the Hückel's rule of aromaticity (4n+2 delocalized π-electrons, where n = 4).

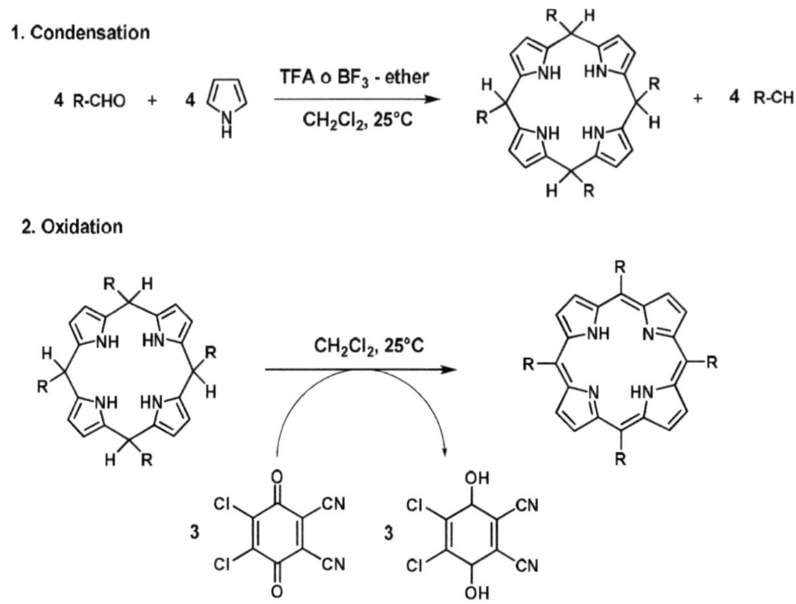

Fig. 4. Two-step one-flask room-temperature synthesis of porphyrins.

Its structure supports a highly stable configuration of single and double bonds with aromatic characteristics that permit the electrophilic substitution reactions typical of aromatic compounds such as halogenation, nitration, sulphonation, acylation, deuteration, formylation. Although this, in the porphyrins there are two different sites on the macrocycle where electrophylic substitution can take place with different reactivity (Milgrom, 1997): positions 5, 10, 15 e 20, called *meso* and positions 2, 3, 7, 8, 12, 13, 17 and 18, called β-pyrrole positions (Fig. 5). The first kind of compounds are widely present in natural products, while the second have no counterpart in nature and were developed as functional artificial models. The activation of these sites depends of the porphyrins electronegativity that can be controlled by the choice of the metal to coordinate to the central nitrogen atoms. For this, the introduction of divalent central metals produces electronegative porphyrin ligands and these complexes can be substituted on their *meso*-carbon. On the other hand, metal ions in electrophylic oxidation states (e.g. Sn IV) tend to deactivate the *meso*-position and activate the β pirrole to electrophylic attack. The chemical characteristics of substituents in β-pyrrole and meso-position determine the water or solvent solubility of porphyrins.

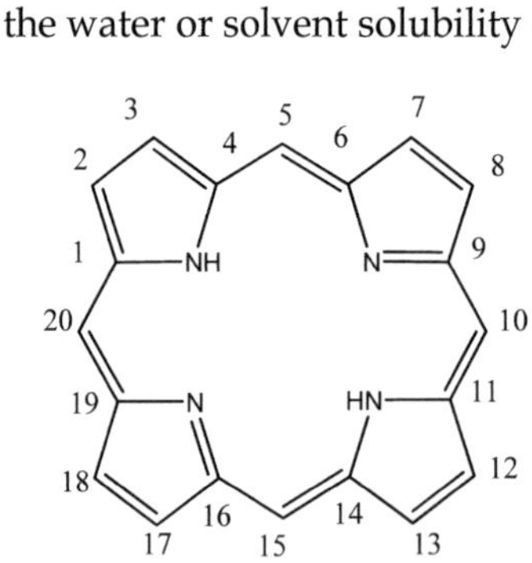

Fig. 5. Porphyrin numeration.

3. Uv-vis spectra of porphyrins

It was recognized early that the intensity and colour of porphyrins are derived from the highly conjugated π-electron systems and the most fascinating feature of porphyrins is their characteristic UV-visible spectra that consist of two distinct region regions: in the near ultraviolet and in the visible region (Fig. 6).

It has been well documented that changes in the conjugation pathway and symmetry of a porphyrin can affect its UV/Vis absorption spectrum (Gouterman, 1961; Whitten et al. 1968; Smith, 1976; Dolphin, 1978; Nappa & Valentine, 1978; Wang et al. 1984; Rubio et al. 1999).

The absorption spectrum of porphyrins has long been understood in terms of the highly successful "four-orbital" (two highest occupied π orbitals and two lowest unoccupied π^* orbitals) model first applied in 1959 by Martin Gouterman that has discussed the importance of charge localization on electronic spectroscopic properties and has proposed the four-orbital model in the 1960s to explain the absorption spectra of porphyrins (Gouterman, 1959; Gouterman, 1961).

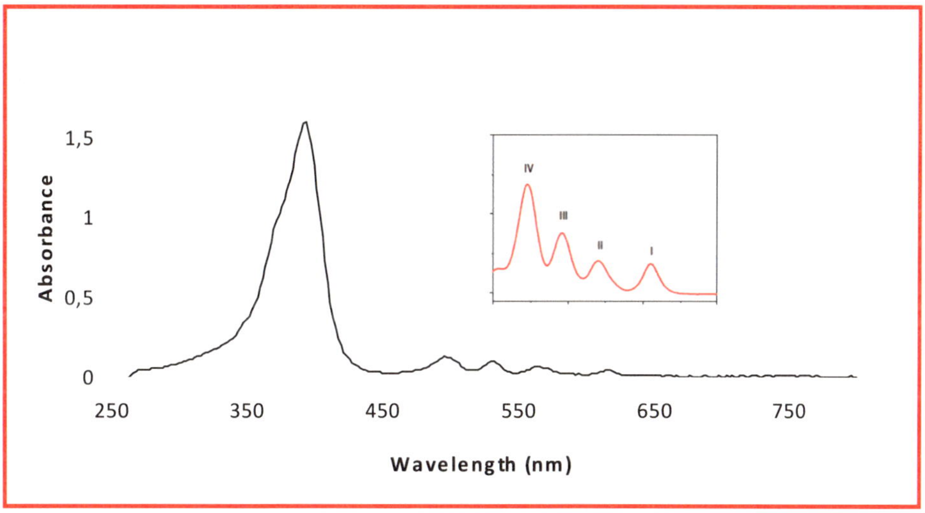

Fig. 6. UV-vis spectrum of porphyrin with in insert the enlargement of Q region between 480-720 nm.

According to this theory, as reported in Figure 7, the absorption bands in porphyrin systems arise from transitions between two HOMOs and two LUMOs, and it is the identities of the metal center and the substituents on the ring that affect the relative energies of these transitions. The HOMOs were calculated to be an $a1_u$ and an $a2_u$ orbital, while the LUMOs were calculated to be a degenerate set of e_g orbitals. Transitions between these orbitals gave rise to two excited states. Orbital mixing splits these two states in energy, creating a higher energy state with greater oscillator strength, giving rise to the Soret band, and a lower energy state with less oscillator strength, giving rise to the Q-bands.

The electronic absorption spectrum of a typical porphyrin (Fig. 6) consists therefore of two distinct regions. The first involve the transition from the ground state to the second excited state (S0 → S2) and the corresponding band is called the Soret or B band. The range of absorption is between 380-500 nm depending on whether the porphyrin is β- or *meso*-substituted. The second region consists of a weak transition to the first excited state (S0 → S1) in the range between 500-750 nm (the Q bands). These favourable spectroscopic features of porphyrins are due to the conjugation of 18 π- electrons and provide the advantage of easy and precise monitoring of guest-binding processes by UV-visible spectroscopic methods (Yang et al. 2002; Gulino et al., 2005; Di Natale et al. 2000; Paolesse & D'Amico, 2007) CD, (Scolaro et al. 2004; Balaz et al. , 2005) fluorescence, (Zhang et al., 2004; Zhou et al., 2006) and NMR spectroscopy (Shundo et al., 2009; Tong et al., 1999).

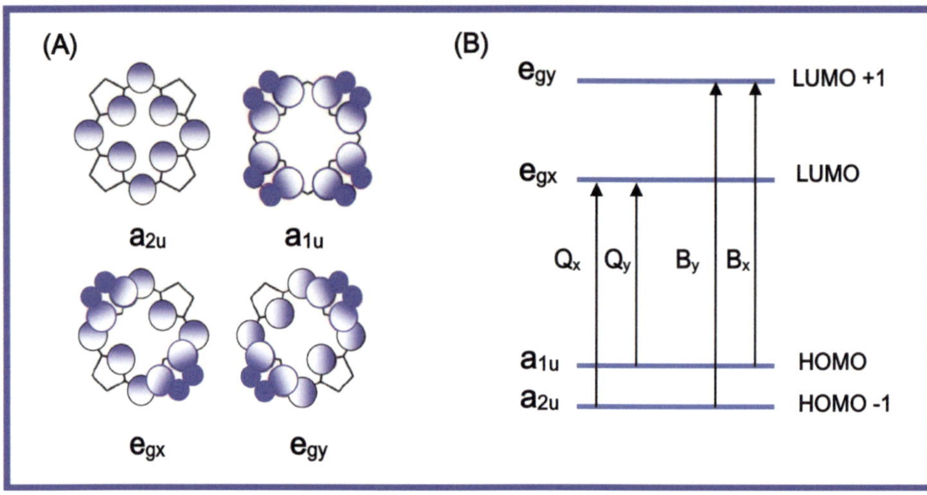

Fig. 7. Porphyrin HOMOs and LUMOs. (A) Representation of the four Gouterman orbitals in porphyrins. (B) Drawing of the energy levels of the four Gouterman orbitals upon symmetry lowering from *D4h* to C2V. The set of e_g orbitals gives rise to Q and B bands.

The relative intensity of Q bands is due to the kind and the position of substituents on the macrocycle ring. Basing on this latter consideration, porphyrins could be classified as *etio, rhodo, oxo-rhodo* e *phyllo* (Prins et al 2001).

When the relative intensities of Q bands are such that IV > III > II > I, the spectrum is said *etio-type* and porphyrins called *etioporphyrins*. This kind of spectrum is found in all porphyrins in which six or more of the β-positions are substituted with groups without π-electrons, *e.g.*, alkyl groups. Substituent with π-electrons, as carbonyl or vinyl groups, attached directly to the β-positions gave a change in the relative intensities of the Q bands, such that III > IV > II > I. This is called *rhodo-type* spectrum (*rhodoporphyrin*) because these groups have a "reddening" effect on the spectrum by shifting it to longer wavelengths. However, when these groups are on opposite pyrrole units, the reddening is intensified to give an *oxo-rhodo-type* spectrum in which III > II > IV > I. On the other hand, when *meso*-positions are occupied, the *phyllo-type* spectrum is obtained, in which the intensity of Q bands is IV > II > III > I (Milgron 1997).

While variations of the peripheral substituents on the porphyrin ring often cause minor changes to the intensity and wavelength of the absorption features, protonation of two of the inner nitrogen atoms or the insertion/change of metal atoms into the macrocycle usually strongly change the visible absorption spectrum.

When porphyrinic macrocycle is protonated or coordinated with any metal, there is a more symmetrical situation than in the porphyrin free base and this produces a simplification of Q bands pattern for the formation of two Q bands.

4. The equilibrium of porphyrins

Neglecting the overall charge of the macrocycle, a monomeric free-base porphyrin H_2-P in aqueous solution can add protons to produce mono H_3-P^+ and dications H_4-P^{2+} at very low pHs, or loose protons to form the centrally monoprotic H-P^- at pH about 6 or aprotic P^{2-} species at pH ≥ 10 (Fig. 8). These chemical forms of porphyrin may exist in equilibrium, depending upon the pH of the solution and can be characterized from the change of the electronic absorption spectrum. The change in spectra upon addition of acid or basic substances can generally be attributed to the attachment or the loss of protons to the two imino nitrogen atoms of the pyrrolenine-like ring in the free-base (Gouterman, 1979; Giovannetti et al, 2010). The N-protonation induced a red-shifts that are consistent with frontier molecular orbital calculations for protonated porphyrins (Daniel et al., 1996).

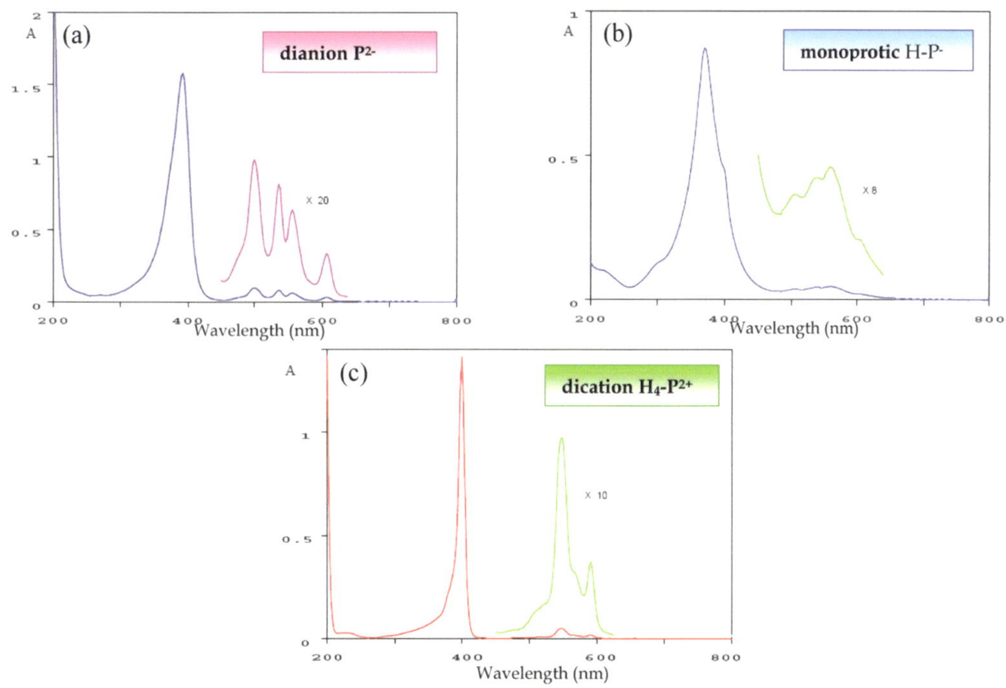

Fig. 8. Typical Uv-vis specrtum of dianion P^{2-} (pH about 10) monoprotic H-P^- (pH about 6) and dication H_4-P^{2+} porphyrin (pH about 1).

Spectrophotometric titration was employed for determining the acid dissociation constants over the inter pH range and change in absorbance with pH can be attributed to the following acid dissociation reactions of porphyrins. Upon addition of acid the spectral pattern of porphyrins changes from the four Q-band spectrum, indicating *D*2h symmetry for free-base porphine, to a two Q-band spectrum for the formation of dications H_4-P^{2+} (Fig. 8 c), indicating *D*4h symmetry, characteristic of porphyrin coordinated to a metal ion through the four N-heteronuclei. In addition, in all cases, the intense Soret band is red-shifted (to an extent dependent on the particular meso-substituents).

5. The reaction of porphyrins with metal ions: Regular and sitting-atop complexes

The metalloporphyrin formation reaction is one of the important processes from both analytical and bioinorganic points of view. The large molar absorption coefficient and the very high stability of porphyrins is valuable for the separation of various kinds of metal ions (Tabata et al.,1998). A variety of metalloporphyrin formation rates are also applicable for the kinetic analysis of metal ions (Tabata & Tanaka, 1991). Also, kinetic studies of metalloporphyrin formation are indispensable in order to understand in vivo metal incorporation processes leading to the natural metalloporphyrins. Generally porphyrins are synthesized in a metal-free form and metal ions are successively inserted.

When the metal ion M^{n+} is incorporated into the porphyrin H_2P to form $MP^{(n-2)+}$, the two amine protons in H_2P are dissociated from the two pyrrole groups as reported in equation (1):

$$M^{n+} + H_2P \leftrightarrow MP^{(n-2)+} + 2H^+ \quad (1)$$

In the formation of metalloporphyrins an marked colour changes with trasformation of the Uv-Vis spectrum especially in the Q zone has been observed. The two Q band obtained are called α and β (Fig. 9). The relative intensities of these bands can be correlated with the stability of the metal complex; in fact when α > β, the metal forms a stable square-planar complex with the porphyrin, in the other case when β > α (e.g. Ni(II), Pd(II), Cd(II)), the metals are easily displaced by protons (Milgron, 1997).

Studies on water soluble and insoluble porphyrins have elucidated aspects of the mechanisms of metal ion incorporation into porphyrins to form metalloporphyrins (Bailey & Hambright, 2003; Hambright et al., 2001; Lavallee, 1987; Funahashi et al., 2001).

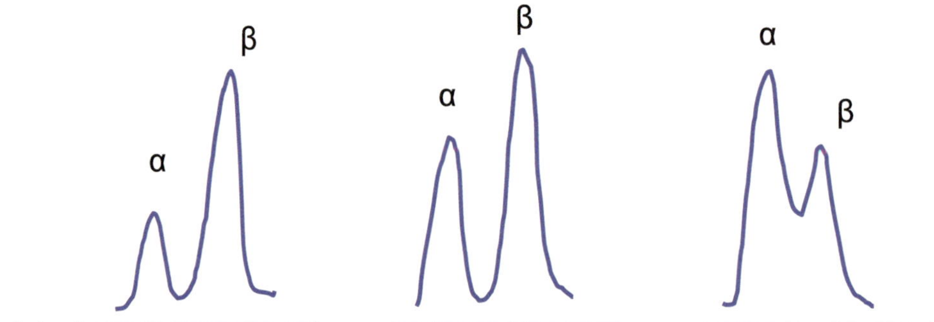

Fig. 9. Q band in the porphyrin metal complexes

The size of the porphyrin-macrocycle is perfectly suited to bind almost all metal ions and indeed a large number of metals can be inserted in the center of the macrocycle forming metalloporphyrins that play key roles in several biochemical processes, due to their central role in photosynthesis, oxygen transport and in various redox reactions (Mathews et al., 2000; Garret & Grisham, 1999; Knör & Strasser, 2005; Lim et al., 2005; Martirosyan et al., 2004; Tovmasyan et al., 2008; Ren et al., 2010; Kawamura et al., 2011).

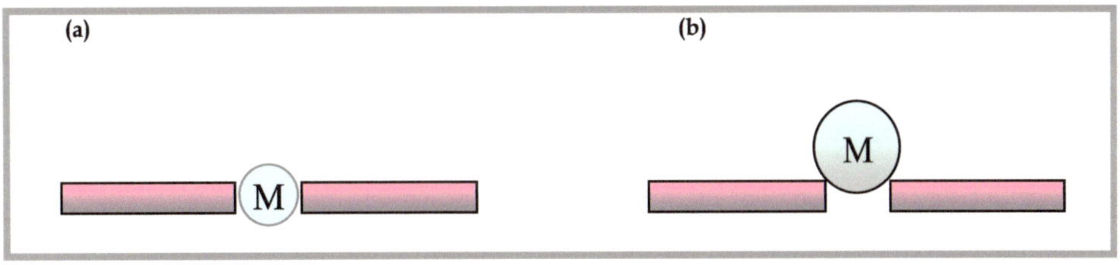

Fig. 10. Schematic representation of (a) regular and (b) SAT metalloporphyrins.

Depending on their size, charge, and spin multiplicity, metal ions (*e.g.* Zn, Cu, Ni, Co, *etc.*) can fit into the center of the planar tetrapyrrolic ring system forming *regular metalloporphyrins* resulting in a kinetically inert complexes (Fig. 10a).

When divalent metal ions (e.g. Co(II), Ni(II), Cu(II)) are chelated, the resulting tetracoordinate chelate has no residual charge. While Cu(II) and Ni(II) in their porphyrin complexes have generally low affinity for additional ligands, the chelates with Mg(II), Cd(II) and Zn(II) readily combine with one more ligand to form pentacoordinated complexes with square-pyramidal structure (Fig. 11a). Some metalloporphyrins (Fe(II), Co(II), Mn(II)) are able to form distorted octahedral (Fig. 11b) with two extra ligand molecules (Biesaga et al., 2000).

Most of the natural metalloporphyrins are of regular type, i.e. their metal centres are located within the plane of the macrocyclic ligand as a consequence of their fitting size. The cationic radii are in the range of 55-80 pm corresponding to the sphere in the porphyrin core surrounded by the four pyrrolic nitrogens. While the symmetry group of the free-base porphyrins is D2h due to the two hydrogen atoms on the diagonally located pyrrolic nitrogens, the coplanar (regular) metalloporphyrins (without these protons) are of higher symmetry (Khan & Bruice, 2003).

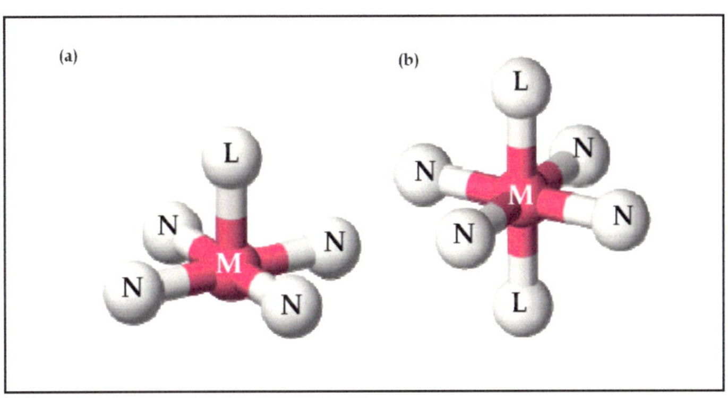

Fig. 11. Schematic pictures of square-pyramidal (a) and octahedral structures (b) (only enclose nitrogen N, metal M and extra ligands L).

If, however, the ionic radius of the metal ions is too large (over ca. 80-90 pm) to fit into the hole in the centre of the macrocycle, they are located out of the ligand plane, distorting it forming *sitting-atop (SAT) metalloporphyrins* (Fig. 10b) that are characterized by special properties (Fleischer & Wang 1960; Barkigia et al., 1980; Liao et al., 2006; Walker et al., 2010) originating from the non-planar structure caused by, first of all, the size of the metal center.

These complexes are kinetically labile and display characteristic structural and photoinduced properties that strongly deviates from those of the regular metalloporphyrins. The latter kind of structure induces special photophysical and photochemical features that are characteristic for all SAT complexes. The symmetry of this structures is lower (generally C_{4v}-C_1) than that of both the free-base porphyrin (D2h) and the regular, coplanar metalloporphyrins (D4h), in which the metal center fits into the ligand cavity.

The rate of formation of in-plane (or normal) metalloporphyrins is much slower than that of the *SAT complexes* because of the inflexibility of porphyrins. In fact, in an SAT complex the distortion of the porphyrin caused by the out-of-plane location of the metal center makes two diagonal pyrrolic nitrogens more accessible on the other side of the ligand due to the increase of their sp^3 hybridization (Tung & Chen, 2000).

Deviating from the regular metalloporphyrins, the SAT complexes, on account of their distorted structure and kinetic lability, display peculiar photochemical properties, such as photoinduced charge transfer from the porphyrin ligand to the metal center, leading to irreversible ring opening of the ligand and dissociation on excitation at both the Soret- and the Q-bands (Horváth et al., 2006). Moreover, the absorption and emission characteristics of these complexes are also significantly deviating from those of the normal (in-plane) metalloporphyrins (Horváth et al., 2006). Also the formation of bi and even trinuclear (bis-porphyrin) complexes has been observed (Lehn, 2002).

In Figure 12 is shown a schematic Energy-level diagram of the frontier orbital of a porphyrin in free-base state (H_2P), in a regular and in a SAT metalloporphyrin.

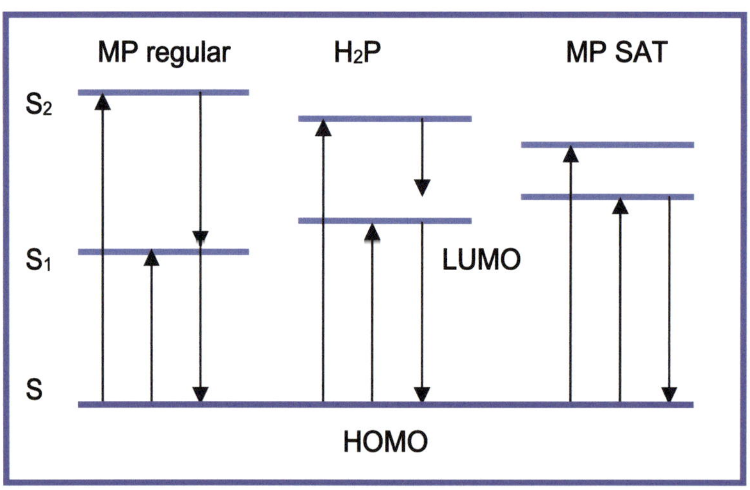

Fig. 12. Simplified energy-level diagram of the frontier orbital of a porphyrin in free-base state H_2P, in a regular and in a SAT metalloporphyrin.

The photoinduced behavior of normal metalloporphyrins have been thoroughly studied for several decades, while the investigation of SAT complexes started in this respect only in the past 8–10 years (Horváth et al., 2004; Valicsek et al., 2008; Valicsek et al., 2009; Valicsek et al., 2007; Huszánk et al., 2005; Huszánk et al., 2007; Valicsek et al., 2011).

Interestingly, in the case of lanthanide ions as metal centers, triple decker porphyrin sandwich complexes were also synthesized and studied (Wittmer &. Holten, 1996).

While the natural porphyrin derivatives are exclusively hydrophobic, some artificial porphyrins having ionic substituents made it possible to prepare water-soluble metalloporphyrins of both regular and SAT type. Kinetically labile complexes are mostly examined in the excess of the ligand.

In the case of metalloporphyrins, however, metal ions are applied generally in excess, especially for spectrophotometric measurements, partly because of the extremely high molar absorbances (mainly at the Soret-bands) of the porphyrins. The formation of kinetically labile SAT complexes, deviating from the regular metalloporphyrins, is an equilibrium process. It can be spectrophotometrically monitored because the absorption and emission bands assigned to ligand-centered electron transitions undergo significant shift and intensity change upon coordination of metal ions.

Special attention was devoted to the reaction of porphyrins with essential metal ions as manganese, iron and chromium show that the most important properties of manganese in complex biological systems is the highly variable oxidation states of the metal from +2 to +5 (Kadish et al. 1999). All these compounds can be easily spectrophotometrically distinguished among them; this is because they have different absorption spectra (Spasojevic & Batinic-Haberle, 2001,) from which is possible to know the oxidation state. Manganese–porphyrin complexes have more extensively studied because were found to be similar to the biologically active compounds (Nakanishi et al. 2000; Meunier, 1992; Perie & Barbe, 1996; Balahura & Kirby, 1994; Haber et al., 2000; Cuzzocrea et al., 2001), and because were also used as catalysts for the oxygenation of alkanes, alkenes and compounds containing nitrogen and sulphur (Mansuy & Momenteau, 1982; Fontecave & Mansuy, 1984). The very important properties that influence the reactivity of the Mn(III)-porphyrin concerns the changes in the oxidation states of Mn in the complexes for its high reactivity with O_2 (Cuzzocrea et al., 2001). Manganese, in the complexes obtained by the reaction of Mn(II) with the porphyrins, has oxidation number +3, so the complex of Mn(II) can be obtained only by reduction, while those of Mn(IV) and Mn(V) for the oxidation of Mn(III)-complexes. Interesting is the reactions of a natural porphyrin, the acid 2,7,12,17 tetrapropionic of 3,8,13,18 tetramethyl-21H, 23H-porphyrin called Coproporphyrin- I (CPI), with manganese (III) that, with different pH and solvent compositions, show the formation of $[Mn^{III}CPI(H_2O)_2]$, $[Mn^{III}CPI(OH)_2]$, $[Mn^{IV}(O)CPI(OH)]$, $[Mn^{V}(O)CPI(OH)]$, $[Mn^{II}CPI(OH)]$ (Fig. 13) with specific Uv-Vis adsorptions as reported in Table 1. (Giovannetti et al., 2010).

5.1 Complexation kinetics

Rates of the complexation of porphyrins with metal ions are very much slower by several orders of magnitude than those of acyclic ligands (Funahashi, S. et al., 1984). Such very slow rates have been discussed in terms of the rigidity of the planar porphyrin framework. The electronic nature of porphyrins, and also the steric accessibility of the bound metal center, can be varied by using electron-donating or with drawing substituents at the meso carbon or in the pyrrolic positions. While such substituent-based changes have been seen to influence the extent of apical ligand binding, as well as the stability of the metal complexes, there is a relatively small effect on the ability to insert cations into the nitrogen core (Lin & Lash, 1995).

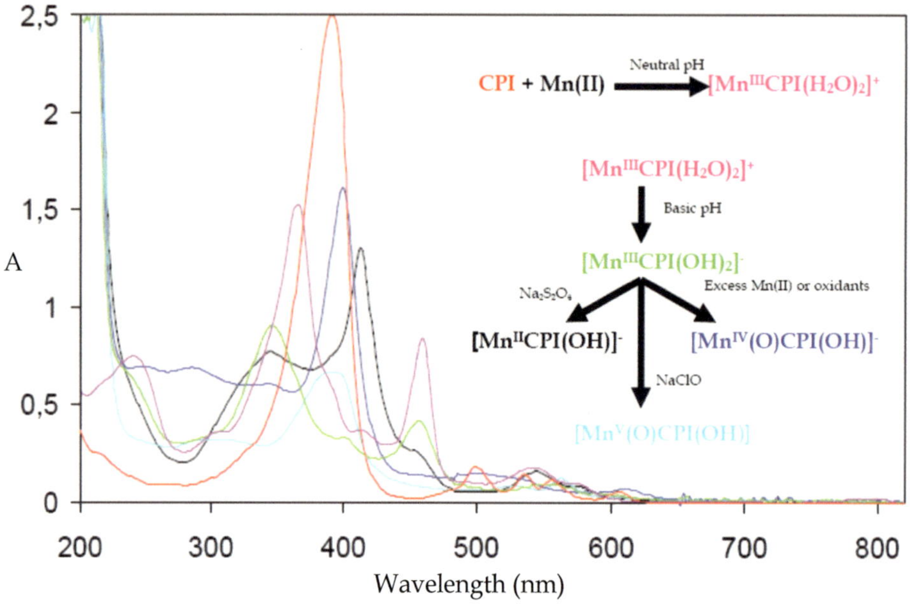

Fig. 13. Uv-Vis adsorption spectra of CPI, $[Mn^{III}CPI(H_2O)_2]$ $[Mn^{III}CPI(OH)_2]$, $[Mn^{IV}(O)CPI(OH)]$, $[Mn^{V}(O)CPI(OH)]$, $[Mn^{II}CPI(OH)]$ with in insert, the experimental conditions for the preparation of all complexes obtained with several reagents.

Complex	Soret band λ (nm)	Q band (β, α) λ (nm)
$[Mn^{II}CPIOH]^-$	414	544, 574
$[Mn^{III}CPI(H_2O)_2]^+$	366, 458	542, 572
$[Mn^{III}CPI(OH)_2]$	348, 458	556
$[Mn^{IV}OCPI(OH)]^-$	400	502, 610
$[Mn^{V}OCPI(OH)]$	384	526, 562

Table 1 Spectral characteristics of Mn–CPI complexes.

In the case of N-substituted porphyrins, in which one of the two hydrogen atoms bound to the pyrrole nitrogen atoms is substituted by an alkyl or aryl group, displacement of the substituent from the porphyrin plane due to its bulkiness causes tilting distortion of the pyrrole rings of the porphyrin as shown by X-ray crystallography (Lavallee & Anderson, 1982; Aizawa et al., 1993). The investigation of the kinetics of the complexation of N-substituted porphyrins with several metal ions, showing that N-alkylporphyrins form metal complexes much faster than corresponding non-N-alkylated porphyrins (Aizawa et al., 1993; Shah et al., 1971; Shah et al., 1971; Anderson & Lavallee, 1977; Lavalle et al., 1978; Anderson et al., 1980; Kuila & Lavallee, 1984; Schauer et al., 1987; Balch et al., 1990; McLaughlin, 1974;

Bain-Ackerman & Lavallee, 1979; Funahashi et al., 1984; Funahashi et al., 1986;Lavallee et al., 1986; Bartczak et al., 1990; Shimizu et al., 1992).

The formation of metalloporphyrins can be accelerated substantially by the use of an auxiliary complexing agent. For example, the rate of complexation of TMPyP with Cu(II) and Mg(II) was accelerated by L-cysteine (Watanabe & Ohmori 1981) and 8-quinolinol (Makino & Itoh, 1981), respectively. The organic ligands with an extended π-electron structure, such as imidazole or bipyridine (Giovannetti et al., 1995; Kawamura et al., 1988; Ishi & Tsuchiai 1987; Tabata & Kajhara, 1989) and L-tryptophan (Tabata & Tanaka, 1988) also show a tendency to accelerate the formation of metalloporphyrins because form intermediate molecular complexes with metal ions and porphyrin reagents. In the presence of tryptophan, the rate of incorporation of Zn(II) to TPPS4 is about 100 times greater than in its absence (Tabata & Tanaka, 1988).

Hence, larger metal ions such as Pb^{2+}, Hg^{2+}, or Cd^{2+} can catalyse the formation of regular metalloporphyrins via generation of SAT complexes as intermediates. This because in the SAT complexes, the distortion caused by the out-of-plane location of the larger metal center, makes two diagonal pyrrolic nitrogens more accessible to another metal ion, even with smaller ionic radius, on the other side of the porphyrin ligand that can easily coordinate to them (Stinson & Hambright, 1977; Inamo et al 2001; Wittmer & Holten, 1996; Tabata & Tanaka 1985; Tung & Chen, 2000; Tabata et al., 1995; Grant & Hambright 1969; .R. Robinson & Hambright, 1992; C. Stinson & Hambright, 1977; Barkigia et al., 1990; Giovannetti et al., 1998).

Since the deformation of the porphyrin ring proved to be the main factor governing the acceleration of the metalloporphyrin formation (Lavallee, 1985; Tabata & Tanaka, 1991), this can also be achieved by substituents at the porphyrin core or at the peripheral ring. Thus, e.g., the peripheral or substituted octabromoporphyrins display a buckled structure due to steric hindrance between the substituents (Bhyrappa & Krishnan, 1991; Mandon et al., 1992; Henling et al., 1993; Brinbaum et al., 1995). Such a deformation profoundly enhanced the reactivity of the porphyrin even towards Hg^{2+} (Nahar & Tabata, 1998), the ionic radius of which is rather large anyway.

6. Aggregation of porphyrins

An increasing interest in recent years is due to supramolecular assemblies of π-conjugated systems for their potential applications in optoelectronic and photovoltaic devices (Schenning & Meijer, 2005).

Molecular aggregates of several dyes have been studied as organic photoconductors (Borsenberger et al.,1978), as markers for biological and artificial membrane systems (Waggoner, 1976), as materials with high non-linear optical properties suitable for optical devices ([Hanamura,1988; Sasaki & Kobayashi, 1993; Wang, 1986; Wang, 1991). Some properties of molecular structure of aggregates permit their use in superconductivity, and other processes (Kobayashi,1992; Schouten et al., 1991; Collman, 1986). Aggregation of small organic molecules to form large clusters is of large interest in chemistry, physics and biology. In nature, particularly in living systems, self-association of molecules plays a very important role; an example is given by molecular aggregates of chlorophyll that have been found to mediate the primary light harvesting and charge-transfer processes in photosynthetic complexes (Creightonet al., 1988; Kuhlbrandt, 1995). In fact, light-harvesting and the primary charge-separation steps in photosynthesis are facilitated by aggregated species, i.e., chlorophylls.

Self-assembly of molecules, driven by not-covalent intermolecular interactions, is a convenient route for manufacturing of new functional materials (Lidzey at al., 2000; Van der Boom et al., 2002; Fudickar et al. 2002; Lagoudakis et al., 2004; Li et al., 2003).

Recently, porphyrin assembly has been used for light-driven energy transduction systems, copying the photophysical processes of photosynthetic organisms (Choi et al.,2004); Choi et al., 2003; Choi et al., 2002; Choi et al., 2001; Luo et al., 2005).

The aggregation and dimerization of porphyrins and metalloporphyrins in aqueous solution have been widely investigated (Borissevitch & Gandini, 1998; Pasternack et al, 1985) and it has been deduced that it is dependent strictly on physical-chemical characteristics, such as, ionic strength, pH and solvent composition; the combination of these factors can facilitate the aggregation processes (Kubat et al., 2003; Giovannetti et al., 2010).

The aggregation of porphyrins, changing their spectral and energetic characteristics, influences their efficacy in several applications thus, it is very important take on detailed informations about the formation dynamic and on the typology of aggregates. In the metal complexation of porphyrins the efficiency reaction is affected by their aggregation (Yusmanov et al, 1996). Several authors have observed that in the photogeneration of H_2O_2 by porphyrins, the efficiency of production was highly dependent on their aggregation state (Komagoe et al, 2006).

The diverse chemical and photophysical properties of porphyrins are in many cases due to their different aggregation mode and, as a result of interchromophoric interactions, perturbations in the electronic absorption spectra of dyes occur. Deviations from Beer's law are often used to investigate the porphyrin aggregation in solution.

Because the aggregates of porphyrins show peculiar spectroscopic properties, the molecular associations of porphyrins were generally investigated using UV–vis absorption and fluorescence spectroscopy (Ohmo, 1993).

The characteristic of porphyrin molecule with 22 π-electrons causes a strong π –π interaction (Van de Craats, & Warman, 2001), facilitating the formation of two structure types: "H-type" with bathochromic shift of B and Q bands and "J-type" with blue shift of B band and red-shift of Q band, with respect to those of monomer.

The J-type aggregates (side-by-side) were formed for transitions polarized parallel to the long axis of the aggregate, while H-type (face-to-face) for transitions polarized perpendicular to it (Fig. 14).

J-aggregates are formed with the monomeric molecules arranged in one dimension such that the transition moment of the monomers are parallel and the angle between the transition moment and the line joining the molecular centers is zero (Bohn, 1993). The strong coupling of monomers results in a coherent excitation with a red-shift relative to the monomer band. *H-aggregates* are again a one-dimensional arrangement of strongly coupled monomers, but the transition moments of the monomers are perpendicular (ideal case) to the line of centers. On the contrary of J-aggregates, the arrangement in H-aggregates is face-to-face. The dipolar coupling between monomers leads to a blue shift of the absorption band (Czikklely et al.,

1970; Nuesch et al., 1995). The H-aggregates are not known to have sharp spectra like the J-aggregates; nevertheless, there are many examples where the spectroscopic blue shift, evident for formation of H-aggregates, was observed.

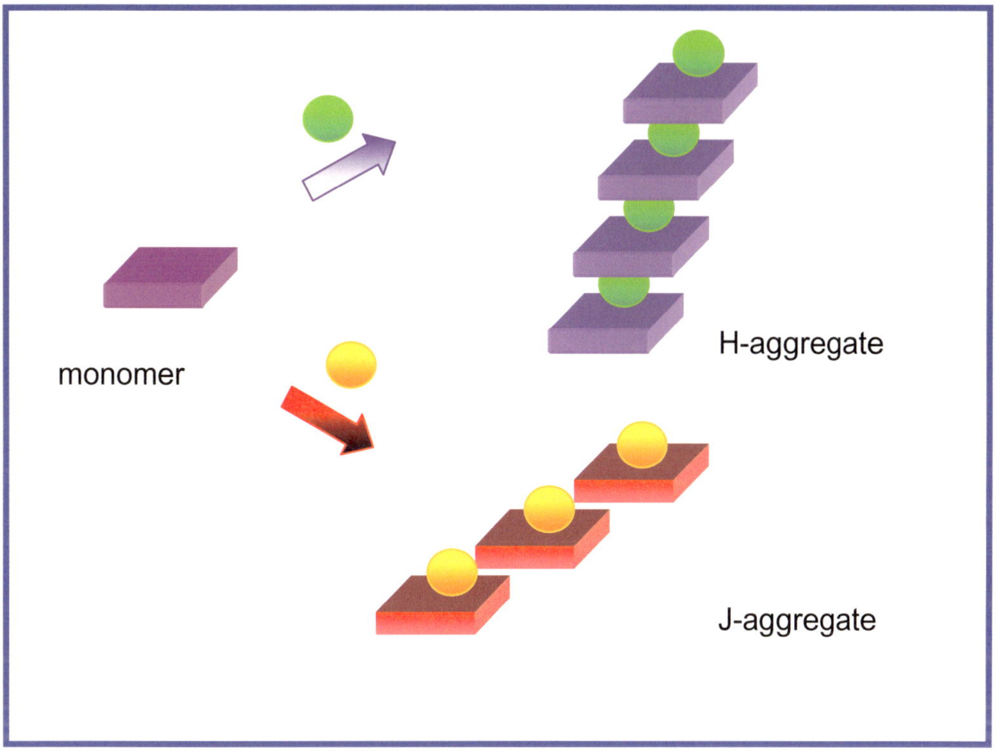

Fig. 14. Schematic representation of H and J aggregates.

These aggregated are of particular interest because the highly ordered molecular arrangement present unique electronic and spectroscopic properties that can be predicted (Fidder et al., 1991; West & Carroll,1966; Furuki et al., 1989; Spano & Mukamel, 1989; Bohn, 1993).

H and J aggregates can be obtained under specific conditions (Misawa & Kobayashi, 1999; Maiti et al., 1995; Kanojk & Kobayashi, 2002; Pasternack et al., 1994; Akins et al., 1994; Ohno et al., 1993; Luca et al., 2005; Luca et al., 2006; Luca et al., 2006). Due to the distinct optical properties, also the control of the formation of H- and J-aggregated states of dyes has attracted much research interest (Maiti et al., 1998; Shirakawa et al., 2003; De Luca et al., 2006; Egawa et al., 2007; Yagai et al., 2008; Gadde et al., 2008; Ghosh et al., 2008; Zhao et al., 2008; Delbosc et al., 2010).

As a result, a wide variety of self-assembled porphyrin structures are highly desirable for practical use, which can be applied to nonlinear optical materials (Collini et al., 2006; Liu et al., 2006; Matsuzaki et al., 2006; Terazima et al., 1997), organic solar cells (Hasobe et al., 2003) and sensor devices (Fujii et al., 2005; Lucaet al., 2007).

In general, the aggregate formation of porphyrins have been studied in solution, and their physicochemical properties can be affected by the ionic strength, nature of the titrating acid, temperature, pH, peripheral substitution and presence of surfactants of ions (Choi et al., 2003; Ohno et al., 1993; Napoli et al., 2004; Kubat et al., 2003; Siskova et al., 2005).

H- and J-aggregates was formed by simply mixing aqueous solutions of two kinds of porphyrins with opposite charges (Xiangqing et al., 2007).

Moreover, self-assembly can be mediated by templates that allows for obtaining aggregates with additional properties, e.g., chiral templates (Koti & Periasamy, 2003; Mammana et al., 2007). In these systems, coupling of strong transition dipoles can result in a perturbations to the electronic absorption spectra of monomer with hyposochromic and bathochromic shift of the monomer Soret band, for H and J aggregates formation respectively (Gourterman et al., 1977; Zimmermann et al., 2003; Scherz & Parson, 1984). The produced splitting is proportional to the magnitude of transition dipole coupling between adjacent molecules.

Although much has been studied about the spectroscopic features and excitonic interactions in molecular aggregates, the detailed information of geometrical structure, especially the molecular orientation, are still the subjects of continuing interests (Nikiforov et al., 2008; Jeukens et al., 2004).

The aggregates of porphyrins have been formed in solutions in the form of fibers, ribbons and tubules by the self-association or aggregation method (Rotomskis, 2004; Fuhrhop, 1993; Giovannetti et al., 2010).

Some H- or J-type aggregates of porphyrins play a role as light harvesting assemblies to gather and transfer energy to the assembled devices, and to obtain a higher incident photon-to-photocurrent generation efficiency (Kamat et al.,2000; Sudeep et al., 2002).

The structural, kinetic, and spectroscopic studies on J- and H-aggregates provide useful information for understanding molecular interactions in aggregation processes.

The kinetics of the formation of the porphyrin aggregate and its structure are sensitive of experimental conditions (Giovannetti et al., 2010). The monomer - aggregated species is a system of multiple equilibria. Spectrophotometric monitoring in the time of the Uv- Vis absorbance permit to obtain information of intermediate species, of type of the aggregate, and of their transformation. For this, for evaluated the polymerization kinetic constants, the concentrations of monomeric [M] and dimeric form [D], can be calculated from the relative absorption maxima at each time. If k_{pol} is the polymerisation kinetic constant, C_M and C_D denoted the initial monomer and dimer concentrations, the reaction rate can be expressed as in equation (2):

$$k_{pol}t = \frac{1}{C_M - C_D} \ln \frac{C_D[M]}{C_M[D]} \tag{2}$$

The plot of the right term of this equation versus t gave good straight lines the slopes of which represented the values of k_{pol}.

The information derived from such studies can help in achieving appropriate design of photoactive aggregates for mimicking light-harvesting natural photosynthetic pigments, photodynamic therapeutic use, and advanced nonlinear optical materials.

7. Conclusions

The porphyrins represent a fascinating world of molecules with sensational properties. Many results have been obtained by careful observation and with detailed studies of their chemical and physical properties due to the use of UV-Vis spectrophotometry between the interpretation of Soret and Q band transformations.

In this contest, the light absorbing power of porphyrins and related compounds should be used in the near future for other many applications and much more can still be studied in the future.

8. References

Acheson, R.M. (1976). *An Introduction to the Chemistry of Heterocyclic Compounds*, 3rd ed; John Wiley & Sons: New York.

Adler, A. D. Longo, F. R. Finarelli, J. D. Goldmacher, J. Assour, J. Korsakoff, L. (1967). *J. Org. Chem.* 32, 476.

Aizawa, S. Tsuda, Y. Ito, Y. Hatano K. Funahashi, S. (1993). *Inorg. Chem.*, 32, 1119.

Akins, D. L. Zhu, H.-R. Guo, C. (1996). *J. Phys. Chem.*, 100, 5420.

Akins, D. L. Zhu, H.-R. Guo, C. (1994). *J. Phys. Chem.*, 98, 3612.

Anderson, H.J. Loader, C.E. Jackson, A.H., (1990) In *Pyrrole,* Jones, R. A. Ed. *The Chemistry of Heterocyclic Compounds*, Vol. 48, John Wiley & Sons: New York 295-397.

Anderson O.P. Lavallee, D.K. (1976). *J. Am. Chem. Soc.*, 98, 4670.

Anderson O.P. Lavallee, D.K. (1977). *J. Am. Chem. Soc.*, 99, 1404.

Anderson O.P. Lavallee, D.K. (1977). *Inorg. Chem.*, 16, 1634.

Anderson, O.P. Kopelove A.B. Lavallee, D.K. (1980). *Inorg. Chem.*, 19, 2101.

Arai Y. & Segawa H. (2011). *J. Phys. Chem. B*, 115, 7773.

Atwood, J.L. Davies, J.E.D. MacNicol, D.D. Vogtle, F. Lehn, J.-M., Eds., *Handbook : Comprehensive Supramolecular Chemistry*, (1996). Pergamon, Oxford.

Bailey, S. L. , Hambright P. (2003). *Inorganica Chimica Acta* , 344, 43.

Bain-Ackerman M.J. Lavallee, D.K. (1979). *Inorg. Chem.*, 18, 3358.

Balaban, T. S. (2005). *Acc.Chem. Res.*, 38, 612.

Balahura, R.J. Kirby, R.A. (1994). *Inorg. Chem.*, 33, 1021.

Balaz, M. Holmes, A.E. Benedetti, M. Rodriguez, P.C. Berova, N. Nakanishi, K. Proni, G., (2005). *J. Am. Chem. Soc., 127*, 4172.

Balch, A.L. Cornman, C.R. Latos-Grazynski L. Olmstead, M.M. (1990), *J. Am. Chem. Soc.*, 112, 7552.

Barkigia, K.M. Berber, M.D. Fajer J., Medforth, C.J. Renner M.W., Smith, K.M. (1990). *J. Am. Chem. Soc.*, 112,

Barkigia, K.M. Fajer, J. Adler, A.D. Williams, G.J.B. (1980). *Inorg. Chem.*, 19, 2057.

Bartczak, T.J. Latos-Grazynski L. Wyslouch, A. (1990). *Inorg. Chim. Acta*, 171, 205.

Beletskaya, I. Tyurin, V. S. Tsivadze, A. Y. Guilard, R. Stern, C. (2009). *Chem. Rev.*, 109, 1659.

Bhyrappa, P. Krishnan, V. (1991) *Inorg. Chem.*, 30, 239.

Biesaga, M. Pyrzynska, K. Trojanowicz, M. (2000). *Talanta,* 51, 209.

Bohn, P. W. (1993). *Annu. Rev. Phys. Chem., 44*, 37.

Borissevitch, I.E. Gandini, S.C. (1998). *J. Photochem. Photobiol. B: Biol.*, 43, 112.

Borsenberger, P. Chowdry, A. Hoesterey, D. Mey, W. (1978). *J. Appl. Phys.*, 44, 5555.

Brinbaum, E.R. Hodge, J.A. Grinstaff, M.W. Schaefer, W.P. Henling, L. Labinger, J.A. Bercaw, J.E. Gray, H.B. (1995). *Inorg. Chem,*. 34, 3625.
Calvin, M. Ball, R. H. Arnoff, S. (1943). *J. Am. Chem. Soc.*, 65, 2259.
Choi, M. Y. Pollard, J. A. Webb, M. A. McHale, J. L. (003). *J. Am. Chem. Soc.*, 125, 810.
Choi, M.-S. Aida, T. Luo, H. Araki, Y. Ito, O. (2003). *Angew. Chem., Int. Ed.*, 42, 4060.
Choi, M.-S. Aida, T. Yamazaki, I. Yamazaki, T. (2001). *Angew. Chem., Int. Ed.*, 40, 194.
Choi, M.-S. Aida, T. Yamazaki, I. Yamazaki, T. (2002). *Chem. Eur. J.*, 8, 2667.
Choi, M.-S. Aida, T. Yamazaki, I. Yamazaki, T. (2004). *Angew. Chem., Int. Ed.* ,43, 150.
Collini, E. Ferrante, C. Bozio, R. Lodi, A. Ponterini, G. (2006). *J. Mater. Chem.* 16, 1573.
Collman, J.P. McDeevitt, J.T. Yee, G.T. Leidner, C.R. McCullough, L.G. Little, W.A. Torrance, J.B. (1986). *Proc. Natl. Acad. Sci. U.S.A.*, 83, 44581.
Creighton, S. Hwang, J. Warshel, A. Parson, W. Norris, J. (1988). *Biochemistry*, 27, 774.
Cuzzocrea, S. Riley D.P., Caputi A.P., Salvemini D. (2001). *Pharmacol. Rev.* , 53, 135.
Czikklely, V. Forsterling, H. D. Kuhn, H. (1970). *Chem. Phys. Lett.*, *6*, 207.
Dailey, H.A. Ed. (1990) *Biosynthesis of Heme and Chlorophylls*, McGrawHill, Inc.: New York.
De Luca, G. Pollicino, G. Romeo, A. Scolaro, L. M. (2006). *Chem. Mater.*, 8, 18.
De Luca, G. Romeo, A. Scolaro, L. M. (2005). *J. Phys. Chem. B*, 109, 7149.
De Luca, G.; Romeo, A.; Scolaro, L. M. *(2006). J. Phys. Chem. B,* 110, 14135.
De Luca, G.; Romeo, A.; Scolaro, L. M. (2006). *J. Phys. Chem. B*, 110, 7309.
Delbosc, N. Reynes, M. Dautel, O. J. Wantz, G. Lere-Porte, Fidder, H. Terpstra, J. Wiersma, D. A. (1991). *J. Chem. Phys*, *94*, 6895.
Delbosc, N. Reynes, M. Dautel, O. J. Wantz, G. Lere-Porte, J.-P. Moreau, J. J. E. (2010). *Chem. Mater.*, 22, 5258.
Di Natale, C. Salimbeni, D. Paolesse, R. Macagnano, A. D'Amico, A., (2000). *Sens. Actuators, B*, *65*, 220
Di Natale,C. Paolesse, R. D'Amico, A., (2007). *Sens. Actuators, B*, *121*, 238.
Dolphin, D. Ed *The Porphyrins,* (1978).Academic, New York,.
Drain, C. M. Alessandro, V. Radivojevic, I. (2009). *Chem. Rev.*, 109, 1630.
Egawa, Y. Hayashida, R. Anzai, J. (2007) .*Langmuir*, 23, 13146.
Elemans, J. A. A. W. Van Hameren, R. Nolte, R. J. M. Rowan, A. E. (2006). *Adv. Mater.*, 18, 1251.
Fleischer, E.B. Wang, J.H. (1960). *J. Am. Chem. Soc.*, 82, 3498.
Fontecave, M. Mansuy, D. (1984). *Tetrahedron Lett.*, 21, 4297.
Fuhrhop, J.H. Bindig, U. Siggel, U. (1993). *J. Am. Chem. Soc.*, 115, 11036.
Fujii, Y. Hasegawa, Y. Yanagida, S. Wada, Y. (2005). *Chem. Commun.* 3065.
Funahashi S., Inada, Y. Inamo, (2001). M. *Anal. Sci.* 917.
Funahashi, S. Ito, Y. Kakıto, H. Inamo, M. Hamada Y. Tanaka, M. (1986). *Mlkrochlm. Actu,* 33.
Funahashi, S. Yamaguchi, Y. Tanaka, M. (1984). *Bull. Chem. Soc. Jpn.*, 57, 204.
Funahashi, S. Yamaguchi, Y. Tanaka, M. (1984). *Inorg. Chem.*, 23, 2249
Furuki, M. Ageishi, K. Kim, S. Ando, I. Pu, L. S. (1989).*Thin Solid Films*, *180*, 193.
Gadde, S. Batchelor, E. K. Weiss, J. P. Ling, Y. Kaifer, A. E. (2008). *J. Am. Chem. Soc.*, 130, 17114.
Ghosh, S. Li, X. Stepanenko, V. Wrthner, F. (2008).*Chem. Eur. J.*, 14, 11343.
Giovannetti R., Alibabaei, L. Petetta, L. (2010).*Journal of Photochemistry and Photobiology A: Chemistry*, 211,108.
Giovannetti, R. Bartocci, V. Ferraro, S. Gusteri, M. Passamonti, P. (1995). *Talanta,* 42, 1913.

Giovannetti, R. Alibabaei, L. Pucciarelli, F. (2010). *Inorganica Chimica Acta*,363, 1561.

Goldberg D.E. Thomas, K.M. (1976). *J. Am. Chem. Soc.*, 98, 913.

Gourterman, M. Holten, D. Lieberman, E. (1977). *Chem. Phys.*, 25, 139.

Gouterman, M. (1961). *J. Mol. Spectroscopy*, 6, 138.

Gouterman, M. (1959). *J. Chem. Phys.*, *30*, 1139.

Grant, C. Hambright, P. (1969). *J. Am. Chem. Soc.*, 91, 4195.

Gulino, A. Mineo, P. Bazzano, S. Vitalini, D. Fragalà, I., (2005). *Chem. Mater.*, *17*, 4043.

Haber, J. Matachowski, L. Pamin, K. Potowicz, J. (2000). *J. Mol. Catal. A*, 162, 105.

Hambright, P. inK.H. Kadish, K.M. Smith, R. Guilard (Eds.), *The Porphyrin Handbook*, (2001). vol. 3 (chap. 18), Academic Press, NewYork,.

Hanamura, E. (1988). *Phys. Rev. B*, 37,1273.

Hasobe, T. Imahori, H. Fukuzumi, S. Kamat P.V., (2003). *J. Mater. Chem.*,13, 2515.

Henling, L.M. Schaeffer, W.P. Hodge, J.A. Hughes, M.E. Gray, H.B. (1993). *Acta Crystallogr.*, 49, 1743.

Hoeben, F. J. M. Jonkheijm, P. Meijer, E. W. Schenning, A. P. H. (2005). *J. Chem. Rev.*, 105, 1491.

Horváth, O. Huszánk, R. Valicsek, Z. Lendvay, G. (2006). *Coord. Chem.Rev.*, 250, 1792.

Horváth, O. Valicsek, Z. Vogler, A. (2004). *Inorg. Chem. Commun.*, 7, 854.

Huszánk, R. Horváth, O. (2005). *Chem. Commun.* 224.

Huszánk, R. Lendvay, G. Horváth, O. (2007). *J. Bioinorg. Chem.*, 12, 681.

Inamo, M. Kamiya, N. Inada, Y. Nomura, M. Funahashi, S. (2001). *Inorg. Chem.*, 40, 5636.

Ishi, H. Tsuchiai, H. (1987) *Anal. Sci.*, 3, 229.

Jeukens, C.R.L.P.N. Lensen, M.C. Wijnen, F.J.P. Elemans, J.A.A.W. Christianen, P.C.M.

Kadish, K.M. Smith, K.M. Guilard, R. (1999). *The Porphyrin Handbook*, Academic Press, S. Diego, CA.

Kahn, K. Bruice, T.C., (2003). *J. Phys. Chem. B*, *107*, 6876.

Kamat, P.V. Barazzouk, S. Hotchandani, S. Thomas, K.G. (2000). *Chem. Eur. J.*, 6, 3914.

Kamat, P.V. Barazzouk, S. Thomas, K.G. Hotchandani, S. (2000).*J. Phys. Chem. B*, 104, 4014.

Kano, H. Kobayashi, T. J. (2002). *Chem. Phys.*, 116, 184.

Kawamura, K. Igarashi, S. Yotsuyanagi, T. (2011). *Microchim. Acta* ,172, 319.

Kawamura, K. Igrashi, S. Yotsuyanagi, T. (1988). *Anal. Sci.*, 4, 175.

Kilian, K. Pyrzynska, K. (2003). *Talanta*, 60, 669.

Knör, G. Strasser, A. (2005). *Inorg. Chem. Commun.* , 9, 471.

Kobayashi, (1992). S. *Mol. Cryst. Liq. Cryst.*, 217, 77.

Komagoe, K. Katsu, T. (2006). *Anal. Sci.*, 22, 255.

Koti, A.S.R. Periasamy, N. (2003). *Chem. Mater.*, 15, 369.

Kubat, P. Lang, K. Prochazkovà, K. Anzenbacher Jr. , P. (2003). *Langmuir*, 19, 422.

Kuhlbrandt, W. (1995). *Nature* , 374, 497.

Kuila D. Lavallee, D.K. (1984). *J. Am. Chem. Soc.*, 106, 448.

Lagoudakis, P.G. de Souza M.M., Schindler F., Lupton J.M., Feldmann, J. Wenus, J. Lidzey, D.G. (2004). *Phys. Rev. Lett.* ,93, 257401.

Lavallee D.K. and Anderson, O.P. (1982). *J. Am. Chem. Soc.*, 104, 4707.

Lavallee, D.K. (1987).*The Chemistry and Biochemistry of N-Substituted Porphyrins, VCH, New York.*

Lavallee, D.K. (1985). *Coord. Chem. Rev.*, 61, 55.

Lavallee, D.K. Kopelove A.B. Anderson, O.P. (1978). *J. Am. Chem. Soc.*, 100, 3025.

Lavallee, D.K. Wite, A. Diaz, ABattioni . J.-P. Mansuy, D. (1986). *Tetrahedron Lett.*, 27, 3521.
Lehn, J.-M., (2002). *Science, 295*, 2400.
Li, G. Fudickar, W. Skupin, M. Klyszcz, A. Draeger, C. Lauer, M. Fuhrhop, J.-H. (2002). *Angew. Chem. Int.* Edit 41, 1828.
Li, L.-L. Yang, C.-J. Chen, W.-H. Lin, K.-J. (2003). *Angew. Chem. Int.* Edit 42, 1505.
Liao, M.S. Watts, J.D. Huang, M.J. (2006). *J. Phys. Chem. A*, 110, 13089.
Lindsey, J. S. Schreiman, I.C. Hsu, H.C. Kearney, P.C. Marguerettaz, A.M. (1987). *J. Org. Chem.* 52, 827.
Lidzey, D.G. Bradley, D.D.C. Armitage, A. Walker, S. Skolnick, M.S. (2000). *Science,* 288, 1620.
Lim, M.D. Lorkovic, I.M. Ford, P.C. (2005). *J. Inorg. Biochem.*, 99, 151.
Lin, Y. Lash, T. (1995).*Tetrahedron Lett.*, 36, 9441
Liu, Z.-B. Zhu, Y.-Z. Chen, S.-Q. Zheng, J.-Y. Tian, J.-G. (2006). *J. Phys. Chem. B*, 110, 15140.
Luca, G.D. Pollicino, G. Romeo, A. Scolaro, L.M. (2007). *Chem. Mater.*, 18, 2005.
Luca, G.D. Romeo, A. Scolaro, L.M. (2005). *J. Phys. Chem. B,* 109, 7149.
Luca, G.D. Romeo, A. Scolaro, L.M. (2006). *J. Phys. Chem. B,* 110, 14135.
Luca, G.D. Romeo, A. Scolaro, L.M. (2006). *J. Phys. Chem. B,* 110, 7309.
Luo, H. Choi, M.-S. Araki, Y. Ito, O. Aida, T. (2005). *Bull. Chem. Soc. Jpn.*, 78, 405.
Ma, S.Y. (2000). *J. Chem. Phys. Lett*, 332, 603,
Maiti, N. C. Mazumdar, S. Periasamy, N. (1998). *J. Phys. Chem. B* 102, 1528.
Maiti, N.C. Ravikanth, M. Mazumdar, S. Periasamy N., (1995). *J. Phys. Chem.*, 99, 17192.
Makino, T. Itoh, J. (1981) *Clin. Chim. Acta,* 111, 1.
Mammana, A. Durso, A. Lauceri, R. Purrello, R. (2007). *J. Am. Chem. Soc.*, 129, 8062.
Mandon, D. Ochsenbein, P. Fischer, J. Weiss, R. Jayaraj, K. Austin, R.N. Gold, A. White, P.S. Brigaud, O. Battioni, P. Mansuy, D. (1992). *Inorg. Chem.*, 31, 2044.
Mansuy, D. Momenteau, (1982). M. *Tetrahedron Lett.,* 2781.
Martirosyan, G.G. Azizyan, A.S. Kurtikyan, T.S. Ford, P.C., (2004). *Chem. Commun.*, 1488.
Matsuzaki, Y. Nogami, A. Tsuda, A. Osuka, A. Tanaka K., (2006). *J. Phys. Chem. A* ,110, 4888.
McLaughlin, G.M. (1974). *J. Chem. Soc., Perkin Trans.* 2, 136.
Medforth, C. J. Wang, Z. Martin, K. E. Song, Y. Jacobsenc, J. L. Shelnutt, (2009). *J. A. Chem. Commun.*, 7261.
Menotti, A. R. (1941). *J. Am. Chem. Soc.*, 63, 267.
Meunier, B. (1992). *Chem. Rev.* , 92, 1411.
Milgrom, L.R. (1997) *The Colours of Life*, OUP, Oxford,;
Misawa, K. Kobayashi, T. (1999). *J. Chem. Phys.*, 110, 5844.
Nahar, N. Tabata, M. (1998). *J. Porphyr. Phthalocyan.*, 2, 397.
Nakanishi, I. Fukuzumi, S. Barbe, J.M. Guilard, R. Kadish, K.M. (2000). *Eur. J. Inorg.Chem.*, 1557.
Napoli, M.D. Nardis, S. Paolesse, R. Graca, M. Vicente, H. Lauceri, R. Purrello, R. (2004). *J. Am. Chem. Soc.* 126, 5934.
Nappa, M. J. S. Valentine, (1978). *J. Am. Chem. Soc.*, 100, 5075
Nikiforov, M.P. Zerweck, U. Milde, P. Loppacher, C. Park, T.-H. Tetsuo Uyeda, H. Therien, M.J. Eng, L. Bonnell, D. (2008). *Nano Lett.*, 8,110.
Nuesch, F.; Gratzel, M. (1995). *Chem. Phys.*, 193, 1.
Ohmo, O. Kaizu, Y. Kobayashi, H. (1993). *J. Chem. Phys.* 99, 4128.
Okada, S.; Segawa, H. (2003). *J. Am. Chem. Soc.*, 125, 2792.

Pasternack, R. F. Schaefer, K. F. Hambright, P. (1994). *Inorg. Chem.*, 33, 2062.
Pasternack, R.F. Gibbs, E.J. Antebi, A. Bassner, S. Depoy, L. Turner, D.H. Williams, A. Laplace, F. Lansard, M.H. Merienne, C. Perrée-Fauvet, M. (1985). *J. Am. Chem. Soc.*, 107, 8179.
Perie, K. Barbe J.M., Cocolios P., (1996). *Soc. Chim. Fr.*, 133, 697.
Porra, R.J. (1997) *Photochem. Photobiol.*, *65*, 492.
Prins, L. J. Reinhoudt, D. N. Timmerman, P., (2001).*Angew. Chem. Int. Ed.*, *40*, 2382
Ren, Q.G. Zhou, X.T. Ji, H.B. (2010). *Chin. J. Org. Chem.* , 30, 1605.
Ribo, J. M. Crusats, J. Farrera, J. A. Valero, M. L. (1994). *J. Chem. Soc., Chem. Commun.*, 681.
Robinson, L.R. Hambright, P. (1992). *Inorg. Chem.*, 31, 652.
Rothemund, P. (1935). *J. Am. Chem. Soc.* 57, 2010.
Rothemund, P. (1936). *J. Am. Chem. Soc.* 58, 625.
Rotomskis, R. Augulis, R. Snitka, V. Valiokas, R. Liedberg, B. (2004). *J. Phys. Chem., B* 108, 2833.
Rowan, A.E. Gerritsen, J.W. Nolte, R.J.M. Maan, J.C. (2004). *Nano Lett.*, 4, 1401.
Rubio, M. Rios, B. O. Serrano-Andres, L. Merchan, M. (1999). *J. Chem. Phys.*, 110, 7202
Sasaki, F. Kobayashi, (1993). S. *Appl. Phys. Lett.*, 63, 2887.
Schauer, C.K. Anderson, O.P. Lavallee, D.K. Battioni J.-P. Mansuy, D. (1987). *J. Am. Chem. Soc.*, 109, 3922.
Schenning, A.P.H.J. Meijer, E.W. (2005). *Chem. Commun.*, 3245.
Scherz, A. Parson, W.W. (1984). *Biochim. Biophys. Acta* , 766, 653.
Schouten, P. Warman, J. De Haas, M. Fox, M. Pan, H. (1991). *Nature*, 353, 736.
Scolaro, L.M. Andrea Romeo, A. Pasternack, R.F., (2004). *J. Am. Chem. Soc.*, *126*, 7178.
Shah, B. Shears B. Hambright, P. (1971). *Inorg. Chem.*, 10, 1828.
Shen, Y. Ryde, U. (2005). *Chem. Eur.* J. ,11 1549.
Shimizu, Y. Taniguchi, K. Inada, Y. Funahashi, S. Tsuda, Y. Ito, Y. Inamo, M. Tanaka, M. (1992). *Bull. Chem. Soc. Jpn.*, 65, 771.
Shirakawa, M. Kawano, S. Fujita, N. Sada, K. Shinkai, S. (2003). *J. Org. Chem.*, 68, 5037.
Shundo, A. Labuta, J. Hill, J.P. Ishihara, S. Ariga, K., (2009). *J. Am. Chem. Soc.*, *131*, 9494.
Siskova, K. Vlckova, B. Mojzes, P. (2005). *J. Mol. Struct.* 744.
Spano, F. C. Mukamel, (1989). S. *Phys. Rev. A*, *40*, 5783.
Spasojevic´ I., Batinic´-Haberle, (2001). I. *Inorg. Chim. Acta*, 317, 230.
Stinson, C. Hambright, P. (1977). *J. Am. Chem. Soc.*, 99, 2357.
Sudeep, P.K. Ipe, B.I. Thomas, K.G. George, M.V. Barazzouk, S. Hotchandai, S. Kamat, P.V. (2002). *Nano Lett.*, 2, 29.
Tabata, M. Kajhara, N. (1989) *Anal. Sci.*, 5, 719.
Tabata, M. Miyata, W. Nahar, N. (1995). *Inorg. Chem.*, 34, 6492.
Tabata, M. Tanaka, M. (1985). *J. Chem. Soc., Chem. Commun.* 42.
Tabata, M. Tanaka, M. (1988).*Inorg. Chem.*, 27, 203.
Tabata, M. Tanaka, M. (1991).*Trends Anal. Chem.*, 10, 128.
Tabata, M. Nishimoto, J. Kusano, K. (1998). *Talanta* 46, 703.
Terazima, M. Shimizu, H. Osuka, A. (1997). *J. Appl. Phys.*, 81, 2946.
Tong,Y. Hamilton, D.G. Meillon, J.-C. Sanders, J.K.M., (1999) *Org. Lett.*, *1*, 1343.
Tovmasyan, A.G. Babayan, N.S. Sahakyan, L.A. Shahkhatuni, A.G. Gasparyan G.H., Aroutiounian, R.M. Ghazaryarn, R.K. (2008). *J. Porphyr. Phthalocya.*, 12, 1100.
Tung, J.Y. Chen, J.-H. (2000). *Inorg. Chem.*, 39, 2120.

Valicsek, Z. Horváth, O. (2007). *J. Photoch. Photobio. A*, 186, 1.
Valicsek, Z. Horváth, O. Lendvay, G. Kikas, Skoric, I.I. (2011). *J. Photoch. Photobio. A*, 218, 143.
Valicsek, Z. Lendvay, G. Horváth, O. (2008). *J. Phys. Chem. B*, 112, 14509.
Valicsek, Z. Lendvay, G. Horváth, O. (2009). *J. Porphyr. Phthalocya.*, 13, 910.
Van de Craats, A.M. Warman, J.M. (2001). *Adv. Mater.*, 12, 130.
Van der Boom, T. Hayes, R.T. Zhao, Y. Bushard, P.J. Weiss, E.A. Wasielewski, M.R. (2002). *J. Am. Chem. Soc.* 124 9582.
Waggoner, A. Membr. J. (1976). *Biol.* , 27, 317.
Walker, V.E.J. Castillo, N. Matta, C.F. Boyd, R.J. (2010). *J. Phys. Chem. A.*, 114, 10315.
Wang, M.-Y. R. Hoffman, B. M. (1984) *J. Am. Chem. Soc.*, 106, 4235.
Wang, Y. (1986). *Chem. Phys. Lett.*, 126, 209.
Wang, Y. (1991). *J. Opt. Soc. Am. B*, 8, 981.
Watanabe, H. Ohmori, H. (1981). *Talanta* 28 774.
West, W. Carroll, B. H. (1966). In *The Theory of Photographic Processes*, 3rd ed. James, T. H., Ed.; The McMillan Company: New York,; Chapter 12.
Whitten, D. G. Lopp, I. G. Wildes, P. D. (1968). *J. Am. Chem. Soc.*, 90, 7196
Wittmer, L.L. Holten, D. (1996), *J. Phys. Chem.*, 100, 860.
Wittmer, L.L. Holten, D. (1996). *J. Phys. Chem.*, 100, 860.
Xiangqing Li, Line Zhang, Jin Mu, (2007). *Colloids and Surfaces A: Physicochem. Eng. Aspects* ., 311, 187.
Yagai, S. Seki, T. Karatsu, T. Kitamura, A. Wrthner, F. (2008). *Angew. Chem.*, 47, 3367.
Yamamoto, S. Watarai, H. (2008). *J. Phys. Chem. C*, 112, 12417.
Yang, R. Wang, K. Long, L. Xiao, D. Xiaohai Yang, X. Tan, W. (2002). *Anal. Chem.*, , *74*, 1088.
Yusmanov, V.E. Tominaga, T.T. Borissevich, L.E. Imasato, H. Tabak, M. (1996). *Magn. Reson. Imaging*, 14, 255.
Zhang, Y. Yang, R. H. Liu, F. Li, K.A., (2004) *Anal. Chem.*, *76*, 7336.
Zhang, Y. Chen, P. Liu, M. (2008). *Chem. Eur. J.*, 14, 1793.
Zhang, Y. Chen, P. Ma, Y. He, S. Liu, M. (2009). *ACS Appl. Mater. Interfaces*, 1, 2036.
Zheng, W. Shan, N. Yu, L. Eang, X. (2008). *Dyes and Pigments* 77, 153.
Zhou, H. Baldini, L. Hong, J. Wilson, A. J. Hamilton, A.D., (2006). ,*J. Am. Chem. Soc.*, *128*, 2421.
Zimmermann, J. Siggel, U. Fuhrhop, J.-H. Roder, B. (2003). *J. Phys. Chem. B*. 107, 6019.

7

Spectrophotometric Methods as Solutions to Pharmaceutical Analysis of β-Lactam Antibiotics

Judyta Cielecka-Piontek[1], Przemysław Zalewski[1],
Anna Krause[2] and Marek Milewski[2]
[1]*Poznan University of Medical Sciences, Department of Pharmaceutical Chemistry*
[2]*PozLab Contract Research Organization at Centre of Transfer of Medical Technologies*
Poland

1. Introduction

Following the discovery of the first analog of penicillin by A. Fleming (1929), the β-lactam antibiotics are still a developing group of chemotherapeutics and are used in treatment of majority of diseases with bacterial etiology. β-lactam antibiotics have a broad spectrum of antibacterial activity, favourable pharmacokinetic parameters and low side effects. In β-lactam therapy two main problems are still current. The increasing resistance of some bacterial strains which implicates necessity to combine the therapy with inhibitors of β-lactamases and other chemotherapeutics. The second problem of therapy of β-lactam antibiotics is their significant instability [1-3]. The analogs from that group are easily degraded in aqueous solutions and in solid state. They are a special group of drugs because parallel to losing the antibacterial efficiency, the strong allergic properties can also appear as a results of their degradation. Therefore in terms of quality control, the stability of β-lactam antibiotics in solutions was widely studied. The evaluation of stability concerned also the studies of their metabolites and intravenous solutions after preparations of pharmaceutical dosage forms. Moreover, the evaluation of concentration changes during storage of substance in solid state was also conducted. As problem of the instability of some β-lactam analogs has been solved their oral administration is possible. An intake of oral formulations is connected with appearance of excipients, which can influence rate of degradation and cause formation of different degradation products.

The common element of chemical structure of all β-lactam antibiotics is five-membered β-lactam ring. Currently, higher significance in treatment have derivatives in which the β-lactam ring is fused to:

- thiazolidine ring in penam analogs,
- 2,3-dihydro-2*H*-1,3-tiazine ring in cephem analogs,
- 2,3-dihydro-1*H*-pyrrole in carbapenem analogs,
- 2,3-dihydrotiazole in penem analogs (Fig .1).

These connections implicate the different intra-ring stress. The presence of sulphur atom and/or double bonds influence on length of bond and intra-molecular angle in molecule of β-lactam analog. Finally for some derivatives, the differences in stability are noticeable. Additionally, the factor distinguishing a stability of derivatives of β-lactam analogs are

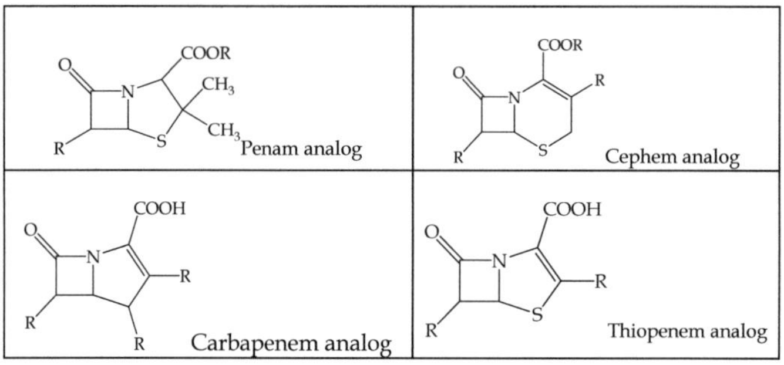

Fig. 1. Chemical structure of penam, cephem, carbapenem and thiopenem nuclei.

chemical structures of substituents at C2, C3, C5, C6, and C7. The amount and type of degradation products of β-lactam antibiotics often depend on affecting factors (solvents, concentration of substance and hydrogen ions, temperature). Moreover, most of the β-lactam antibiotics obtained by chemical synthesis or fermentation contain impurities being remnants of the process. In the development of analytical methods for the determination of β-lactam antibiotics, selectivity is a fundamental validation parameter. A reliable, selective method is expected to allow separation and determination of parental substance in the presence of related ones. Current International Conference on Harmonization (ICH) guidelines require the development of analytical methods permitting analysis in the presence of related products (Q1A-R2) [4]. These requirements are restrictions but also challenges during the development of analytical methods for the determination of β-lactam antibiotics. The problem of the overlapping of the "background" originating from related products (impurities, degradation products and metabolites) and/or the presence of other active substances in a sample (inhibitors of β-lactamases, other drugs) was solved during the determination of β-lactam antibiotics by using chromatographic techniques (high-performance liquid chromatography, thin layer chromatography). On the other hand, search of new solutions and analytical methods, especially being in accordance with the "*green chemistry*" concept, is very important and up-to-date. Analytical methods based on determination of spectrophotometric properties of β-lactam analogs are a developing tools in their analysis. Non-destructive investigations of β-lactam analogs, did not producing residues, were reported in fields of many spectrophotometric methods. A few methods of determination of β-lactam analogs were developed by using infrared spectrophotometry enriched by chemometric procedures [5-6]. Most of all analytical methods for the determination of β-lactam analogs were developed in range of visible and ultraviolet radiations. Desired, selective signals were possible to obtain by application of following techniques:

- as spectrophotometric methods
  - direct spectrophotometry
  - direct spectrophotometry enriched by chemometric procedures
  - derivative spetrophotometry
  - derivative spectrotometry enriched by chemometric procedures
- as visible spectrophotometric methods
  - measurement of absorption of species being a result of reaction between analyte and derivatizating reagent
  - measurement of absorption of species being a result of reaction between degradation products of analyte and derivatizating reagent (Table 1).

Derivative	Ultraviolet region			Visible region	
	1	2	3	1*	2*
Analog of penam	✓	✓	✓	✓	✓
Analog of cephem					
I generation	✓	✓	✓	✓	✓
II generation		✓		✓	✓
III generation	✓	✓	✓	✓	✓
IV generation		✓			
Analog of carbapenem	✓	✓	✓		
Analog of penam		✓			

1. direct spectrophotometry enriched by chemometric procedures
2. derivative spetrophotometry
3. derivative spectrotometry enriched by chemometric procedures
1.* measurement of absorption of species being a result of reaction between analyte and derivatizating reagent
2.* measurement of absorption of species being a result of reaction between degradation products of analyte and derivatizating reagent

Table 1. Possibilities of application of visible and ultraviolet spectrophotometric determinations for analysis of β-lactam antibiotics in the period of time 1994–2011.

2. Spectrophotometric methods for determination of β-lactam antibiotics

2.1 Direct spectrophotometry

Spectra of β-lactam antibiotics recorded by using direct spectrophotometry do not have desired selectivity due to the presence of related products. A comparison of sharp zero-order spectra and/or value of absorption maxima for some β-lactam analogs with ones obtained for CRS (*chemical reference substance*) is recommended by pharmacopeias for an their identification [7] . Lack of desired absorbing species in chemical structure of penam analog often do not allow to apply direct spectrophotometry even for qualitative studies of substance of high purity.

Paradoxically, the significant instability of analogs can sometimes solve this problem due to formation of degradation products that can absorb ultraviolet radiation permitting determination of parental substance.

Significant susceptibility of β-lactam analogs to degradation in basic medium was reported during analysis of cephem analogs. It was confirmed that formation of piperazine-2,5-dione derivative, peak at 340 nm, was possible via intra-molecular nucleophilic attack of the primary amine from the side chain on β-lactam ring (pH = 11 was required) (Fig. 2) [8].

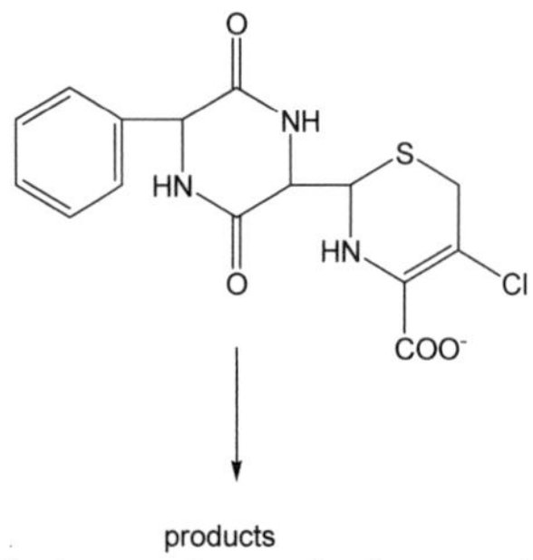

The degradation products of cefactor - piperazine-2,5-dione derivative.

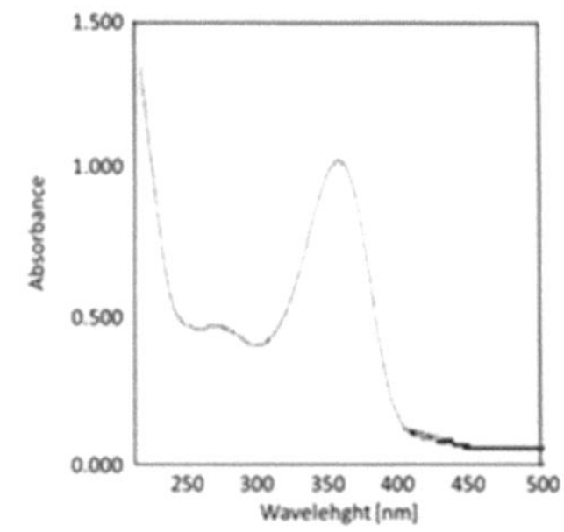

The spectrum of piperazine-2,5-dione derivative of cefaclor.

Fig. 2. Chemical structures of degradation products of cefaclor (1.0 mmol/l) formed at pH 11.0 and its spectrum [8].

The degradation of penam analogs in acidic conditions was also a base for spectrophotometric determination. As it is shown in Fig. 3, different pathways of degradation (including enzymatic one) can lead to obtaining absorbing species in the range of ultraviolet radiation. As a results of chemical degradation of penam analog in acidic conditions, the penicilloic acid, penillic acid and penicillenic acid are formed and absorb the ultraviolet radiation in the range 320–360 nm, respectively [9]. While during the enzymatic degradation under the influence of penicillin acylase, D-4-hydroxyphenylglycine (D-HPhG) and 6-aminopenicillanic acid are formed. Then the D-HPhG was catalyzed by D-phenylglycine aminotransferase to form L-glutamate and hydroxybenzoylformate which strongly absorb UV light at 335 nm [10].

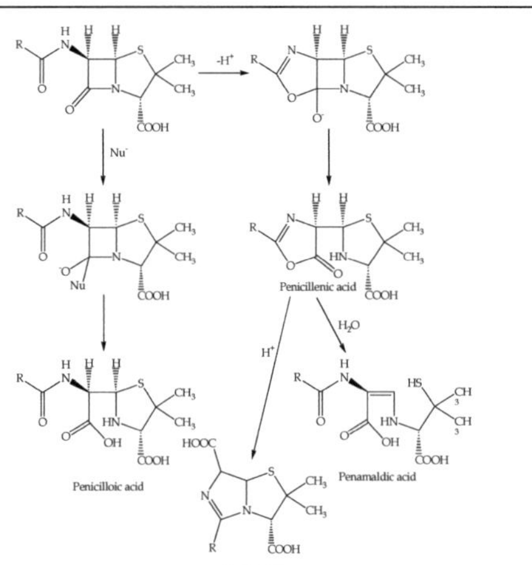

Chemical degradation of penicillin in acidic conditions

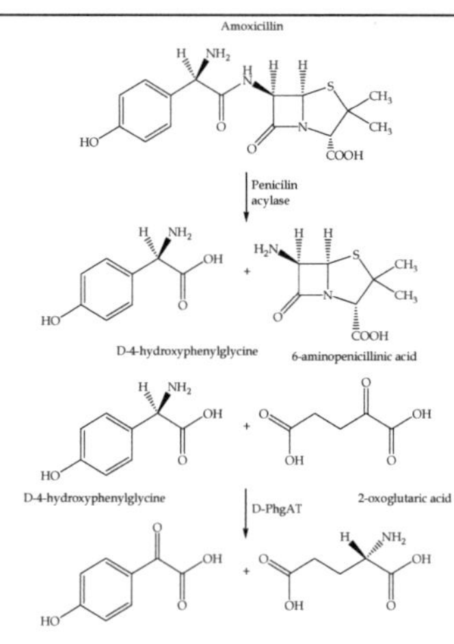

Enzymatic degradation of amoxicillin using bienzymatic factors

Fig. 3. The pathways of obtaining of absorbing degradation products of penam analog [9-10].

2.2 Direct spectrophotometry enriched by chemometric procedures

The other way of improving the selectivity of direct spectrophotometry for the determination of β-lactam antibiotics is the enrichment of data analysis by chemometric procedures. A literature review revealed the application of the following determinations of β-lactam antibiotics enriched by chemometric procedures that solved the problem of spectral overlap without additional separation techniques at the stage of sample preparation, were used:

- a separation of analog of cephem in the presence of impurities originating from synthesis (e.g., cephalexin in the presence of 7-aminocephalosporanic acid and acid-induced degradation products) using H-point standard additions method (HPSAM) [11]
- determination of analog of penam in the presence of other drugs (e.g., amoxicillin in the presence of diclofenac) using partial least squares (PLS) regression analysis [12]
- determination of analog of cephem in the presence of alkali-induced degradation products using full spectrum quantitation (FSQ) (e.g., cefotaxime, ceftazidine, ceftiaxome, in the presence of degradation products) [13].

Each chemometric method relies on different tools of regression analysis of multicomponent system permitting simultaneous determination of two or more components.

The determination of β-lactam analyte in the presence of known and unknown inferences was possible by the application of HPSAM procedure, where analyte concentration is calculated from the following equation:

$$\frac{(A_0\text{-}b_0)+(A'\text{-}b)}{M(\lambda_1)-M(\lambda_2)} = -C_X + \frac{(A'\text{-}b)}{M(\lambda_1)-M(\lambda_2)} \tag{1}$$

where b_0 and A_0 are the absorbance values for β-lactam analyte, b and A′ ones for the interferent, at λ_1 and λ_2 and $M(\lambda_1)$, $M(\lambda_2)$ are slopes of plots at selected wavelengths.

In PLS technique, analytical sensibility was defined as $\gamma = \frac{SEN_k}{\|\sigma_r\|}$ where $SEN_k = \frac{1}{\|b_k\|}$, σ_r is a value estimated from standard deviation of blank samples, b_k value is a vector of the regression coefficient for the *k* analytes and *k* is a number of components in a mixture.

The FSQ technique during a determination of β-lactam antibiotics applies Fourier pre-processing of the entire absorption spectra of the individual β-lactam analogs with their degradation products at variable concentration to calculate matrix calibration coefficients.

2.3 Derivative spectrophotometry

A derivative spectrophotometry using derivatives of absorbance with respect to wavelength (first $\frac{dA}{d\lambda} = f(\lambda)'$, second $\frac{d^2A}{d^2\lambda} = f(\lambda)''$; third $\frac{d^3A}{d^3\lambda} = f(\lambda)'''$; respectively) is a suitable tool for overcoming the overlapping spectra problem in analysis of many β-lactam analogs. Possibility of application of derivative spectrophotometry with zero-crossing point is widely used in analysis of all β-lactam analogs. The direct correlation between order of used derivative spectrophotometry and similarities of chemical structures of nuclei of β-lactam analogs has not been observed, e.g., both second-derivative and first-derivative were developed for cephem analogs including the same nuclei (Fig. 4).

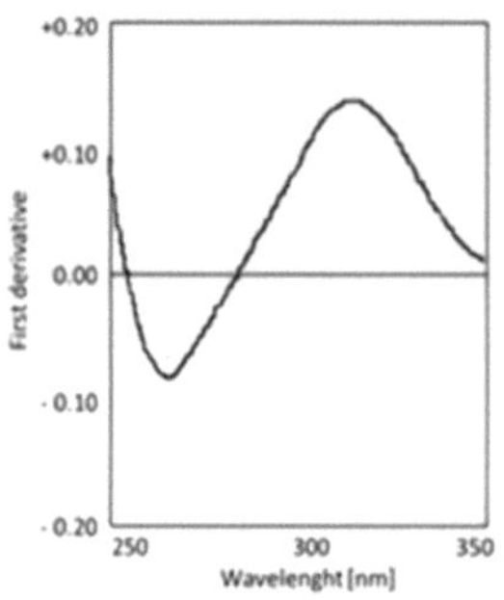

First-derivative spectrum of cefprozil (1 mg%) in 0.1 M HCl.

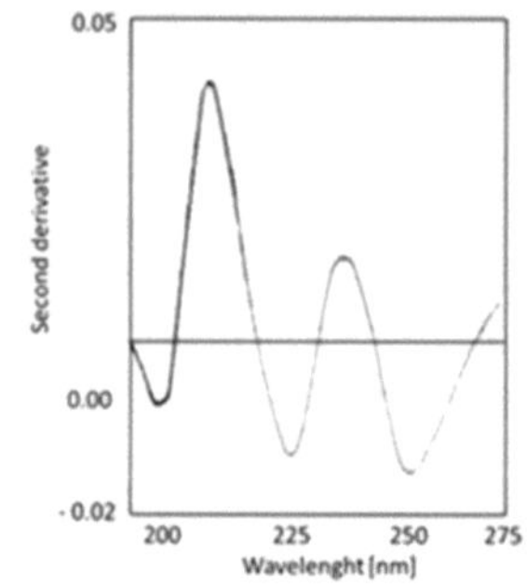

Second-derivative spectrum of cefepime (20 μg/ml) in water.

Fig. 4. The application of derivative spectrophotometry for analysis of cephem analogs [15-16].

The application of derivative spectrophotometry for determination of β-lactam antibiotics was used in the following areas:

- a separation and determination of penam/cephem analogs and inhibitors of β-lactamase in aqueous solution (e.g., determination of ampicillin sodium in the presence of sulbactam sodium; determination of cefsulodin in the presence of clavulanic acid) [14-15]
- a separation and determination of cephem/carbapenem analog and excipients used in parenteral pharmaceutical dosage forms (e.g., determination of cefepime in the presence of *L*-argininie) [16]
- a separation and determination of cephem analog and its degradation products (e.g., determination of cefprozil in the presence of its degradation products) [17]
- a separation and determination of cephem analog and related compounds from the synthesis (e.g., determination of triethylammonium salt of cefotaxime in the presence of 2-mercaptobenzothiazole) [18]
- a separation and determination of penam/cephem/carbapenem analogs in biological matrix (e.g., determination of amoxicillin, cefuroxime, imipenem in urine) [19].

The separation of often structurally very similar species (e.g., two analogs of cephem, cephem analog and its impurities from synthesis or carbapenem analog and its degradation products) was possible by using derivative spectrophotometry (Fig. 5).

It was proved that the derivative spectrophotometry can be recommended as a method for routine control analysis of pharmaceutical preparation of β-lactam antibiotics. Derivative spectrophotometry ensured the rapid analysis of parenteral dosage forms and also removed a "*background*" excipients in oral pharmaceutical dosage forms.

The special potency of derivative spectrophotometry was possibility of its usage in determination of β-lactam analogs in biological matrix. In this case, to meet the requirements of analytical methods, the selectivity had to be extended in regard with interference of biological endogenous components. It was noticed that the selective determination of penam/cephem/carbapenem in the presence of metabolites (open-ring degradation product) and endogenous substance of urine was possible to achieve.

<table>
<tr><th colspan="2">Chemical structures of components of mixture</th><th>Derivative spectra</th></tr>
<tr><td colspan="3">The separation of β-lactam analog with the same nuclei</td></tr>
<tr><td>Cephalothin
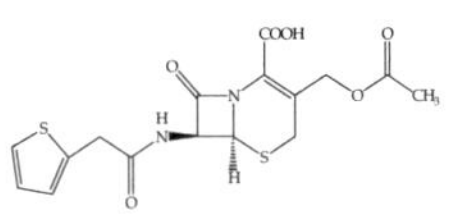</td><td>Cefoxitin
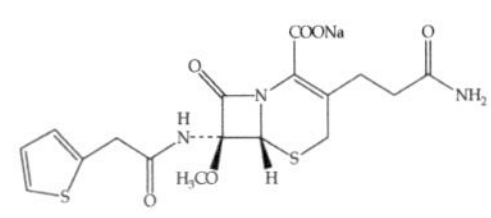</td><td>First-derivative of spectrum of cephalothin CL (20.0 μg/ml) and cefoxitin CX (20.0 μg/ml) and mixtures of each component M)
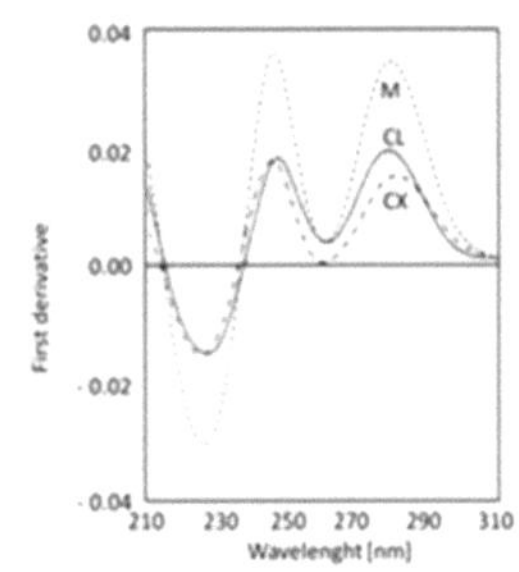
</td></tr>
<tr><td colspan="3">The separation of β-lactam analog and impurties from synthesis</td></tr>
<tr><td>Triethylammonium salt of cefotaxime
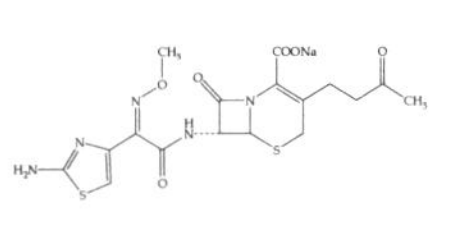</td><td>2-mercaptobenzothiazole
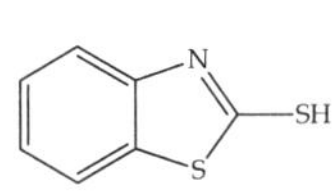</td><td>First derivative spectrum of triethylammonium salt of cefotaxime A (40 μg/ml) , 7-aminocephalosporanic acid B (40 μg/ml), S-(2-benzothazolyl)2-amino-α-(methoxyimino-4-thiazoleethanethioate) C (0.03 μg/ml).
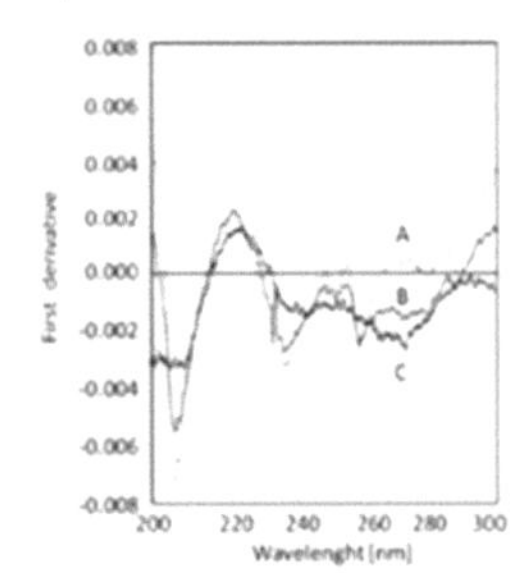
</td></tr>
<tr><td colspan="3">The separation of β-lactam analog and degradation products</td></tr>
<tr><td>Biapenem
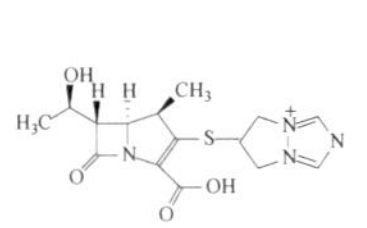</td><td>Open-ring hydrolysis products of biapenem
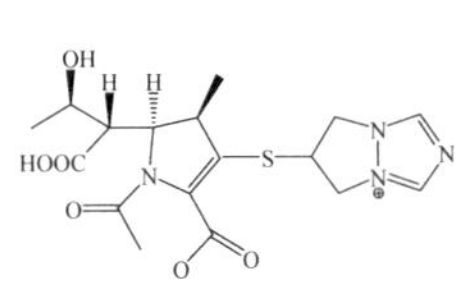</td><td>First-derivative spectra of biapenem during degradation at 313 K: in HCl
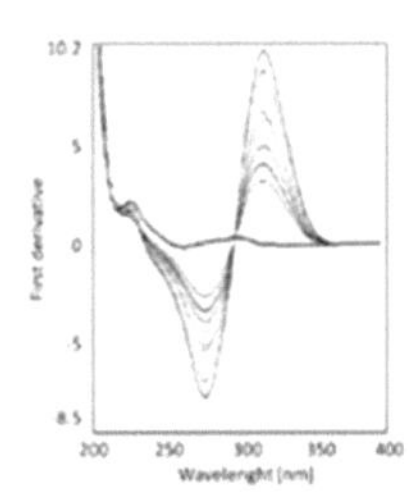
</td></tr>
</table>

Fig. 5. Separation of some β-lactam analogs using derivative spectra [18, 20-21].

2.4 Derivative spectrophotometry enriched by chemometric procedures

The application of chemometric procedures coupled with derivative spectroscopy permits achievement of higher selectivity in determination of β-lactam antibiotics. Currently, chemometric procedures based on the estimated ratio of spectra derivative for the selective determination of β-lactam analogs are the most common. It was proved that the application of the ratio of different-order spectra derivatives permitted the separation of binary and tertiary mixtures of β-lactam antibiotics [22]. During the determination of concentrations of three components (e.g., penicillin-G sodium, penicillin-G procain and dihydrostreptomycin sulphate salts) in a mixture the equation describing the ratio spectra derivative spetrophotometry is as follows:

$$\frac{d(A_{a+b,\lambda}/A_{a,\lambda^0})}{d\lambda} = C_b\frac{d(k_{b,\lambda}/A_{a,\lambda^0})}{d\lambda} + C_c\frac{d(k_{c,\lambda}/A_{a,\lambda^0})}{d\lambda} \quad (2)$$

where $A_{a+b+c,\lambda}$ is the absorbance of the ternary mixture of *a, b* and *c* at wavelength λ, A_{a,λ^0} is the absorbance of pure component at wavelength λ, C_b and C_c – are the concentrations of *b* and *c*, $k_{b,\lambda}$ and $k_{c,\lambda}$ are the products of the molar absorption coefficient of *b* at wavelength λ and the thickness of the absorption cell. Equation 2 is divided by C_b while divisor can be any component of ternary mixture (Fig. 6):

$$\frac{d(A_{a+b,\lambda}/A_{a,\lambda^0})}{d(A_{b,\lambda^0}/A_{a,\lambda^0})} = \frac{C_b}{C_b^0} + (C_c d\frac{d(k_{c,\lambda}/A_{a,\lambda^0})}{d(A_{b,\lambda^0}/A_{a,\lambda^0})}) \quad (3)$$

Equation 3 is drawn:

$$J = C_c\, d(d(\frac{d(k_{c,\lambda}/A_{a,\lambda^0})}{d(A_{b,\lambda^0}/A_{a,\lambda^0})}))/d\lambda \quad (4)$$

Finally, after the next derivation J (as the left side of equation 3), is proportional to the C_c value and can be used to determine concentration of component in the ternary mixture (when A_{a,λ^0} and A_{b,λ^0} are fixed) [23].

Depending on chemometric procedure, the selective determination of following analogs was possible:

- a separation and determination of carbapenem and degradant (e.g., the determination of ertapenem and its degradant) when the substraction technique was used [24-25]
- a separation and determination of penam and cephem analogs (e.g., the determination of penicillin-G, penicillin-G procain in the presence of dihydrostreptomycin sulphate salts or the determination of cefotaxime and cefadroxil), cephem analog and inhibitor of β-lactamases (e.g., the determination of cephradine and clavulanic acid) and carbapenem analog and degradation products (e.g., the determination of meropenem and its degradant) when ratio spectra of derivative with all orders were used [26-29]
- a separation and determination of carbapenem and degradation products (e.g., the determination of ertapenem) when the substraction technique was used [30]
- a separation and determination of carbapenem and degradation products (e.g., the determination of ertapenem) when the Kraise's method was used [31]
- a separation and determination of penam analogs (e.g., the determination of ampicillin and flucloxacillin) when multivariate methods (classical least squares and principle component regression) were used.

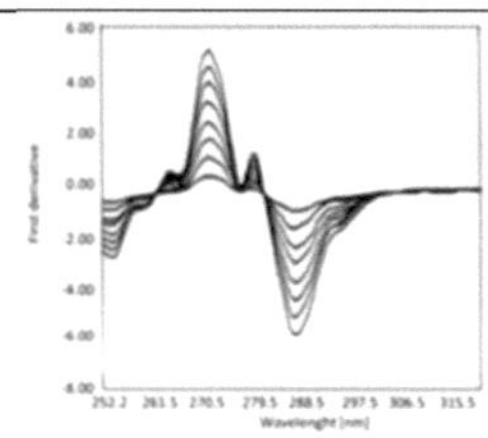

First order derivative ratio spectra of dicloxacillin sodium 50.0–400.0 mg/l using of 68.47 μg/ml ampicillin sodium as a divisor [26]

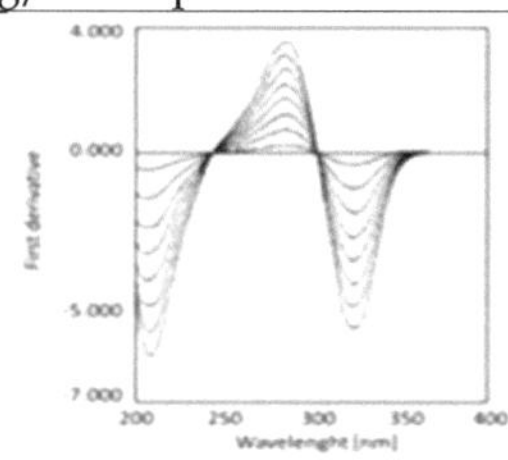

First order derivative ratio spectra of meropenem 4.0–60.0 μg/l using of 32 μg/ml degradate as a divisor [29]

Fig. 6. The application of ratio spectra of derivative spectrophotometry in analysis of β-lactam antibiotics.

3. Visible spectrophotometric methods for determination of β-lactam antibiotics

The β-lactam analogs itselfs do not absorb in visible region of radiation. However, many visible spectrophotometric methods were developed for the determination of β-lactam antibiotics using the effect of formation of "species" giving signals in visible region as the result of chemical derivatization (Fig. 7).

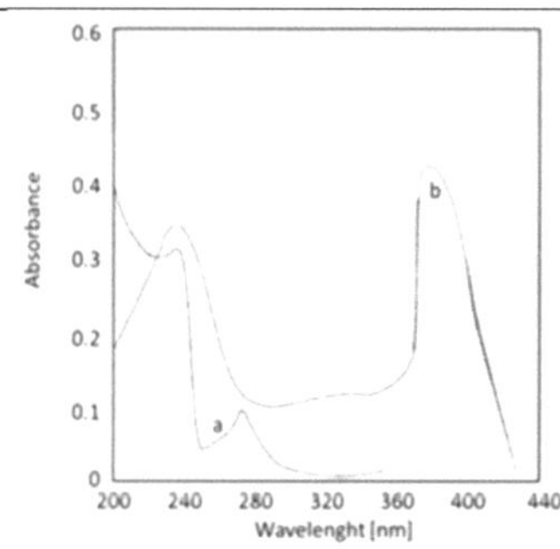

Absorption spectra of amoxicillin trihydrate
(a) 30 μg/ml; (b) after reaction with Fe (III) [31]

Fig. 7. The application of derivatization for detetmination of penam analog.

Formation of "species" absorbing visible radiation can be a result of reactions of chemical reagents with:

- β-lactam analog

- degradation product of some β-lactam analog.

Above-mentioned methods were described only for the determination of penam and cephem analogs. The species, absorbing visible radiation, used in analysis of β-lactam analogs were formed as a result of the following reactions:

- redox
- complexation of metals
- complexation on the base of charge-tranfer process
- formation of ion pairs
- coupling with specified reagents.

3.1 Visible spectrophotometric methods based on redox reactions

The methods based on the selective oxidation were reported for penam and cephem analogs containing phenolic substitutes at C6 and C7, respectively. These methods permitted also selective determination of β-lactam analogs in the presence of excipients being in their pharmaceutical preparations. The application of oxidation properties of iron ions was used in analysis of a huge number of β-lactam analogs:

- when, as the result of direct reaction with Fe(III) in acidic medium, yellow coloured products (λ_{max} = 397 nm) were produced (cefoperazone sodium, cefadroxil monohydrate, cefprozil anhydrous, amoxicillin trihydrate). A possible mechanism of reaction is presented in Fig. 8 [32].

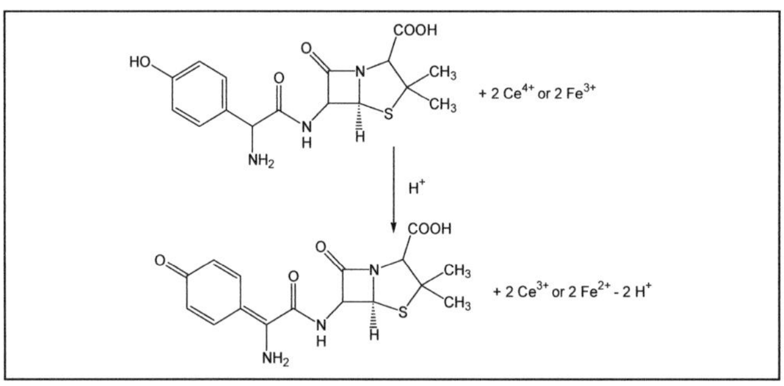

Fig. 8. The mechanism of reaction of amoxicillin and Fe(III)[32].

- when, as the result of indirect reaction, red complex Fe-(o-phen)$_{2/3}$ (λ_{max} = 510 nm) was produced. This complex is formed between *o*-phenanthroline and Fe(II) which previously was reduced from Fe(III) as the result of oxidation of β-lactam analogs in alkali medium [33].

The reduction of oxidized quercetin by cephem analogs was used in development of visible spectrophotmetric method for determination of β-lactam analogs. Quercetin is a flavonol (3,5,7,3′,4′- pentahydroxyflavone) which is oxidized by *N*-bromosuccinimide giving reddish green colour (λ_{max} = 510 nm). As the result of reduction of oxidized form of quercetin by cephem analog fade colour was observed (Fig. 9). This colour is the result of formation of *o*-quinone derivative of quercetin under the mild oxidants [34].

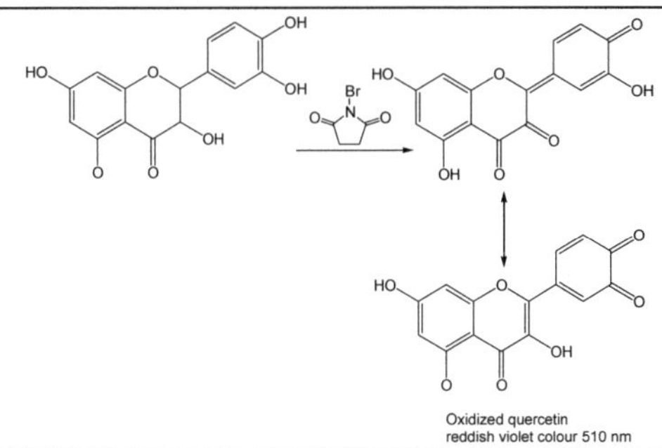

Fig. 9. The mechanism of reduction of quercetin oxidation with *N*-bromosuccinimide [34].

Also, another reagent which is able to oxidize some sulphur atoms present in compounds such as cefotaxime and cefuroxime, is 1-chlorobenzotiazol. As the result of reaction of cephem analogs and 1-chlorobenzotiazol, a product with yellow colour is formed, absorbing radiation at λ_{max} = 298 nm. The suggested possible reaction pathways and absorption spectra are shown in Figure 10 [35].

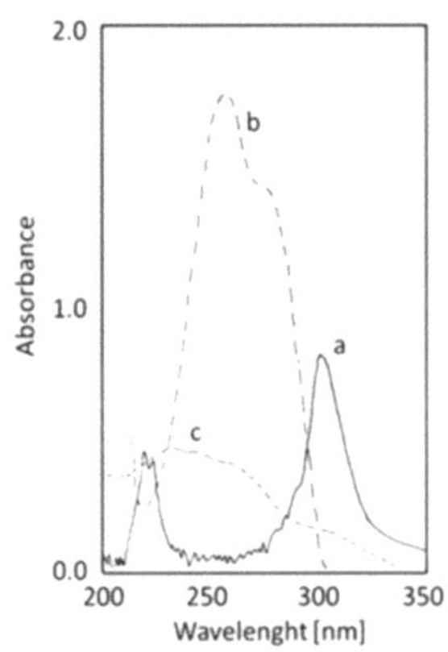

Absorption spectra of 65 μg/ml cefotaxime sodium (c), 1-chlorobenzotiazol (0.002 M) (b) and their reaction product (a).

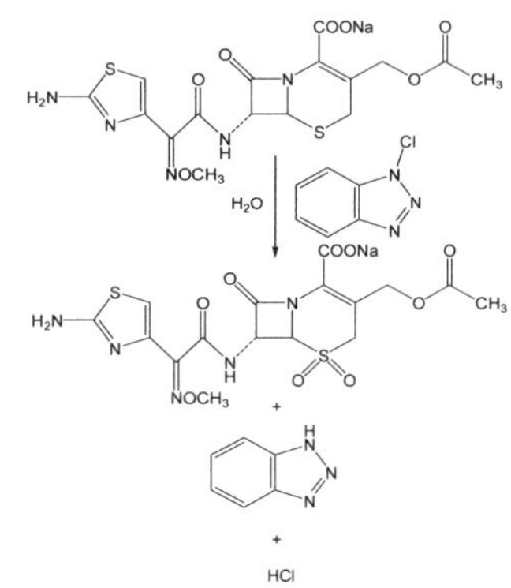

The schemes of reaction between cefotaxime sodium and 1-chlorobenzotiazol

Fig. 10. The mechanism of reaction of cephem analog and 1-chlorobenzotiazol [35].

In indirect spectrophotometry, redox properties of iodine were used in determination of penam analog (amplicillin, penicillin V, amoxicillin, cloxacillin) and cephem analogs (cefadroxil, ceftezoxime). The method based on formation of hypoiodite, from excess of iodine (which did not react with β-lactam analog) under alkaline conditions. Hypoiodite reduced the intensity of wool fast blue colour (5,9-dianilo-7-phenyl 4,10-disulphbenzo[α]phenazinium hydroxide) by disruption of phenazine chromophore [36].

Especially, for cefadoxil many visible spectrophotometric methods based on redox reactions were proposed (Fig. 11). These methods were based on the reaction with different oxidizing reagents:

- 4-aminoantipyrine in the presence of potassium hexacyanoferrate(III) in alkaline medium (λ_{max} = 505 nm [37]
- 3-methyl-2-benzothiazolinone hydrozone hydrochloride in the presence of ceric ammonium sulphate (λ_{max} = 410 nm) [38]
- 4-aminophenazone in the presence of potassium hexacyanoferrate (III) (λ_{max} = 510 nm)
- 2,6-dichloroquinone-4-chlorimide (Gibb's reagent) (λ_{max} = 620 nm) [39]
- *N*-bromosuccinimide or *N*-chlorosuccinimide in alkali medium (λ_{max} = 395 nm) [40]
- sodium persulfate in alkaline medium (λ_{max} = 350 nm) [41]

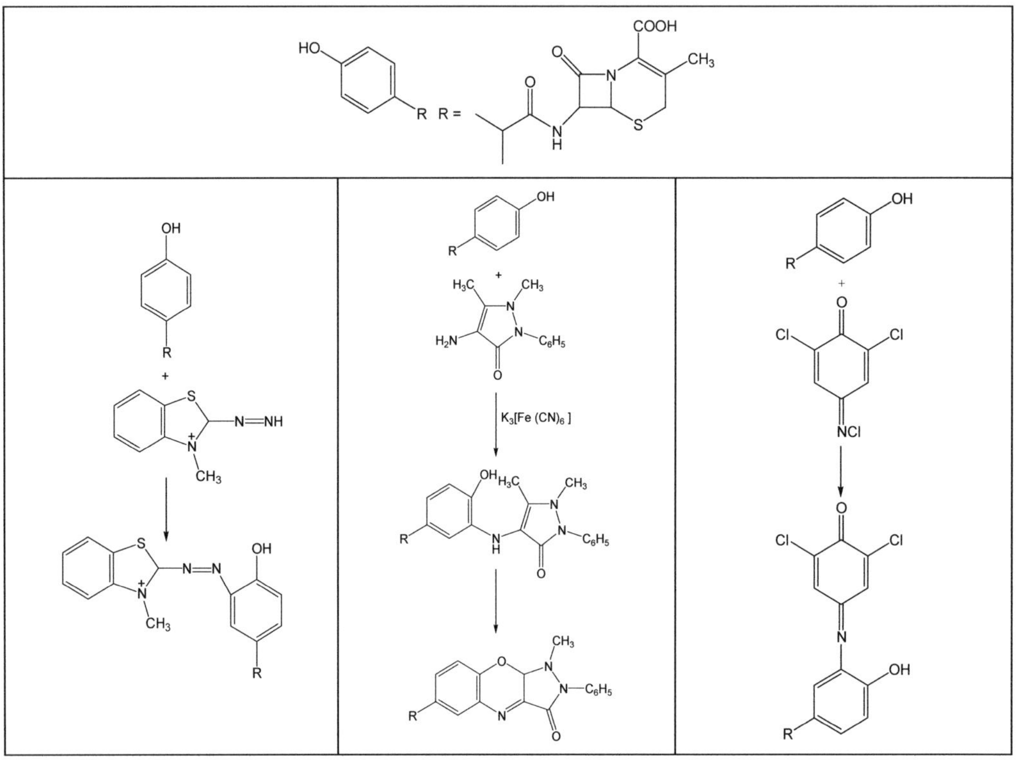

Fig. 11. Schemes of some oxidation reactions of cefadoxil [39].

3.2 Visible spectrophotometric methods based on formation of complex with metal

The formation of colored complexes as a consequence of the interactions between metal ions and analytes (penam and cephem analogs) resulted from:

- a direct reaction between β-lactam analogs and metal ions
- a reaction between the degradation products of β-lactam analogs and metal ions.

Direct complexation of β-lactam analogs and metal ions with formation of yellowish-brown chelate complex was possible due to the presence of suphur atoms in the β-lactam ring and the thiazole ring. Cephem analogs (cefpodoxime, ceftizoxime, ceftazidime, ceftiaxone and cefixime) gave with palladium(II) ions, absorbing complex, in the presence of sodium lauryl sulphate as surfactant, in the range 300–500 nm [42]. Β-lactam analogs containing the phenolic ring with free *ortho* position to the hydroxyl group (amoxicillin trihydrate, cefperazone sodium, cefadroxil monohydrate, cefprozil anhydrous) reacted with nitrous acid forming the nitroso derivatives. They were capable of tautometric interconversions to form colored complex in the presence of copper(II) ions (Fig. 12). The stoichiometric ratios (nitroso derivative to copper(II)) were determined by the Job's method 2:1 [43].

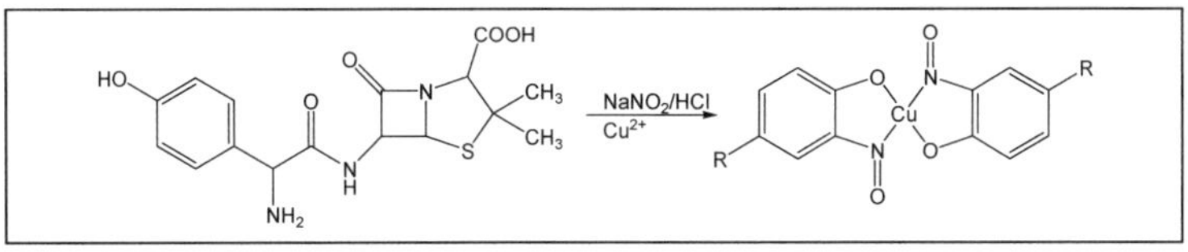

Fig. 12. Scheme of the reaction of nitrous acid, Cu(II) with phenolic β-lactam analogs [43].

The determination of cephalexin, cefixime, ceftriaxone, cefotaxime based on the Bent-French method, in which degradation products of β-lactam analogs with metal ions form the colored complexes, was developed. Hydroxamic acids formed by hydroxiaminolysis of cephem analogs (1:3), formed complexes with iron (II) ions (Fig. 13) [44].

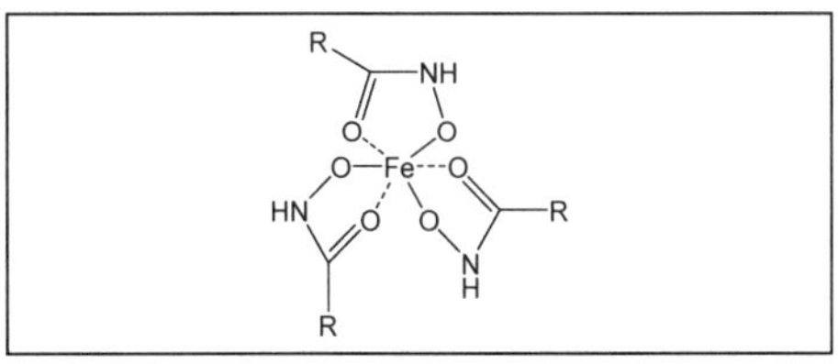

Fig. 13. The scheme of hydroxamic acid-iron(III) [44].

3.3 Visible spectrophotometric methods based on formation of charge-transfer complex

Some drugs, including penam and cephem analogs, are electron donors. Therefore, they form charge-transfer complexes with compounds that are σ- and п-accepptors of electrons. The wavelengths at which the absorption maxima of charge-transfer complexes of β-lactam antibiotics were measured depended on what of reagent was used the acceptor:

- *p*-chloranilic acid gives coloured complex species at 520–529 nm during analysis of cefotaxime sodium, cefuroxime sodium, ceftazidime pentahydrate, cephalexin monohydrate, cefotaxime sodium, cephradine, cephaloridine sodium, cefoperazone sodium (ratio 1:1); cephalotin sodium, cefixime, cefprozil anhydrous, cefazolin sodium, cephapirin (ratio 1:2); cefaclor (ratio 1:4) [45]
- *p*-nitrophenol, 2,4-dinitrophenol, 3,5-dinitrosalycilic acid, picramic acid and picric acid give greenish yellow complexes at 446, 435, 442, 473 and 439 nm, respectively during determination of flucloxacillin (ratio 1:1) [46]
- 7,7,8,8-tetracyanoquinodimethane (TCNQ) gives coloured complex species at 838–843 nm during analysis of cefotaxime sodium, cefuroxime sodium, cephapirin sodium, cefazoline sodium, cephalexin monohydrate, cefadroxil monohydrate, cefoperazone and ceftazidime (ratio 1:1)[47]
- 2,3-dichloro-5,6-dicyano-*p*-benzo-quinone (DDQ) gives coloured complex species at 460 nm during analysis of cephapirin sodium, cefazoline sodium, cephalexin monohydrate, cefadroxil monohydrate, cefoperazone and ceftazidime (ratio 1:1) [47]

- iodine gives coloured complex species at 838–843 nm during analysis of cephapirin sodium, cefazoline sodium, cephalexin monohydrate, cefadroxil monohydrate, cefoperazone and ceftazidime (ratio 1:1) [47].

During formation of charge-transfer complexes between *p*-chloranilic acid and d-donors electrons (A) from group of cephem analogs (D) in polar solvents the radical anion is formed.

$$D + A \rightleftharpoons \underset{\text{complex}}{(D\text{---}A)} \xrightleftharpoons{\text{polar sovent}} \overset{\bullet +}{D} + \overset{\bullet -}{A}$$

Electron transfer from donor to the acceptor moiety occurred with the formation of intensely coloured radical ions with high molar absorptivity value. The formation of charge-transfer complex of some cephem analogs (D) with iodine in 1,2-dichloroethane (J) is observed with the change of colour from violet to lemon yellow:

$$D + J_2 \rightleftharpoons \underset{\text{outer complex}}{D\text{---}\overset{\oplus}{J}\,\overset{\ominus}{J}} \rightleftharpoons \underset{\text{inner complex}}{\left[D\text{---}\overset{\oplus}{J}\right]^{\ominus}} + J^{\ominus} \xrightleftharpoons{J_2} \underset{\text{tri-iodide ion}}{J_3^{\ominus}}$$

Only in 1,2-dichloroethane, formation of tri-iodide ion pair (inner complex), showing two absorption maxima at 290 nm and 364 nm, was possible. This complex originated from an early intermediate outer complex D---J_2. While the interactions of some cephem analogs with DDQ and TCNQ took place according to the following simple relationship:

$$D + A \rightleftharpoons \underset{\text{complex}}{(D\text{---}A)}$$

3.4 Visible spectrophotometric methods based on formation of ion pair

In visible spetrophotometric analysis of penam and cephem analogs, their ability to form ion-pair was also used. The penam analogs contacting the tertiary amine group (ampicillin, dicloxacillin, flucloxacillin, amoxicillin) and Mo(V)-thiocyanate binary complex in hydrochloric acid give coloured ion-pair formation absorbing at λ = 467 nm (Fig. 14) [48].

$$Mo(VI) \underset{\text{ascrobic acid}}{\rightleftharpoons} Mo(V) \xrightleftharpoons{6SCN^-} Mo(SCN)_6^-$$

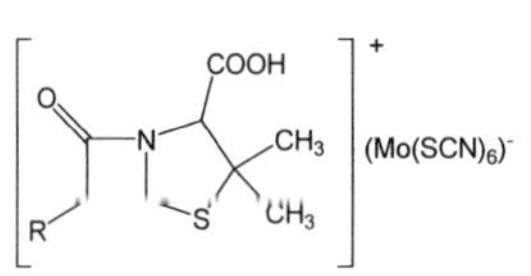

Fig. 14. Scheme of Mo(V)-thiocyanate-β-lactam ion-pair [48].

Some cephem analogs (cephaprin sodium, cefuroxime sodium, cefotaxime sodium, cefoperazone sodium, cefadroxil, ceftazidime, cefazolin sodium and cefaclor) can be determined spectrophotometrically based on formation of ion-pair complex with ammonium reinecekate. In acidic medium at 25 ± 2°C, as the reaction products, the complex, absorbing at 525 nm, was formed according to the scheme [49]:

cephem analog hydrochloride + ammonium reineckate → cephem reineckate + ammonium chloride

3.5 Visible spectrophotometric methods based on coupling with specified reagents

The analysis of phenolic derivatives of penam and cephem analogs were possible by measurement of absorption species formed as a result of reations with specified reagents.

Diazo coupling of β-lactam analogs was conducted with the following compounds:

- benzocaine in thrietylamine medium for determination of cefadroxil and amoxicillin. Stoichiometric ratio of formed species was 1:1 with peaking at λ=455 nm and λ=442 nm, respectively [50] .
- electron-deficient polinitro derivatives for determination of amoxicillin, cefoperazone, cefadroxil, cefprozil. Complexes of the Meisenheimer type were formed [51].

Suggested mechanisms of coupling reaction of phenol derivative of β-lactam analogs were presented in Fig. 15.

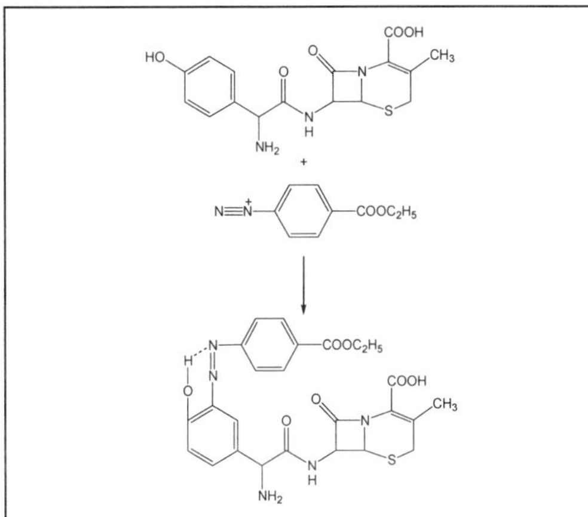

Reaction of diazo coupling of cefadroxil

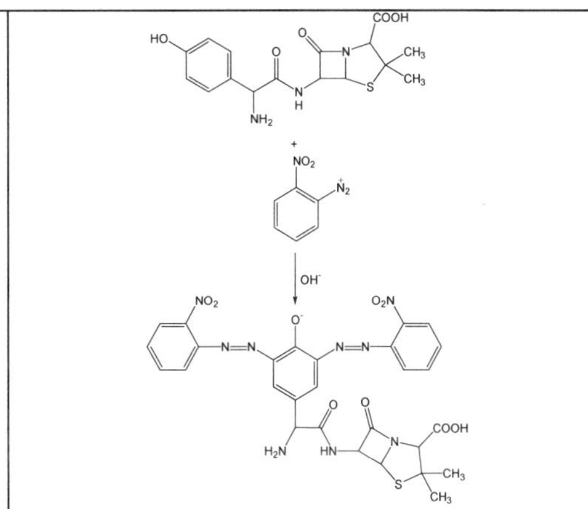

Reaction of diazo coupling of amoxicillin

Fig. 15. Schemes of reactions of coupling of β-lactam analogs [50-51].

For determination of phenolic derivative of β-lactam analog (cefadroxil) measurement of absorption of formed product in the reaction between it and 4-aminoantipyrine in the presence of alkaline potassium hexacyanoferrate(III) at 510 nm was also proposed (Fig. 16). Potassium hexacyanoferrate(III), being oxidant in this reaction, yielding *N*-substituted quinine imines and in the result was responsible for formation of red-colored antipyrine dye. Additionally, a sequential injection analysis (SIA) spectrophotometric procedure for the determination was reported [52].

1,2-naphthoquinone-4-sulfonic acid is the reagent permitting the nucleophilic substitution reaction in area of amino group of penem (amoxicillin) and cephem (cephalexin) analogs (Fig. 17). The stoichiometric ratio of these species was 1:1 and they absorb at λ = 463 nm [53].

The extension of the methodology for determination of cephalexin by using the H-point standard additions method (HPSAM) and the generalized H-point standard additions methods (GHPSAM) (after solid phase extraction cartridges) permitted also its analysis in urine [53].

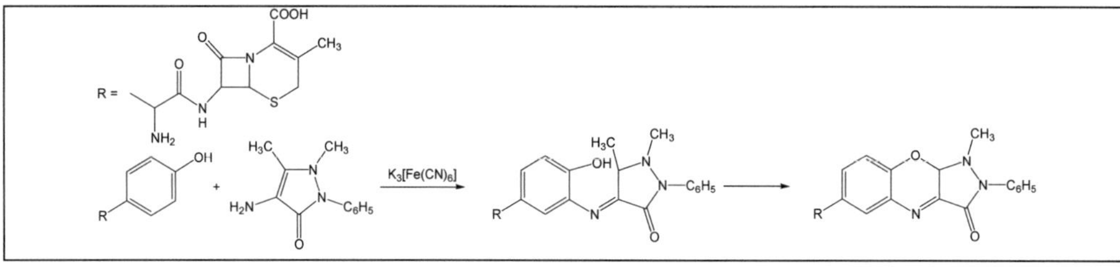

Fig. 16. The reaction mechanism of cefadroxil with 4-aminoantipyrine in the presence of alkaline $[Fe(CN)_6]^{3-}$

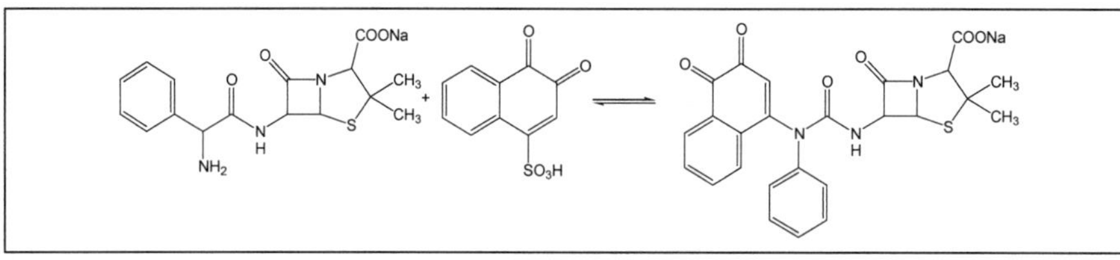

Fig. 17. The reaction mechanism of ampicillin sodium and 1,2-naphthoquinone-4-sulfonic acid [53].

3.6 Visible spectrophotometric methods based on degradation products of β-lactam analogs

The significant instability of β-lactam analogs was also exploited in the spectrophotometrical determination of β-lactam analogs in the visible region.

As the intermediate stages of determination, depending on the affecting factors, the following were present:

- degradation products typical of an acidic environment
- degradation products characteristic of a basic environment.

The formation of degradation products of β-lactam analogs in conditions of a basic hydrolysis was also the first stage during development of visible spectrophotometric methods. As the result of reactivity of degradation products formed in basic medium with some reagents, the determination of the following β-lactam analogs was possible:

- cephem analogs (cefadroxil, cefotaxime), when coupling factor were *N,N*-diethyl-*p*-phenylenediamine sulfate and Fe(III) (λ = 670 nm) or *p*-phenylenediamine dihydrochloride and Fe(III) (λ = 597 nm) [54]
- cephem analogs (cefotaxime, ceftriaxone, cefradine) when reducing factor was potassium iodate (required acidic medium) and a result of the reaction was colour change of leuco crystal violet under the influence of formed iodine (λ = 588 nm) [55]
- cephem analogs (cefotaxime sodium) when coupling factor was 1,10-phenanthroline and ferric chloride (λ = 520 nm) [56]
- penam analogs (amoxicillin, ampicillin) and cephem analogs (cephalexin, cephradine) when reducing factor of formed hydrolyzed products was J_2 (required acidic medium) (λ = 460 nm) [57].

The significant expansion of possibilities for the developed analytical method was the usage of flow injection analysis (FIA).

Similarly, in the case of determination of penam analysis during acidic hydrolysis (1.0 M HCl), formation of non-absorbing degradation products was the intermediate stage of their analysis. Complex which is necessary for achievement of spectrophotometric signals, was formed between penicillamine and palladium(II) chloride, peak at 334 nm (Fig. 18) [58].

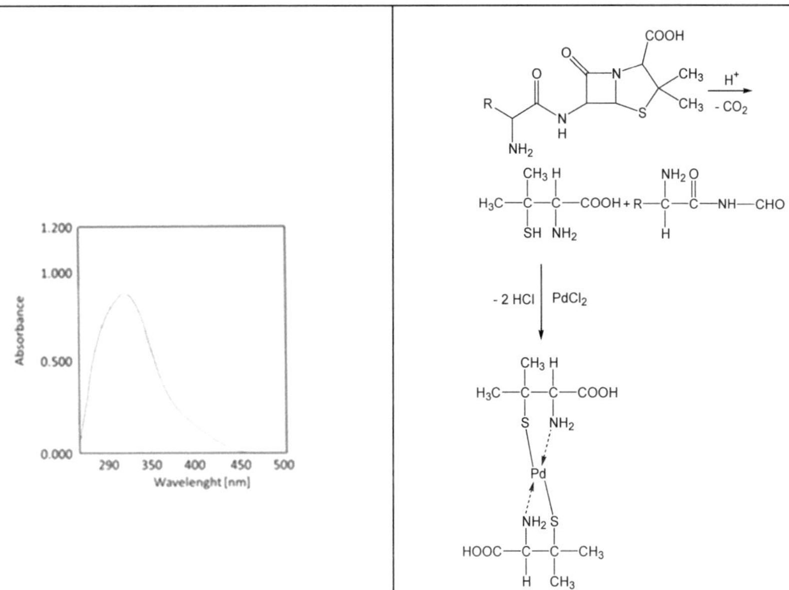

Absorption spectrum of the reaction product of palladium(II) chloride with hydrolysis product of ampicillin.	The proposal ot hte reaction pathway between the hydrolysed penicillins na palladium(II)

Fig. 18. Absorption spectra based on reaction of degradation products of penam analogs and palladium (II) and suggested mechanism of the reaction [58].

In conditions of acidic hydrolysis, determination of cephem analogs took place. Vanadium(IV) after reduction from vanadium(V), reacted with forming degradation products. Colored complexes of some cephem analogs (cephalexin, cephaprine sodium, cefazolin sodium, cefotaxime) were found, absorbing at 515, 512, 518 and 523 nm, respectively (Fig. 19) [59]

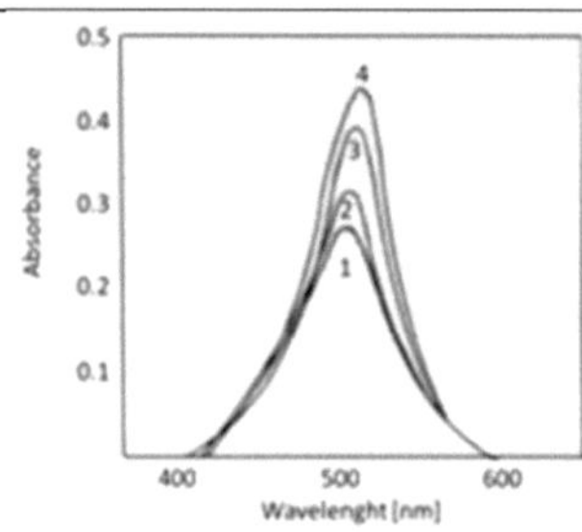

Absorption spectra of the red coloured products of 20 μg/ml of cephalexin (1), cephaprine sodium (2), cefazolin sodium (3) and cefotaxime (4) using 5 ml 0.1% vanadophosphoric acid in 2M H_2SO_4.

Fig. 19. Absorption spectra based on reaction of degradation products of cephem analogs and vanadium(IV) [59].

4. Conclusion

Spectrophotometric determinations of drugs in Vis-UV region are ones from simpler and cheaper methods used in quantitative pharmaceutical analysis. The key problem in application of these methods in analysis of multicomponents mixture is achievement of desired selectivity. Above-mentioned procedures shown that using various methods of chemical derivatization (significant for analysis in Vis region) or ones based on mathematical extends (significant for analysis in UV region) the selective determination of such labile drug as β-lactam antibiotics was possible. Taking into consideration variability of stability of some β-lactam analogs (in the presence of impurities, degradation products, inhibitors of β-lactamases, excipients and other drugs) and pathways of their degradation, confirmation of selectivity of method and values of other validation parameters for each analogs is required. It is important that analytical method used as alternative, referential one should not use spectrophotometric properties of compound as the base of detection. Recently, the development of chemometric procedures, as the complementary method for spectrophotometric ones, is often supported by suitable, high level software and equipment of apparatus allowing to assume that analysis based on calculated algorithms meet all validation criteria and can replace these methods which require pre-treatment of samples. Moreover, these methods permit multiple determination of the same sample and can be used in routine control analysis of intravenous pharmaceutical dosage forms. Additionally, proposed conditions of determination using spectrophotometric methods are in accordance with all criteria identified as the trends of *"green chemistry"*. A enrichment of visible spectrophotometric methods by the application of FIA and SIA permit accept of challenges of modern industrial-scale pharmaceutical analysis [60]. It can be expected that spectrophotometric methods enriched by chemometric procedures will become an important field of pharmaceutical analysis including also labile drugs.

5. Acknowledgment

The authors thank the State Committee for Scientific Research, Poland, for project N N405 683040.

6. References

[1] Cielecka-Piontek J. Michalska K. Zalewski P. Jelińska A. Recent Advances in stability studies of carbapenems. Current Pharmaceutical Analysis 2011;7 213-227.

[2] El-Shaboury S. Saleh G. Mohamed F. Rageh A. Analysis of cephalosporin antibiotics. Journal of Pharmaceutical and Biomedical Analysis.2007;45 1–19.

[3] Cielecka-Piontek J. Michalska K. Zalewski P. Zasada S. Comparative review of analytical techniques for determination of carbapenems. Current Analytical Chemistry 2012; 8 91-115.

[4] ICH. Stability testing of new drug substances and products. In: Proceedings of International conference on Harmonization. Geneva: IFPMA; 2000.

[5] Parisotto G. Ferrao M. Furtado J. Molz R. Determination of amoxicillin content in powdered pharmaceutical formulations using DRIFTS and PLS. Brazilian Journal of Pharmaceutical Sciences 2007;43(1) 89-95.

[6] Tabelbpour Z. Tavallaie R. Agmadi S. Abdollahpour A. Simultanesous determination of penicillin G salts by infrared spectroscopy: Evaluation of combining orthogonal signal correction with radial basis function-partial least squares regression. Spectrochimica Acta Part A 2010;76 452–457.

[7] European Pharmacopoeia 7th ed. 2010

[8] Ivama V. Rodrigues L. Guaratini C. Zanoni M. Spectrophotometric determination of cefaclor in pharmaceutical preparations. Quimica Nova 1999;22(2) 1–6.

[9] Deshpande A. Baheti K. Chatterjee N. Degradation of β-lactam antibiotics. Current Science 2004;78(12) 1684-1695.

[10] Rojanarata T. Opanasopit P. Ngawhirunpat T. Saehuan Ch. Wiyakrutta S. Meevootisom V. A simple sensitive and green bienzymatic UV-spectrophotometric assay for amoxicillin formulations. Enzyme and Microbial Technology 2010;46 292–296.

[11] Campins-Falco P. Sevillano-Cabeza A. Gallo-Martinez L. Bosch-Reig F. Monzo-Mansanet I. Comparative Study on the Determination of Cephalexin in its Dosage Forms by Spectrophotometry and HPLC with UV-vis Detection. Microchimica Acta 1997;126 207–215.

[12] Cantarelli M. Pellerano R. Marchevsky E. Camina J. Simultaneous Determination of Amoxicillin and Diclofenac in Pharmaceutical Formulations Using UV Spectral Data and the PLS Chemometric Method. Analytical Sciences 2011;27 73–78.

[13] Abdel-Hamid M. FSQ spectrophotometric and HPLC analysis of some cephalosporins in the presence of their alkali-induced degradation products. Il Farmaco 1998;53 132–138.

[14] Mahgoub H. Aly F. Uv-spetrophotometric determination of ampicillin sodium and sulbactam sodium in two-component mixtures. Journal of Pharmaceutical and Biomedical Analysis 1998;17 1273–1278.

[15] Murillo J. Lemus J. Garcia L. Simultaneous determination of the binary mixtures of cefsulodin and clavulanic acid by using first-derivative spectrophotometry. Journal of Pharmaceutical and Biomedical Analysis 1995;13(6) 769–776.

[16] Rodenas V. Parra A. Garcia-Villanova J. Gomez M. Simultaneous determination of cefepime and L-arginine in injections by second-derivative spectrophotometry. Journal of Pharmaceutical and Biomedical Analysis 1995;13 1095–1099.

[17] Daabees H. Mahrous M. Abdel-Khalek M. Beltagy Y. Emil K. Spectrophotometric determination of cefprozil in pharmaceutical dosage forms in urine and in the

presence of its alkaline induced degradation products. Analytical Letters 2001;34(10) 1639–1655.

[18] Nuevas L. Gonzalez R. Rodriguez J. Hoogmartens J. Derivative spectrophotometric determination of the thriethylammonium salt of cefotaxime in presence of related compound from the synthesis. Journal of Pharmaceutical and Biomedical Analysis 1998;18 579–583.

[19] Forsyth R. Ip D. Determination of imipenem and cilastatin sodium in Primaxin® by first order derivative ultraviolet spectrophotometry. Journal of Pharmaceutical and Biomedical Analysis 1994;12 1243-1248.

[20] Murillo J. Lemus J. Garcia L. Analysis of binary mixtures of cephalothin and cefoxitin by using first-derivative spectrophotometry. Journal of Pharmaceutical and Biomedical Analysis 1996;14 257-266.

[21] Cielecka-Piontek J. Lunzer A. Jelińska A. Stability-indicating derivative spectrophotometry method for the determination of biapenem in the presence of its degradation products. Central European Journal of Chemistry 2011;9 35–40.

[22] Mohamed A. Salem S. Maher E. Chemometrics-assisted spectrophotometric determination of certain β-lactam antibiotics combinations. Thaiwan Journal Pharmaceutical Sciences 2007;31 9–27.

[23] Lin Z. Liu J. Chen G. A new method of Fourier-transform smoothing with ratio spectra dervative spectrophotometry. Fresenius Journal Analytical Chemistry 2001;370 997–1002.

[24] Zając M. Cielecka-Piontek J. Jelińska A. Development and validation of UV spectrophotometric and RP-HPLC methods for determination of ertapenem during stability studies. Chemia Analityczna 2006;51 761–768.

[25] Cielecka-Piontek J. Jelińska A. The UV-derivative spectrophotometry for the determination of doripenem in the presence of its degradation products. Spectrochimica Acta Part A 2010;77 554–557.

[26] Morelli B. Determination of ternary mixtures of antibiotics by ratio-spectra zero-crossing first- and third-derivative spectrophotometry. Journal of Pharmaceutical and Biomedical Analysis 1995;13(3) 219-227.

[27] Morelli B. Derivative spectrophotometry in the analysis of mixtures of cefotaxime sodium and cefadroxil monohydrate. Journal of Pharmaceutical and Biomedical Analysis 2003;32 257-267.

[28] Murillo J. Lemus J. Garcia L. Application of the ratio spectra derivative spectrophotometry to the analysis of cephradine and clavulanic acid in binary mixtures. Fresenius Journal Analytical Chemistry 1993;347 114–118.

[29] Elragehy N. Abdel-Moety E. Hassan N. Rezk M. Stability-indicating determination of meropenem in presence of its degradation product. Talanta 2008;77 28–36.

[30] Zając M. Cielecka-Piontek J. Jelińska A. Development and validation of UV spectrophotometric and RP.HPLC methods for determination of ertapenem during stability studies. Chemia Analityczna 2006;51 761–768.

[31] Hassan N. Abdel-Moety E. Elragehy N. Rezk M. Selective determination of ertapenem in the presence of its degradation product. Spectrochimica Acta Part A 2009;72 915-921.

[32] Salem H. Saleh G. Selective spectrophotometric determination of phenolic β-lactam antibiotics. Journal of Pharmaceutical and Biomedical Analysis 2002;28 1205-1213.

[33] Al-Momani I. Spectrophotometric determination of selected cephalosproins in drug formulations using flow injection analysis. Journal of Pharmaceutical and Biomedical Analysis 2001;25 751-757.

[34] Saleh G. El-Shaboury S. Mahomed F. Rageh A. Kinetic spectrophotometric determination of certain cephalosporins using oxidized quercetin reagent. Spectrochimica Acta Part A 2009;73 946-954.

[35] Ayad M. Shalaby A. Abdellatef H. Elsaid H. Spectrophotometric determination of certain cephalosporins though oxidation with cerium(IV) and 1-chlorobenzotriazole. Journal of Pharmaceutical and Biomedical Analysis 1999;20 557-564.

[36] Sastry Ch. Rao S. Naidu P. Strinivas K. New spectrophotometric method for the determination of some drugs with iodine and wool fast blue BL. Talanta 1998;45 1227-1234.

[37] Feng S. Jiang J. Fan J. Chen X. Sequential injection analysis with spectrophotometric detection of cefadroxil and amoxicillin in pharmaceuticals. Chemia Analityczna 2007;52(83) 83-92.

[38] Sastry Ch. Rao K. Prasad D. Determination of cefadroxil by three simple spectrophotometric methods using oxidative coupling reactions. Mikrochimica Acta 1997;126 167-172.

[39] Makchit J. Upalee S. Thongpoon Ch. Liawruangrath B. Liawruangrath S. Determination of cefadroxil by sequential injection with spectrophotometric detector. Analytical Sciences 2006;22 591-597.

[40] Salem G. Two selective spectrophotometric methods for the determination of amoxicillin and cefadroxil. Analyst 1996;121 641-645.

[41] Helaleh M. Abu-Named E. Jamhour R. Spectrophotometric determination of selected cephalosporins. Acta Poloniae Pharmaceutica 1998;55(2) 87-91.

[42] Walily A. Gazy A. Belal S. Khamis E. Quantitative determination of some triazole cephalosporins through complexation with palladium (II) chloride. Journal of Pharmaceutical and Biomedical Analysis 2000;22 385-392.

[43] Salem H. Selective spectrophotometric determination of phenolic β-lactam antibiotics in pure forms and in their pharmaceutical formulations. Analytica Chimica Acta2004;515 333-341

[44] Eric A. Karljikovic-Rajic K. Vladimirov S. Zivanov-Stakic D. Spectrophotometric determinations of certain cephalosporins using ferrihydroxamate method. Spectroscopy Letters 1997;30(2) 309-313.

[45] Saleh G. Askal H. Darwish I. El-Shorbagi A. Spectroscopic analytical study for the charge-transfer complexation of certain cephalosporins with chloroanilic acid. Analytical Sciences 2003;19 281-287.

[46] El-Mammli M. Spectrophotometric determination of flucloxacillin in pharmaceutical preparations using some nitrophenols as a complexing agent. Spectrochimica Acta Part A 2003;59 771-776.

[47] Saleh G. Askal H. Radwan M. Omar M. Use of charge-transfer complexation in the spectrophotometric analysis of certain cephalosporins. Talanta 2001;54 1205-1215.

[48] Mohamed G. Spectrophotometric determination of amipicillin dicluxacillin flucloxacillin and amoxicillin antibiotic drugs: ion-pair formation with

molybdenum and thiocyanate. Journal of Pharmaceutical and Biomedical Analysis 2001;24 561-567.
[49] Salem H. Askal H. Colourimetric and AAS determination of cephalosporins using Reineck's salt. Journal of Pharmaceutical and Biomedical Analysis 2002;29 347-354.
[50] El-Ashry S. Belal F. El-Kerdawy M. Wasseef D. Spectrophotometric determination of some phenolic antibiotics in dosage forms Mickrochimica Acta 2000;135 191–196.
[51] Freitas S. Silva V. Araujo A. Canceicao M. Montenegro M. Reis B. Paim P. A multicommuted flow analysis method for the photometric determination of amoxicillin in pharmaceutical formulations using a diazo coupling reaction. Journal of Brazilian Chemistry Society 2011;22(2) 279–285.
[52] Xu L. Wang H. Xiao Y. Spectrophotometric determination of amipicillin sodium in pharmaceutical products using sodium 12 naphtoquinone-4-sulfonic as the chromogentic reagent. Spectrochimica Acta Part A 2004;60 3007-3012.
[53] Gallo-Martinez L. Sevillano-Cabeza A. Campins-Falco P. Bosch-Reig F. A new derivatization procedure for the determination of cephalexin with 12-naphtoquinone 4-sulphonate in pharmaceutical and urine samples using solid-phase extraction cartridges and UV-visible detection. Analytica Chimica Acta 1998;370 115–123.
[54] Metwally F. Alwarthan A. Al-Tamimi S. Flow-injection spectrophotometric determination of certain cephalosporins based on the formation of dyes. Il Farmaco 2001;56 601–607.
[55] Buhl F. Szpilkowska-Sroka B. Spectrophotometric determination of cephalosporins with Leuno crystal violet Chemia Analityczna 2003;48(145) 145–149.
[56] Rao G. Kumar K. Chowdary P. Spectrophotometric methods for the determination of cefotaxime sodium in dosage forms. Indian Journal of Phamraceutical Sciences 2001;63(2) 161–163.
[57] Al-Momani I. Flow-injection spectrophotometric determination of amoxicillin cephalexin ampicillin and cephradine in pharmaceutical formulations. Analytical Letters 2004; 37(10) 2099–2110.
[58] Belal F. El-Kerdawy M. El-Ashry S. El-Wasseef D. Kinetic spectrophotometric determination of ampicillin and amoxicillin in dosage forms. Il Farmaco 2000;55 680–686.
[59] Amin A. Shama S. Vanadophosphoric acid as a modified reagent for the spetrophotometric determination of certain cephalosporins and their dosage forms Monatsefte fur Chemie 2000;131 313–319.
[60] Tzanavaras P. Themelis D. Review of recent applications of flow injections spectrophotometry to pharmaceutical analysis. Analytica Chimica Acta 2007;588 1–9.

8

Identification, Quantitative Determination, and Antioxidant Properties of Polyphenols of Some Malian Medicinal Plant Parts Used in Folk Medicine

Donatien Kone[1,2], Babakar Diop[2], Drissa Diallo[3], Abdelouaheb Djilani[1] and Amadou Dicko[1]
[1]*Université Paul Verlaine-Metz/LCME. 1, Metz*
[2]*Université de Bamako/Faculté des Sciences et Techniques, BP*
[3]*INRSP/Département de Médecine Traditionnelle, Bamako,*
[1]*France*
[2,3]*Mali*

1. Introduction

In biological systems, during the cellular respiration, reactive oxygen species (ROS) like hydroxyl radical (•OH), superoxide anion ($\bullet O_2^-$) and hydrogen peroxide (H_2O_2) are generated, as the natural consequence of oxidation reactions (Tarnawski et al., 2005). Reactive oxygen species (ROS) damage living cells causing lipid, protein, and DNA oxidation (Shukla et al., 2010). They are involved in the development of various diseases such as diabetes, rheumatic disorders (Luximon- Ramma et al., 2002), aging, cancer, cardiovascular or neurodegenerative disorders (Ju et al., 2004; Tarnawski et al., 2005), malaria and gastric ulcer (Gülçin et al., 2006).

The interest in searching natural antioxidants has recently increased. These natural products could be used in food or in medicinal materials to replace synthetic antioxidants which are about to be restricted owing to their side effects such as carcinogenesis (Gülçin et al., 2006). Many medicinal plants contain large amounts of antioxidants, such as polyphenols, which can play an important role in adsorbing and neutralizing free radicals, in quenching singlet and triplet oxygen, or in decomposing peroxides. The compounds that are responsible of antioxidant activity could be used for the prevention and treatment of free radical-related disorders (Gomez-Caravaca et al., 2006). Indeed, the consumption of antioxidants prevents different diseases such as neurological degeneration, inflammatory disorders, coronary diseases, aging and cancers (Djeridane et al., 2006). Hence, the studies on natural antioxidants have gained increasingly greater importance.

A large number of different plants have been studied as new sources of natural antioxidants (Cakir et al., 2003; Lee et al., 2000; Kumaran & Karunakaran, 2007; Muanda et al., 2009). For the first time, we report here the antioxidant properties of extracts from the following six Malian folk medicine plants, which were previously studied for their biological activities:

Anogeissus leiocarpus (DC.) Guill. et Perrot (Combretaceae), *Cissus populnea* Guill. et Perr. (Vitidaceae), *Mitragyna inermis* (Willd.) O. Ktze. (Rubiaceae), *Terminalia macroptera* Guill. et Perrott (Combretaceae), *Vepris heterophylla* R. Let. (Rutaceae) and *Zizyphus mucronata* Willd. (Rhamnaceae). These plants were selected for their traditional used in the treatment of inflammatory diseases such as: malaria, oedema, arthritis, rheumatism, ulcer, gingivitis, conjunctivitis (Burkill, 2000; Malgras, 1992; Arbonnier, 2002; Inngjerdingen et al., 2004). Vonthron-Sénécheau et *al.*, (2003) reported the *in vitro* antiplasmodial activity of the extracts of the leaves of *Anogeissus leiocarpus*. Geidam et al., (2004) reported evidence-proved similar effects of the aqueous stem bark extract from *Cissus populnea* on some serum enzymes in alloxane induced diabetic rats. They have attributed hypoglycaemic properties to these extracts. Aqueous extract from *Mitragyna inermis* has been used by traditional healers for the treatment of various diseases, particularly for hepatic illness, malaria and hypertention. Recently, studies by Ouédraogo et *al.*, (2004) demonstrated the hypotensive, cardiotropic and vasodilatory properties of bark aqueous extract from *Mitragyna inermis*. To identify new antimalarial compounds, Conrad et *al.* (1998) selected *Terminalia macroptera* for an antiplasmodial screening. Moulis et al., (1994) studied the volatile constituents of the leaves of *Vepris heterophylla*. They found that among thirty-three compounds - that were identified by capillary GC - the main constituents were geijerene and pregeijerene. Recently, Mølgaard et *al.*, (2001) reported good activity of the extracts of root bark from *Zyzyphus mucronata* which were tested *in vitro* against tapeworms and schistosomules.

To our knowledge, there are no previous reports concerning *in vitro* antioxidant activities of these plant part extracts. The purposes of this study were to determine the total phenolic and the total flavonoid contents, to evaluate their antioxidant activities using 2, 2'-azino-bis (3-ethylbenzothiazoline)-6-sulfonic acid (ABTS) and 2, 2-diphenyl-1-picrylhydrazyl (DPPH) tests, and finally to identify and to quantify some polyphenolic compounds by using a RP-HPLC coupled to an UV detector.

2. Materials and methods

2.1 Plant material

Professor N'Golo Diarra performed the plant taxonomy in the "Departement of Traditional Medecine" in Bamako, and voucher specimens were deposited in its herbarium as *Anogeissus leiocarpus* (DC.) Guill. et Perrot (Combretaceae) Ref. N° 1559, *Cissus populnea* Guill. et Perr. (Vitidaceae) Ref. N°1368, *Mitragyna inermis* (Willd.) O. Ktze. (Rubiaceae) Ref. N°1394, *Terminalia macroptera* Guill. et Perrott (Combretaceae) Ref. N°1617, *Vepris heterophylla* R. Let. (Rutaceae) Ref. N°2444 and *Zizyphus mucronata* Willd. (Rhamnaceae) Ref. N°2499. The plant material was collected around the district of Bamako in December 2005.

2.2 Apparatus

The HPLC analyses were performed with a Waters 600E coupled to a Waters 486 UV visible tunable detector (SPD-M10Avp) and a Reverse Phases C18 symmetry analytical Alltech Intertsil ODS- 5 µm 4.6mm x 150 mm column. In addition, spectrophotometer analyses were carried out with a Cary 50 scan UV- Visible apparatus (UV Mini 1240).

2.3 Chemical reagents

Standards: catechin and gallic acid, 3,4 dihydroxybenzoic acid (protocatechuic acid), chlorogenic acid, rutin were purchased from Across Organics (France). *p*-coumaric acid, isovitexin and quercetin 3-ß-D-glucoside were obtained from Fluka Chemical Company (France).

Aluminium chloride (AlCl3), ascorbic acid, 2-2′-azino-bis (3-ethylbenzothiazoline-6-sulfonic acid) diammonium salt (ABTS), PBS buffer, 2-2′-azobis (2-methylpropionamidine) dichloride (AAPH), Folin-Ciocalteu's phenol reagent, sodium carbonate (Na_2CO_3), caffeic acid and sodium nitrite ($NaNO_2$), stable free radical DPPH were purchased from Sigma Chemical Company (France). All commercial standards and reagents were of the highest analytical grade.

2.4 Preparation of extracts

Each plant material was dried in a dark ventilated room for 5–7 days. The different parts of the plants (leaves, root barks, and stem barks) were ground to powder and sifted in a sieve (0.750µm).

The extraction of the samples was performed by the ultrasound-assisted method (Kim et al., 2002, 2003). The air-dried plant material (10 g) was extracted using 100 mL of 80% aqueous methanol, the mixture of freeze-dried powder and 80% aqueous methanol was sonicated for 20 min with continual nitrogen gas purging. The mixture was filtered through Whatman N°2 filter paper and rinsing with 50 mL of 100% methanol. The extraction of the residue was repeated using the same conditions and the two filtrates were combined and transferred into a 1 L evaporating flask with an additional 50 mL of 80% aqueous methanol. The solvent was evaporated using a rotary evaporator at 40 °C. The remaining extract concentrate was first dissolved in 50 mL of 100% methanol and diluted to a final volume of 100 mL using distilled deionized water (ddH_2O). The mixture was centrifuged at 1500g for 20 min and stored at -4°C until analyses were performed.

2.5 Determination of the total phenolic and of the total flavonoid contents

The concentration of total phenolics was measured by the method described by (Kim et al., 2003). Briefly, an aliquot (1 mL) of appropriately diluted extracts or standard solutions of gallic acid was added to a 25 mL volumetric flask containing 9 mL of ddH_2O. A reagent blank was prepared using ddH_2O. One mililiter of Folin & Ciocalteu's phenol reagent was added to the mixture and shaken. After 5 min, 10 mL of 7% Na_2CO_3 solution were added and the solution was then immediately diluted to volume (25 mL) with ddH2O and mixed thoroughly. After an incubation of 90 min at 23 °C, the absorbance versus prepared blank was read at 750 nm (Cary 50 Scan UV-Visible apparatus). Total phenolic contents of plant parts are expressed as mg of gallic acid equivalents (GAE) / g dry weight. All samples were analyzed at least in triplicate.

Total flavonoids were measured by a colorimetric assay that was developed by Zhishen et al. (1999). We can add to a 10 mL volumetric flask containing 4 mL ddH_2O either 1 mL of aliquot of appropriately diluted sample or 1 mL of a standard solution of catechin. At zero time, 0.3 mL 5% $NaNO_2$ was added to the flask. After 5 min, 0.3 mL 10% $AlCl_3$ was added.

After 6 min, 2 mL of 1 M NaOH was added to the mixture. Immediately, the reaction flask had to be diluted to volume by the addition of 2.4 mL of ddH2O and thoroughly mixed. Absorbance of the mixture was determined at 510 nm (Cary 50 Scan UV-Visible apparatus) versus prepared water blank. Total flavonoids of plant parts are expressed as mg / g dry weight of catechin equivalents (CE). Samples were analyzed at least in triplicate.

2.6 Determination of total antioxidant activity

Various methods have been introduced for the measurement of the total antioxidant capacity (Delgado-Andrade et al., 2005; Gülçin et al., 2006). In this study antioxidant activity was estimated by the method previously described (Kim et al., 2002, 2003). The *in vitro* antioxidant activities have been determined in two antioxidant tests. Among the different methods permitting to evaluate the antioxidant activities, these two simple stable radical chromogens have a high level of sensitivity and allow the analysis of a large number of samples in a timely fashion (Kim et al., 2002). It was reported that a single method is not enough to evaluate the antioxidant capacity of most of the complex natural products (Ozgen et al., 2006). Antioxidant capacity is expressed as mg of vitamin C equivalent (mg VCE) per g dry weight.

2.7 ABTS radical anion scavenging activity

ABTS radical anions were used according to the method of (Kim et al., 2003). In brief, 1.0 mM of 2, 2′-azobis (2-amidino-propane) dihydrochloride (AAPH), a radical initiator, was mixed with 2.5 mM ABTS in phosphate-buffered saline (pH 7.4) and the mixed solution was heated in a water bath at 68 °C for 13 min. The resulting blue-green ABTS solution was adjusted to the absorbance of 0.650 ± 0.020 at 734 nm with additional phosphate-buffered saline. 20 μl of sample were added to 980 μL of the ABTS radical solution. The mixture incubated in a 37°C water bath under restricted light for 10 min. A control (20 μL 50% methanol and 980 mL of ABTS radical solution) was run with each series of samples. The decrease of the absorbance at 734 nm was measured (Cary 50 Scan UV-Visible apparatus) at an endpoint after 10 min. Total antioxidant capacity of plant parts is expressed as mg / g of dry weight of vitamin C equivalents (VCEAC). The radical stock solution had to be freshly prepared and all measurements of the tested samples were repeated at least three times.

2.8 DPPH radical scavenging activity

The DPPH radical scavenging activity was determined according to the method of Kim et *al.*, (2002). The DPPH radical (100 μM) was dissolved in 80% of aqueous methanol. The plant extract solutions, 0.1 mL, were added to 2.9 mL of the methanolic DPPH solution and the mixture was vigorously shaken and was kept at 23 °C in the dark for 30 min. The decrease of the absorbance of the resulting solution was monitored at 517 nm (Cary 50 Scan UV-Visible apparatus) after 30 min. A control, which consists of 0.1 mL of 50% aqueous methanol and 2.9 mL of DPPH solution, was prepared. The DPPH radical scavenging activity of plant extracts is expressed as mg/g of dry weight of vitamin C equivalents (VCEAC). This measure was taken after 30 min reaction time. The radical stock solution had to be daily prepared and the tests were repeated at least three times.

2.9 HPLC analysis

The HPLC analyses were conducted with a Water 600 Pump apparatus. This apparatus was equipped with a quaternary solvent delivery system, a Rheodyne injector with 20µL sample loop and a UV detector Waters 486 Tunable which was fixed at 280 nm. Throughout this study, Alltech Intertsil ODS-5 C18 reversed phase column (150 mm, 4.6 mm, 5µm particle size) was used. The flow rate of the mobile phase was of 1 mL / min and the gradient elution was adapted from (Nakatani et al., 2000; Bouayed et al., 2007). The solvent composition and the gradient elution program are reported in the table 1.

Times (min)	%A	%B	%C
0	100	0	0
5	65	12	23
11	0	15	85
29	0	22	78
36	0	25	75
42	0	25	75
55	0	35	65
60	0	50	50
70	0	100	0
75	100	0	0

Solvent composition: A =50 mM $NH_4H_2PO_4$ at pH 2.60; B = 80% acetonitrile, 20 %A and C = 200 mM O-phosphoric acid at pH 1.50

Table 1. HPLC solvent gradient elution program

Standards of five phenolic acids and two flavonoids were dissolved in 50% MeOH to make a concentration of 0.5; 0.25; 0.125 and 0.10 mg/mL. The plant part extracts and standards solutions were filtered through 0.45-µm olefin polymer (OP) syringe-tip filters. Then, phenolic acids and flavonoids present in the extracts were identified by matching the retention time against their corresponding standard. In this study, the standards used for comparison were gallic acid, protocatechuic acid, chlorogenic acid, caffeic acid, *p*-coumaric acid, isovitexin and quercetin-3-β-D-glucoside (Figure 1). Quantitative analysis was made according to the linear calibration curves of standards compounds. Three replications were made at least for each standard and plant extract.

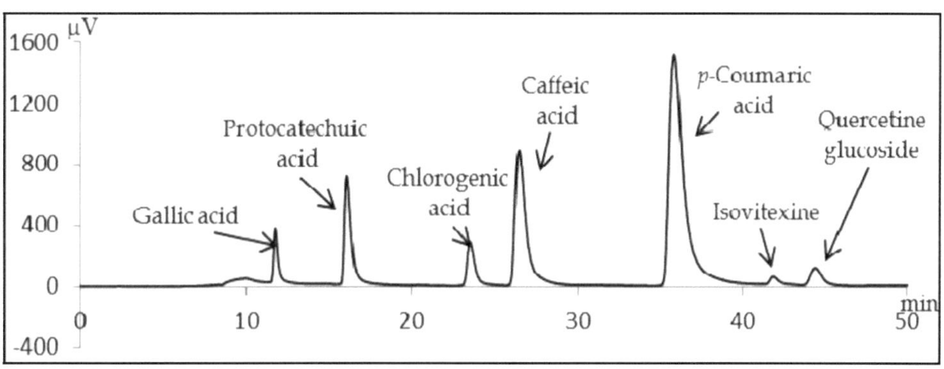

Fig. 1. Chromatogram of standards (1mg/ml)

3. Results

3.1 Total phenolics and total flavonoids

The results of the colorimetric analysis of total phenolics expressed as Gallic Acid Equivalents (GAE) and those of total flavonoids expressed as Catechine Equivalents (CE) are given in the table 2.

Plants name	Plant parts	Total phenolics (mg GAE)	Total Flavonoids (mg CE)
A. leiocarpus	L	223.1 ± 0.2	38.9 ± 1.7
	TB	26.5 ± 0.4	10.3 ± 0.3
C. populnea	RB	76.4 ± 1.1	27.6 ± 1.2
M. inermis	TB	19.5 ± 0.7	11.1 ± 1.3
T. macroptera	TB	48.5 ± 1.3	14.2 ±1.4
	RB	219.6 ± 0.4	33.1 ± 1.3
V. heterophylla	L	51.5 ± 0.5	9.3 ± 0.9
Z. mucronata	L	52.2 ± 0.5	14.4 ± 0.8
	RB	19.3 ± 0.6	9 ± 1.6

L= leaves; TB= trunk barks; RB= root bark.Total phenolics expressed as gallic acid equivalent (GAE), total flavonoids expressed as catechin equivalent (CE).
Values are means of triplicate determination ± standard deviation.
The total phenolic compounds which are present in plant materials were ranged from 19.3 ± 0.6 to 223.1 ± 0.2 mg GAE / g dry weight and the total amount of flavonoids varied from 9 ± 1.6 to 38.9 ± 1.7 mg CE / g dry weight.

Table 2. Total phenolic and total flavonoid contents of the plant parts

3.2 ABTS and DPPH radical-scavenging activity

ABTS and DPPH tests were conducted to evaluate the antioxidant properties of plant part extract on their stable free radicals in comparison to the antioxidant activity of vitamin C, the corresponding results were collected in the figure 2.

The antioxidant activity using ABTS varied from 39 mg to 468 mg VCE per g dry weight. The overall antioxidant capacity of plant parts in VCEAC, which was evaluated by ABTS assay, was in the following order: *T. macroptera* root bark > *A. leiocarpus* leaves > *T. macroptera* trunk bark > *C. populnea* root bark > *A. leiocarpus* trunk bark > *Z. mucronata* leaves > *V. heterophylla* leaves > *M. inermis* trunk bark > *Z. mucronata* root bark.

The antioxidant activity using DPPH ranged from 21 mg to 361 mg VCE per g dry weight. The overall antioxidant capacity of plant parts in VCEAC which was evaluated by DPPH assay decreased in the following order: *T. macroptera* root bark > *A. leiocarpus* leaves >*T. macroptera* trunk bark > *C. populnea* root bark > *Z. mucronata* leaves ≈ *A. leiocarpus* trunk bark > *Z. mucronata* root bark > *V. heterophylla* leaves >*M. inermis* trunk bark.

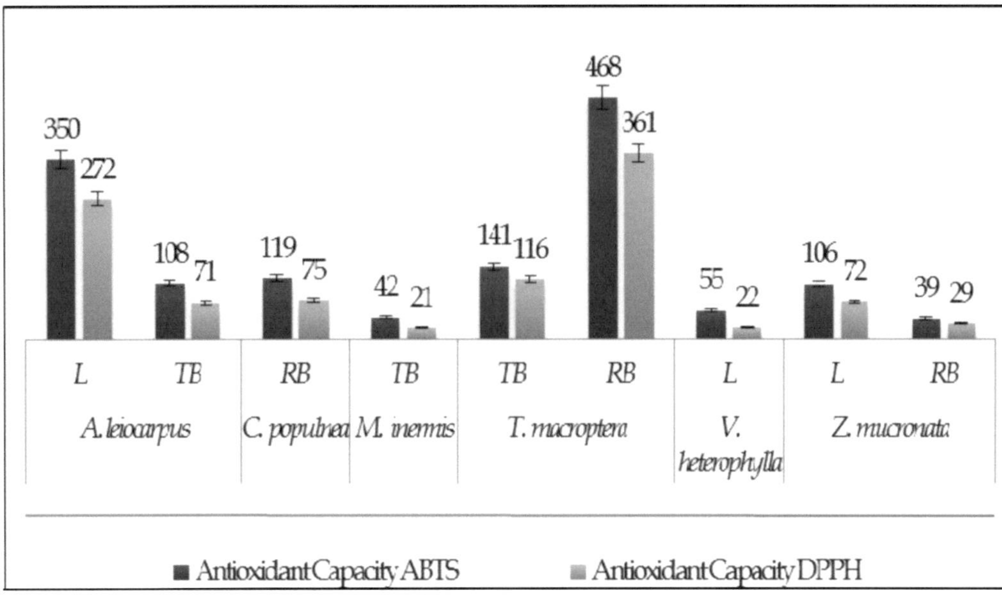

L= leaves; TB= trunk barks; RB= root barks
Values are means of triplicate determination ± standard deviation.

Fig. 2. Vitamin C equivalent antioxidant capacity (VCE mg/g dry weight).

3.3 Analysis of polyphenolic composition in plant part extracts using HPLC

The RP-HPLC results which are summarised in table 3, show that the protocatechuic acid (792.00 - 7.93 mg / 100g dry weight), the *p*-coumaric acid (1833.56 - 4.2 mg / 100g dry weight), the gallic acid (69.00 - 5.14 mg / 100g dry weight) and the chlorogenic acid (2286.08 - 62.09 mg / 100g dry weight) were encountered in high concentration, whereas the caffeic acid (149.86 -43.46 mg / 100g dry weight), the isovitexin (182.22 - 31.46 mg / 100g dry weight) and the quercetin-3-β-D-glucoside (83.53 - 70.89 mg /100g dry weight) were found in low concentration. It should be noted that the trunk and the root barks of *T macroptera* and the leaves of *V. heterophylla* contain the greatest number of compounds similar to the standards, whereas none of the standards was detected for the root barks of Z. *mucronata*.

4. Discussion

The leaves of *A. leiocarpus* had the highest total phenolic contents, which was 4- fold higher than those of the leaves of *V. heterophylla*. The lowest total phenolic and total flavonoid contents were found in the leaves of *V. heterophylla*. In the trunk barks of *T. macroptera*, the total phenolics and the total flavonoids were ranked first, followed by *A. leiocarpus* and then the *M. inermis* ones. The phenolic and flavonoid compounds were found in highest concentration first in the root barks of *T. macroptera* followed by the root barks of *C. populnea*. In contrast, it appeared that the lowest amount of flavonoids was found for the root barks of *Z. mucronata*. The total phenolic and the total flavonoid contents of the root barks of *T. macroptera* were respectively 11-fold and 4-fold greater than those of *Z. mucronata*.

Plant	Plant parts	Gallic Acid	Protocatechuic Acid	Chlorogenic Acid	Caffeic Acid	*p*-coumaric Acid	Isovitexin	Quercetin
Anogeissus leiocarpus	L	49.1 ± 3.5	67.6 ± 0.3	2286 ± 80	nd	nd	nd	nd
	T B	5.1 ± 0.2	35.5 ± 0.2	nd	nd	221 ± 20	nd	nd
Cissus populnea	R B	nd	26 ± 1.2	62.1 ± 0.5	nd	4.2 ± 0.2	nd	70.9 ± 0.5
Mitragyna inermis	T B	5.9 ± 0.16	53.2 ± 1.2	nd	nd	7.5 ±0.6	nd	nd
Terminalia macroptera	T B	16.4 ± 2.2	7.9 ± 1.3	67 ± 2	nd	59.3 ± 2.2	182 ± 42	nd
	R B	69 ± 1.4	792 ± 7	113.4 ± 1.4	149.9 ± 2.6	1833 ± 4.6	nd	nd
Vepris heterophylla	L	7.5 ± 0.44	10.1 ± 0.7	79 ± 1.7	nd	31.9 ± 2.15	31.5 ± 1.1	nd
Zizyphus mucronata	L	nd	53 ± 0.4	683 ± 20.1	43.5 ±1.4	nd	nd	nd

L= leaves; TB= trunk barks; RB= root barks; nd= not detected.

Table 3. Concentrations of flavonoids and phenolic acids in the medicinal traditional plant parts (mg / 100g of dry material)

The results obtained by ABTS and DPPH tests show that the antioxidant activity order for these different plant parts was approximately similar in both assays. However, the antioxidant capacity using DPPH compared to the one obtained by ABTS essay was underestimated about 33%. Arnao, (2000) and Delgado-Andrade et *al*., (2005) report the same occurrence and they explain that the DPPH is only dissolved in alcoholic media. In contrast, the ABTS radicals being solubilised in aqueous and in organic media the antioxidant activity measured is due to the hydrophilic and lipophilic nature of the compounds. In addition, at 515 nm near the visible region where the antioxidant activity is measured, interferences occur with the DPPH coloration.

In this study, we found that phenolic compounds are the major contributors to the antioxidant activity, since total phenolics and antioxidant activity showed a good correlation with a correlation coefficient of R^2=0.9208. However, we note that the trunk barks of *A. leiocarpus* exhibit a high antioxidant activity and a low level of total phenol antioxidant. The value of correlation coefficient between total flavonoids and antioxidant activity was R^2 = 0.752 only.

These results showing good antioxidant activity of these plant parts are particularly interesting since the antioxidant agents would induce analgesic, anticarcinogenic, anti-inflammatory, antithrombotic, immune modulating and anti-atherogenic effects (Djeridane et al., 2006).

The results of HPLC analysis were in accordance with those previously reported in the literature. The phytochemical investigations of the different parts of *T. macroptera* led to the isolation of several *C*- and *O*-glycosyl flavones, chlorogenic acid, quercetin, gallic acid (Silva et al., 2000). Chyau et *al.*, (2006) identified 3,4-dihydroxybenzoic acid (protocatechuic acid), *p*-coumaric acid, gallic acid from the leaves of *T. catappa*. Moreover, gallic acid was also present in the trunk barks of *A. latifolia* (Govindarajan et al., 2004). Protocatechuic acid (3,4-dihydroxybenzoic acid) was found in *Mitragyna rotundifolia* (Kang & Hao, 2006). Ojekale et *al.*, (2006) have reported the presence of flavonoids in *C. populnea*. In this study, the presence of isovitexin in the leaves of *V. heterophylla* was identified and quantified.

5. Conclusion

This study permits to evaluate the amount of phenolics, flavonoids and their total antioxidant activity linked to six traditional medicinal plants. Antioxidant activity varied greatly among the different plant parts and was highly correlated with the polyphenolic contents. We take an interest in the leaves of *A.Leocarpus* and in the root barks of *T.Macroptera*, since they exhibited important antioxidant activities and could be attractive sources of natural antioxidants. Moreover, this comparative study permits to identify and determine by RP-HPLC, five individual phenolic acids and two flavonoids that are mainly at the origin of the antioxidant activity in the studied plant parts.

6. Acknowledgement

The authors are thankful to the Service de Cooperation d'Actions Culturelles (SCAC) of the French embassy in Mali for its financial support.

7. References

Arbonnier M. (2002). Arbres, arbustes et lianes des zones sèches d'Afrique de l'Ouest. Paris: 2nd Ed. CIRAD -UICN

Arnao MB. (2000). some methodological problems in the determination of antioxidant activity using chromogen radicals: a practical case. Trends in Food Science & Technology, Vol. 11, pp. 419–421

Bouayed J, Rammal H, Dicko A, Younos C, Soulimani R. (2007). Chlorogenic acid, a polyphenol from *Prunus domestica* (Mirabelle), with coupled anxiolytic and antioxidant effects. Journal of the Neurological Sciences, Vol. 262, pp. 77–84

Burkill HM. (2000).Useful plants of West Tropical Africa, second ed. Royal Botanic Gardens, Kew, London

Cakir A, Mavi A, Yıldırım A, Duru ME, Harmandar M, Kazaz C. (2003).Isolation and characterization of antioxidant phenolic compounds from the aerial parts of Hypericum hyssopifolium L. by activity-guided fractionation. Journal of Ethnopharmacology, Vol. 87, pp. 73–83

Chyau CC, Ko PT, Mau JL. (2006). Antioxidant properties of aqueous extracts from Terminalia catappa leaves. LWT, Vol. 39, pp. 1099-108

Conrad J, Vogler B, Klaiber I, Roos G, Walter U, Kraus W. (1998). Two triterpene esters from Terminalia macroptera bark. Phytochemistry, Vol. 48, pp. 647-650.

Delgado-Andrade C, Rufiaän-Henares JA, Morales FJ. (2005). Assessing the antioxidant activity of melanoidins from coffee brews by different antioxidant methods. J. Agric. Food Chem., Vol. 53, pp. 7832-7836

Djeridane A, Yousfi M, Nadjemi B, Boutassouna D, Stocker P, Vidal N. (2006). Antioxidant activity of some Algerian medicinal plants extracts containing phenolic compounds. Food Chemistry, Vol. 97, pp. 654-660

Geidam MA, Adoga GI, Sanda FA. (2004). Effects of aqueous stem bark extract of Cissus populnea on some serum enzymes in normal and alloxan induced diabetic rats. Pakistan Journal of Biological Sciences, Vol. 7, pp. 1427- 1429

Gómez-Caravaca AM, Gómez-Romero M, Arráez-Román D, Segura-Carretero A, Fernández-Gutiérrez A. (2006). Advances in the analysis of phenolic compounds in products derived from bees. Journal of Pharmaceutical and Biomedical Analysis, Vol. 41, pp. 1220-1234

Govindarajan R, Vijayakumar M, Rao CV, Shirwaikar A, Mehrotra S, Pushpangadan P. (2004). Healing potential of *Anogeissus latifolia* for dermal wounds in rats. *Acta Pharmaceutica*, Vol. 54, pp. 331-338

Gülçin I, Mshvildadze V, Gepdiremen A, Elias R. (2006). Screening of antiradical and antioxidant activity of monodesmosides and crude extract from Leontice smirnowii tuber. Phytomedicine, Vol. 13, pp. 343-351

Inngjerdingen K, Nergard CS, Diallo D, Mounkoro PP, Paulsen BS. (2004). An Ethnopharmacological survey of plants used for wounds healing in dogoland, Mali, West Africa. Journal of Ethnopharmacology, Vol. 92, pp. 233-244

Ju EM, Lee SE, Hwang HJ, Kim JH. (2004). Antioxidant and anticancer activity of extract from Betula platyphylla var. japonica. Life Sciences, Vol. 74, pp. 1013-1026

Kang W, Hao X. (2006). Triterpenoid saponins from *Mitragyna rotundifolia. Biochemical Systematics and Ecology*, Vol. 34, pp. 585-587

Kim DO, Lee KW, Lee HJ, Lee CY. Vitamin C Equivalent Antioxidant Capacity (VCEAC) of Phenolic Phytochemicals. (2002). J. Agric. Food Chem., Vol. 50, pp. 3713-3717

Kim DO, Seung WJ, Lee CY. (2003). Antioxidant capacity of phenolic phytochemicals from various cultivars of plums. Food Chemistry, Vol. 81, pp. 321-326

Kumaran A, Karunakaran R. (2007). Activity-guided isolation and identification of free radical-scavenging components from an aqueous extract of Coleus aromaticus. Food Chemistry, Vol. 100, pp. 356-361

Lee KY, Weintraub ST, Yu BP. (2000). Isolation and identification of a phenolic antioxidant from Aloe barbadensis. Free Radical Biology & Medicine, Vol. 28, pp. 261-265

Luximon-Ramma A, Bahorun T, Soobrattee MA, Aruoma OI. (2002). Antioxidant activities of phenolic, proanthocyanidin, and flavonoid components in extracts of Cassia fistula. J. Agric. Food Chem., Vol. 50, pp. 5042-5047

Malgras D. Arbres et arbustes guérisseurs des savanes maliennes. (1992). Ed. Karthala et ACCT, Paris-France

Moulis C, Fouraste I, Keita A, Bessiere JM. (1994). Composition of leaf essential oil from Vepris heterophylla. Flavour and Flagrance Journal, Vol. 9, pp. 35 -37

Mølgaard P, Nielsen SB, Rasmussen DE, Drummond RB, Makaza N, Andreassen J. (2001). Anthelmintic screening of Zimbabwean plants traditionally used against schistosomiasis. Journal of Ethnopharmacology, Vol. 74, pp. 257- 264

Muanda F, Koné D, Dicko A, Soulimani R, Younos C. (2009). Phytochemical composition and antioxidant capacity of three Malian medicinal plant parts. Evidence-based Complementary and Alternative Medcine. (eCAM), Vol. 2011, Article ID 674320. Doi: 10.1093/ecam/nep 109, pp. 1-8

Nakatani N, Kayano S, Kikuzaki H, Sumino K, Katagiri K, Mitani T. (2000). Identification, quantitative determination, and antioxidative activities of chlorogenic acid isomers in prune (Prunus domestica L.). J. Agric. Food Chem., Vol. 48, pp. 5512-5516

Ojekale AB, Lawal OA, Lasisi AK, Adeleke TI. (2006). Phytochemisty and spermatogenic potentials of aqueous extract of *Cissus populnea* (Guill. and Per) stem bark. *The Scientific World Journal*, Vol. 6, pp. 2140–2146

Ouédraogo S, Ranaivo HR, Ndiaye M, Kaboré ZI, Guissou IP, Bucher B, Andriantsitohaina R. (2004). Cardiovascular properties of aqueous extract from Mitragyna inermis. Journal of Ethnopharmacology, Vol. 93, pp. 345–350

Ozgen M, Reese RN, Tulio JR AZ, Scheerens JC, Miller AR. (2006). Modified 2, 2-Azino-bis-3-ethylbenzothiazoline-6-sulfonic Acid (ABTS) Method to measure antioxidant capacity of selected small fruits and comparison to ferric reducing antioxidant power (FRAP) and 2, 2′-Diphenyl-1-picrylhydrazyl (DPPH) methods. J. Agric. Food Chem., Vol. 54, pp. 1151–1157

Silva O, Gomes ET, Wolfender JL, Marston A, Hostettmann K. (2000).Application of High Performance Liquid Chromatography coupled with Ultraviolet spectroscopy and Electrospray Mass Spectrometry to the characterisation of Ellagitannins from Terminalia macroptera Roots. Pharmaceutical Research, Vol. 17, pp. 1396-1401

Shukla KK., Mahdi AA., Ahmad MK., Jaiswar SP., Shankwar SN. & Tiwari SC. (2010). Mucuna pruriens reduces stress and improves the quality of semen in infertile men. Evidence-based Complementary and Alternative Medicine, Vol. 7, No. 1, pp. 137-144

Tarnawski M., Depta K., Grejciun D. & Szelepin B. (2006). HPLC determination of phenolic acids and antioxidant activity in concentrated peat extract - a natural immunomodulator. Journal of Pharmaceutical and Biomedical Analysis, Vol. 41, pp. 182–188.

Vonthron-Sénécheau C, Weniger B, Ouattara M, Tra Bi F, Kamenan A, Lobstein A, Brun R, Anton R. (2003). In vitro antiplasmodial activity and cytotoxicity of ethnobotanically selected Ivorian plants. Journal of Ethnopharmacology, Vol. 87, pp. 221–225

Zhishen J, Mengcheng T, Jianming W. (1999). The determination of flavonoid contents in mulberry and their scavenging effects on superoxide radicals. Food Chemistry, Vol. 64, pp. 555–559

Section 3

FT-IR Spectroscopy

9

Organic Compounds FT-IR Spectroscopy

Adina Elena Segneanu, Ioan Gozescu*,
Anamaria Dabici, Paula Sfirloaga and Zoltan Szabadai
National Institute for Research and Development in Electrochemistry and Condensed Matter, Timisoara (INCEMC-Timisoara)
Romania

1. Introduction

General spectral range of electromagnetic radiation with a wavelength greater than 750 nm (i.e. with the number of wavelength below 13000 cm^{-1}) bears the name of the domain infrared (IR). In this field samples absorb electromagnetic radiation due to transitions of vibration of the structure of molecules, molecular transitions in vibrations crystalline network (if the sample is in the solid state of aggregation) or due to transitions of molecular rotation. Subdomain of spectral wavelengths between 2500 - 50000 nm (respectively the wave numbers 4000 to 200 cm^{-1}) bears the name of the *middle infrared domain.*

From the point of view of analytical control of medicinal products, this domain is the most used. At the base of absorption is being generated electromagnetic radiation in this area spectral transitions are the vibrations of individual molecules or of crystalline network (if the sample examined is solid). Show effects such transitions caused by the vibrations of individual molecules provides information about molecular structure of the sample examined, and show effects such crystalline network to identify a particular forms of crystallization of the substance of interest.

The most frequent use of the absorption spectrophototometry in the middle infrared field lies in the identification of substances through molecular vibration. The wavelength (i.e, the wave numbers) of the of the absorption band are characteristic chemical identity of the substance in question. The intensity of the absorption bands allows quantitative analysis of the samples but, unlike in the ultraviolet and visible, in the infrared field diffuse radiation is much more refreshing, and for this reason quantitative determination infrared, are affected by notable errors.

From the standpoint of analytical use, the spectra of molecular vibration is enjoying increased popularity in comparison to the study of the crystal latice's vibrations. A molecule may be considered to be a vibrator with more than one degree of freedom, able to execute more modes of vibration. In each mode of vibration every atom in the molecule oscillates about their own position of equilibrium. Such oscillations have different amplitudes for

---

* Corresponding Author

different atoms of the molecule, but at a certain mode of vibration, each atom in the molecule oscillates with the same frequency. In other words, the frequency of oscillations of the atoms in molecule is characteristic of a particular mode of oscillation of the molecule.

A molecule composed of N atoms has several possible modes of oscillation. In each mode of oscillation (in principle) all the atoms of molecule perform periodic shifts around level position with a frequency of oscillation mode which is a feature of the assembly. Because each of the N atoms can run periodic shifts in 3 perpendicular directions each other, the assembly of N atoms can have 3N ways of motion. But, those displacements that correspond to moving molecule as a whole (not deform the geometry of the molecule) and movements, which correspond to entire molecules rotation about an axis (also without deforming the molecule's geometry), do not represent actual oscillation (associated with actual deformation of the molecule).

These displacements (3 in number) and rotations around the three orthogonal axis (also 3 in number) are eliminated of the total number of atomic movements possible. Therefore, a molecule is, in general, (3N - 6) distinct modes of oscillation and in each of these (3N - 6) modes of oscillation each atom oscillates with frequencies characteristic individual modes of vibration. A special case represents molecules whose structures are linear, because in these cases the inertia of the molecule, in relation to the axis flush by molecule, it is practically zero. For this reason, in the case of a linear molecules consisting of N atoms, the number of modes of vibration is 3N - 5.

2. The vibration of a diatomic molecule

For an understanding of the vibrations of a polyatomic molecule, should be first a preliminary analysis of the oscillations of a molecule composed of two atoms linked by covalent binding. Such a molecule, with N= 2 atoms, shows N = 3x2 - 5 = 1 modes of vibration. The steering as defined by the covalent binding of the two atoms is the only special steering, it is ordinary to accept that atoms will move (in a periodic motion) after direction of the covalent connection. Assembly oscillation may be considered in relation to several systems of reference. It may choose as origin of the system of reference the center of gravity of the diatomic assembly.

In this case the both atoms perform periodic shifts in relation to this reference point. The mathematical analysis of oscillations is advantageous to place the reference origin in one of atoms. In this case, however, in the place of mass m_A and m_B of the atoms of the molecule A-B are used *reduced mass* (noted with μ) of the assembly dimolecule. Reduced mass is calculated from the **mA** and **mB** of atoms of the dimolecule assembly in accordance with following relationship.

$$\frac{1}{\mu}=\frac{1}{m_A}+\frac{1}{m_B} \tag{1}$$

Rigorous deduction of the relationship (1) can be found in literature on the subject. During the oscillation, the kinetic energy, $\mathbf{E_c}$, and potential energy $\mathbf{E_p}$ of the assembly are varying, periodically. If the system does not radiate energy to environment, or do not accept energy from the environment, then the amount of E_c and E_p remains constant during oscillation. Potential energy is dependent on the single variable of the diatomic system (namely, the deviation of the **Δr** inter-atomic distance to **r0**) which is variable in time. Potential energy dependence of the **Δr** (i.e. lengthening the deformation of the diatomic molecule) is expressed, in the harmonic approximation, of the relationship (2).

$$E_p = \frac{1}{2} \cdot k \cdot \Delta r^2 = \frac{1}{2} \cdot k \cdot (r - r_0)^2 \qquad (2)$$

In the relationship (2) the coefficient 'k' is *constant of force*, size that characterises the strength of inter-atomic connection in the molecule. On the basis of expression (2) the potential energy of the diatomic assembly, using the mechanics in this quantum mechanics, may deduct quantified values ('allowed') of diatomic oscillator.

These values of energy 'allowed' shall be calculated on the basis of the expression (3) by substituting for the number of quantum vibration (**n**vib) integers numbers (0, 1, 2, . .

$$E_{vib}(n_{vib}) = E_c + E_p = h \cdot v_0 \cdot \left(n_{vib} + \frac{1}{2} \right) \qquad (3)$$

The expression (3) shows that the energy $\mathbf{E}_{vib}$ (the sume of the kinetic energy $\mathbf{E}_c$ and potential energy $\mathbf{E}_p$) has a state of vibration allowed to diatomic system depends on the number of vibration quantum $\mathbf{n}_{vib}$.

The lower value of energy (in the fundamental vibration's state diatomic system) is obtained by replacing **n**vib= 0 in the relationship (3). In the relationship (3) h is the size Planck constant. (6,626075 x 10^{-34} Js). If diatomic molecule fundamental changes from the vibration (**n**vib = 0) in the state of vibration excited immediately above (**n**vib = 1), then change of energy **ΔEvib(0→1)** is expressed by the relationship (4).

$$\Delta E_{vib}(0 \rightarrow 1) = h.v0 \qquad (4)$$

This value to change the vibration energy determines how often (or the number of wavelength) at which diatomic molecule shows preferential absorption of radiation.

In principle, diatomic molecule can pass from the fundamental ($\mathbf{n}_{vib}$ = 0) in a excited state (for example, corresponding $\mathbf{n}_{vib}$ = 2) but, those quantum transitions in which the number is changing more than one establishment are prohibited by the rules of selection.

Rigorous justification of the rules of selection is treated in detail in literature on the subject.

Preferred frequency ($\mathbf{v}_0$), the favorite number of wave **n**vib = 0 to which a small diatomic molecule absorbs radiation (hence to which generates a strip of absorption) as the transition (0→1), is expressed quantitatively the relationship (5)

$$v_0 = \frac{1}{2 \cdot \pi} \cdot \sqrt{\frac{k}{\mu}} \quad ; \quad \tilde{v}_0 = \frac{1}{2 \cdot \pi \cdot c} \cdot \sqrt{\frac{k}{\mu}} \qquad (5)$$

In the relationship (5) 'C' is the speed of propagation of electromagnetic radiation in a vacuum. At the harmonic approximation, the dependency **Δr** is sinusoidal. But in the case of molecules, the potential energy is dependent on the momentary deflection **Δr** of the system in a manner more complicated, so the approximation describes successfully the harmonic oscillations limited to a diatomic molecules. As a result of difficulties with mathematical order but the description of molecular oscillations, especially in the case poliatomice molecules, it accepts harmonic approximation."

$$E_p = D_e \cdot \left[1 - e^{-\beta \cdot (r - r_0)} \right]^2 \tag{6}$$

Figure 1 represents the dependency of potential energy E_p of a diatomic molecules to the momentary distance (r) in a approximately more faithful than the harmonic (based on parabolic dependence). In a more or less accurate in the description diatomic vibration of molecules, energy dependence potential (E_p) by the distance inter atomic (r) is described by a function of type Morse (6) in place of a relationship of type (2).

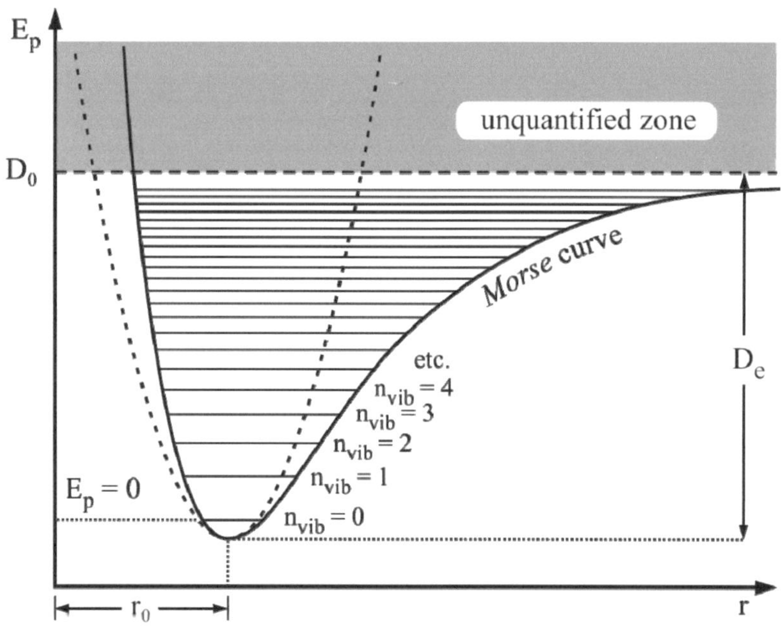

Fig. 1. Dependency of potential energy Ep of a diatomic molecule to the momentary distance

In the function (6.) the coefficient **β** depends on the mass reduced (μ) of the assembly diatomic, in accordance with relationship (7).

$$\beta = \nu_0 \cdot \sqrt{2 \cdot \pi^2 \cdot \mu \cdot D_e} \tag{7}$$

Continuous curve in figure 1 graphically represents the function (6). Morse function curve is compared with the curve corresponding harmonic approximation (parable with the interrupted curve).

The horizontal lines, arranged on the inside of the cavity Morse curve, shows the values allowed (quantifiable) of the energy of vibration of the assembly diatomic. Advanced to a deformation of the length of connection inter atomic, the energy potential of deformation tends toward a limit value (**D_0**) over which the energy of deformation of the assembly shall cease to be quantified.

In an approximation more accurate, 'anharmonic', in the phenomenon of vibration, the amounts permitted of the energy of oscillation are expressed a relationship similar to (3), with the difference that the anarmonic approximation. Status of vibration energy depends on the binomial **n_{vib}** quantum number after an expression of the degree 2 in relation:.

$$E_{vib}(n_{vib}) = E_c + E_p = h \cdot \nu_0 \cdot \left(n_{vib} + \frac{1}{2}\right) - h \cdot \nu_0 \cdot x \cdot \left(n_{vib} + \frac{1}{2}\right)^2$$
$$\nu_0 = \frac{1}{2 \cdot \pi} \cdot \sqrt{\frac{k}{\mu}} \tag{8}$$

The coefficient 'x' in the relationship (8) characterized quantitatively anarmonicity of molecule diatomic vibration, i.e. the drift behavior system from the model of harmonic vibration.

3. Potential energy dependence of the inter atomic distance of a diatomic molecule in Morse potential energy approximation

In inharmonic approximation of the vibration of diatomic molecules of the selection rule, relating to the variation in **n_{vib}** allowed for the quantum number, it is not so strict as in the case described harmonics. The model does not exclude the possibility inharmonic transitions between the status of vibration to which variation **n**vib quantum number to be 2,3 , etc. , in practice IR spectrophotometry.

Transitions associated with variations in higher than the unit are called harmonics of the upper fundamental transition (i.e., the transition that starts at the same lower status, but for which **Δn_{vib}** = 1).

The appearance of the absorption bands assigned to upper harmonics inherent in spectra are observed frequently in IR (especially in the case polyatomic molecules), but as a rule occur with intensities that are smaller than corresponding fundamental bands.

Strips of the upper harmonics associated with fundamental tape appear at frequencies (or wave numbers) which are approximately multiples whole frequency (or the wave-number) fundamental.

Another practical consequence of the inharmonicity of vibration of molecules is the rise of the inter-combination bands in the IR absorption spectra.

These bands of absorptions are observed at frequencies equal to the sum or the difference between two frequencies or fundamental frequency of a fundamental and a harmonic one. By cause of bands of combination appear various normal modes of oscillation of the molecule. The high harmonics and the bands of combination in IR absorption spectra cause considerably complications in their interpretation.

4. Vibration of polyatomic molecules

In the case the vibration diatomic molecules atoms can oscillate just in the direction of connection covalent) binding atoms. In the case of molecules consisting of several atoms (N atoms) the description of the assembly oscillations, even in harmonic approximation, is significantly more complicated. In principle, each atom in the structure molecule can execute, independently of the other atoms of the same molecule, the three-way oscillation linear independent (after three axes orthogonal coordinate attached each of atoms).

Therefore, the N atoms can run periodic shifts after 3N directions. In other words, a polyatomic molecule, consisting of N atoms, has 3N degrees of freedom of movement of constituent atoms. From this number, not all directions of movement of individual atoms correspond to deformations of real three-dimensional structure of the molecule, as oscillations in which each constituent atom is moving at the same time in the same direction, with the same amplitude and phase. They are equivalent to move whole molecules (translation molecule) without deforming them.

Also, movements that the N atoms can be synchronized in such a way that the assembly rotation movements correspond to atomic molecule as a whole, without causing structural deformation. Whereas entire molecule can be translated into 3 directions linear independent of each other and can rotate around a three axis-oriented perpendicular to each other. Of the total number of 3N directions of movement are deleted atomic 6 shifts (that is 6 degrees of freedom) in order to obtain the number (3N - 6) for detailed rules for the movement of atoms, therefore the same number of modes of oscillation actual molecular structure. .

It must be remembered, however that if molecule is not free, but is linked to a structure with crystalline comparable forces with those which act between atoms of molecule, then moving molecule as a whole from its position of equilibrium means a deformation, But, in this case is not of the molecule, instead of the structure supra molecular (crystalline lattice) in which molecule is a constituent.

In the particular case of a molecules polyatomic (with N atoms) having structure linear (all the atoms constituents are willing co-linear), the number of actual oscillation of the molecule is equal to (3N - 5). In this case is deleted of the total number of 3N possible directions of displacement 3. Degrees of freedom correlated with translation no deformation of the entire molecule and only 2 degrees of freedom corresponding to rotation molecule around two mutually orthogonal axis and perpendicular to the longitudinal axis of the molecule.

The explanation lies in the fact that in this case of the inertia of the molecule to the third axis of rotation (the flush to the longitudinal axis of the molecule) is vertical, so that rotation around this axis is virtually builds up kinetic energy.

At each of the (3N - 6) (i.e. (3N - 5)) possible ways of real oscillation, in principle each of constituents of atoms vibrating molecule running around their own positions of equilibrium. But in a particular mode of oscillation, individual atoms oscillate after other directions and with other amplitudes, but with the same frequency (feature mode of vibration in question). Directions and the amplitudes individual travel, and the frequency (common) of oscillation are characteristic mode of vibration.

The vibrating atoms in a molecule polyatomic can be described as a function of internal coordinates instead of cartesian coordinates. Such are eliminated those movements which correspond translations to atoms and rotation molecule without deformation. Internal co-ordinates of a molecules polyatomic can be defined in different ways. How to define the most frequently involves the covalent connection between pairs of atoms connect (l_{ab} for atoms a and b), Angles between connections covalent) binding centered on an atom common (αabc for atoms a and c bound by common atom (b) and diedral angles θabcd (the angle of the plans P and R containing three connections between four **atoms covalent) binding) (Figure 2).**

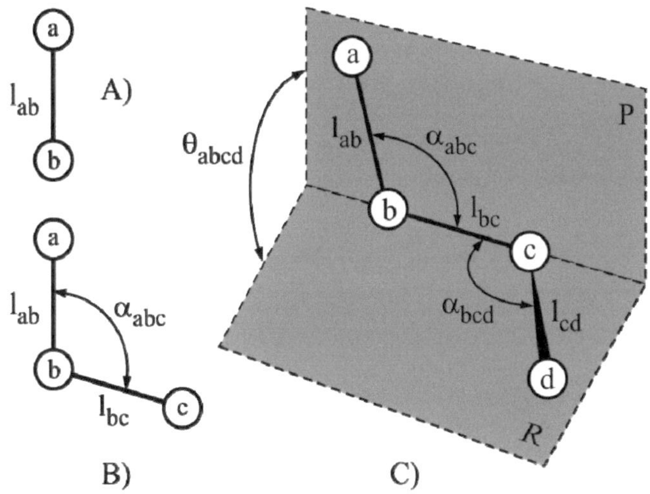

Fig. 2. Define Internal coordinates l , α and θ

Each vibration mode of the molecule can be described as a time-dependent periodic variation of all the internal coordinates. For a specific molecular structure, consisting of N atoms, the whole structure runs ((3N-6) or (3N-5)) modes of preferential oscillations, involving in (3N-6) (or (3N-5)) ways the internal coordinates of the molecule. These "preferential" types of oscillations represent the normal oscillations of the molecule. In each normal type of oscillation, all the internal coordinates of the molecule oscillate at a common frequency (in principle), specific for that type of oscillation.

In each normal mode of oscillation, the internal coordinates are involved in varying degrees. For a normal oscillation is characteristic of the internal coordinate of the molecule (e.g., a covalent bond length) is more involved than the others, then it may be said (with some tolerance) that normally oscillation that the respective oscillation type is characteristic for the respective internal coordinate (e.g., the length of the covalent bond).

Figure 3 represents the fine vibration structure of the fundamental electronic state and of the first excited electronic states, for the case of a hypothetical triatomic molecule. This type of molecule has $3 \cdot 3\text{-}6=3$ normal vibration modes. Oscillation modes are also represented in Figure 3, indicating the direction and the direction of the relative shift of the individual atoms in one of the half period of oscillation. Each electronic state consists in a number of vibration states characterized by the vibration quantum numbers $\mathbf{n_{vib}}$. For each normal way of vibration (indicated by "1", "2" and "3" in Figure 3), the vibrations in the electronic states may be differently arranged. In other words, for different normal types of vibrations, the fundamental electronic state is differently "split" in substrates of vibration. In practice, the most probable vibrational transition occurs between the vibration states corresponding to the quantum numbers n_{vib}=0 and n_{vib}=1 for each normal vibration (transitions represented in Figure 3 by bold arrows). There is only a diminished probability to also occur, for every type of normal vibration, transitions between the vibration states corresponding to the quantum numbers n_{vib}=0 and n_{vib}=2 (transitions depicted in Figure 3 by dashed arrows). This kind of transition at a normal vibration type generates its first superior harmonic.

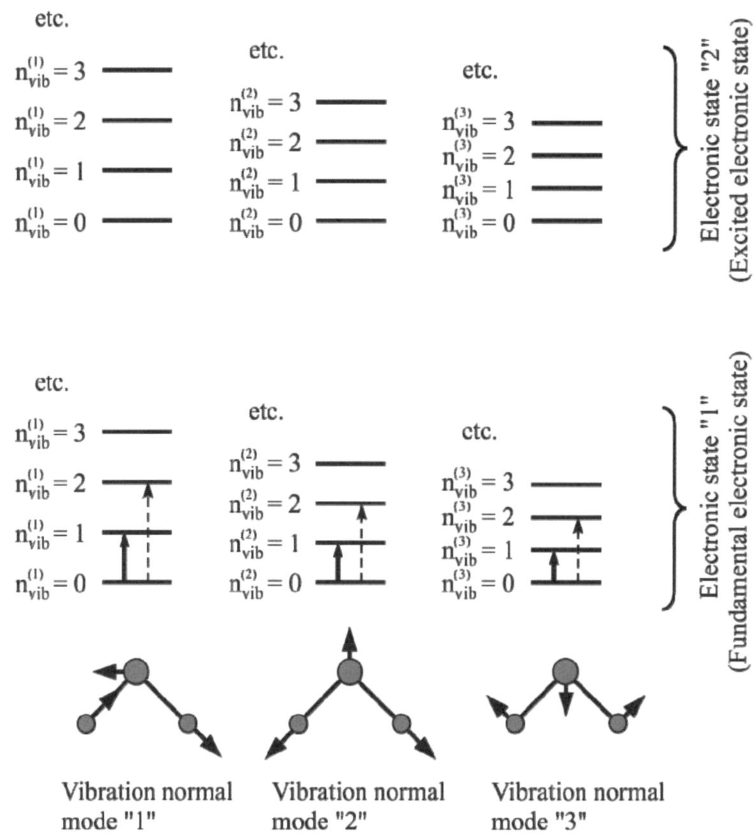

Fig. 3. Fine vibration structure of the fundamental and first excited electronic states

The transition between the vibrational states is initiated by the absorption of radiation with proper frequency (or wave number). The transition is visualized, in principle, by the appearance of the absorption band in the absorption spectrum of the analyzed substance. The incident radiation on the sample. Probability that is able produce transitions between the vibrational states of the fundamental electronic state), is located in the infrared (IR) domain. In principle, the IR absorption spectrum of a molecule contains a number of absorption bands that is equal to the normal modes of vibration of the molecule in question. Thus, in the case illustrated in Figure 3, the molecule would present the first absorption band corresponding to the normal mode of vibration no. "1" at the highest value of the wave number (the energy gap is maximum), the second absorption band corresponding to the normal mode of vibration no. "2", at an intermediate wave number (the energy gap is intermediate) and the third absorption band corresponding to the normal mode of vibration no."3", at the lowest value number of the wave number (the energy gap is the smallest). The superior indices of the vibration quantum numbers that are present in Figure 3 $\left(n_{vib}^{1}, n_{vib}^{2}, n_{vib}^{3}\right)$ refers to the serial number of the respective normal vibration mode. The above made statement made, that each normal vibration type generates an absorption band in the IR spectrum of the analyzed substance, is true only in principle. It is common that not all the normal vibration types of a molecule (3N-5 or 3N-6, for an N atoms molecule) to present absorbtion bands.

If the covalent bonds in a molecule (or a molecular fragment) have comparable force constants or the masses of the atoms involved in the covalent bonds are close, then those atoms are involved, to some extent, in all possible normal modes of vibration of the molecule (or molecular fragment). In these situations, the individual internal coordinates (bond lengths or individual angles) do not independently vibrate; their vibrations are coupled, generating the appropriate number (N) of normal vibrations.

If one of the atoms linked by a covalent bond has a much smaller mass than the other atom (for example, the C-H, N-H, O-H, S, H, P-H bonds), the reduced mass of the ensemble of bound atoms is almost equal to the mass of the lighter atom and in this case the oscillation of the bond length is quasi-independent from the oscillation of the rest of the molecule. It may be said, in this case, that the absorption band associated with the normal mode of vibration affects the covalent bond, and it is characteristic of the presence of that chemical bond in the molecule structure.

On the other hand, if two atoms in a molecule are connected by a significantly stronger covalent bond then the other covalent bonds of the molecule (e.g., the isolated double or triple bond, as in one of the cases: >C=C<, >C=O, >C=N-, -N=N-, -C-C-), than the vibration of the respective bond can be also considered quasi-independent from the oscillation of the rest of the molecule. In this case, we can say that in one of the normal modes of vibration of the molecule practically occurs only the elongation oscillation of that bond, so the optical absorption band, generated in the IR spectrum by the normal mode of vibration, is characteristic for the presence of that covalent bond

Often, a group of atoms (or even a functional group) have their "own" normal ways of vibrations, quasi-independent of the rest of the molecule vibrations. In these situations, the normal "own" modes of the group of atoms is manifested in the IR absorption spectrum as a corresponding number of "group characteristic bands". An example is the characteristic band of the amidic functional group.

5. Practical aspects of infrared spectrophotometric analysis

The IR absorption spectrum is the graphical representation of a measure of energy depending on a measure of wave of the involved radiation. IR practice has established the use of the wave number (reciprocal of the wave length and proportional to the frequency of the radiation) and of the percentage transmission (T%) or absorbance (A), as related to the energy of the radiation.

There are significant differences between the UV-Vis and IR absorption spectra. IR spectra, even those of samples in condensed states, are generally characterized by a large number of well defined, sharp bands, with easily localizable positions. Therefore, IR spectra are useful for the fast, non-destructive identification of the chemical substances, and it is extremely unlikely that two substances that are chemically different to have, accidentally, identical IR spectra. Instead, the quantitative determinations in the IR spectral range are more difficult because diffuse radiation in the IR spectrophotometers is greater than in the UV-Vis spectrophotometer, so the error sources affect the results of quantitative determinations more than in the UV or Vis range.

Another difference between the two spectral areas consists in the transparency (and usability) of auxiliary materials (glass, optics, etc.). Spectra of liquid samples are recorded using similar cells to those used in UV-Vis spectrophotometry, except that glass walls are made of specific transparent materials (NaCl, KBr, KCl, ZnSe, As_2S_3, KRS-5, and others). The thickness of the sample in IR spectrophotometry are usually much smaller (0.05 mm - 1 mm) than those found in the UV-Vis absorption spectrophotometry (1 - 10 mm).

The IR absorption spectra can be recorded for solid, liquid or gaseous samples. The most common presentation state of samples in drug control is solid state. Commonly practiced method for obtaining IR spectra of substances or mixtures of pharmaceutical interest in solid form, consists in incorporate them into a solid, microcrystalline medium (for exemple potassium bromide). This method of sample preparation is called "inclusion in the tablet (or pill) of potassium bromide." To achieve such a compressed is triturated a small amount medium (1-2 mg) of solid interest with 200-250 mg of potassium bromide microcrystalline. Potassium bromide used for this purpose must be high purity (purity "for spectroscopy") and dried before use for several hours at 180 ° C. The triturating of solid mixture containing the substance of interest and potassium bromide, is running in agate mortar medium (glass or porcelain mortar is not appropriate). After the grain sufficiently fine, solid mixture is placed in a special mold will compress high pressure medium (about 10 ton-force) with a hydraulic press. Before applying pressure, air is removed from the stencil to prevent inclusion of air microbubbles in solid mass during pressing, that may produce microfissure in mass tablet at the end pressure . During pressing, the potassium bromide microcrystalline are sinterising forming a solid transparent, optically homogeneous. A compressed pellet carried out in ideal conditions is transparent without opaque area. In the spectral range 4000 - 300 cm^{-1} potassium bromide shows very good transparency, which is why this mode is used preferentially for sample preparation. Whereas, the included technique of sample in a potassium bromide matrix keeps crystallization form of the sample solid, the IR spectrum obtained by compression in potassium bromide is dependent from the crystallization of the sample. For substances that shows polymorphic, the IR absorption spectrum of solid samples , included in compressed potassium br An essential difference between the UV-Vis spectra registration procedure and the IR field is that in the UV-Vis domain, the solvents absorbtion is insignificant, so their absorption can be compensate, in the case of IR domain all used solvents presents its own band absorption, sometimes even more powerful, that in these areas of the spectrum of energy received by the detector is too small to differentiate the strong absorption of the solvent from the sample absorption that exceeding only in small extent the solvent absorbtion. For this reason, in the IR absorption spectra of the solutions are frequently areas where the IR radiation detector is inactive, and the signal recorded in these areas is irrelevant. To view the sample spectrum in these areas, is repeat the recording spectrum in another solvent that is transparent.omide, allows the identify the crystallization form of the interest substance. Another technical detail is related by the fact that in IR absorption domain the solvents presents absorption bands, sometimes quite strong. Some IR spectrophotometers operating in double beam mode (similar to the double beam spectrophotometers used in the spectral UV-Vis). In the case of UV-Vis spectrophotometers is introduced, in the right reference optical path, a vat filled with pure solvent, and the right of the second optical path is introduced a vat of the same thickness, filled with solution (solvent and solute). The electronics parts of the spectrophotometer compare the absorbances of vats located in the two optical paths and subtract the absorbance of the solvent, located in the reference route, from solution absorbances located in the route of sample. Because the interest substances absorbance is marked in UV-Vis, and the absorbance of solvent is insignificant, the difference between the absorbances associated with two optical route is almost always positive.

In IR domain, where the absorbances of dissolved substances are comparable with those of solvents. It is easily understood that if a certain place in the spectrum (the number of wavelengths $\tilde{\upsilon}$) the solvent has a strong absorption band (molar absorptivity large solvent $\boldsymbol{\varepsilon}_0(\tilde{v})$ and the solute does not absorb significantly at that wavenumbers ($\boldsymbol{\varepsilon}(\tilde{v})$ small), then, the same layer thickness "d", the absorbance values measured in the two opticalroute ($\mathbf{A_{ref}}(\tilde{v})$ for route reference and $\mathbf{A}(\tilde{v})$ for solution route), are expressed by the relations (9).

$$\begin{aligned} A_{ref}(\tilde{v}) &= d\cdot\varepsilon_0(\tilde{v})\cdot C_0 \\ A(\tilde{v}) &= d\cdot\varepsilon_0(\tilde{v})\cdot C_0^* + d\cdot\varepsilon_0(\tilde{v})\cdot C \end{aligned} \tag{9}$$

In the (9) relation, $\mathbf{C_0}$ şi $\mathbf{C^*_0}$ represents molar concentration of pure solvent in the two cells (placed in reference route and in the sample route respectively), and C is the molar concentration of solute in the cells placed in the route sample. It is obvious that $\mathbf{C^*_0 < C_0}$ because in the sample cells is in addition to solvent, a quantity of solution. The signal recorded by spectrometry, $\mathbf{\Delta A}(\tilde{v})$, to the number of wave $(\tilde{v})$, is the difference between the both absorbance of relationship (9)

$$\Delta A(\tilde{v}) = A(\tilde{v}) - A_{ref}(\tilde{v}) = d\cdot[\cdot\,\varepsilon0(\tilde{v})\cdot C^*_0 - C_0) + \varepsilon(\tilde{v})\cdot C] \tag{10}$$

The first term in the right side of the parenthesis right above relationship is negative, because $\mathbf{C^*_0 < C_0}$. If $\varepsilon_0(\tilde{v})$.is significantly higher than $\boldsymbol{\varepsilon}(\tilde{v})$, then it can happen that $\mathbf{\Delta A}(\tilde{v})$ to have a negative value for the number of wave $\tilde{v}$. Obviously, such a negative value of absorbance is an artifact, without spectrophotometric real significance. To overcome this problem, manifested in solutions recording spectra, are used in the reference route a cell with variable thickness. The solution is placed in a cell of fixed thickness"**d***", lower than thickness"**d**", in the reference route. The absorbances $\mathbf{A^*_{ref}}(\tilde{v})$ and $\mathbf{A}(\tilde{v})$, and the difference of absorbances, $\mathbf{\Delta A^*}(\tilde{v})$, associated with the two optical route, the new working conditions, are given by the relations (11).

$$\begin{aligned} A^*_{ref}(\tilde{v}) &= d^*\cdot\varepsilon_0(\tilde{v})\cdot C_0 \\ A(\tilde{v}) &= d\cdot\varepsilon_0(\tilde{v})\cdot C_0^* + d\cdot\varepsilon(\tilde{v})\cdot C \end{aligned} \tag{11}$$

If we choose suitable variable thickness of the cell in the reference route, the expression $\Delta A^{*}(\tilde{v}) = A(\tilde{v}) - A^{*}ref(\tilde{v}\sim) = \varepsilon_0(\tilde{v})\cdot(d\cdot C^{*}0 - d^{*}\cdot C0) + d\cdot\varepsilon(\tilde{v})\cdot C$

If we choose suitable variable thickness of the cell in the reference route, the expression **($d.C^{*}_0$ - $d^{*}.C_0$)** is null for any value of wavenumber (as the expression does not depend on the number of wavelengths), so the choice of suitable thickness **d*** resolve the problem reported for the entire spectrum. In this case the absorbance of the solute depends on its concentration, according to the relation BLB.

$$\Delta A^{*}(\tilde{v}) = A(\tilde{v}) - A_{ref}(\tilde{v}) = d\cdot\varepsilon(\tilde{v})\cdot C \qquad (12)$$

Because IR absorption spectra are generated by transitions between vibrational states of sample molecules and the frequency of normal modes vibration depends (in addition to the force constants associated with deformations of the molecule) of the masses of atoms, it is expected that the replacement of atoms in molecular structure sample with different isotopes of the respective atoms (isotopic marking of the molecule) to induce dramatic changes of the IR absorption spectrum of the sample.

By isotopic marking in the known positions of molecule and by confronting these changes with changes in IR absorption band positions, significant conclusions can be drawn on whether the different atoms are involved in the normal modes of vibration of the molecule. If an atom of molecule is replaced by its heavier isotope, then IR absorption bands is moving to lower wavenumbers. The most significant movement is found in these absorption bands corresponding to normal vibration modes which involves mostly the replaced atom with heavier isotope.

The biggest relative change in mass of an atom by isotopic substitution is made for replacement of the hydrogen atom (isotope 1H) with deuterium (2H isotope). It follows that by sample deuterating, the IR absorption bands associated with chemical bonds in which one of the atom is hydrogen, suffer very significant movement toward smaller wave numbers.

Isotopic displacement of absorption bands is useful and for choice of suitable solvent in those cases where the IR absorption spectrum should be recorded in solution. Because of own absorption, some solvent (eg chloroform, $HCCl_3$) can not be used except in those domain where this is sufficiently transparent. The spectral regions in which the chosen solvent substantially absorbed are not used. But if using deuterated solvent (e.g. deuterocloroform, $DCCl_3$), this have unusable areas at other wavenumbers. Original solvent (undeuterated) and deuterated solvent presents identical dissolution properties, but are complementary with respect to transparency in the IR spectral range.

6. Aspects of construction and specific features of operating mode for Fourier transform spectrophotometers (FTIR)

Old design spectrophotometers work similar with those for UV-Vis domain, i.e. are composed of radiation source, monochromator designed to select a desired wavelength radiation, the sample chamber and the radiation detector. In IR domain, diffuse radiation presents more serious problems then in ultraviolet and visible domain. Thus, in IR domain, the ratio of useful signal and noise is more disadvantageous. The new concept of Fourier transform IR spectrophotometers meant an important step in achieving spectra with high quality even for difficult samples where the spectrophotometers with traditional construction proved to be powerless.

Construction scheme and specific features of operating mode for a Fourier Transform Infrared Spectrophotometer (FTIR - "F*ourier* **T**ransform **I**nfra**R**ed") are presented in Figure 4. It is noted that the optical assembly has no monochromator, which is replaced by a Michelson interferometer type. The polychromatic radiation from **LS** (light source) source is transmitted through concave mirror $\mathbf{M_1}$ (mirror 1) and radiation divider $\mathbf{BS_2}$ (beam splitter 2) to sample **S** (sample). After crossing the sample, the radiation reach the radiation divider $\mathbf{BS_1}$ (beam splitter 1) that divides the flow of radiation in two tracks: one for the mirrors $\mathbf{M_2}$ (mirror 2) and another for the mirrors $\mathbf{M_3}$ (mirror 3). Mirrors $\mathbf{M_2}$ and $\mathbf{M_3}$ turn back the radiation to the radiation divider $\mathbf{BS_1}$. Reaching its surface, radiations which had different routes, merge and produce a interference phenomena. The only constructive element moving while recording the spectrum is the set of $\mathbf{M_3}$ mirrors. If the mirrors $\mathbf{M_3}$ are in position **A**, the optical path difference, corresponding to the two optical paths is null, thus the radiations which turn back on the surface of the radiation divider produces an interference maximum. By translation of $\mathbf{M_3}$ mirrors, optical path difference δ between the two routes is changed progressive.

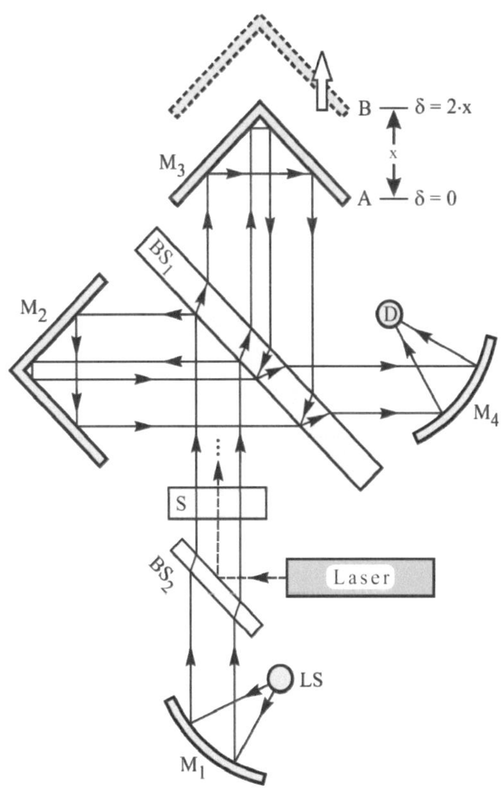

Fig. 4 Construction and operating scheme for Fourier Transform Infrared Spectrophotometer

It can be shown that the detector **D**, which measures the intensity of interference as a function of **M_3** mirrors position (so depending on the optical path difference δ between the two routes) records an interferogram which depends on inverse Fourier transform of emission spectrum of the source **LS** and on inverse Fourier transform of transparency (transmission) spectrum of the sample **S** (sample). After Fourier transform of detector **D** signal and some additional mathematical operations on detector signal the transmission (or optional absorption spectrum) spectrum of the sample **S** in known form is obtained.

Figure 4 shows the typical interferogram recorded by the detector (representing the light flux, which reached the detector, as a function of the optical path difference δ associated with **BS_1 - M_2 - BS_1** and **BS_1 - M_2 - BS_1**).

To know the exact positions of absorption maxima in the IR spectrum of the sample, the position of mirrors **M_3** must be known exactly commensurable with the radiation wavelength in each moment of this whole movement. Therefore, together with the radiation of source **LS**, it is sent another radiation, this time the radiation is monochromatic, coming from a laser emitting in the visible (usually red radiation) or near infrared (often with a wavelength of 1064 nm) range. The interferogram produced by monochromatic laser radiation is practically a sinusoidal function. This sinusoidal signal, also noted by the detector **D**, is superimposed on the signal generated by the sample. By tracking the interference maximum and minimum (sinusoidal type) of the laser radiation, it can be indicated the current location of the mirrors **M_3**, in each moment of the recording operation, with an accuracy comparable to the wavelength of laser radiation.

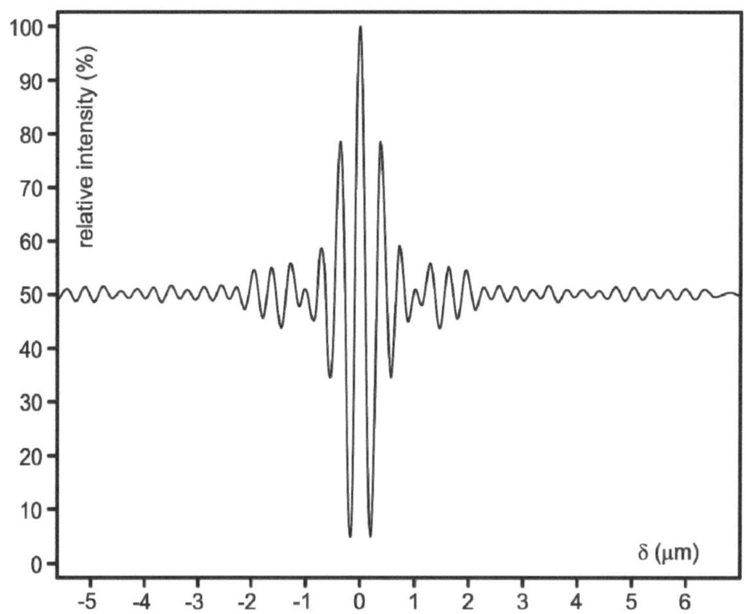

Fig. 5. The typical interferogram recorded by the detector

Recording of an IR spectrum of a sample based on Fourier Transform method has many advantages:

a. using a spectral interferometer ensures the achievement of a resolution much higher than that offered by spectrophotometers with dispersion or a high signal / noise for a given resolution;
b. lack of slits in the optical assembly of the Fourier transform spectrophotometers removes a series of disadvantages related to the fact that for spectrophotometers with dispersion the optical image of the input slit is deformed due to dispersive optical element (prism or diffractive optical network);
c. the signal / noise ratio achieved in Fourier transform spectrophotometers it is more advantageous with several magnitude orders compared with dispersion spectrophotometers;
d. because of the signal / noise ratio advantage, recording of an interferogram (a single displacement of mirrors $\mathbf{M_3}$ from position corresponding to $\delta = 0$ to a position with a extreme δ value) can be achieved in a very short time (less one second) reason why, within a reasonable time, interferogram recording operation can be repeated for several times, followed by the mediation of the obtained signals.

The last aspect is particularly important for difficult samples, which absorb infrared radiation in a very advanced position. For these samples, a single interferogram recording, with all the inherent advantages of Fourier multiplexing technique, signal / noise ratio is often unsatisfactory. In these cases, overlapping a larger number of records (the number can be **N**), followed by calculating the arithmetic average of the records, significantly improves the signal / noise ratio. In theory errors can be demonstrated that the overlap (acquisition) of N records, followed by mediation of the obtained interferogram, the signal / noise ratio is improved by a factor equal to $\sqrt{N}$ in comparison with a single record case. Thus, and in IR spectra (obtained by the Fourier transform of the interferogram) the signal / noise ratio is improved by acquisition of spectra.

The electrical signal of the detector is digitized with an appropriate electronic interface and data (pairs of wavenumber values vs. absorbance or wavenumbers vs. transmission percentage) are stored in a file created by a computer. The stored data can be later processed in different ways. Thus, you can add or subtract algebraic different spectra, can make corrections of baseline, can reduce noise by techniques different than acquisition (e.g. "signal smoothing") or it can be presented the absorbance derived as a function of scanned size (wavenumber). Derivation of the original spectrum as a function of wavenumer often surprises some details of the IR spectrum which are harder to observe in its original form (absorbance vs. wavenumber).

7. Organic compounds

Elucidation of the molecular structure is especially important in organic chemistry. An analytical method for the identification of functional groups from organic compounds uses one of the most physical properties of a chemical compound: the infrared absorption spectrum. Compared with other physical properties: melting point, refractive index, or specific gravity which offer only a single point of comparison with other substances, the IR spectrum of a specific compound, gives a multitude of important information (position of bands, band intensity). The intensity is indicative of the number of a particular group contributing to absorption.

It is well known that molecules absorb a unique set of IR light frequencies, because the frequency of vibration involved depends on the masses of atoms involved, the nature of the bonds and the geometry of the molecule. A molecule absorbs only those frequencies of IR light that match vibrations that cause a change in the dipole moment of the molecule. Each organic molecule, with the exception of enantiomers, has a unique infrared spectrum. This is because symmetric structures and identical groups at each end of one bond will not absorb in the IR range. The spectrum has two regions. The *fingerprint* region is unique for a molecule and the *functional group* region is similar for molecules with the same functional groups.

The entire spectral pattern is unique for a given compound. The bands that appear depend on the types of bonds and the structure of the molecule.

In a complicated molecule many fundamental vibrations are possible, but not all are observed movements which do not change the dipole moment for the molecule or the those which are so much alike that they coalesce into one band.

IR is usually preferred when a combination of qualitative and quantitative analysis is required. It is often used to follow the course of organic reactions allowing the researcher to characterize the products as the reaction proceeds.

For the analysis, the samples can be liquids, solids, or gases. The only molecules transparent to IR radiation under ordinary conditions are monatomic and homonuclear molecules such as Ne, He, O_2, N_2, and H_2. One limitation of IR spectroscopy is that the solvent water is a very strong absorber and attacks NaCl sample cells.

Computerized spectra data bases and digitized spectra are widely used today especially in research, chemistry, medicine, criminology, etc

8. Interpretation of spectra

Identification of a molecular structure from the IR spectrum can be realized using information from correlation tables and absorbances from the functional group region of the spectrum and comparison of the obtained spectrum with those of known compounds or obtain a known sample of a suspected material.

A preliminary examination of a spectrum use requires the examination of two important spectrum areas: functional group region (**4000-1300 cm^{-1}**) and the **909-650** cm^{-1} region. The characteristic stretching frequencies for important functional groups such as OH, NH, and C=O occur in this portion of the spectrum. The absence of absorption in the assigned ranges for the various functional groups can usually be used as evidence for the absence of such groups from the molecule. The absence of absorption in the **1850-1540** cm^{-1} region excludes a structure containing a carbonyl group.

Strong skeletal bands for aromatics and heteroaromatics fall in the **1600-1300** cm^{-1} region of the spectrum. These skeletal bands arise from the stretching of the carbon-carbon bonds in the ring structure. The lack of strong absorption bands in the **909-650** cm^{-1} region generally indicates a aliphatic structure. Aromatic and heteroaromatic compounds display strong out-of-plane C-H bending and ring bending absorption bands in this region. The intermediate portion of the spectrum, **1300-909** cm^{-1} is usually correspond to the fingerprint region. The absorption pattern in this region is complex, with bands originating in interacting vibrational modes. Absorption in this intermediate region is probably unique for every molecular species. For example, in the cases of hydrocarbons, organic compounds classified as saturated or unsaturated based on the absence or presence of multiple bonds, the energy of the infrared light absorbed by a C-H bond depends on the hybridization of the hybrid orbital, in the order of sp^3>sp^2>sp. The sp^3-hybridized C-H bonds in saturated hydrocarbons absorb in the 2850-3000 cm^{-1} region. The sp^2-hybridized C-H bonds from alkenes absorbs at **3080** cm^{-1}. A sp-hybridized C-H bond in a molecule, alkyne absorbs infrared at 3320 cm^{-1}. Another classification of hydrocarbons can be made based on absorptions due to the carbon-carbon bond. Carbon-carbon bond strength increases in the order of single>double>triple. Saturated hydrocarbons all contain carbon-carbon single bonds that absorb in the 800-1000 cm^{-1} region. But, unsaturated hydrocarbons also contain carbon-carbon single bonds that absorb in this same region. So, this interval can not be considered as fingerprint region because most organic compounds have carbon-carbon single bonds.

The alkanes give an IR spectrum with relatively few bands because there are only CH bonds that can stretch or bend.

The next table present the characteristic group frequencies of organic molecules.

Class	**Group**	**Wavenumber (cm^{-1})**
Hydrocarbons		
Alkane	C-H	2850-3000
	C-C	800-1000
Aromatic	C-H	3000-3100
	C=C	1450-1600
Alkene	C-H	3080-3140
	C=C	1630-1670
Alkyne	C-H	3300-3320
	C-C	2100-2140
Oxygen Compounds		
Alcohol	O-H	3300-3600
	C-O	1050-1200
Ether	C-O	1070-1150
Aldehyde	C=O	1720-1740
	C-H	2700 -2900
Carboxylic Acids	C=O	1700-1725
	O-H	2500-3300
	C-O	1100-1300
Ester	C=O	1735-1750
	C-O	1000-1300 (2 bands)
Ketone	C=O	1700-1725
Acyl halides	C=O	1785-1815
Anhydrides	C=O	1750;1820 (2 bands)
	O-C	1040-1100
Amides	C=O	1630-1695

Class	**Group**	**Wavenumber (cm-1)**
	N-H	1500-1560
Isocyanates,Isothiocyanates, Diimides, Azides, Ketenes	-N=C=O, -N=C=S -N=C=N-, $-N_3$, C=C=O	2100-2270
Nitrogen compounds		
Amines	N-H	3300-3500
	C-N	1000-1250
	NH_2	1550-1650
	NH_2 & N-H	660-900
Nitriles	C≡N	2240-2260
Oxidized Nitrogen Functions		
Oxime **(=NOH)**	O-H	3550-3600
	C=N	1665± 15
	N-O	945± 15
Amine oxide (N-O)	aliphatic	960± 20
	aromatic	1250± 50
N=O	nitroso	1550± 50
	nitro	1530± 20;1350± 30
Alkyl bromide	C-H	667
Sulfur compounds		
Thiols	S-H	2550-2600
Esters	S-OR	700-900
Disulfide	S-S	500-540
Thiocarbonyl	C=S	1050-1200
Sulfoxide	S=O	1030-1060
Sulfone	S=O	1325± 25; 1140± 20
Sulfonic acid	S=O	1345
Sulfonyl chloride	S=O	1365± 5;1180± 10
Sulfate	S=O	1350-1450
Phosphorous compunds		
Phosphine	P-H	2280-2440 950-1250
Phosphonic acid	(O=)PO-H	2550-2700
Esters	P-OR	900-1050
Phosphine oxide	P=O	1100-1200
Phosphonate	P=O	1230-1260
Phosphate	P=O	1100-1200
Phosphoramide	P=O	1200-1275
Silicon compounds		
Silane	Si-H	2100-2360
	Si-OR	1000-1110
	$Si-CH_3$	1250± 10

Table 1. Schematic representation of the Infrared Group Frequencies classification

9. References

E.O. Brigham: "The Fast Fourier Transform", Prentice-Hall, Inc., 1974

P.L. Polavarapu: "Vibrational Spectra: Principles and Applications with Emphasis on Optical Activity", Elsevier Science B.V. 1998;

Y. Wang, R. Tsenkova, M. Amari, F. Terada, T. Hayashi, A. Abe, Y. Ozaki: 'Potential of Two-Dimensional Correlation Spectroscopy in Analysis of NIR Spectra of Biological Fluids. I. Two-Dimensional Correlation Analysis of Protein and Fat Concentration-Dependent Spectral Variations of Milk", Analusis Magazine, 1998:26, M64-M69

R.N. Bracewell: "The Fourier Transform and Its Applications", McGraw Hill, 2000

D. Baurecht, U.P. Fringeli: "Quantitative Modulated Excitation Fourier Transform Infrared Spectroscopy", Review of Scientific Instruments, 2001:72, 3782-3792

M-J. Paquet, M. Laviolette, M. Pézolet, M. Auger: "Two-Dimensional Infrared Correlation Spectroscopy Study of the Aggregation of Cytochrome C in the Presence of Dimyristoilphosphatidylglycerol", Biophysical Journal, 2001:81, 305-312

Y. Kauppinen, J. Partanen: "Fourier Transforms in Spectroscopy", Wiley-VCH Verlag Gmbh, 2001

J.M. Chalmers, P.R. Griffiths (Editors): "Handbook of Vibrational Spectroscopy. Theory and Instrumentation", John Wiley & Sons Ltd., 2002

M. Cho: "Two-Dimensional Optical Spectroscopy", CRC Press, 2009

P.W. Atkins, Physical Chemistry. 2nd Ed. San Francisco: W.H. Freeman and Company, 1982. Discussion of vibrational spectra from a quantum mechanical view.

B.W. Cook, K. Jones. A Programmed Introduction to Infrared Spectroscopy. New York: Heyden & Son Inc., 1972. Excellent resource for the beginning spectroscopist.

R. T. Morrison, R. N. Boyd. Organic Chemistry. 5th Ed. Boston: Allyn and Bacon, Inc., 1987. Provides a brief description of spectroscopy. Includes relevant IR spectra for each family of organic compounds.

R. L., Shriner, R. C., Fuson, D. Y., Curtin, T. C. Morrill. The Systematic Identification of Organic Compounds. 6th Ed. New York: Wiley, 1980. Contains a brief section on IR spectroscopy. Mainly a text for identification of compounds by chemical tests.

R. M., Silverstein, G. C., Bassler, T. C. Morrill. Spectrometric Identification of Organic Compounds. 4th Ed. New York: Wiley, 1981. Description of mass spectrometry, IR spectrometry, 1H NMR spectrometry, 13C spectrometry, and UV spectrometry.

A. Lee Smith,. Applied Infrared Spectroscopy: Fundamentals, Techniques, and Analytical Problem-Solving. New York: Wiley, 1979. Comprehensive treatment of IR spectroscopy. Includes history, instrumentation, sampling techniques, qualitative and quantitative applications.

G. Socrates, Infrared Characteristic Group Frequencies. 2nd Ed. Chichester: Wiley, 1994. A comprehensive reference of correlation tables.

A.Streitweiser, Jr., C. H. Heathcock. Introduction to Organic Chemistry. 2nd Ed. New York: Macmillan Publishing Co., Inc., 1981. Introductory organic text with a section on IR spectroscopy. Includes spectroscopic information as each family is presented.

10

Application of Infrared Spectroscopy in Biomedical Polymer Materials

Zhang Li, Wang Minzhu, Zhen Jian and Zhou Jun
State Food and Drug Administration Medical Device Supervising and Testing Center of Hangzhou Zhejiang Insitute for the Control of Medical Device, Hangzhou, China

1. Introduction

Infrared Spectrum (IR) is mainly used to study molecular structure and composition in substances and thus is also called molecular spectrum. When the sample is exposed to infrared light with continuously changing frequency, the molecule absorbs irradiation of certain frequencies and is subject to vibration or rotation, thus to cause the change of dipole moment. The molecule's transition from normal state to excited state weakens the intensity of the corresponding transmitted light in the absorption region, then the infrared software is used to obtain the IR spectrum. Started in 1970, Fourier Transform Infrared Spectroscopy (FTIR) was of high resolution and high scanning speed. It was not only limited to middle infrared (MIR), Spectrum ranges ultraviolet to far infrared section with the assistance of the beam splitter. The main direction of modern analysis, study and development is to combine technology of FTIR with that of the other instrument. For example, FTIR-TGA (Thermogravimetry Analysis) can be used to obtain thermogravimetric curve as well as the IR spectrum of the weight loss material, thus to determine the real composition of vapor generated in the various weight loss stages and the decomposition process.

Ultraviolet and visible absorption spectrum are usually used to study unsaturated organic matter, especially the organic compounds with conjugated system. However, infrared spectroscopy mainly studies chemical compounds with change of dipole moment during vibration. Thus, almost all organic compounds, except single atoms and homonuclear molecule, absorb in the infrared spectrum region. Except for optical isomers, some high polymer of high molecular weight and compounds with slight difference in molecular weight, two compounds of different structure are unlikely to have the same infrared spectrum. Wave number position, number of wave peaks of infrared absorption band and the intensity of absorption band indicate the characteristics of the molecular structure, and thus can be used to identify the structural composure of the unknown objects or its chemical groups. The absorption intensity of the absorption band is related to the contents of the molecular composition or the chemical groups and can be used for quantitative analysis and purity identification. Infrared spectrum analysis has distinctive characteristic and can be used to test gas, liquid and solid samples. Lie other analysis methods, it can be used for qualitative and quantitative analysis and is one of the effective methods for chemical compound identification and molecular structure elucidation.

2. Summary of biomedical polymer materials

Biomedical polymer material is an important component of biological material and is a remarkable functional polymer. It involves in physics, chemistry, biochemistry, medicine, pathology subjects. The synthetic polymer materials and the organisms (natural polymer) have a very similar chemical structure, which determines their similarity in performance and enables high molecular polymer to meet the many complicate and rigorous functional requirements for medical products. Most metal and inorganic materials are incapable in this respect. Presently, high molecular polymers are widely used in medical products.

2.1 Presently there are two types of high molecular polymers: Non-biodegradable and biodegradable polymer materials

Non-biodegradable biomedical polymer material is widely used in manufacturing of adhesives, coating, and artificial lens as well as for repair of the many soft and hard tissues and organs of human body such as ligament, tendon, skin, blood vessel, artificial organs, bone and teeth. Most non-biodegradable biomedical polymer materials have no biological activity and are difficult to bond firmly with tissues and thus may result in toxicity and allergic reaction. Biodegradable biomedical polymer material is mainly used in temporary execution and replacement of tissue and organ functions in clinic or be used as the medicine controlled-release system and delivery carrier, absorbable surgical suture and wound dressings. It is readily biodegradable and the degradation products can be excreted through metabolism. Thus, it has no negative effect to tissue growth. Presently, it has become the key focus in development of biomedical polymer materials.

Non-biodegradable Polymer Materials	Biodegradable Polymer Material
Silicone Rubber Polyethylene	Polyvinyl alcohol Modified natural polysaccharides
Polyacrylate	Protein
Polytetrafluoroethene -PTFE, etc.	(Back spinning) synthetic polyester
Dacron	Polypeptide-polyolefine, silk ossein
Carbon-graphite fiber, etc.	PLA

Table 1. Category of Biomedical Polymers

However, biomedical polymer is an interdisciplinary subject and has varying types of classification based on different purposes and practices. For example, it can be categorized based on the source and application purpose of the medical polymer or based on the influence of living tissues to the materials. Presently there is no uniform standard for the classification.

2.2 Biomedical polymer material

Polymer material is generally composed of high polymers and low molecular weight substances. High polymers are divided into homopolymer, copolymer, blends and oligomer; Low molecular weight materials include: 1) additives: regulator, chain transfer agent, terminator and emulsifier; 2) additives: plasticizers, stabilizer, filler and colorant etc.; 3) unreacted monomer, residual catalyst, etc.

Compared with ordinary organic substance, polymer-a long chain connected by multiple monomer units through covalent bonds—features diversified and designable structure, with very high mechanical strength and non-fixed molecular weight. Polymer is partially crystallized or non-crystallized and provides the properties of elastomer and fluid.

Polymer Material	Chain structure	Repetitive unit structure
		Molecular weight and its distribution
		Copolymer composition and sequence distribution
		Stereospecificity
		Branching and Cross-linking
	Aggregation structure	Crystallization
		Orientation
		Physical condition
		Heterogeneous structure and morphology

Table 2. Polymer materials

2.3 Basic requirements of biomedical polymer and common materials

Biomedical polymer is directly applied to human body and is closely related to human health. Therefore, the materials used for clinic should be strictly controlled, otherwise, it may cause adverse effect instead of life saving. Below are requirements on properties and performances of the biomedical polymer:

1. Resistance to biological aging: polymer for long-term implantation should have good biological stability.
2. Physical and mechanical stability: different strength, elasticity, size, stability, fatigue resistance and wear resistance for different application.
3. Easy for processing and molding
4. Proper materials with reasonable cost
5. Convenient for sterilization.

Requirements on body effect of biomedical polymer

1. Non-toxic, i.e. chemically inert
2. No pyrogenic reaction
3. Non carcinogenic
4. Non-teratogenic
5. Does not cause allergic reaction or interfere with the body's immune mechanism
6. Does not damage adjacent tissue or cause calcified deposition on material surface.
7. Excellent blood compatibility without causing coagulation when contacting with blood.

Requirements for biomedical polymer production and processing: besides the strict control on biomedical polymer itself, matters harmful to human body shall also be prevented during material production; the purity of the raw materials used in biomedical polymer synthesis shall be strictly controlled, no harmful matter is allowed and the content of heavy mental shall be within the limit; additive processing shall meet medical standard; the production environment should meet proper standard for cleanliness.

The commonly used biomedical polymer materials include Polytetrafluoroethene, polyurethane, polyvinyl chloride, silicone rubber, polypropylene, polysiloxane gel, poly methyl acrylate, chitin derivatives and Polymethylmethacrylate.

2.4 Biomedical polymer material study content

Structure-property relationship: different materials have different properties; the same kinds of materials have different properties; the property of the material not only relevant to its composition but more importantly is relevant to its structure. Moreover, its basic performance and processing performance are sometimes inconsistent with its operational performances.

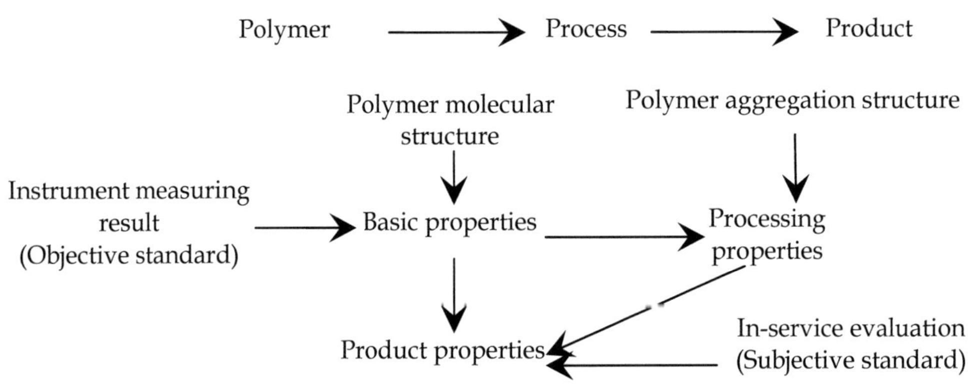

Table 3. Structure-property relationship of Polymer Material

2.4.1 Polymer chain structure study

1. Polymer-based band

Most bands in the spectra are characterized by small molecule band with structure similar to repetitive units, i.e. elemental bands; also there are some unique absorbing bands different from the small molecular organism that belongs to polymer-based bands. Conformation bands; Stereoregularity bands; Conformational regularity band; Crystal band

Such bands are of great significance to the study of polymer.

2. Determination of chemical composition
3. Determination of degree of branching
4. Determination of copolymer composition
5. Determination of the sequence distribution of the copolymer
6. Influence from additives

Adding modifier into certain materials may obtain new material with different performances. But the difference of the infrared spectra maybe small.

2.4.2 Change of polymer materials

Monomer—polymer—Products—Application

2.4.3 Analyses on polymer material

Development of new materials, assimilation of introduced technology and gradual internationalization; discrimination of true and false materials

3. Characteristics of polymer IR spectrum

Infrared spectroscopy is the most effective method to identify the various polymers and additives in polymer material analysis. The main advantages of the infrared spectroscopy include: 1) it does not cause damage to the sample under analysis; 2) it may analyze organic and inorganic compounds of various physical states (gas, liquid and solid) and various exterior forms (elastic, fibrous, thin film, coating and power form); 3) well-developed molecular vibration spectroscopy (the basis of IR spectrum) makes it easy to understand the explanation of the IR spectrum of the compound; 4) a large number of standard infrared spectrogram for various kinds of chemical compounds have already been published in the world and can be referred to in spectra analysis. With application of computer and establishment and improvement of spectral database, the spectral identification will be easier and conclusion will be more reliable. Chemical compounds of different chemical structures have infrared absorption spectroscopy of different characteristics, There is no completely identical spectroscopy except for some isomers. Moreover, each absorption band (band) in the infrared spectrogram represents a certain vibration type of a certain atomic group or radical in the chemical compound. Their vibration frequency (corresponding to the wave numbers of the absorption band on the spectrum) is directly related to the mass and chemical bond strength of the atom in the atomic group or radical. They are consequently subject to changes of proximity structure and different influences of chemical environment. As each molecule of the polymer contains a great number of atoms, it may be considered that the polymer spectrum would be extremely complicate with considerable number of normal vibration. But this is not the case, IR spectra of some polymers are more simple than that of the monomers. This is because polymer chain is made of many repetitive units and each repetitive unit has basically same bond force constant with roughly similar vibration frequency. Moreover, due to limitation of strict selection law, only part of the vibration has infrared activity. The explanation of infrared spectrum of polymer must take into account the molecular chain structure and the aggregate structure of the concerned polymer. Different structural characteristics correspond to different absorption bands: ① absorption bands: reflects the chemical composition of polymer structural unit, connection type of monomers, branching or cross-linking and sequence distribution. ② Conformation bands: such bands are related to the certain conformation of some radical in the molecular chain and have different representations in different morphologies. ③ Stereoregularity bands: these bands are related to the structure of the molecular chain and are therefore identical in the various morphologies of the same high polymer. ④ Conformational regularity bands: these kinds of bands are generated as result of the mutual action of the adjacent radicals in the molecular chain. ⑤ Crystal bands: the bands are formed as a result of the interaction between the adjacent molecular chains in the crystal.

4. Preparation of samples

In order to obtain high quality infrared spectra, a proper method should be selected for sample preparation according to the characteristics of the sample. Close attention should be paid to the following in sample preparation: 1) Purified and separated sample. IR spectrum of mixed sample will be shown in bands of various components and may mislead judgment for the sample. 2) Free of moisture. The infrared spectroscopy is very sensitive to moisture. If there is any trace of moisture in the sample, bands with intensive OH radical characteristic will be appeared near 3400cm^{-1}, affecting the identification of N-H, C-H bonds. 3) Proper concentration and thickness of the sample. In good infrared spectra, most of the absorption bands have transmittance of between 20 to 80%. In especially low concentration or thin thickness, weak absorption bands and band of medium strength will disappear, resulting in inaccurate absorption spectrum. On the contrary, the excessive strong absorption will result in flat peak with no maximum value of the peak or indistinct double peaks.

4.1 Solid sample preparation technology

Common solid samples include polymer and some organic compounds. The following methods are usually adopted for sample preparation:

1. Pressed halide disk method (KBr Pressed Disk Method), mix a quantitative amount of samples (accurately measured for quantitative analysis) And some ground and screened KBr powder in the agate mortar in proper proportion and pulverize them completely until there is no obvious particles in the blends. Please be noted: KBr requires the use of analytical reagent or higher. Qualified KBr should be no absorption in mid-infrared region. Moisture absorption will result in moisture absorption peak. In addition, KBr powder tends to absorb the vapor in the air and should be dried before use. Screen the KBr powder with 200-mesh sieve, then dried under 120°C and put the powder in the dryer for use. The temperature should not be too high(greater than120°C), otherwise it will decompose (see spectra 1).
2. Paste Method

Sample Preparation

Put the fully grounded powder sample in the agate mortar, add 1-2 drops of medium by dropper and blend evenly. Use stainless steel spatula to scoop out the evenly grounded sample, paste it on a window slice, compress with another window slice to a proper thickness and then measure it.

Notes:

All the media used in the method are organic matter and can absorb in certain range. The absorption positions of the media are as shown below:

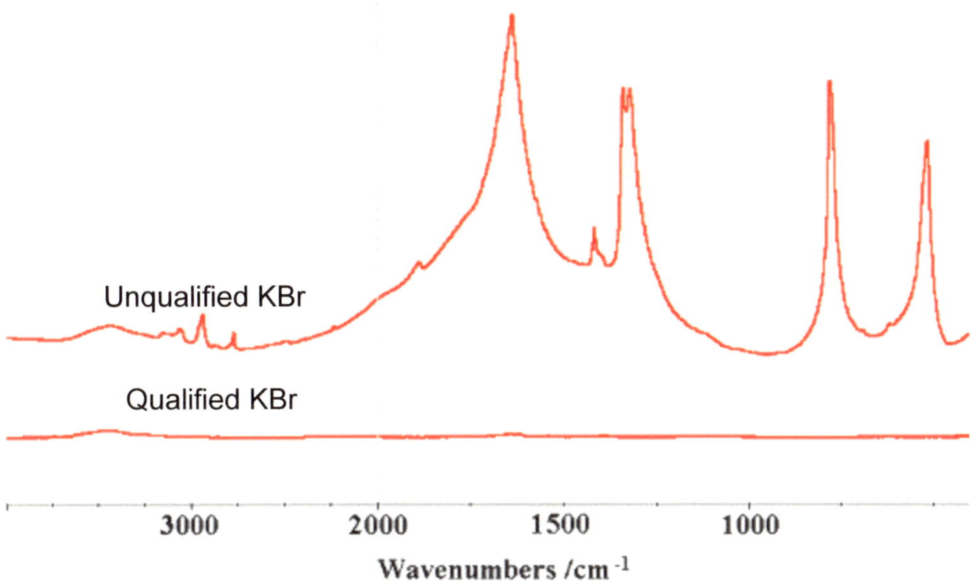

Fig. 1. Absorption requirement of qualified KBr

Name of media	Absorption peak location and assignment
Paraffin oil (long-chain alkane)	3000-2850cm^{-1} (V_{C-H2}, V_{C-H3}) 1468 cm^{-1}, 1397 cm^{-1} (δ_{C-H2}, δ_{C-H3})
Fluorocarbon oil (perfluoroparaffin)	With C-F absorption of different intensity in 1400-1500 cm^{-1}
Hexachlorobutadiene (chloroalkene)	With C=C and C-Cl absorption in 1500-16010 cm^{-1}, 1150-1200 cm^{-1}, 600-1000 cm^{-1}

Table 4. The absorption positions of the media

The IR spectra of above three media have complementary. Complete infrared spectrogram of the sample can be obtained by at least two media. In this consideration, the influence of the media spectrogram should be deducted from spectrum analysis of the sample to baseline.

3. Solution film casting method

Sample preparation method

Dissolve the sample in an appropriate volatile solvent to prepare a solution with concentration of around 2-5%, apply the solution evenly on the watch glass and glass sheet, or drop it directly on the infrared wafers (KBr, NaCl, BaF etc), then solvent volatilize to obtain sample film.

Application Scope: mainly used in sample analysis with both film-forming and solvent dissolving.

Notes:

The selection of solvent and the concentration of the sample prepared play an important role in obtain even and pure film.

Principle for Solvent Selection:

All solvents have infrared spectrum absorption band. Commonly used solvent oil include CCl_4, $CH\ Cl_3$, CS_2 and n-hexane.

Select solvent with low boiling point instead of using solvent with high boiling point and strong polarity, otherwise, the film thickness will be uneven as result of the fast solvent volatilization.

The solubility of the sample in the solvent should be sufficient high. The solution concentration can be regulated.

4. Hot-press film making method

Hot-press device: composed of hydraulic membrane machine, heating die and temperature control device.

Hot-press temperature setup for various common polymers

Sample preparation method

Place a piece of aluminum foil in the core of the die, put the sample on the aluminum foil and then place another piece of aluminum foil on the sample; put on the upper module and then place the die on the tablet press. Control the pressure within 1000-3000Kpa/cm^2. The heating temperature and time for pressing shall be controlled to the extent that there is no thermal decomposition or other chemical changes. After pressing for about 1min, take down the die from the hydraulic press (wear insulating glove to avoid burn), cool to room temperature and then unpack the die to take out the sample tablet. Reduce sample volume if the film is too thick. Reduce temperature if the film turns yellow or has bubbles. Uneven film is caused by low heating temperature, too short heating time or small pressure. Reselect film making condition to remake the film in case of any of the above circumstances.

Application Scope: applicable for non-oxidizing and non-degradable thermoplastic or inorganic polymer materials near the softening point or melting point.

Notes:

Select appropriate hot pressing temperature and pressing time according to the nature of the analyzde sample in order to obtain hot press film in line with infrared transmittance while not destroying structure performance of the sample. Take protective measure against burn when operating in high temperature.

General solid sampling technique can be used in the infrared analysis. However, improvement has been made in actual operation based on the features of the polymer. Mr.

Yang Rui from Tsinghua University has made a detailed research and analysis (in 2011) as below:

Thermoplastic resin: dissolve flow casting film, hot-pressing film or dissolve coated tablet.

Thermosetting resin: such as curing epoxy resin and phenolic resin. Use clean steel file to file sample powder and then use KBr pellet.

Mild cross-link polymer: sample that only swells instead of dissolving in solvent. Grind with KBr in swelling (with solvent) condition, then dry the solvent and pulverize the tablet.

Fiber sample: Filament with diameter of less than 10 microns can be neatly arranged (or be cut into piece) to determine the transmission spectra with KBr pellet. Filament with larger diameter or non-filament fiber should be entangled on the aluminum tablet or be squashed for determination with ART.

Polymer	Temperature /°C	Polymer	Temperature /°C
High-density polyethylene Low density linear polyethylene	170	Nylon 11	220
Low density polyethylene	150	Nylon 66	280
Polypropylene	200	Metaformaldehyde	190
Polystyrene	130	Makrolon	260
PMMA	260	Polybutylene terephthalate	290
PVC	190	Polybutylene terephthalate	250
Nylon 6	250	Teflon	360

Table 5. Reference Temperature for Hot-pressing Film of Common Polymers

Figure 2 shows PMMA infrared spectrum from different sample preparation methods. The PMMA infrared spectrum from powder pulverized tablet and hot-pressing film are nearly same on wavenunbers, and slightly different on wave intensity. But there is another additional infrared eigen wave (at 760 cm^{-1}) for PMMA infrared spectrum from solution film casting method (chloroform resolving).

4.2 Liquid sample preparation technology

4.2.1 Material of window slice used for organic liquid testing

KBr and NaCl are the most commonly used window slice materials used for determination of IR spectrum of the organic solution. But KBr is widely used than NaCl. KBr is the most suitable window slice material for testing organic liquid in middle infrared.

Rinse the KBr wafer immediately after testing. As KBr is non-dissolvable with in soluble, remove the organic solvent from the surface of the wafer by anhydrous ethanol instead of clean water, then dry with crystal-tipped tissue.

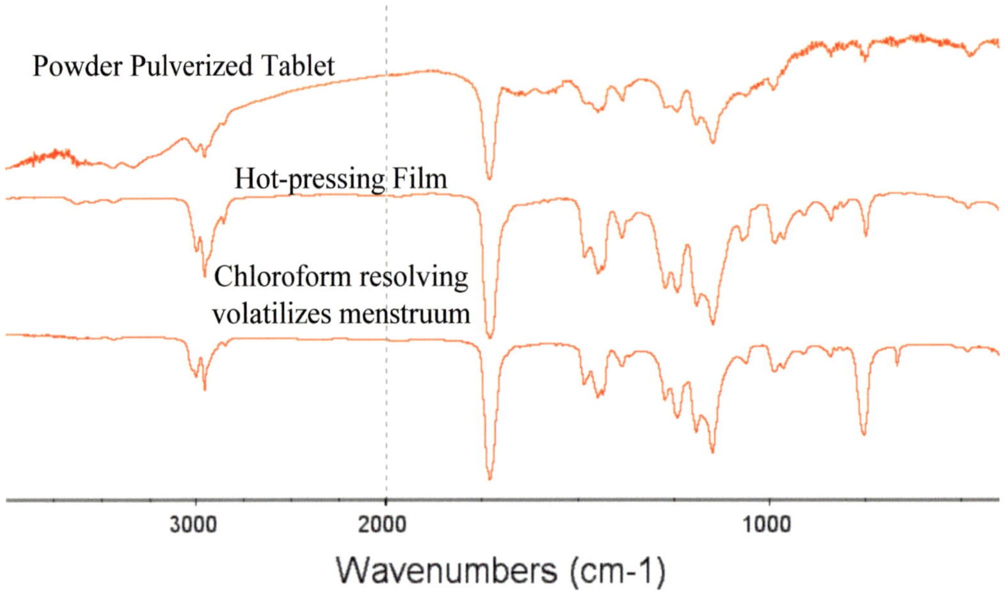

Fig. 2. PMMA Infrared Spectrum from Different Sample Preparation Methods

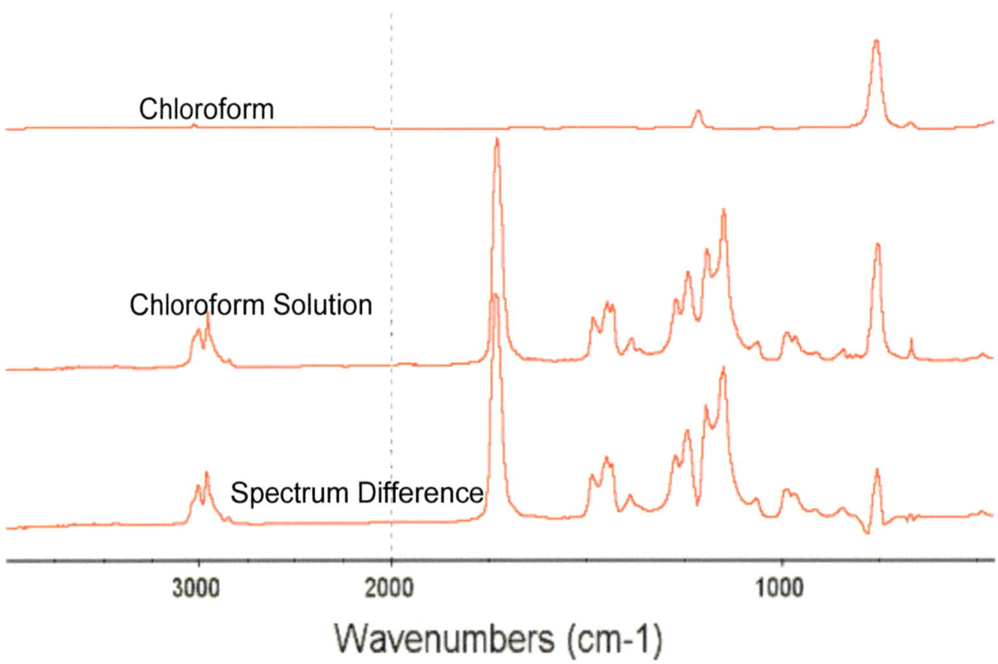

Fig. 3. PMMA Chloroform Solution and Difference Spectrum of Chloroform

4.2.2 Window slice for testing aqueous solutions

The most commonly used Window slice materials for IR spectroscopy of aqueous solution samples is BaF2 wafer, followed by CaF2 wafer. Though BaF_2 wafer is non-dissolvable in water, it can be dissolved in acid and ammonium chloride and can react with phosphate and sulfate to generate barium phosphate or barium sulfate respectively, thus erode the surface of the wafer. When testing metal salt solution, the metal ion will exchange with barium ion and thus erode wafer surface.

4.2.3 Preparation technology for liquid samples with different boiling points

1. Liquid with low boiling points

As the sample has low boiling point, a sealed tank should be used to prevent evaporation of sample. The thickness of the liquid membrane is decided by the nature of the sample. The stronger the polarity, the thinner the tanker is.

2. Liquid with high boiling point and low viscosity

Liquid sample with high boiling point, good flow property and low viscosity may be dropped between the two window slices of the removable liquid tank, then compress the automatically formed even liquid film for measurement.

3. Sample with high boiling point and high viscosity

For qualitative analysis of plasticizer and pyrolysates used in the samples with high boilling point and high viscosity, such as grease, polymers, apply a small amount of sample on KBr window slice by stainless spatula, and scrape evenly, then put it on the sample rack for measurement.

5. Analysis of infrared spectroscopy in biomedical polymer material

The key difference between medical polymer material and other polymer materials is that the former has both medical functionality and biocompatibility and resorts to chemical or physical means to achieve polymeric modification of polymer materials. Fourier Transform Infrared Spectroscopy (FT-IR) is an effective method to analyze polymer materials and its modification.

5.1 The transformation degree of dental composite resin after polymerization may directly affect the biological, chemical and mechanical strength. The most urgent problem in the application and development of the composite resin material is to remove various unfavorable factors that may affect solidification of the composite resin to maximize its transformation degree. FTIR spectroscopic technology may comprehensively study the polymerization of chemical curing and visible-light curing composite resin and the influences of various factors to the degree of polymerization as well as the relationship between transformation degree of the composite resin and the various indirect indicators. The existing study on FTIR indicates: Of different brands of dental composite resin available in market, the transformation degree of visible-light curing resin is superior to that of the chemical-curing resin; the double bond transformation degree and the mechanical properties of the resin have positive correlation with the contents of the catalyst and the reductive and have negative correlation with the contents of the inhibitors.

5.2 As the polymer surface properties affect the cohesiveness, wettability and biocompatibility of the polymer and its actual application, the study focusing on improving polymer properties through polymer surface modification. The quantitative analysis of the polymer surface composition is the important basis for the property study. Attenuated total reflectance infrared spectroscopy (ART-FTIR) technology is one of the most effective methods to test the information of the material construction of the polymer surface without interference of the test samples. Just as transmission infrared spectrum, ATR-FTIR provides information [1-5] about chemical structure, three-dimensional structure, molecular orientation and hydrogen bond of the material. The existing study[6] uses attenuated total reflectance Fourier transform infrared spectroscopy (FTIR-ATR) to test the surface composition of polyethylene glycol/ polyethylene blends (PEG/PE) film. With the corresponding characteristic peak intensity ratio as the basis for the quantitative determination, ATR correction procedures, and NMR-FTIR correction equation[6~7] : Y=0.346 48-1.336 55X+1.268 37X^2,, are used for quantitative analysis of the relative composition of the polyethylene glycol chain and the polyethylene chain on the blend surface to obtain a better result of the reproducibility and comparability. The quantitative analysis of the relative composition of polyethylene glycol chain and the polyethylene chain on the surface layer of the blend film is achieved through working curve method. See figure as below:

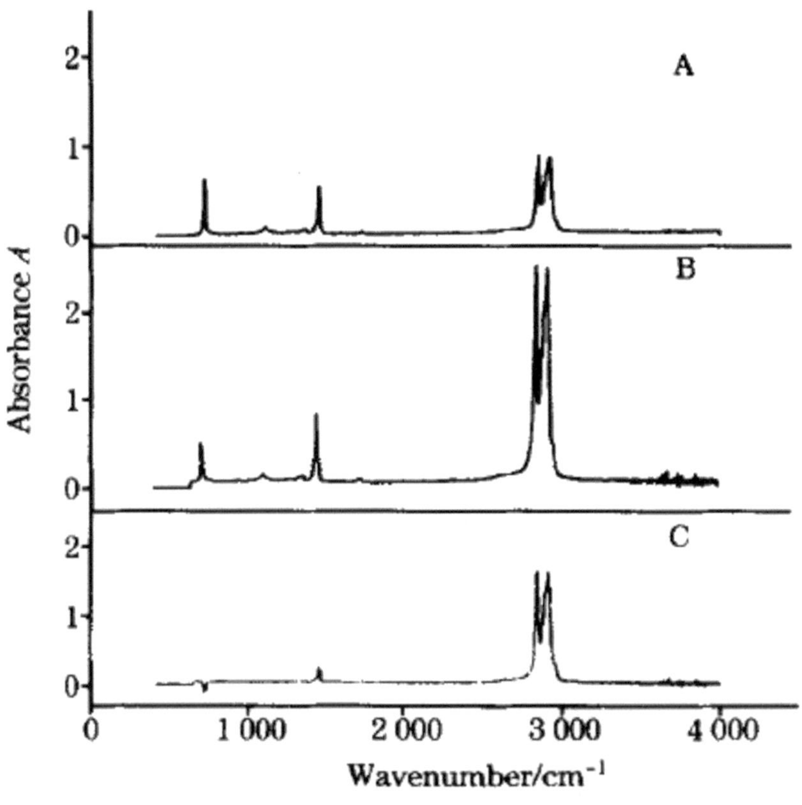

A. ATR spectrumwithout calibration; B. Corrected ATR spectrum;
C. The difference spectrum between spectra A and B

Fig. 4. ATR spectra of blend of PEG (2000) and LLDPE

5.3 Another key focus of study is to use combined infrared spectroscopy and computer technology to make quantitative analysis of the chemical structure of the auxiliary materials added to the medical polymer, such as additive, adhesives and plasticizer. Spectrum subtraction technology can be used to identify the additives in the high polymer products. Medical infusion devices are made of conventional polymer material polyvinyl chloride (PVC) and 2-ethylhexyl phthalate (DEHP) is added to plasticize rigid polyvinyl chloride (PVC), with additive dosage of 40-60%. Study has verified that DEHP can enter human body through venous transfusion, respiratory tract and skin and bring damage to human health. This has become focus of academic research and disputes and has attracted attention from media. Though DEHP's toxicity and carcinogenicity has been fully confirmed in experimental animals, its adverse effect in human body is still controversial. Using infrared spectroscopy subtract technology to analyze PVC infrared spectrogram of PVC and the infrared spectrogram of plasticized PVC may determine the kernel of material construction of the plasticizer. FTIR spectrum subtraction may also be used in polymer end-group analysis, polymer oxidation and degradation reaction analysis and inter-molecular analysis.

Below is the infrared spectra of traditional bis (2-ethylhexyl) phthalate (DEHP) plasticized rigid PVC and PVC materials used in medical infusion equipment.

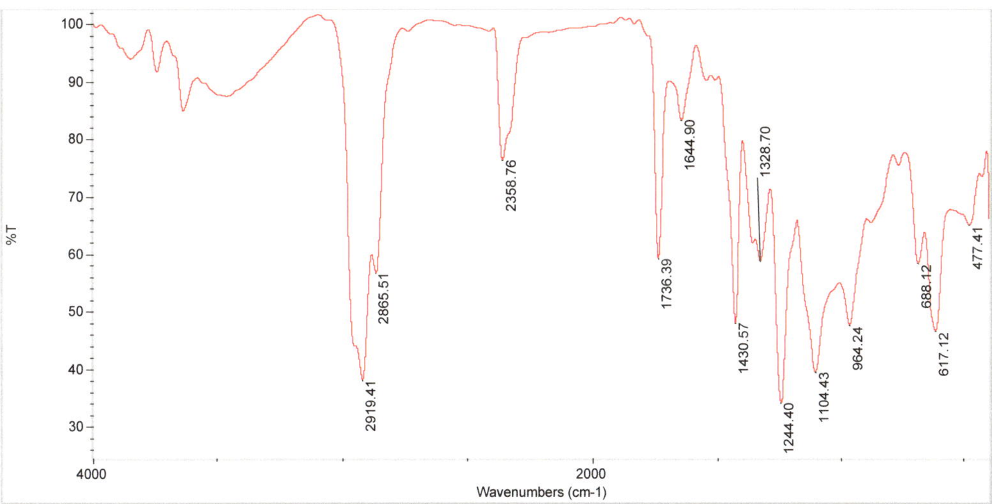

Fig. 5. Infrared spectrogram of PVC and DEHP

5.4 Polymer materials in ophthalmology

Contact lenses are the most common polymer product in ophthalmology. The basic requirements for this type of materials are: ①excellent optical properties with a refractive index similar to cornea; ② good wettability and oxygen permeability; ③ biologically inert, degradation resistant and not chemically reactive to the transfer area; ④ with certain mechanical strength for intensive processing and stain and precipitation prevention. The common used contact lens material includes poly-β-hydroxy ethyl methacrylate, poly-β-hydroxy ethyl methacrylate-N-vinyl pyrrolidone, poly-β-hydroxy ethyl methacrylate, Poly-β-hydroxy ethyl methacrylate - methyl amyl acrylate and polymethyl methacrylate ester-N-vinyl pyrrolidone, etc. The artificial cornea can be prepared by silicon rubber, poly methyl acrylate, poly-casein or other thin films. The main body of the artificial lens can be made of polymethacrylate. The researchers have attached increasingly great importance to the qualitative analysis of polymer materials for this kind of medical equipments. As explicitly specified in YY0477-2004 "Orthokeratology Using Gas Permeable Rigid Contact Lens", infrared spectrum analysis is adopted to determine the components of the lens materials. The following drawing is the infrared spectrum analysis of this material: point 2961cm-1 is the methyl characteristic peak of methyl acrylate; points 1104 cm-1 and 1046 cm-1 are characteristic absorption peaks of siloxane, points 1730 cm-1,1227 cm-1 and 1199 cm-1 are ester peaks of methyl acrylate; points 893 cm-1 and 7556 cm-1 are the structure characteristic peaks of polymethacrylate; also it is worth knowing that carbonyl peak on point 1769 cm-1 is the characteristic peak of fluoro-alkylated methyl acrylate.

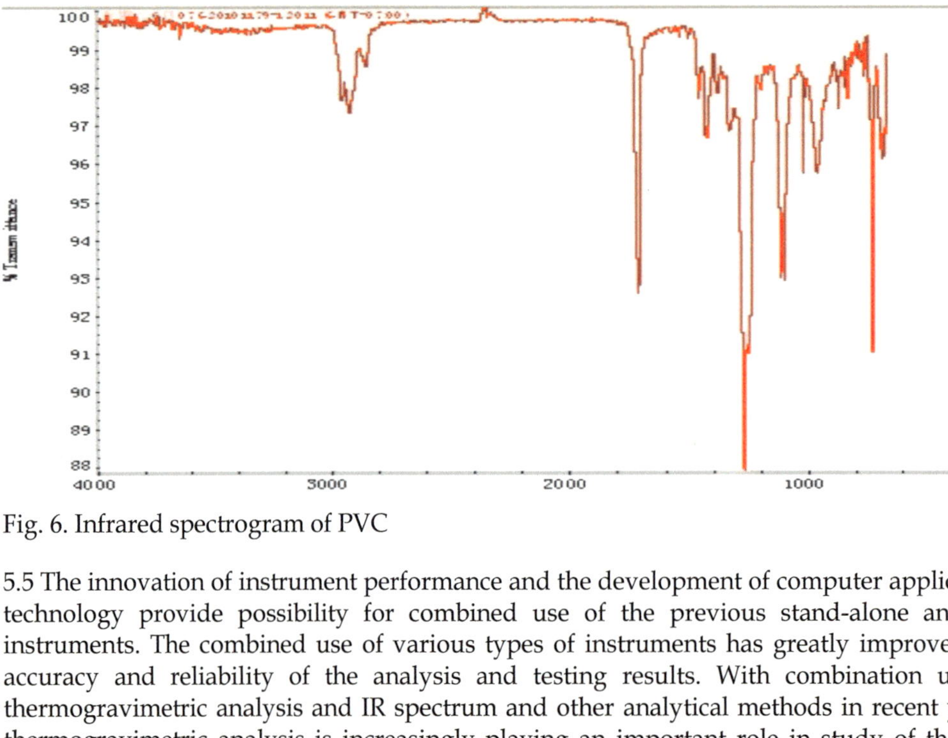

Fig. 6. Infrared spectrogram of PVC

5.5 The innovation of instrument performance and the development of computer application technology provide possibility for combined use of the previous stand-alone analysis instruments. The combined use of various types of instruments has greatly improved the accuracy and reliability of the analysis and testing results. With combination use of thermogravimetric analysis and IR spectrum and other analytical methods in recent years, thermogravimetric analysis is increasingly playing an important role in study of thermal behavior in chemical materials. In comparison with traditional thermogravimetric analysis method, TGA-IR spectrum combined analysis can directly and accurately determine the various physical-chemical change during the heating proves and identify the chemical composition of decomposition or degradation products during the during the weight loss process. Thus, it has been a key experimental method in studying thermostability and thermal decomposition (degradation) process of various inorganic, organic and polymer materials and proves a promising prospect in respect of thermal performance analysis of the materials. As Fourier transform infrared spectrometer is of high noise-signal ration and high precision, it may detect the slight intensity change and frequency shift of the infrared bands in the sample before and after the heat treatment. Thus to provide structural differences of polymer film for three different thermal stages of high molecular polymer, from high-elastic state slow cooling, high-elastic state quenching to heat treatment below temperature Tg. The study shows the conformation changes of PVC films with different thermal histories in heating process. Meanwhile, in FTIR measurement, the sample subjected to heat treatment below Tg temperature occurred sudden change of conformation in the temperature range corresponding to enthalpy absorption peak of differential scanning calorimetry (DSC).

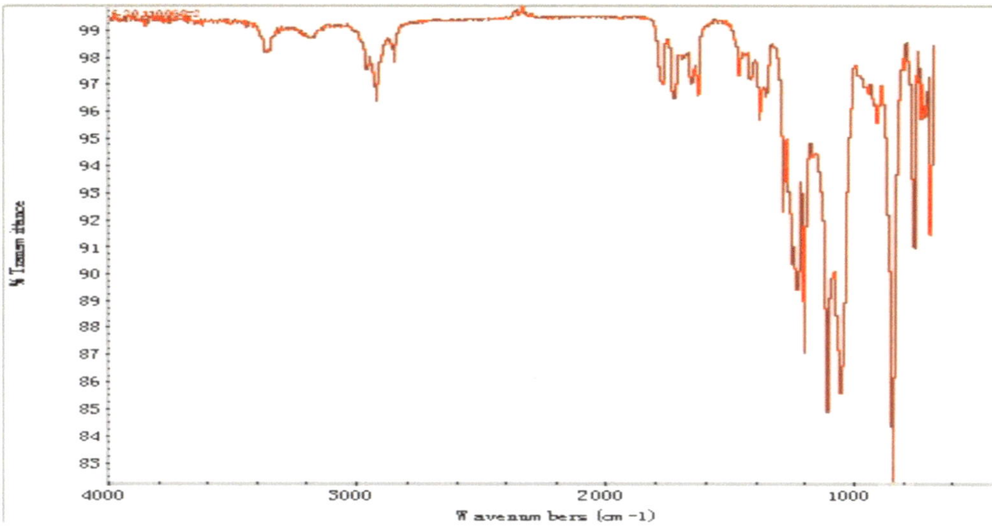

Fig. 7. Spectrum of Orthokeratology Using Gas Permeable Rigid Contact Lens

6. Prospect

FTIR is becoming widely used in the field of medical polymer materials, especially for quantitative analysis of the material properties. At present, the study on application of infrared spectroscopic technology or combined application of IR spectrum with other technologies in polymer material, to be still growing. With rapid development of scientific technology, the research in IR spectrum is further deepened. IR spectrum is not only widely used in polymer materials but is also widely used in pharmaceuticals, foods and environmental science. The gradually improved infrared detection method may be used on vivo analysis of pathological tissue and greatly contribute to the rapid and accurate diagnosis of the diseases. Meanwhile, the disease mechanism and progression maybe clarified through analysis of infrared spectrogram of tissues or cells.

7. References

[1] Yang Qun, Wang Yi-lin, Yao Jie, et al. Spectroscopy and Spectral Analysis, 2006, 26(12): 2219.

[2] Jiang Zhi, Yuan Kai-jun, Li Shu-fen, et al. Spectroscopy and Spectral Analysis, 2006, 26(4): 624.

[3] Bergberiter D E, Srinivas B. Macromolecules, 1992, 25: 636.

[4] Lee K W, Kowalczyk S P, Shaw J M. Macromolecules, 1990, 23: 2097.

[5] Francis M, Mirabella J R. Appl. Spectrosc. Rev., 1985, 21: 45.

[6] Qian Hao, Zhu Ya-fei, XU Jia-rui Spectroscopy and Spectral Analysis, 2003, 23 (4): 708-713
[7] Chen Jia-xing, Joseph A. Gardella Jr. Appl. Spectroscopy, 1998, 52(3): 361.

Section 4

Fluorescence Spectroscopy

11

Laser Fluorescence Spectroscopy: Application in Determining the Individual Photophysical Parameters of Proteins

Alexander A. Banishev*
Department of Physics and Astronomy, University of California, Riverside, CA, USA

1. Introduction

This work was initiated by the problem of investigating the photophysical properties of complex protein molecules and performing the diagnostics of such molecules in water environment. At the present time, fluorescence spectroscopy (fluorimetry) is widely used to study complex organic compounds (COC) (Lakowicz, 1999). Together with spectrophotometry these methods form the basis for fast and nondestructive diagnostics of COC in the natural environment, i.e. they present the diagnostic methods *in vivo* and *in situ*. However, the conventional (linear) fluorescence spectroscopy methods can not provide complete information on fluorescent objects under study because of insufficient selectivity (fluorescence bands of most COC are broad and structureless at room temperature).

The capabilities of fluorescence spectroscopy can be enhanced by using the methods of laser fluorimetry, in particular nonlinear laser fluorimetry (Fadeev et al., 1999). This method allows one to get information on the molecular level and determine the photophysical parameters of molecules (absorption cross section, lifetime in excitation state, intersystem crossing and energy transfer rates, etc.). Furthermore, the parameters can be measured *in vivo* and *in situ* in the absence of *a priori* information, which is necessary for conventional spectroscopic methods (for example, molecular concentration (Banishev et al., 2009)).

The diagnostics of protein complexes is an intricate problem if a molecule contains more than one absorption/fluorescent center (Permyakov, 1992). The problem becomes much more complex if, in addition, the protein specimen (ensemble of molecules) is a mixture of several chemically nonidentical types of molecules (subensembles) which cannot be separated, i.e. their partial concentrations are unknown. The second situation is typical for the special kind of proteins, namely, fluorescent proteins (FPs) (Piatkevich et al., 2010). The solutions of FPs are usually mixtures of several types of molecules, which are chemically different and have their own set of photophysical properties (Verkhusha et al., 2004). In this case for unambiguous interpretation of experimental data it is necessary to make simultaneous measurements of a large number of parameters, i.e. to simultaneously apply (or, better, synthesize) several spectroscopic methods.

*Corresponding Author

In this chapter, a new approach based on the simultaneous use of nonlinear laser fluorimetry, spectrophotometry and conventional fluorimetry methods is presented. The approach allows us to *in vivo* determine the individual photophysical parameters of fluorophores in multi-fluorophore protein complexes. The approach has been applied for investigation of the photophysical properties of the protein molecules of different complexity. Two classes of proteins have been chosen, namely, serum albumins (by the examples of human and bovine serum albumins) and fluorescent proteins (by the example of monomeric red FP mRFP1). The following new results are presented.

i. The photophysical parameters such as (a) true absorption cross section (at 266 nm) of tryptophan and intersystem crossing rate in single-tryptophan-containing protein human serum albumin and (b) true absorption cross section (at 266 nm) of tryptophans and rate of energy transfer between them in two-tryptophan-containing protein bovine serum albumin have been determined.
ii. The complete solution of the task of determining the photophysical parameters of all mRFP1 spectral forms is given. The mechanism of photophysical processes in the spectral forms under their excitation by UV radiation (at 266 nm) has been clarified.
iii. The study of the influence of a single amino acid substitution in mRFP1 protein on individual photophysical parameters of the chromophore (a heterogroup[1] responsible for light absorption and fluorescence in the visible wavelength range) of fluorescent spectral form is performed. The 66th amino acid residue (glutamine 66) has been chosen as a position to be replaced. This residue participates in formation of the chromophore and, as was shown in (Banishev et al., 2009), its substitution by polar serine or cysteine changes the spectral and photophysical properties of the resultant mutant of the mRFP1. In the present work this study has been extended. The optical properties of new variants of mRFP1 with polar (asparagine, histidine) and non-polar (alanine, leucine, phenylalanine) substitution have been investigated. It was found that the individual extinction coefficient of the chromophore and the position of the steady-state spectra of the proteins with polar substitution correlate with the volume of the substituted amino acid residue at position 66. The explanation of this effect is given.

Except for this key target, the methodological task has been put, namely, to demonstrate the unique capabilities of the nonlinear laser fluorimetry method (which is not, so far, well known in a wide circle of opticians) on the specific object.

2. The method of nonlinear laser fluorescence spectroscopy

The fluorescence signal from fluorophores of complex organic compound (COC) under powerful laser excitation is represented as the nonlinear function of the number of detected fluorescence photons N_{Fl} (or fluorescence intensity I_{Fl}) on the photon fluxes F of pumping radiation (Filipova et al., 2001). The dependence $N_{Fl}(F)$ is called fluorescence saturation curve, its typical view is represented in the Fig. 1(a). There are several reasons for that nonlinear dependence: the non-zero lifetime of organic molecules in excited state; intercombination conversion; intermolecular interactions including singlet–singlet annihilation, etc.

[1] The heterogroup is called a chromophore, even when it produces fluorescence, in other words, it is a fluorophore.

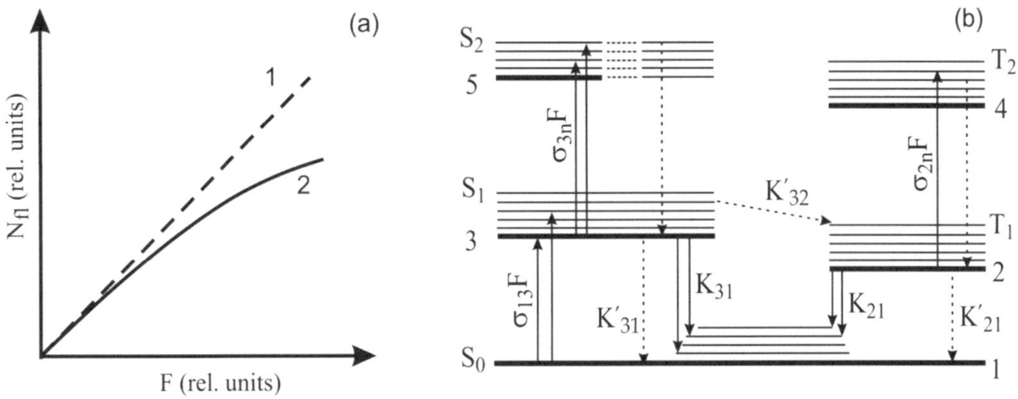

Fig. 1. (a) The $N_{Fl}(F)$ dependencies (see text): (1) in the absence of fluorescence saturation, and (1b) when saturation appears (for most COC at $F > 10^{23}$ cm^{-2}s^{-1} (Fadeev et al., 1999)). (b) Photophysical processes in COC (Lakowicz, 1999), without accounting for intermolecular interactions. Solid and dotted vertical lines are radiation and radiationless transitions respectively, S_i are singlet states and T_i are triplet states.

The parameters of saturation curves depend on photophysical characteristics of molecules fluorophores, so that such characteristics can be extracted from these curves after resolving an inverse problem (Fadeev et al., 1999). This is a basement of nonlinear laser fluorimetry as a method for investigation of photophysical properties of COC. To solve the inverse problem, we should first calculate (either analytically or numerically) the theoretical saturation curves by using the fluorescence response formation model of an ensemble of fluorescent molecules under their excitation by laser radiation. In present work two models have been used: the conventional model of fluorescence response formation and the model of localized donor-acceptor (LDA) pairs.

The conventional model (Banishev et al., 2008a, 2009) in describing the fluorescence response is represented as a system of equations that describes the kinetics of concentration of COC molecules at the corresponding energy states (Fig. 1(b)). In the case of monomolecular solutions of non-interactive organic compounds, we must give priority to the following photophysical parameters when defining the saturation curve:

- the absorption cross section $\sigma \equiv \sigma_{13}$, which determines the probability of a molecule transition from the ground singlet state S_0 (level 1) to the first excited state S_1 (level 3) stimulated by a photon flux with the density F;
- the full lifetime $\tau \equiv \tau_3$ of the molecule in the S_1 state (the fluorescence decay time);
- the quantum yield $\eta_T = K'_{32}/K_3$ of the molecule transition to the lower triplet state T_1 (level 2) due to the intersystem crossing, where $K_3 \equiv \tau^{-1} = K_{31} + K'_{31} + K'_{32}$; K_{31} and K'_{31} are the rates of radiative and radiationless transitions from S_1 to S_0; K'_{32} is the rate of the transition from S_1 to T_1.

The model of the fluorescence response, which takes into account processes pointed out above, can be described by the following set of kinetic equations (Fadeev et al., 1999):

$$\frac{\partial n_1(t,\vec{r})}{\partial t} = -F(t,\vec{r})\cdot\sigma\cdot[n_0(t,\vec{r})-n_3(t,\vec{r})-n_2(t,\vec{r})]+(K_3-K_{32}^{'})\cdot n_3(t,\vec{r})$$
$$\frac{\partial n_3(t,\vec{r})}{\partial t} = F(t,\vec{r})\cdot\sigma\cdot[n_0(t,\vec{r})-n_3(t,\vec{r})-n_2(t,\vec{r})]-K_3\cdot n_3(t,\vec{r})$$
$$\frac{\partial n_2(t,\vec{r})}{\partial t} = K_{32}^{'}\cdot n_3(t,\vec{r}) \qquad (1a)$$
$$n_0 = n_1 + n_2 + n_3,$$

where n_0 is the total concentration of molecules; n_3, n_2, and n_1 are concentrations of molecules in the S_1, T_1 and S_0 states, respectively; $F(t,\vec{r})$ is the photon flux density of exciting radiation at the coordinate point $\vec{r}$ at instant of time t. The rest of parameters are defined above. In model (1a), the transition from T_1 to S_0 is neglected. This assumption is valid if the light pulse duration (t_p) is much less than the lifetime in the T_1 state, i.e. t_p is much less then $(K_{21}+K'_{21})^{-1}$. For pulse lasers often used in laser fluorescence spectroscopy, the t_p is ~10 ns, and this condition is fulfilled.

The conventional model (1a) describes the processes in a system in the absence of interaction between molecules. If there is the interaction and an ensemble of fluorophores generating the fluorescence response consists of subensembles of the donor and acceptor molecules, then the conventional approach is reduced to two systems of kinetic equations, i.e. separately for each subensemble. The term describing the energy transfer is in this case a "cross term" that connects these two systems of equations. Such model, based on separate mathematical description of two subensembles, is able to describe the fluorescence response when each molecule of the donor is surrounded by a large number of the acceptor molecules onto which the energy transfer can occur (Agranovich & Galanin, 1982) (i.e. the possibility that the donor molecule and all the locally surrounding it acceptor molecules simultaneously stay in the excitation state is excludes). The situation like this is typical for a concentrated binary solution of single-fluorophore molecules (for example, dye solutions with the concentration higher than 10^{-4} M) or for complexes with high local concentration (Fadeev et al., 1999), such as phytoplankton. The energy transfer process in that case is called the intermolecular one.

If there is a donor–acceptor pair within a single molecule (i.e. we have a molecule with a LDA pair), the situation is possible when the donor and the acceptor are simultaneously in an excited state. Therefore, the description of the energy transfer in the framework of a conventional scheme is impossible and the model (1a) should be modified. Let us note that the molecular objects with LDA pair are finding more and more wide applications at present time. Commonly, systems of this kind are constructed artificially from pairs of organic compounds, for example, from dye molecules (Srinivas et al., 2001) or FP macromolecules (Truong & Ikura , 2001). In (Banishev et al., 2008b) a fluorescence response formation model of an ensemble of LDA pairs has been suggested by the author. The model makes it possible to describe the energy transfer inside a LDA pair, disregarding the energy transfer between the pairs. The main idea of this approach consists in the following. Let us introduce a notion of the *collective states* of a LDA pair (Fig. 2); each of these states simultaneously describes both the donor state and the acceptor state:

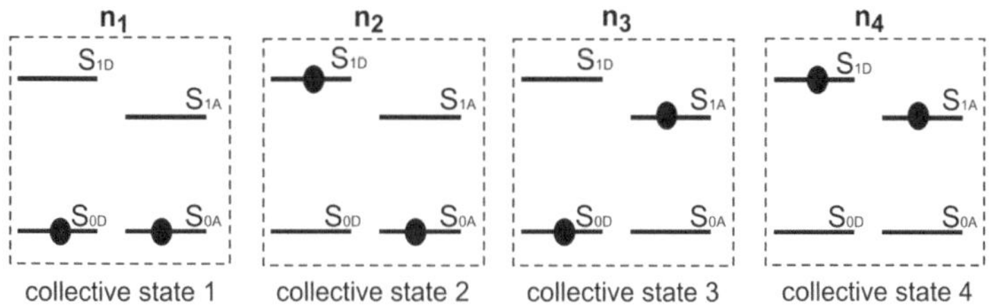

Fig. 2. The collective states of the LDA pair (nonmetering the process of singlet-singlet annihilation and intersystem crossing). The S_{0D}, S_{1D} and S_{0A}, S_{1A} are the energy levels of the donor and acceptor respectively.

- Collective state 1: the donor and the acceptor are in the ground state S_{0D} and S_{0A}, respectively; the concentration of such molecules is denoted as $n_1 \equiv n_1(t,\vec{r})$.
- Collective state 2: the donor is in the first excited singlet state S_{1D}, and the acceptor is in the state S_{0A}; the concentration of such molecules is denoted as $n_2 \equiv n_2(t,\vec{r})$.
- Collective state 3: the donor is in the state S_{0D}, and the acceptor is in the first excited singlet state S_{1A}; the concentration of such molecules is denoted as $n_3 \equiv n_3(t,\vec{r})$.
- Collective state 4: the donor and the acceptor are in the first excited singlet state S_{1D} and S_{1A}, respectively; the concentration of such molecules is denoted as $n_4 \equiv n_4(t,\vec{r})$.

As a result, there is no need to describe the fluorescence response from a sub-ensemble of donor and acceptor molecules separately, i.e. it is unnecessary to create two systems of equations (one for a donor sub-ensemble and one for an acceptor sub-ensemble) similar to (1a), as it takes place in the conventional approach. Instead of this, the system of equations that describes the populations of the collective states can be written. The dynamics of variation in the concentrations of these four collective states of the LDA pair (nonmetering the process of intersystem crossing in molecules of the donor and acceptor) is mathematically described by the following system of kinetic equations:

$$\begin{aligned}
&\frac{\partial n_1}{\partial t} = -F(t,\vec{r})\cdot(\sigma_D+\sigma_A)\cdot n_1 + \frac{n_2}{\tau_D} + \frac{n_3}{\tau_A} \\
&\frac{\partial n_2}{\partial t} = -F(t,\vec{r})\cdot\sigma_A\cdot n_2 - \frac{n_2}{\tau_D} - K_{DA}\cdot n_2 + F(t,\vec{r})\cdot\sigma_D\cdot n_1 + \frac{n_4}{\tau_A} \\
&\frac{\partial n_3}{\partial t} = -F(t,\vec{r})\cdot\sigma_D\cdot n_3 - \frac{n_3}{\tau_A} + F(t,\vec{r})\cdot\sigma_A\cdot n_1 + \frac{n_4}{\tau_D} + K_{DA}\cdot n_2 + K_{SS}\cdot n_4 \\
&\frac{\partial n_4}{\partial t} = F(t,\vec{r})\cdot\sigma_A\cdot n_2 + F(t,\vec{r})\cdot\sigma_D\cdot n_3 - \frac{n_4}{\tau_D} - \frac{n_4}{\tau_A} - K_{SS}\cdot n_4 \\
&n_1+n_2+n_3+n_4 = n_0,
\end{aligned} \tag{1b}$$

where τ_D, τ_A and σ_D, σ_A are the lifetime and absorption cross section of the donor (denoted by D) and acceptor (denoted by A) as it defined above; K_{DA} is the rate of the energy transfer from the excited donor to the unexcited acceptor; K_{SS} is the rate of energy transfer from the excited donor to the excited acceptor (singlet–singlet annihilation (Fadeev et al., 1999); $F(t,\vec{r})$ is the photon flux density (see Eqs. (1a)); and n_0 is the total concentration of molecules containing a LDA pair. In this model, the following photophysical parameters are presented: the absorption cross section and the excited state lifetime of the donor and acceptor, the energy transfer rates K_{DA} and K_{SS}.

By solving systems (1a) and (1b) numerically, one can find the concentration of the fluorescent molecules in the excited state and calculate the number of fluorescence photons N_{Fl}, emitted from the volume V after the action of the laser pulse (Filipova et al., 2001; Fadeev et al., 1999). The theoretical saturation curve for the model (1a) can be calculated from following equation:

$$N_{Fl}(\lambda) = K_{31} \cdot \int_V d\vec{r} \int_{-\infty}^{+\infty} n_3(t,\vec{r})dt \tag{2a}$$

For the model (1b), for the donor (2b) and the acceptor (2c) curves, respectively:

$$N_{Fl}^{D}(\lambda) = \tau_D^{-1} \cdot \eta_D \cdot \int_V d\vec{r} \int_{-\infty}^{+\infty} (n_2(t,\vec{r}) + n_4(t,\vec{r}))dt \tag{2b}$$

$$N_{Fl}^{A}(\lambda) = \tau_A^{-1} \cdot \eta_A \cdot \int_V d\vec{r} \int_{-\infty}^{+\infty} (n_3(t,\vec{r}) + n_4(t,\vec{r}))\, dt \tag{2c}$$

where η_D and η_A are the fluorescence quantum yield, which is defined as the ratio of the radiation decay rate of S_1 state to the sum of all rates of S_1 state decay (i.e. $\eta \equiv K_{31}/K_3$), of the donor and acceptor, correspondingly; λ is the fluorescence registration wavelength. Other symbols are defined in Eqs. (1a).

In considered model (1a), the fluorescence saturation is caused by a finite lifetime τ and by intercombination conversion. In model (1b), due to the finite fluorescence lifetime and due to the saturation of the energy transfer channels. Let us note that the model (1b) could be also supplemented with the intersystem crossing mechanisms, but preliminary experiments showed that at the given parameters of the laser radiation the process for albumins and mRFP1 is small compared to the mechanisms under study and contributes little to fluorescence saturation. Therefore, this mechanism has been excluded to increase the stability of the inverse problem solution (details and mathematical basement of inverse problem solution of nonlinear laser fluorimetry can be found elsewhere (Boychuk at al., 2000). For the same reason the induced processes from the excited states (two-photon absorption or photoizomerization, etc) have been excluded.

As was mention above, the photophysical parameters of fluorophores (σ, K'_{32} and τ in the model (1a) and τ_D, τ_A, σ_D, σ_A, K_{DA} and K_{SS} in the model (1b)) can be determined from the dependence $N_{Fl}(F)$, by solving the inverse problem. However, in experiments, it is convenient to normalize the number of detected fluorescence photons N_{Fl} to the reference signal (will denote as N_{Ref}), which can represent a part of exciting radiation directed to the reference channel of the detection system by a beamsplitter or a Raman scattering signal from water molecules (Fadeev et al., 1999). In this case, one has to deal with the dependence $[\Phi(F)]^{-1}=N_{Ref}/N_{Fl}$ (which is also called a saturation curve, $\Phi(F)$ is the fluorescence parameter) rather than $N_{Fl}(F)$. According to the practical experience such normalization also helps to increase the stability of the inverse problem solution. In the absence of saturation, Φ stops being dependent on photon flux density F and tends to a constant which is denoted by Φ_0 ($\Phi_0 \equiv lim|_{F\to 0}[\Phi(F)]$). If in measurements of Φ_0 the Raman scattering band of water molecules as a reference signal is used, then it is possible to find the fluorescence quantum yield of a complex organic compound (Filipova et al., 2001).

For the reasons pointed out in (Banishev et al., 2008a), the same laser fluorimeter has been optimized for measuring the nanosecond fluorescence decay (the kinetic mode of the fluorimeter operation). The curve represents the dependence of the number $N_{Fl}(t_{del})$ of fluorescence photons in the detector gate (with wide t_g) on the gate delay time t_{del} with respect to a laser pulse. For the model (1a) an expression for kinetic curve can be written as:

$$N_{Fl}(t_{del}) = K_{31} \cdot \int_V d\vec{r} \int_{-t_g/2+t_{del}}^{t_g/2+t_{del}} n_3(t,\vec{r})dt \quad (3)$$

The t_{del} changes discretely and proportionally to the detector gate step: $t_{del}=i \times t_{step}$, where i is a number of the detector gate step. Similar expressions can be written for (2b, c). Gate position at which its centre coincided with the laser pulse maximum was taken as the zero delay (t_{del}=0). This was detected by the maximum of water Raman line (Banishev et al., 2006).

By solving the inverse problem, the fluorescence lifetime τ of a fluorophore can be determined independently from the dependence $N_{Fl}(t_{del})$. In the experiment the fluorescence signal is measured in relative units. For comparison of the experimental data with the theoretical ones it is necessary to normalize the obtained experimental curve to the fluorescence intensity at some fixed time delay. This procedure, the fluorimeter capabilities in the kinetic mode and the corresponding theory can be found elsewhere (Banishev et al., 2006). The difference of such variant of kinetic fluorimetry from the conventional time-resolved fluorimetry is that the fluorescence is excited by a pulse with rather long duration (~10 ns), and for fluorescence registration an optical gated multichannel analyser is used. Whereas, the conventional time-resolved fluorimetry (Lakowicz, 1999) is based on the analysis of fluorescence decay curves after the excitation pulse, whose duration is much shorter than the lifetime of a fluorophore in the excitation state (picoseconds).

For determination of the photophysical parameters from experimental curves (solution of the inverse problem) the variation algorithm (Banishev et al., 2008a) was used. It is based on the procedure of minimizing the functional of the residue between the experimental curves and curves calculated from models (1a) or (1b) by varying the photophysical parameters (Boychuk at al., 2000).

It is necessary to point out two distinctive features of nonlinear laser fluorimetry: (i) as the method implies detection of fluorescence photons (see Eqs. (2)), the photophysical parameters derived from the saturation curve relate only to a fluorescent molecule of COC; (ii) information on the concentration of fluorescent molecules is not used in deriving the photophysical parameters from the saturation curve (Banishev et al., 2009; Filipova et al., 2001). Thus, the method allows one to determine individual photophysical parameters of a molecule in the case when the following complex situation takes place: (i) the sample under study is a multicomponent ensemble of molecules, the absorption bands of its subensembles overlapping (i.e. when the sample is excited, all the subensembles absorb light); (ii) the concentrations of molecules from the subensembles in the mixture are *a priori* unknown.

3. Experimental

3.1 Nanosecond laser fluorescence spectrometer

A homemade laser fluorimeter (Fig. 3) has been built for the experiments. The fluorimeter consist of a laser source, optical elements for light conversion, a fiber-optic cable, a cuvette and an optical multichannel analyser.

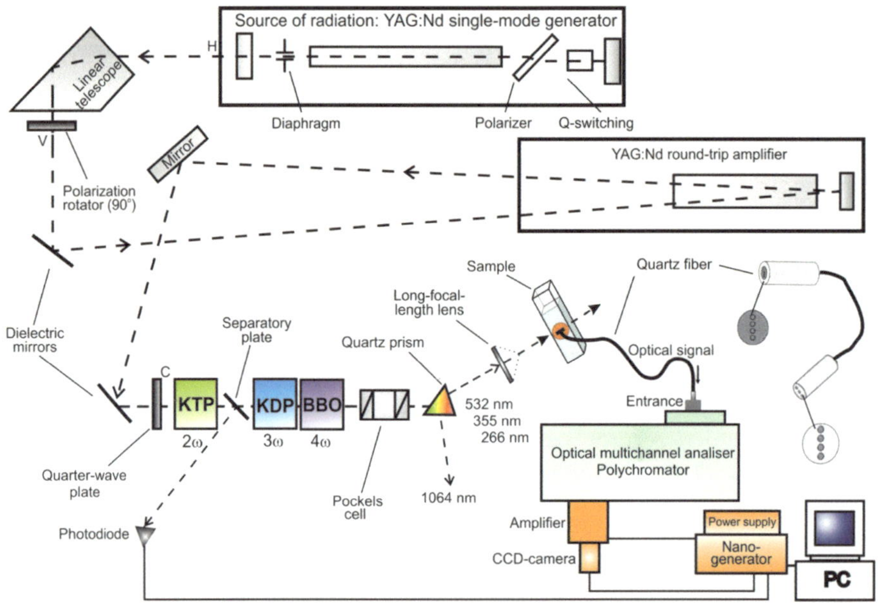

Fig. 3. Nanosecond laser fluorescence spectrometer scheme (for details see text). The letters H, V and C denote the horizontal, vertical and circular light polarization.

3.1.1 Laser excitation system

The pulsed single mode Nd:YAG laser (the fundamental wavelength is 1064 nm) with the set of nonlinear crystals for generating the 2nd (532 nm), the 3d (355 nm) and the 4th (266 nm) harmonics of the fundamental radiation was used. At a pulse repetition frequency of 10 Hz, the average pulse energy was: 5 mJ (532 nm), 2 mJ (355 nm) and 0.7 mJ (266 nm). The 2-mm-diameter Nd:YAG crystal with the diaphragm (the diameter is 1.5 mm) on the outlet face was used to obtain the light generation in a single mode regime (single transverse mode). Q-switching was made by the electro-optical shutter, working based on the Pockels effect. Between the Nd:YAG crystal and the shutter, the polarizer, transmitting light only with horizontal polarization, was placed. Thus, at the generator output, we had a single mode beam with horizontal polarization. After the generator, the light beam passed through the linear telescope, which also worked as a rotating prism. The main purpose of this component was to decrease the beam divergence in the horizontal plane, which is necessary for effective conversion of the fundamental frequency to its harmonics. Then, the laser beam was sent to the 90 degree polarization rotator, which changed the horizontal polarization to vertical one (required for reducing energy losses on rotating mirrors). Passed through the rotator and reflected from the rotating mirror beam was directed through the round-trip amplifier, which was composed of the Nd:YAG crystal (5 mm in diameter) and the rear mirror. Then the beam arrived at the polarizer, which reflects only vertical polarization. The reflected beam was directed through the quarter-wave plate for changing the vertical polarization to the circular one. That improves the efficiency of frequency-doubling in a KTP crystal. The KDP and BBO nonlinear crystals were used for generating the 3d and 4th harmonics. After the frequency conversion, the radiation of the 4th (or the 3d), 2nd and the fundamental harmonics was transmitted through the quartz prism for their spatial divergence in the horizontal plane. The continuous adjustment of the laser intensity during the saturation curves measurement was carried out by the Pockels cell, which was placed in the beam way right after the KTP crystal. The cell consists of the electro-optical component (DKDP crystal) and two Glan prisms. Changing the voltage on the DKDP, one can adjust the radiation output intensity at the outlet from the Pockels cell.

3.1.2 Registration system

The laser radiation was focused on the cuvette with the sample solution by a long-focal-length lens (the focal length is 20 cm). For collecting the fluorescence photons, the light-guide cable (the length is 5 m), consisting of seven quartz fibers (the diameter of each fiber is 600 micron), which were laid out in a row (like a slit) at both ends, was used. The cable inlet was fixed at the cuvette side, and the outlet was clasped to the polychromator entrance slit; the sample solution was excited by transmitted laser radiation. As a detector of radiation, the optical multichannel analyser (OMA) was used. The optical chamber (DeltaTech, Scientific Park of MSU) of the analyser consists of the electro-optical converter based on a gated microchannel plate (MCP), CCD matrix, and optical device for transferring an image from the MCP to the CCD matrix (a pixel size is $11\times11\mu m^2$). The chamber was fixed to the polychromator (MUM without the output slit, reciprocal linear dispersion is 0.15 nm per channel) optical output. The multichannel analyser was connected to the PC. As a result, optical image in the polychromator output slit plane could be obtained as a 2D picture on the PC monitor. The software installed on the PC allowed the OMA to operate both in the continuous and gated modes. When working in gated mode, a part of light was sent to the silica photodiode PD-265 (the building-up time of the leading edge is less than 2 ns). The photodiode was connected to the nanogenerator triggering inlet (the trigger level is 0.6 V). The gating of the MCP was implemented by high-voltage pulses from the nanogenerator (the amplitude is 800 V). The detector gate delay time could be adjusted through the nanogenerator over the range of 50 ns (the dead time of the detector) to 1200 ns with the step t_{step}=2.5 ns. Exactly because of the dead time of the registration system, when it operates in the gated mode (i.e. in the case of kinetic curves measurements), the light-guide cable as the dead time compensator (the optical delay line) was used. The detector gate width t_g could be varied from 10 to 1200 ns; in our experiments it was set to 10 ns.

3.1.3 Laser radiation parameters

The laser pulse duration at the wavelengths of 532 and 266 nm were fitted well by Gaussian function with the full width at half maximum (t_p) of 12 and 10 ns, respectively. When measuring (i) the kinetic curves, the laser beam diameter was 3 mm and (ii) the saturation curves, the beam was focused to a spot with a diameter from 600 to 800 μm, depending on the protein under investigation. The values of the photon flux density F were determined according to the equation $F^{-1}=E^{-1}\hbar\omega\cdot S\cdot t_p$, where $\hbar\omega$ is a photon energy, E is average pulse energy, S is cross section area of a laser beam, t_p is a pulse duration. The photon flux density F_{max} at a maximum of the saturation curve was measured before each experiment. Then, the photon flux density was gradually decreased. The F value at each point was obtained by the value of the Raman scattering signal from water molecules N_{RS}: $F=F_{max}\cdot N_{RS}/N^{max}_{RS}$, where N^{max}_{RS} is the Raman scattering signal at $F=F_{max}$. In nonlinear laser fluorimetry experiments, the fluorescence saturation curves were measured in the following ranges of F: (i) $5\times10^{23}\div7\times10^{25}$ $s^{-1}cm^{-2}$ when the excitation wavelength was 532 nm and (ii) $2\times10^{24}\div5\times10^{25}$ $s^{-1}cm^{-2}$ when excited at the wavelength of 266 nm.

3.2 Picosecond time-resolved and steady-state spectral measurements

The fluorescence lifetime measurements were performed with a streak camera (Agat SF 3M, VNIIOFI, Russia). A Nd:YAG laser with excitation wavelengths of 532 and 266 nm (the second and fourth harmonic of fundamental radiation) was used as a light source. The laser radiation parameters of the fluorimeter were as follows (for 532 nm): pulse energy 160 μJ, duration 20 ps (fwhm), beam diameter 5 mm. The error in determining the fluorescence lifetimes in time intervals of several nanoseconds did not exceed 5 %. In addition to the laser equipment, the Cary 100 spectrophotometer (Varian, Inc., USA) and the Cary Eclipse spectrofluorimeter (Varian, Inc., USA; slits width was 5 nm) were used for optical density measurements and fluorescence registration, respectively.

4. The object

4.1 Albumins

In this work, the solutions of human serum albumin (HSA) (>96%, Sigma) and of bovine serum albumin (BSA) (>98%, MP Biomedicals) in a phosphate buffer (0.01 M, pH 7.4) have been used. The proteins concentrations were 10^{-5} (absorption spectra measurement) and 10^{-9} M (fluorescence measurement at the nanosecond laser fluorimeter). All of the experiments were performed at a temperature of 25±1 ^{0}C. The structure and biological functions of HSA and BSA can be found in (Peters, 1996). Tryptophan, tyrosine, and phenylalanine (with relative contents of 1:18:31 in HSA and 2:20:27 in BSA) are the absorption groups in these proteins (as in many other natural proteins). The tyrosine fluorescence in HSA and BSA (as in many other natural proteins) is quenched due to the effect of adjacent peptide bonds, polar groups (such as CO, NH_2), and other factors, and phenylalanine has a low fluorescence quantum yield (0.03) (Permyakov, 1992). Therefore, the fluorescence signal in these proteins is determined mainly by tryptophan groups. In that case the fluorescence, registered in nonlinear and kinetic laser fluorimetry measurements, correspond to tryptophan residues (this fact will be used in Section 6.1).

As described in (Peters, 1996), HSA and BSA have similar structure and amino acid sequences that differ insignificantly by location of some certain amino acids. However, HSA contains one tryptophan residue in the protein matrix (Trp-214), and BSA contains two residues (Trp-212 and Trp-134). Trp-212 in BSA and Trp-214 in HSA have a similar microenvironment and, hence, their spectral properties are similar (Eftink et al., 1977). Tryptophans of BSA are not spectrally identical due to the stronger integration of Trp-212 into the protein's structure and the more hydrophobic environment of Trp-212 in comparison with Trp-134. The distance between tryptophans in BSA is about 3.5 nm. This fact makes the intramolecular energy transfer between them using the Forster resonance energy transfer (FRET (Valeur, 2002)) mechanism possible.

4.2 Red fluorescent proteins

Fluorescent proteins are a class of proteins that have a distinguishing property of forming their chromophore without involvement of any additional cofactors and ferments (autocatalytic reaction), except for molecular oxygen. In recent years, FPs have gained enormous popularity as genetically encoded fluorescence markers that enable to visualize a broad range of biological processes in cells and tissues. The most popular for practical applications are FPs whose fluorescence is shifted to the red (red FPs) and whose molecules are monomers (Piatkevich et al, 2010). The mRFP1 protein possesses these properties (Cambel et al., 2002), which made it an object of the research.

In this work the red FP mRFP1 and its mutants at 66 amino acid residue (glutamine 66) have been used. The seven mutants with substitution of the glutamine 66 for the serine (protein mRFP1/Q66S[2]), cysteine (mRFP1/Q66C), asparagine (mRFP1/Q66N), histidine (mRFP1/Q66H), alanine (mRFP1/Q66A), leucine (mRFP1/Q66L) and phenylalanine (mRFP1/Q66F) have been created. The method for fabrication and purification of proteins is described in (Vrzheshch et al., 2008). All the experiments were performed in a 0.06 M phosphate buffer at a temperature of 25±1 °C. The Bradford method (McCluskey, 2003) was used to determine initial concentration of the proteins. The proteins concentration was: (1) 10^{-9} M when measured fluorescence saturation curves; (2) $5.5\times10^{-10} \div 3\times10^{-10}$ M (depending on protein) when measured fluorescence quantum yield; (3) 10^{-6} M when measured steady-state spectra and fluorescence life-times on a picoseconds laser fluorimeter.

The formation of a fluorescent molecule of red FPs is a complicated process (usually called maturation) consisting of several stages (Verkhusha et al., 2004; Strack et al., 2010). At some stages, intermediate protein forms are produced, which remain in the resultant specimen (solution) of protein. In other words, the solution of red FPs is an ensemble of protein molecules consisting of several chemically non-equivalent subensembles (a mixture of different spectral forms of FPs). At neutral pH, the solution of mRFP1-like proteins contains three different spectral forms, namely, the mature form (will denote below as R form), the immature form with protonated chromophore (GH form) and the immature form with deprotonated chromophore (G form). The detailed information about formation of the chromophore of red FPs and analysis of composition of the samples can be found elsewhere (Verkhusha et al., 2004; Strack et al., 2010). According to a widely used terminology (Verkhusha et al., 2004), I'll also call the mature form as a red form and the immature forms as a green form in this work. The spectral forms can be detected in an absorption spectrum as the bands with maxima in the blue (360-420 nm, GH form), green (450-540 nm, G form) and red (550-600 nm, R form) spectral ranges. It is known (Pakhomov et al., 2004) that the balance between GH and G forms of FP can be disturbed with the changing of the solution acidity or as a result of light (from the UV, blue, or green spectral ranges) influence on the

[2] In this work, the name of each mRFP1 mutant is designated like mRFP1/Q66#, where Q66 is the abbreviation of the glutamine (single letter code is Q) at position 66, which is substituted. In place of the symbol #, the single letter code of amino acid the glutamine is substituted for is written down. The list of the amino acids abbreviation can be found elsewhere (Zamyatin, 1972).

protein. Irradiation of the protein molecules by the light from the green spectral range provokes a conversion of G form molecules to the molecules of GH form; as a result, the protein solution will represent a mixture of only two forms, namely, GH and R. In current work, for conversion the radiation of an Ar laser (LG-106-M1) at 488 nm (300mW) was used. An increment of the protein solution pH to 9 leads to a practically total transfer of the GH form molecules to the G form molecules (in that case, the protein solution is a mixture of the G and R forms). These procedures will be used as a way to decrease the amount of simultaneously existing forms in the mRFP1 solution when the method of the nonlinear fluorimetry will be implemented (see Section 6).

5. Steady-state fluorescence and absorption spectroscopy

5.1 Human and bovine serum albumins

The absorption spectra of the proteins and (for comparison) the corresponding equimolar solutions (solution of tryptophan, tyrosine, and phenylalanine at the same ratio as they are contained in protein) are shown in Fig. 4. It is seen that the parameters of the absorption bands of proteins do not coincide with the corresponding parameters for the equimolar solutions.

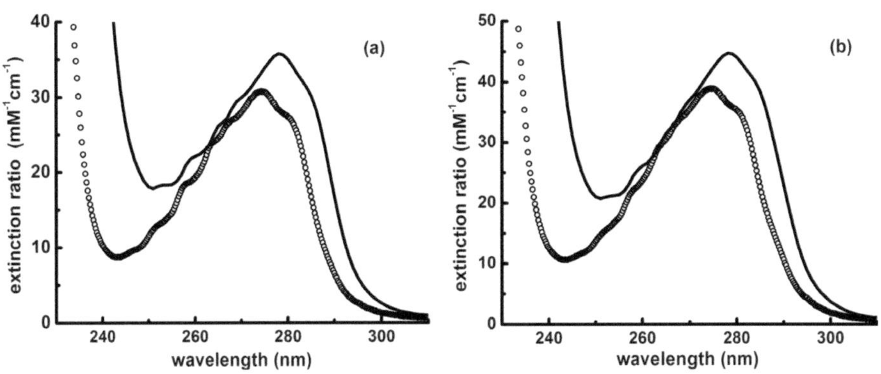

Fig. 4. Solid line is an absorption spectrum of (a) human and (b) bovine serum albumin. Circles are absorption spectrum of an equimolar solution with relative contents of tryptophan, tyrosine, and phenylalanine as (a) 1:18:31 and (b) 2:20:27.

The measurements of the fluorescence emission spectra of the proteins (data not showed) revealed that the fluorescence of both proteins is blue shifted relative to the tryptophan fluorescence (353 nm) in a buffer solution (Banishev et al., 2008a). This is due to a decrease in the tryptophan environment polarity in the proteins. The maximum of the HSA fluorescence (332 nm) is blue shifted in comparison with BSA (342 nm). Since the fluorescence spectrum of the tryptophan residues reflects the polarity of their nearest environment, and since the properties of the environments of Trp-212 in BSA and Trp-214 in HSA are similar (Eftink et al., 1977), such a shift can be related to the fact that BSA contains tryptophan Trp-134 located in the environment with a higher polarity (in comparison with Trp-212). Thus, the total fluorescence spectrum of BSA is red shifted. This result will be necessary for choosing the registration wavelength in measuring the acceptor and donor fluorescence when the nonlinear and kinetic curves will be measured (Section 6.1).

5.2 Monomeric red fluorescent protein and its mutants at residue 66

One can see from the absorption spectra of the mRFP1 (Fig. 5(a)) that in the wavelength range from 370 to 650 nm, there exist three absorption bands, which are explained by the presence of three spectral forms in the solution (Verkhusha et al., 2004), i.e. R, G and GH forms (the corresponding absorption maxima at 584, 503 and 380 nm). One can also see that the absorption bands of G and R forms overlap. Excitation of fluorescence in the absorption band of each form indicates that G and GH forms do not fluoresce, the R form fluoresces with maxima at 607 nm. The excitation of the protein solution by irradiation at a wavelength of 270 nm (tryptophan absorption band) leads to the appearance in the signal spectrum not only of an UV band (maximum at 330 nm), which corresponds to the tryptophan fluorescence in the protein matrix, but also of a band in the visible region of wavelengths (maximum at 607 nm) corresponding to fluorescence of the chromophore mRFP1 R form (Fig. 5(b)). The chromophore of GH and G forms is non-fluorescent (Campbell et al., 2002).

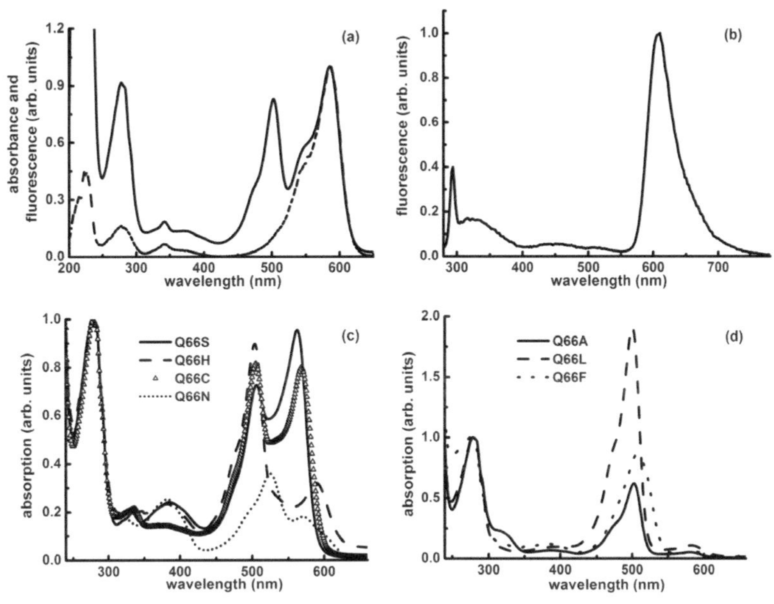

Fig. 5. (a) Absorption (solid) and fluorescence excitation (dotted, registration at 607 nm) spectra of mRFP1. The spectra are normalized to the signal at 584 nm. (b) Emission of mRFP1 (excitation by 270 nm). (c) Absorption spectra of the mutants with polar and (d) non-polar substitutions at position 66. The spectra are normalized to optical density at 270 nm.

Absorption spectra of the variants of mRFP1 with mutation at position 66 are shown in the Fig. 5(c, d).The absorption spectra structure of the mutants is similar to one for mRFP1. One can see that for all proteins in the wavelength range from 370 to 650 nm, the spectrum contains three bands with the maxima at 360-420 nm, 450-540 nm and 550-600 nm depending on the protein. These absorption peaks are corresponding to the GH, G and R forms. There only signal in the red spectral range were observed in the fluorescence emission and excitation spectra (i.e. the G form was found to be non-fluorescent) for all proteins except mRFP1/Q66F. For mRFP1/Q66F, the additional band in fluorescence excitation (maxima at 502 nm) and emission (maxima at 512 nm) spectra was detected, which attributed to the chromophore of G form. The spectral characteristics of the proteins are presented in the Table 1.

Note that the presence of the three protein forms can be qualitatively seen in the absorption spectrum (Fig. 5). However, as is described in (Banishev et al., 2009), the quantitative determination of the individual photophysical parameters of their chromophore with the help of only conventional methods are problematic. This is explained by the fact that the preparative separation of the forms is rather difficult and, as a result, it is hard to find their partial concentrations. At that rate, for example, for calculating the molar extinction coefficient (or absorption cross section) of chromophore from absorption spectra, the total protein concentration (total concentration of all forms) is used. As a result, the extinction coefficients are artificially underestimated (Kredel at al., 2008). At present, the only method that used for determining the individual extinction coefficient of chromophore of each spectral form can be found in (Ward, 2005). However, as it was pointed out in (Kredel et al., 2008), the values measured by the method are inaccurate in case of red FPs. Although, the procedure which enables to reduce the experimental errors has been proposed by (Kredel et al., 2008), the problem remains still topical.

Protein	Residue at 66	Green (G) form	Red form		
		$\lambda^{max}{}_{abs}$	$\lambda^{max}{}_{abs}$	$\lambda^{max}{}_{ex}$	$\lambda^{max}{}_{em}$
mRFP1	Glutamine (Gln)	503	584	584	607
mRFP1/Q66S	Serine (Ser)	506	562	561	579
mRFP1/Q66C	Cysteine (Cys)	505	568	568	588
mRFP1/Q66N	Asparagine (Asn)	525	570	570	604
mRFP1/Q66H	Histidine (His)	504	588	588	618
mRFP1/Q66A	Alanine (Ala)	500	582	578	605
mRFP1/Q66L	Leucine (Leu)	500	582	577	613
mRFP1/Q66F	Phenylalanine(Phe)	507	591	595	624

Table 1. The position (nm) of the maximum of the absorption, fluorescence excitation and emission spectra of the proteins.

6. Nonlinear fluorescence spectroscopy of proteins

6.1 Determination of photophysical parameters of single- and double-tryptophan-containing proteins

The emission spectra at several values of photon flux density *F* are shown in Fig. 6(a, b). In the figures, the band with a maximum value of *F* (4×10^{25} $cm^{-2}s^{-1}$) corresponds to the maximum value of the Raman scattering signal from water molecules. One can see that the maximum of the human serum albumin (HSA) fluorescence band does not change its position when *F* is changed. This is due to the fact that HSA contains one saturating fluorophore. However, the bovine serum albumin (BSA) fluorescence band is blue shifted (from 340 to 335 nm) when *F* is increased, owing to the fact that BSA contains two fluorophores in different environments (therefore, with different spectral properties), which exhibit different degrees (factors) of saturation. Taking into account the blue shift of the HSA fluorescence spectrum (in comparison with the BSA fluorescence spectrum) and the similarity of the properties of Trp-214 in HSA and Trp-212 in BSA (see Section 5.1), one can assume that, in the system of two tryptophans of BSA, Trp-212 serves as the donor of the energy (the energy transfer occurs via Forster mechanism), and Trp-214 is the acceptor (i.e., its fluorescence spectrum is presumably shifted towards long wavelengths).

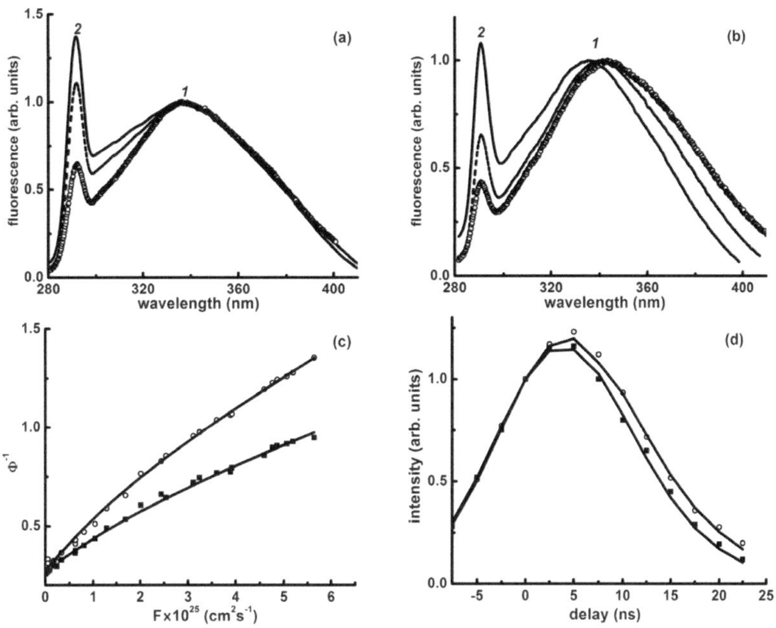

Fig. 6. (a) HSA and (b) BSA emission spectra at several values of photon flux density (see text); (1) is fluorescence and (2) is water Raman scattering band. (c) Saturation and (d) kinetic curves for BSA; fluorescence was registered at 390 (squares) and 310 (circles) nm. Lines are plotted using model (1b) and Eqs. (2b, c) for parameters from the Table 2.

The kinetic curves (see Eq. 6) and the fluorescence saturation curves of BSA are shown in Fig. 6(c, d). For BSA, the saturation curves depend on the registration wavelength in the wavelength range 310–390 nm. This is due to the fact that the BSA fluorescence band is a superposition of the bands of two tryptophans possessing different spectral properties. A similar difference in the curves for HSA is negligible. For the determination of the photophysical parameters of HSA fluorophore from fluorescence saturation and kinetic curves, the model (1a) and Eq. (2a) have been used. The same calculation procedure was done for BSA experimental curves, but with using the model (1b) and Eqs. (2b,c); the fluorescence signal was measured at 310 nm (when registered the fluorescence saturation and kinetic curves of the donor) and 390 nm (for similar curves of the acceptor). The resulting values of the parameters of protein fluorophores are presented in the Table 2.

As one can see from the Table 2 that the values of photophysical parameters σ and τ of Trp-214 in HSA and Trp-212 in BSA are similar. This result should have been expected based on a comparison of the structures of these proteins. The rates of energy transfer in BSA from excited donor to unexcited acceptor (K_{DA}) and to excited acceptor (K_{SS}) are small in comparison with the rate of intramolecular relaxation (τ^{-1}). This can be due to following reasons: (i) in BSA, the tryptophan residues in the D–A pair are located at a distance that is insufficient for a noticeable energy transfer between them. According to the data on the BSA structure (Peters, 1996), the distance between two tryptophans in the molecule is about 3.5 nm. For comparison, the Forster radius for the energy transfer between free tryptophans ranges from 0.6 to 1.2 nm (depending on the solvent). (ii) perhaps, the mutual orientation of the transition dipoles of fluorophores impedes the energy transfer (Lakowicz, 1999).

Protein	Parameters	Tryptophan residues	
		Trp-214	
HSA	τ, ns $\sigma(\lambda_{ex}=266)$, cm^2 K'_{32}, s^{-1}	4.5 ±0.5 (1.3±0.1)×10^{-17} <10^{-7}	
		Trp-134 (donor)	*Trp-212 (acceptor)*
BSA	τ, ns $\sigma(\lambda_{ex}=266)$, cm^2 K_{DA}, s^{-1} K_{SS}, s^{-1}	6.2±0.5 (3±0.3)×10^{-17} <10^{-7} <10^{-7}	5±0.5 (1±0.1)×10^{-17} <10^{-7} <10^{-7}

Table 2. Photophysical parameters of fluorophores (tryptophan residues) in HSA and BSA.

It is significant that the values of the absorption cross section determined using the method of nonlinear fluorimetry are true values for fluorophores and they are obtained without *a priori* information about the contribution of other groups into absorption at a specific wavelength and about the concentration of fluorophores. This is a unique feature of nonlinear fluorimetry. As mentioned above, there are three absorption groups in proteins (tryptophan, tyrosine and phenylalanine) and the absorption spectra of these three amino acids are overlapping. In that situation it is hard to separate the contribution of each amino acid group from total protein absorption without any preparative action on a molecule. For that reason, the absorption cross section of the tryptophan residue in protein is assumed to be equal to the absorption cross section of free tryptophan in solution (Pace et al., 1995), because it is supposed that this parameter is weakly dependent on the environment. However, a comparison of the equimolar solution and protein solution absorption spectra (Fig. 4) shows that these spectra do not coincide for HSA and BSA. Therefore, it is clear that such an assumption is just estimation and for diagnostics of the state of a protein it is necessary to be able to determine the true photophysical parameters of tryptophan merged in protein matrix. That has been done in this Section. The absorption cross section for tryptophan in aqueous solution is equal to 1.6×10^{-17} cm^2 (Banishev et al., 2008a) (at 266 nm); now, it can be compared with the true values of the absorption cross section of tryptophan residues in native proteins (see Table 2). The abilities of the approach are not limited by albumins.

6.2 Determination of photophysical parameters of mRFP1 protein under UV excitation

According to three-dimensional structure (PDB ID1g7k) of the mRFP1, there is a tryptophan at a distance of about 15Å from the protein chromophore, which could be a potential partner (the donor) for inductive FRET to the protein chromophore. Thereby, the tryptophan and the chromophore form a LDA pair inside a molecule of the FP.

As a preliminary step in determination of the full set of the mRFP1 photopysical parameters, the analysis of fluorescence decay at picoseconds excitation has been done. The excitation wavelength was 266 nm in order to match the acceptor (tryptophan) absorption band. The acceptor fluorescence decay under excitation of an ensemble of donor-acceptor pairs by δ-pulse is described (Valeur, 2002) as:

$$I_A(t) \sim B \cdot \exp(-t/\tau_A) - A \cdot \exp(-t/\tau_{D+A}) \tag{4}$$

where, $B=A+[A^*]_0$, $[A^*]_0$ is the excited acceptor molecules concentration at a time point $t=0$ (immediately after the excitation pulse is over); $A=[D^*]_0 \cdot K_{DA}/(1/\tau_{D+A}-1/\tau_A)$; $[D^*]_0$ is the excited donor molecules concentration at a time point $t=0$; $\tau_{D+A}=(1/\tau_D+K_{DA})^{-1}$ is the fluorescence lifetime of the donor in the presence of the acceptor; other designations are given above.

The fluorescence decay signal was measured at picoseconds time-resolved fluorimeter described in Section 3.2. The excitation wavelength was 266 nm and the signal registration was done in the range of the R form chromophore fluorescence (the chromophores of others forms are non-fluorescent). The experimental time dependence was fitted by function (4), as a result, the lifetimes τ_A, τ_{D+A}, and the partial contributions of B and A components of the fluorescence decay curve were obtained. The K_{DA} cannot be determined only from the fluorescence decay curve, because in this case it would be necessary to remove the acceptor and measure the fluorescence lifetime of the donor (in the absence of the acceptor). As it will be shown below, the nonlinear fluorimetry method makes it possible to resolve this problem for the native protein without any preparative action. The processing of the fluorescence decay curve of the mRFP1 allowed us to determine the lifetime values, τ_A=1.6 ns and τ_{D+A}=0.24 ns. The values of B and A in (4) were found to be practically equal. It is indicative of the absence of the direct excitation of the acceptor; therefore, in the equation system (1b) one can assume σ_A=0 (under excitation at 266 nm) and for this reason, the acceptor fluorescence (under this wavelength excitation) is a result of the energy transfer from the tryptophan to the chromophore.

The next step was to measure and analyse the fluorescence saturation curves with excitation at the wavelength of 266 nm. To improve the stability of the inverse problem solution, the process of singlet-singlet annihilation from the model (1b) was excluded. At pH 7.4 (the acidity at which the protein was initially produced), the mRFP1 solution is represented as a sum of three subensembles of molecules, each having its own set of photophysical parameters, and the dynamics of the populations of the collective states is described by its own system of equations similar to (1b). The number of the photons of the tryptophan fluorescence is calculated from Eq. (2b), where the sum under the integral is the population of the collective states n_2 and n_4 for all three forms. The fluorescence of the R form is calculated from Eq. (2c), where the populations of the collective states n_3 and n_4 of this form are present. It is difficult to resolve the inverse problem of nonlinear fluorimetry in such a situation, because the number of unknown parameters is too large. But, as was mentioned in Section 4.2, there are the techniques that allow reducing the number of the simultaneously present forms to two. In that case if the portion of the concentration of the R form is c_R and that of the second form (for example, G) is c_G, then the numbers of the fluorescence photons of the donor and acceptor (from the unit of volume) are:

$$N_{Fl}^{D}(\lambda)=\tau_D^{-1}\cdot\eta_D\cdot\int_{-\infty}^{+\infty}[c_R\cdot(n_2{}^R(t,r)+n_4{}^R(t,r)\ )+c_G\cdot(n_2{}^G(t,r)\ +n_4{}^G(t,r)\ )]dt \qquad (5a)$$

$$N_{Fl}^{A}(\lambda)=\tau_A^{-1}\cdot\eta_A\cdot\int_{-\infty}^{+\infty}c_R[n_3{}^R(t,r)+n_4{}^R(t,r)]\ dt \qquad (5b)$$

where the symbols R and G denote the R and G form molecules; n_2, n_3, and n_4 are the collective states (see (1b)); η_D and η_A are the fluorescence quantum yields (see (2b,c)); and c is the relative concentration of the forms in total molecules concentration. The amount of the R form molecules c_R can be found from the algorithm that described in the next Section.

In this connection, the following procedure has been realized:

a. The value of the protein solution pH was set near 9; in this situation, the protein solution contains only protein molecules of the R and G forms. After that, two saturation curves were taken under excitation at the wavelength of 266 nm and fluorescence registration at 330nm (the fluorescence saturation curve of the donor, i.e., the molecules of the tryptophan contained in the protein matrix of the R and G forms) and at 607nm (the fluorescence saturation curve of the acceptor, i.e., the chromophore only in the protein molecules of the R form). Resolving the inverse problem, in which fluorescence response formation is described by two systems of the kind of (1b) for the populations of the collective states of the LDA pairs in macromolecules of R and G forms, and the values of the parameters τ_{D+A}, τ_A, σ_A=0 for the R form of the protein are considered to be known (see above), the values of K_{DA} (for the R and G forms) and τ_A (for the G form) have been determined.
b. The protein sample was irradiated with the Ar laser at a wavelength of 488 nm (the pH value is near to a neutral one) and made only the R and GH forms present in the solution. After that, I used the same procedure of measurement and calculation of the saturation curves and determined the following parameters: K_{DA} (for R and GH forms of protein) and τ_A (for the GH form of the protein). The values of the K_{DA} for the R form in both cases coincided in error limits of the experiment.

The experimental dependences $\Phi^{-1}{}_D(F)$ and $\Phi^{-1}{}_A(F)$ for case (a) can be found in (Shirshin et al., 2009). When measured the saturation curves $\Phi^{-1}{}_D(F)$ and $\Phi^{-1}{}_A(F)$, the intensity in the spectra of the first and second orders of the RS valence band of water molecules (the wavelengths are 291 and 582 nm, correspondingly) were used as a reference signal; the excitation was at 266 nm. Having performed this procedure, the photophysical parameters of FP mRFP1 under UV excitation (266 nm) have been determined (see Table 3).

Let us discuss the main results of this Section. First of all, it is interesting to compare the true value of the absorption cross section obtained for tryptophan in the FP mRFP1 with the values for tryptophan in an aqueous solution (1.6×10^{-17} cm^2 (Banishev et al., 2008)), human serum albumin (1.3×10^{-17} cm^2), and bovine serum albumin (σ_D=1×10^{-17} cm^2 and σ_A=3×10^{-17} cm^2), which were determined in previous Section. One can see that the values for tryptophans in proteins are different and do not equal to the value for a free tryptophan, as is often assumed. I want to emphasize that the lifetime of the excited state of the donor (tryptophan) τ_D in the absence of the acceptor (chromophore) has been obtained without the removal of the acceptor (as it is often supposed when determine the energy transfer efficiency in LDA pairs by conventional methods (Valuer, 2002)). For mRFP1, this value can be compared with ones for the free tryptophan (τ_D=2.8 ns (Banishev et al., 2008a)), HSA (τ_D=4.5 ns) and BSA (τ_D=5 and τ_A=6.2 ns). Simultaneously, the excited state lifetime values of the chromophores of the GH and G form have been obtained (τ_A in the Table 3), although the chromophores of these forms are non-fluorescent. The obtained results show that the high volume (E=0.89) of the energy transfer efficiency from the tryptophan to the chromophores in all three forms of the protein is of special scientific and practical interest. This permits employing mRFP1 as a promising fluorescence indicator that makes use of its own inner LDA pair (an alternative is the preparation of such pairs of two proteins (Srinivas et al., 2001; Truong & Ikura, 2001).

Parameter	Values for R form	Values for GH form	Values for G form
$\sigma_D(\lambda_{ex}=266)$, cm^2	$(1\pm0.2)\times10^{-16}$	$(1\pm0.2)\times10^{-16}$	$(1\pm0.2)\times10^{-16}$
$\sigma_A(\lambda_{ex}=266)$, cm^2	0	not defined	not defined
K_{DA}, s^{-1}	$(3.7\pm0.7)\times10^9$	$(7.8\pm1)\times10^9$	$(2.5\pm0.7)\times10^9$
E	0.89	0.94	0.84
τ_A, ns	3±0.15	1.9±0.4	1.7±0.4
τ_D, ns	2.1±0.5	2.1±0.5	2.1±0.5

Table 3. The photophysical parameters of the LDA pairs in mRFP1 by UV excitation. In the table: (1) $\sigma_D(\lambda_{ex}=266)$ and $\sigma_A(\lambda_{ex}=266)$ are the absorption cross section of the donor (tryptophan) and the acceptor (chromophore) at the wavelength of 266 nm; (2) K_{DA} and $E\equiv K_{DA}/(K_{DA}+1/\tau_D)$ are the rate and efficiency of the energy transfer from the excited donor to the unexcited acceptor; (3) τ_D and τ_A are the excited state lifetimes of the donor (in the absence of the acceptor) and the acceptor.

6.3 Influence of a single amino-acid substitution on the individual photophysical parameters of the fluorescent form of the mRFP1 protein

In this Section, the influence of a single amino acid substitution in mRFP1 at position 66 on optical characteristics of the chromophore of fluorescent spectral form (R form) is performed. For that purpose, the method of nonlinear laser fluorimetry was realized in the version when the protein fluorescence is excited by the wavelength of 532 nm (i.e. the only protein chromophore was excited). All technical details of the procedure can be found in (Banishev et al., 2009).

At first, the photophysical parameters of R form chromophore of the proteins were determined. The σ and K'_{32}, defined in Section 2, have been determined from fluorescence saturation curve for each of the eight protein samples. Because the solution of each of the eight proteins contains mature (red) and immature (green) form, the only fluorescence in the red spectral range (from 550 nm) was detected (to obtain the parameters only for R form chromophore). The typical view of the measured fluorescence saturation curves can be found in (Banishev et al., 2009). To simplify the inverse problem solution, the fluorescence lifetime τ was measured independently with the picosecond laser fluorimeter (excitation at 532 nm). It was found that for all protein samples the fluorescence decay best fit by a single-exponential dependence (Banishev et al., 2009). Solving the inverse problem of nonlinear fluorimetry for each saturation curve at given τ, the σ and K'_{32} for each protein sample have been defined. Note that in this scheme of nonlinear laser fluorimetry the 532-nm laser pulses were used for exciting fluorescence, and, hence, σ is the absorption cross section of the protein R form at 532 nm, i.e., $\sigma \equiv \sigma_R^{(532)}$.

At the second stage, the partial concentration of the mature and immature species in the resultant solution of each mutant has been determined. As was said above, the equilibrium between the GH and G forms of FPs can be shifted under the action of external factors. Using this property of red FPs, it is possible to find the ratio of concentrations of all forms. However, the only the red fluorescence of R form is useful for practical application (Piatkevich et al., 2010). The green form is the by-products of maturation and supposed to be absent in ideal case. For that reason the measurement procedure has been simplified and the concentration of the R form and the total concentration of the GH and G forms were determined.

Indeed, given above-mentioned assumptions, one can write the following system of equations:

$$\begin{aligned} &\frac{\Phi_0^{(570)}}{\Phi_0^{(532)}} \cdot \frac{\sigma_{RS}^{(570)}}{\sigma_{RS}^{(532)}} = \frac{\sigma_R^{(570)}}{\sigma_R^{(532)}} \\ &C_R \cdot \sigma_R^{(570)} = 2.3 D^{(570)} l^{-1} \\ &C_R \cdot \sigma_R^{(532)} + C_G \cdot \sigma_G^{(532)} = 2.3 D^{(532)} l^{-1} \\ &C_R + C_G = C_0, \end{aligned} \tag{6}$$

where C_R, C_G are the concentrations of the R form and the total concentration of the G and GH forms in the solution (in cm^{-3}); C_0 is the total concentration of protein molecules determined by conventional methods (McCluskey, 2003); $\sigma_R^{(570)}$ and $\sigma_R^{(532)}$ are the individual absorption cross section of the chromophore of fluorescent form at 570 and 532 nm; $\sigma_R^{(570)}$ is integral absorption cross section of the chromophore of green form; $D^{(570)}$, $D^{(532)}$and $\sigma_{RS}^{(570)}$, $\sigma_{RS}^{(532)}$ are the optical density of the protein solution and Raman scattering cross section of water (Filipova et al., 2001) at 570 and 532 nm, respectively.

The first equality in system (6) reflects the fact that the quantum yields (expressed in terms of the fluorescence parameter Φ_0 (Filipova et al., 2001)) upon excitation of the protein solution at 532 and 570 nm are the same. The second and third equalities are the optical density (determined from the absorption spectrum of the proteins) written in terms of the concentration of protein molecules absorbing light at 570 and 532 nm and in terms of their absorption cross section. The wavelengths of 570 and 532 nm were chosen for reason mentioned in (Banishev et al., 2009). In system (6) the sought-for quantities are C_R, C_G, $\sigma_R^{(570)}$, $\sigma_G^{(532)}$, while experimentally measured values are $\Phi_0^{(570)}/\Phi_0^{(532)}$, $D^{(570)}$, $D^{(532)}$, l, C_0, $\sigma_R^{(532)}$ (the latter found by means of nonlinear laser fluorimetry). After the $\sigma_R^{(570)}$ was found, one can find the maximum value of the individual absorption cross section $\sigma_R^{(max)}$ of the R form chomophore, or the extinction coefficient $\varepsilon_R^{(max)}$, which is more convenient for comparison with data from literature. This value can be calculated using the absorption spectrum and relation $D^{(max)}/D^{(570)}=\varepsilon_R^{(max)}/\varepsilon_R^{(570)}$, where $D^{(max)}$ is the optical density at the maximum of the absorption band of the R form. The results for the eight samples are given in the Table 4.

One can see from the Table 4 that at the absorption maximum of the R form of the mRFP1 $\varepsilon_R^{(max)}$=(215±40) mM^{-1}cm^{-1}, which is drastically (four times) larger than the value published in (Campbell et al., 2002). This difference is due to the fact that the (Campbell et al., 2002) calculated the extinction coefficient using the total protein concentration (and, therefore, found the integral extinction coefficient) rather than the partial concentration (as in our case). As a result, the determination of the partial concentration of fluorescent molecules allowed us to find the individual extinction coefficient of the chromophore of the R form.

Protein	$\varepsilon_R^{(max)}$ (mM^{-1}cm^{-1})	C_R/C_0**, %	η	η_T*
mRFP1	215±40	26±6	0.24±0.03	0.01±0.01
mRFP1/Q66S	85±13	34±6	0.20±0.04	0.05±0.02
mRFP1/Q66C	135±20	17±6	0.19±0.04	0.01±0.01
mRFP1/Q66N	133±15	9±4	0.17±0.03	0.02±0.02
mRFP1/Q66H	230±27	8±4	0.13±0.03	0.07±0.02
mRFP1/Q66A	171±16	2±2	0.19±0.04	0.05±0.02
mRFP1/Q66L	240±40	2±2	0.12±0.03	0.06±0.02
mRFP1/Q66F	142±18	2±2	0.04±0.03	0.12±0.03

Table 4. Individual photophysical parameters of the R form and its fraction in the protein sample. Note: τ, η and η_T are the fluorescence lifetime, fluorescence quantum yield and quantum yield to the triplet state (converted from K'_{32}), respectively. Other parameters are defined after the system of Eqs. (6). * Determined from the fluorescence saturation curve; ** determined by solving system of Eqs. (6).

In the general case, the determination of photophysical parameters of FPs with the help of integral characteristics of the sample is incorrect, which can be proved by several examples. By using the dynamic difference method, (Kredel et al., 2008) obtained the individual extinction coefficient 143 mM^{-1}cm^{-1} for the chromophore of the R form of mPlum, which is also drastically larger than other published values ranging from 22 to 41 mM^{-1}cm^{-1} (Shcherbo et al., 2007) for this protein. It is interesting to note that (Gross et al., 2000) have earlier reported a similar value of 150 mM^{-1}cm^{-1} for the R form chromophore of red FP DsRed (the table value for this protein is assumed to be 75 mM^{-1}cm^{-1}). In their approach, the amount of immature species was deduced from mass spectroscopic analysis. Another example can be found in (Strack et al., 2010). Using the method described in (Ward, 2005), (Strack et al., 2010) got an assessed value of 123 mM^{-1}cm^{-1} for DsRed.T7, which is close to that obtained by (Kredel et al., 2008). From these examples one can see that the values of the individual extinction coefficients of the R form chromophore are close, as it is expected to be, because the chromophores of these proteins are considered to be chemically identical. One can assume a minor disagreement due to chromophore orientation change relative to protein matrix or composition of its closest environment (this can explain the difference in the extinction values for mPlum, DsRed and DsRed.T7 determined by (Kredel et al., 2008; Gross et al., 2000) and (Strack et al., 2010)). However, the published values of the extinction of red FPs with the same chromophore drastically vary depending on the protein: 75 mM^{-1}cm^{-1} per a polypeptide chain for the DsRed, 120 mM^{-1}cm^{-1} per a polypeptide chain for tdimer2(12) (Campbell at al., 2002), 22 mM^{-1}cm^{-1} for mPlum (Shcherbo et al., 2007), 50 mM^{-1}cm^{-1} for mRFP1 (Campbell at al., 2002) and 90 mM^{-1}cm^{-1} for mStrawberry (Shu et al., 2006). Discrepancies follow directly from the content of immature form in the protein samples.

In the measurements I obtained 215 $mM^{-1}cm^{-1}$ for mRFP1, which is larger than the value for DsRed and mPlum. However, it should be noted that I did not take into account the photochemical processes (photoionisation, photobleaching, etc. (Banishev et al., 2008a) in the model of fluorescence response generation (1a). The efficiency of these processes in the protein samples under study may be different. When efficient enough, the photochemical processes may contribute noticeably to fluorescence saturation. In this case their emission can result in the saturation curve giving an overstated value of the absorption cross section and, therefore, overstated quantity of $\varepsilon_R^{(max)}$. On the other hand, as it was mentioned in Section 5.2, the method applied by (Kredel et al., 2008) for determining the individual extinction coefficient of mPlum is not accurate enough in the case of red FPs (the method is well adapted only for GFP-like FPs). Therefore, the obtained value of 143 $mM^{-1}cm^{-1}$ can be underestimated and the precise value of the chromophore extinction is greater.

As it was shown earlier (Banishev et al., 2009), for the R form of the proteins mRFP1, mRFP1/Q66S and mRFP1/Q66C, the position of the maximum of absorption, fluorescence excitation and emission bands depends on the substituted amino-acid residue at position 66 and positively correlates with the volume of this residue: the maximum moves to the red range with increasing the volume of the residue. A similar correlation was described for the individual extinction coefficient of the R form chromophore, i.e. a higher extinction coefficient corresponds to a larger volume of the residue. The results for the new mutants mRFP1/Q66N, mRFP1/Q66H, mRFP1/Q66A, mRFP1/Q66L and mRFP1/Q66F are presented in Fig. 7(a, b). In the same figure the results obtained in (Banishev et al., 2009) for mRFP1, mRFP1/Q66S and mRFP1/Q66C are plotted. The values of the amino acids volume were taken from (Zamyatin, 1972). One can see that the dependence of the position of steady-state spectra (at maximum) of two new mutants (mRFP1/Q66N and mRFP1/Q66H) on the volume of amino-acid residue at position 66 has the same trend as described in (Banishev et al., 2009). The same can be observed for the individual extinction coefficient (but not for the integral one). There are no such dependences for characteristics of the proteins mRFP1/Q66A, mRFP1/Q66L and mRFP1/Q66F.

The results can be explained in the following way. It is known that formation of R form chromophore of red FPs goes through formation of a double bond between the Cα and N atoms of the 66th amino acid residue. Since dehydrogenation of a bond between Cα and N atoms involves the carbonaceous framework of the 66th amino acid residue in the system of conjugation, then the changes in the side radical of this residue can lead to the changes in the spectral and photophysical properties of the new mutants.

In the case of a polar amino acid at position 66 (serine, cysteine, asparagines and histidine), its side radical can form hydrogen bonds with the side radicals of glutamine-42, glutamine-213 and glutamate-215 (see Fig. 7 (c, d)). These radicals, in turn, belong to the protein shell (the β-barrel) and are rather rigidly bonded to it (Khrameeva et al., 2008). A change in the geometry of the side radical at position 66 will in this case cause a change in the geometry of the chromophore imidazolidine ring, because the interaction of the chromophore with the glutamine-42 and glutamine-213 through the hydrogen bond network can be distorted and a new bond with glutamate-215 can be formed. As a result, the chromophore tilt- and twist-angles (the pictures with the explanation of the angles can be found in (Piatkevich et al., 2010)) will change and the chromophore coplanarity will be distorted. In (Piatkevich et al., 2010) it was shown on the basis on x-ray diffraction data that the chromophore planarity is directly connected with the optical properties (the steady-state spectra positions, fluorescence quantum yield, etc.) of red FPs. The deviations from chromophore coplanarity are responsible for the changes in the optical characteristics for mCherry and mStrawberry (Shu et al., 2006). The interrelation between the optical properties of the monomeric red FPs and the geometry of their chromophores was also confirmed by the molecular dynamics simulations conducted for mRFP1 mutants with single polar amino acid substitutions at position 66 (Khrameeva et al., 2008). The simulations have shown that the substitutions have an influence on the torsion angles in the phenolic and imidazolidine rings of the chromophore as well as on the torsion angles in the regions of connection between these rings and chromophore attachment to β-barrel. It was predicted that the volume of the amino acid residue at position 66 can correlate with the optical characteristics of the mutants. The experimental results presented in this Section are consistent with the results of simulations performed by (Khrameeva et al., 2008).

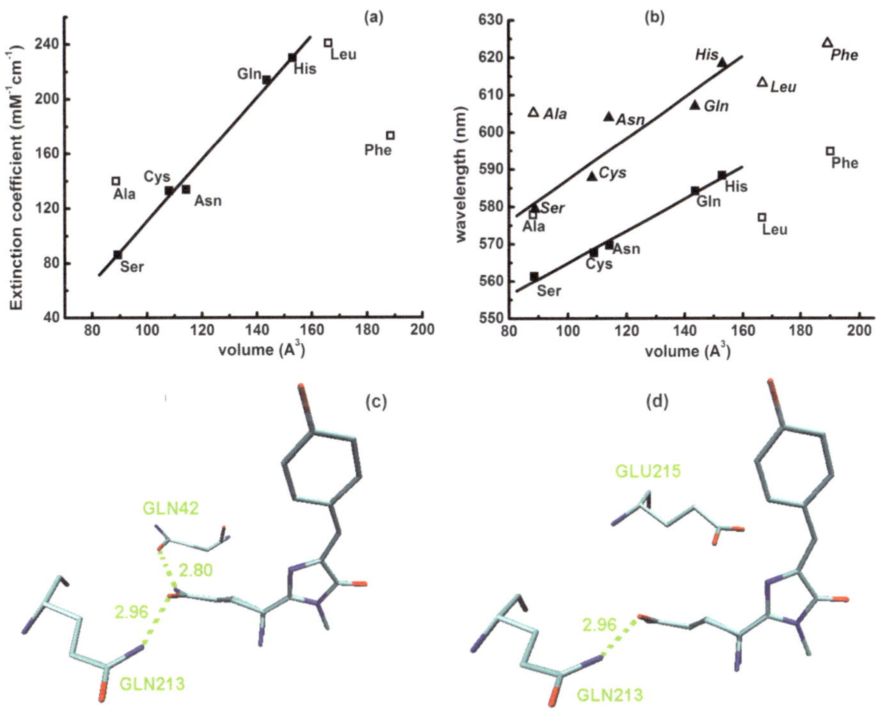

Fig. 7. The dependence of the (a) extinction coefficient, (b) position of absorption/ excitation (squares) and fluorescence emission (triangles) maximum of the protein R form on the volume of 66th amino acid residue. The solid and hollow scatters are polar and non-polar amino acids, respectively. (c) and (d) are schematic diagrams of chrmophore environment, showing the residues location at positions 213, 42 and 213 , 215, respectively. Hydrogen bonds are shown in dashed lines, labeled with lengths in angstroms. Glu is glutamate.

For the mutants with non-polar groups at residue 66 (alanine, leucine and phenylalanine) there is no correlation effect. The non-polar substitutions lead to breakage of hydrogen bonds between the 66th residue of the chromophore and glutamine-42 and glutamine-213. Formation of a new bond between the chromophore and the glutamate-215 is unlikely because of non-polarity of substituted residues. Therefore, there is no defined correlation between the chromophore geometry (consequently, the volume of the substituted amino acid residue) and the optical properties of the proteins.

At the present time the red FPs whose molecules are monomers are of particular interest (Piatkevich et al., 2010) as fluorescent markers. Attempts to find new variants of red FPs in order to improve their properties (higher brightness, photo and pH stability, etc.) are performed. However, the interrelation between optical or photophysical parameters and structural properties of FPs, which is necessary for development of these studies, is rather unclear. A method for prediction of properties of FPs based on their structure is still not developed. This problem might be solved by analysis of properties of mutant proteins with point mutations. Therefore, the results obtained in this Section can be used to tackle the general problem of the development of an algorithm, which could provide the prediction of the spectral properties of FPs based on their structures. The data will also be useful for revealing promising positions for directed mutagenesis.

7. Conclusion

In current work the approach based on the simultaneous use of nonlinear laser fluorimetry, spectrophotometry and conventional fluorimetry methods has been applied for investigation of the photophysical properties of the protein molecules of different complexity. The full set of photophysical parameters of the fluorophores (tryptophan residues) of human and bovine serum albumins has been determined. The photophysical processes in the spectral forms of the red FP mRFP1 under UV (266 nm) and visible (532 nm) irradiation are described quantitatively. The individual photophysical parameters of the new mutants of the mRFP1 protein (a single substitution at the 66 amino acid position) were determined. It was shown that the individual extinction coefficient of the red chromopore of the proteins correlate positively with the volume of the substituted amino acid residue at position 66 (for polar substitution). A similar correlation has been described for the position of the maximum of the absorption, fluorescence excitation and emission spectra: the position of the maximum moves to the red with increasing the volume of the residue. In addition, the partial concentration of the fluorescent spectral form in the resultant solution of each FP variant has been determined.

8. Acknowledgment

The author is grateful to Prof. Victor Fadeev for providing the ability to work in his group and for great help in mastering the fluorescence spectroscopy methods. The author also thanks Evgeny Vrzheshch for the samples of the red FPs and for valuable discussions.

9. References

Agranovich, V.M. & Galanin, M.D. (1982). *Electronic Excitation Energy Transfer in Condensed Matter*, Elsevier Science Ltd, ISBN 978-0444863355, North-Holland

Banishev, A.A.; Vrzhechsh, E.P. & Shirshin, E.A. (2009). Application of laser fluorimetry for determining the influence of a single amino-acid substitution on the individual photophysical parameters of a fluorescent form of a fluorescent protein mRFP1. *IEEE J. Quantum Electron.*, Vol. 39, No. 3, pp. 273-278, ISSN 10637818

[a]Banishev, A.A.; Shirshin, E.A. & Fadeev, V.V. (2008). Determination of photophysical parameters of tryptophan molecules by methods of laser fluorimetry. *IEEE J. Quantum Electron*, Vol. 38, No. 1, pp. 77-81, ISSN 10637818

[b]Banishev, A.A.; Shirshin, E.A. & Fadeev, V.V. (2008). Laser fluorimetry of proteins containing one and two tryptophan residues. *Laser Physics*, Vol. 18, No. 7, pp. 861–867, ISSN 1054-660X

Banishev, A.A.; Maslov, D.V. & Fadeev, V.V. (2006). A nanosecond laser fluorimeter. *Instrum. Exp. Tech.*, Vol. 49, No. 3, pp. 430–434, ISSN 0020-4412

Boychuk, I.V.; Dolenko, T.A.; Sabirov, A.R.; Fadeev, V.V.; Filippova, E.M. (2000). Study of the uniqueness and stability of the solutions of inverse problem in saturation fluorimetry. *IEEE J. Quantum Electron*, Vol. 30 , No. 7, pp. 611-616, ISSN 10637818

Campbell, R.E.; Tour, O.; Palmer, A.E.; Steinbach, P.A.; Baird, G.S., Zacharias, D.A. & Tsien, R.Y. (2002). A monomeric red fluorescent protein. *Proc. Natl. Acad. Sci. USA*, Vol. 99, pp. 7877–7882, ISSN 1091-6490

Eftink, M.R.; Zajicek, J.L. & Ghiron, C.A. (1977). A hydrophobic quencher of protein fluorescence: 2,2,2-trichloroethanol. *Biochim Biophys Acta.*, Vol. 491, No. 2, pp. 473–481, ISSN 0006-3002

Fadeev, V.V.; Dolenko, T.A.; Filippova, E.M. & Chubarov, V.V. (1999). Saturation spectroscopy as a method for determining the photophysical parameters of complicated organic compounds. *Opt. Commun.*, Vol. 166, pp. 25-33, ISSN 0030-4018

Filipova, E.M.; Fadeev, V.F.; Chubarov, V.V.; Dolenko, T.A.& Glushkov, S.M. (2001). Laser fluorescence spectroscopy as a method for determining humic substance. *Appl. Spectrosc.*, Vol. 36, No. 1, pp. 87-117, ISSN 0003-7028

Gross, L.A.; Baird, G.S.; Hoffman, R.C.; Baldridge, K.K. & Tsien, R.Y. (2000). The structure of the chromophore within DsRed, a red fluorescent protein from coral. *Proc. Natl. Acad. Sci. USA*, Vol. 97, pp. 11990–11995, ISSN 1091-6490

Khrameeva, E.E.; Drutsa, V.L.; Vrzheshch, E.P.; Dmitrienko, D.V. & Vrzheshch, P.V. (2008). Mutants of monomeric red fluorescent protein mRFP1 at residue 66: structure modeling by molecular dynamics and search for correlations with spectral properties. *Biochemistry*, Vol. 73, No. 10, pp. 1082-1095, ISSN 0006-2979

Kredel, S.; Nienhaus, K.; Oswald, F.; Wolff, M.; Ivanchenko, S.; Cymer, F.; Jeromin, A.; Michel, F. J.; Spindler, K.D.; Heilker, R.; Nienhaus, G.U & Wiedenmann, J. (2008). Optimized and far-red-emitting variants of fluorescent protein eqFP611. *Chem. Biol.*, Vol. 15, pp. 224–233, ISSN 1074-5521

Lakowicz, Joseph R. (1999). *Principles of Fluorescence Spectroscopy (2nd edition)*, Kluwer/Plenum Publishers, ISBN 0-306-46093-9, New York

McCluskey, K. (2003). The fungal genetics stock center: from molds to molecules. *Adv. Appl. Microbiol.*, Vol. 52, pp. 245–262, ISSN 1365-2672

Pace, C.N.; Vajdos, F. & Fee, L. (1995). How to measure and predict the molar absorption coefficient of a protein. *Protein Sci.*, Vol. 4, pp. 2411, ISSN 1469-896x

Pakhomov, A.A.; Martynova, N.Y.; Gurskaya, N.G.; Balashova, T.A. & Martynov, V.I. (2004). Photoconversion of the chromophore of a fluorescent protein from Dendronephthya sp. *Biochemistry*, Vol. 69, pp. 901–908, ISSN 0006-2979

Permyakov, E.A. (1992). *Luminescent Spectroscopy of Proteins,* CRC Press Inc, ISBN 978-0849345531, Boca Raton, USA

Peters, Jr., T. (1996). All *About Albumin: Biochemistry, Genetics, and Medical Applications,* Academic Press, ISBN 978-0125521109, San Diego, USA

Piatkevich, K.D.; Efremenko, E.N.; Verkhusha, V.V. & Varfolomeev S.D. (2010). Red fluorescent proteins and their properties. *Russ. Chem. Rev.*, Vol. 79, No. 3, pp. 243-258, ISSN 1468-4837

Shcherbo, D.; Merzlyak, E.M.; Chepurnykh, T.V.; Fradkov, A.F.; Ermakova, G.V.; Solovieva, E.A.; Lukyanov, K.A.; Bogdanova, E.A.; Zaraisky, A.G.; Lukyanov, S. & Chudakov, D.M. (2007). Bright far-red fluorescent protein for whole-body imaging. *Nat. Methods,* Vol. 4, pp. 741–746, ISSN 1548-7091

Shirshin, E.A.; Banishev, A.A. & Fadeev, V.V. (2009). Localized donor–acceptor pairs of fluorophores: determination of the energy transfer rate by nonlinear fluorimetry. *JETP Letters,* Vol. 89, No. 10, pp. 475–478, ISSN 0021-3640

Shu, X.; Shaner, N.C.; Yarbrough, C.A.; Tsien, R.Y. & Remington, S.J. (2006). Novel chromophores and buried charges control color in mFruits. *Biochemistry,* Vol. 45, No. 32, pp. 9639–47, ISSN 1520-4995

Srinivas, G.; Yethiraj, A. & Bagchi, B. (2001). FRET by FET and dynamics of polymer folding. *J. Phys. Chem. B,* Vol. 105, pp. 2475–2478, ISSN 1520-5207

Strack, L.R.; Strongin, D.E.; Benjamin, L.M.; Glick, S. & Keenan, R.J. (2010). Chromophore formation in DsRed occurs by a branched pathway. *J. Am. Chem. Soc.*, Vol. 132, pp. 8496–8505, ISSN 0002-7863

Truong, K. & Ikura, M. (2001). The use of FRET imaging microscopy to detect protein—protein interactions and protein conformational changes in vivo. *Curr. Opin. Struct. Biol.* Vol. 11, pp. 573–578, ISSN 0959-440X

Valeur, B. (2002). *Molecular Fluorescence: Principles and Applications,* Wiley-VCH, ISBN 978-3527299195, Weinheim, Germany

Verkhusha, V.V.; Chudakov, D. M.; Gurskaya, N. G.; Lukyanov, S.; Lukyanov K.A. (2004) Common pathway for the red chromophore formation in fluorescent proteins and chromoproteins. *Chem. Biol.*, Vol. 11, pp. 845–854, ISSN 1074-5521

Vrzheshch, E.P.; Dmitrienko, D.V.; Rudanov, G. S.; Zagidullin, V.E.; Paschenko, V. Z.; Razzhivin, A.P.; Saletsky, A.M. & Vrzheshch, P. V. (2006). Optical properties of the monomeric red fluorescent protein mRFP1. *Moscow Univ. Phys. Bull.*, Vol. 63, No. 3, pp. 109-112, ISSN 0096-3925

Ward, W.W. (2005). *Biochemical and Physical Properties of Green Fluorescent Protein. In Green Fluorescent Protein: Properties, Applications and Protocols (2nd Edition),* Wiley John & Sons , ISBN 9780471736820, New Jersey, USA

Zamyatin, A.A. (1972). Protein Volume in Solution. *Prog. Biophys. Mol. Biol.*, Vol. 24, pp. 107-123, ISSN 0079-6107.

Protein structure available from Protein Data Bank, www.pdb.org, ID 2vad

12

Current Achievement and Future Potential of Fluorescence Spectroscopy

Nathir A. F. Al-Rawashdeh
United Arab Emirates University, Department of Chemistry, UAE
Jordan University of Science and Technology, Department of Chemistry, Jordan

1. Introduction

Spectrofluorometric methods of analysis are the most commonly analytical techniques and continue to enjoy wide popularity. The wide availability of the instrumentation, the simplicity of procedure, sensitivity, selectivity, precision, accuracy, and speed of the technique still make the spectrofluorometric methods attractive. These features make fluorescence spectroscopy an attractive technique as compared to other forms of optical spectroscopy or other analytical techniques such as chromatography and electrophoresis. Fluorescence spectroscopy has been used widely as a tool for quantitative analysis, characterization, and quality control in the pharmaceutical, environmental, agricultural, nanotechnology and biomedical fields.

The emission of light from an excited electronic state of a molecular species is called luminescence. The discovery and characterization of luminescence begun from the 15th century. In 1506 Nicolas Monardes was the first to describe the bluish opalescence of the water infusion from the wood of a small Mexican tree. In 1612 Galileo described the emission of light (phosphorescence) from the famous Bolognian stone, which discovered by Vincenzo Casciarolo, a Bolognian shoemaker. Galileo wrote: "It must be explained how it happens that light is conceived into the stone, and is given back after some time, as is childbirth". Even though, some of the first scientific reports of luminescence appeared in the middle of the 18th century. In 1845 Sir J.F.W. Herschel reports on an experiment he did twenty years earlier. Herschel made the first observation of fluorescence from quinine sulfate (quinine: (*R*)-(6-methoxyquinolin-4-yl)-((2*S*, 4S, 8R)- 8-vinylquinuclidin-2-yl)methanol, $C_{20}H_{24}N_2O_2$, quinine absorbs in the UV region), he observed that an otherwise colorless solution of quinine in water emitted a blue color under certain circumstances. Herschel concludes that a species in the solution, "exert its peculiar power on the incident light" and disperses the blue light. The experiment can be repeated simply by observing glass of tonic water exposed to sunlight. Often a blue glow is visible at the surface (Rendell, 1987).

The phenomenon of fluorescence was known by the middle of the nineteenth century. British scientist Sir George G. Stokes first made the observation that the mineral fluorspar exhibits fluorescence when illuminated with ultraviolet light, and he coined the word "fluorescence". In 1852, Sir G.G. Stokes studied the same compound (quinine) that has been used by Herschel and found that the fluorescing (*emitted*) light has longer wavelengths than the excitation (*absorbed*) light, a phenomenon that has become to be known as the Stokes shift. Stokes' paper demonstrated the fundamental property of fluorescence, which simply can be summarized as a photon of ultraviolet radiation collides with an electron in a simple atom, exciting and elevating the electron to a higher energy level. Subsequently, the excited electron relaxes to a lower level and emits light in the form of a lower-energy photon (*higher wavelength*) in the visible light region. In 1871 Adolph Von Baeyer, a German chemist, synthesized the fluorescent dye, fluorescein. In 1880 Edmund Bequerel showed that certain metal ion complexes emit radiation with a very long decay time (Rendell, 1987).

Jabłoński and others developed a modern theoretical understanding of Stokes observation some 70 years later. In the 1920s and 1930s Jabłoński investigated polarized light and fluorescence and was able to show that the transition moments in absorption and emission are two different things. Thus the foundation for the concept of anisotropy was laid. For that and other accomplishments Jabłoński has been referred to as the father of fluorescence and his work has had a major impact on the theoretical understanding of photophysics (Rendell, 1987; Lakowicz, 2006).

How and why do certain molecules known as fluorophores or fluorescent molecules (such as: dyes, polyaromatic hydrocarbon, or heterocyclic,...etc.) emit different colors of light?. Briefly the answer for this question is that some molecules are capable of being excited, via absorption of light energy, to a higher energy state, also called an excited state. The energy of the excited state, which cannot be sustained for long, "decays" or decreases, resulting in the emission of light energy. This process is called *fluorescence*. To "fluoresce" means to emit light via this process. A fluorophore is a molecule that is capable of fluorescing due to the presence of certain chromophores within a molecule. In its ground state, the fluorophore molecule is in a relatively low-energy, stable configuration, and it does not fluoresce. When light from an external source hits a fluorophore molecule, the molecule can absorb the light energy. If the energy absorbed is sufficient, there are multiple excited states or energy levels that the fluorophore can attain, depending on the wavelength and energy of the external light source. Since the fluorophore is unstable at high-energy configurations, it eventually adopts the lowest-energy excited state, which is semi-stable. The length of time that the fluorophore is in excited states is called the *excited lifetime*, and it lasts for a very short time, ranging from 10^{-15} to 10^{-9} seconds. Next, the fluorophore rearranges from the semi-stable excited state back to the ground state, and the excess energy is released and emitted as light. The emitted light is of lower energy, and thus longer wavelength, than the absorbed light. This means that the color of the light that is emitted is different from the color of the light that has been absorbed. Emission of light returns the fluorophore to its ground state. The fluorophore can absorb light energy again and go through the entire process repeatedly (Lakowicz, 2006).

The cyclical fluorescence process, shown in Figure 1, can be summarized as: 1. Excitation of a fluorophore through the absorption of light energy, the excitation wavelength is usually the same as the absorption wavelength of the fluorophore; 2. A transient excited lifetime with some loss of energy, during this period, the fluorophore undergoes conformational changes and is also subject to possible interactions with its molecular environment, with two important consequences: first, the energy of higher excited state is partially dissipated as a heat, yielding a relaxed lowest singlet excited state from which fluorescence emission originates, second, not all the molecules initially excited by absorption (stage 1) return to the ground state by fluorescence emission, as other processes such as collisional quenching, fluorescence energy transfer, and intersystem crossing may also depopulate the first excited state ; and 3. Return of the fluorophore to its ground state, accompanied by the emission of light. The light energy emitted is always of a longer wavelength (*lower energy*) than the light energy absorbed, due to the energy dissipation during the transient excited lifetime, as shown in Step 2. Consequently, the ratio of the number of fluorescence photons emitted (stage 3) to the number of photons absorbed (stage 1) represents the *fluorescence quantum yield* (Lakowicz, 2006).

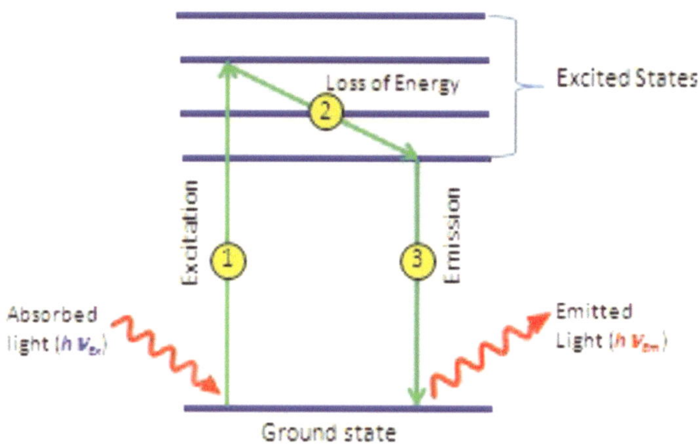

Fig. 1. The Jablonski diagram illustrates the three stages involved in the creation of an excited electronic singlet state by optical absorption and subsequent emission of fluorescence

A fluorophore can repeatedly undergo the fluorescence process—in theory, indefinitely. This is extremely useful, because it means that one fluorophore molecule can generate a signal multiple times. This property makes fluorescence a very sensitive technique for visualizing microscopic samples—even a small amount of the stain can be detected. In reality, however, the fluorophore's structural instability during the excited lifetime makes it susceptible to degradation. High-intensity illumination can cause the fluorophore to change its structure so that it can no longer fluoresce—this is called photobleaching. When a fluorescent sample, such as a slide with mounted tissue, is photobleached, the fluorophores are no longer promoted to an excited state, even when the required light energy is supplied (Lakowicz, 2006).

Now that we've introduced the general process of fluorescence, let's take a look at the basic properties of the light spectrum and its importance in fluorescence. The visible spectrum (Figure 2) is composed of light with wavelengths ranging from approximately 380 nanometers to 750 nanometers.

Light waves with shorter wavelengths have higher frequency and higher energy. Light waves with longer wavelengths have lower frequency and lower energy. As we stated before, an excited fluorophore emits lower-energy light than the light it absorbed. Therefore, there is always a shift along the spectrum between the color of the light absorbed by the fluorophore during excitation, and the color emitted.

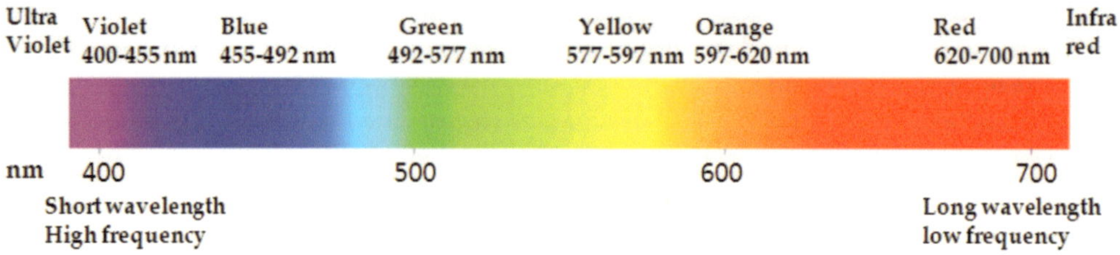

Fig. 2. Visible light spectrum

For example, let's say that we have a tube that contains a particular fluorescent dye. If we shine 480 nanometer light at the dye solution, some of the fluorophore molecules will become excited. However, the majority of the molecules are not excited by this wavelength of light. As we increase the excitation wavelength, say to 520 nanometers, more molecules are excited. However, this is still not the wavelength at which the proportion of excited molecules is maximal. For this particular dye, 550 nanometers is the wavelength that excites more fluorophores than any other wavelength of light. At wavelengths longer than 550 nanometers, the fluorophore molecules still absorb energy and fluoresce, but again in smaller proportions. The range of excitation wavelengths can be represented in the form of a fluorescence excitation spectrum, which looks like the spectrum shown in Figure 3.

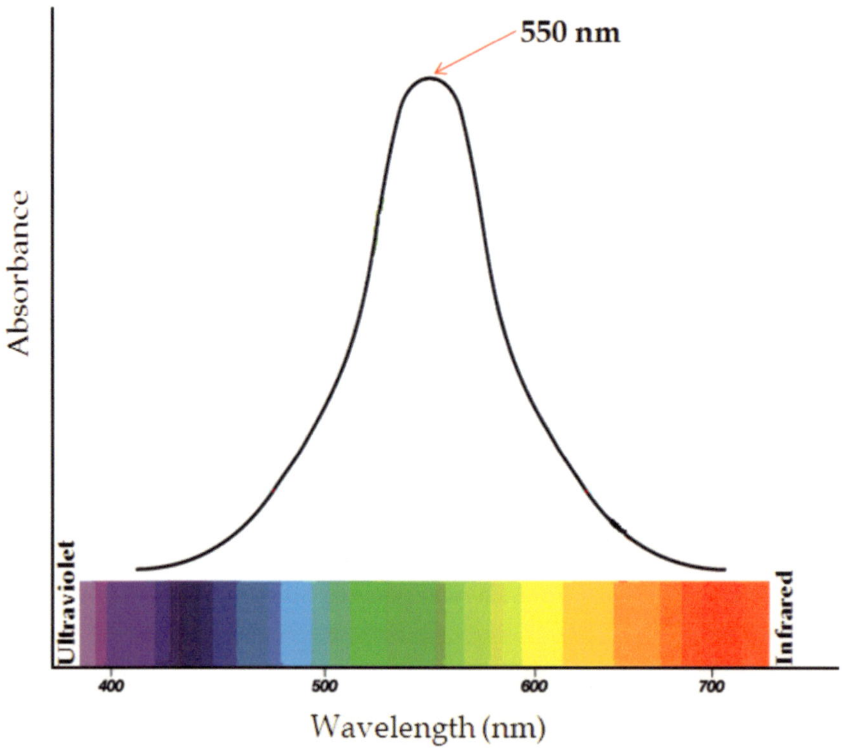

Fig. 3. Excitation spectrum of dye solution recorded at λ_{em} = 570 nm, the excitation wavelength maximum at 550 nm.

A fluorescent molecule absorbs light over a range of wavelengths—and every chemical molecule has a characteristic excitation range. However, some wavelengths within that range are more effective for excitation than other wavelengths. This range of wavelengths reflects the range of possible excited states that the fluorophore can achieve. Thus for each fluorescent molecule, there is a specific wavelength—the excitation maximum—that most effectively induces fluorescence. Now let's look at the light that is emitted by the fluorophore molecules when they are excited at the optimal excitation wavelength. Just as fluorophore molecules absorb a range of wavelengths, they also emit a range of wavelengths. There is a spectrum of energy changes associated with these emission events. When we excite the previously described dye solution at its excitation maximum, 550 nanometers, light is emitted over a range of wavelengths. A molecule may emit at a different wavelength with each excitation event because of changes that can occur during the excited lifetime, but each emission will be within the range. Although fluorophore molecules emit same intensity of light, the wavelengths and therefore the colors of the emitted light are not homogeneous. However, a larger population of molecules has most intensely fluorescence at 570 nanometers (Lakowicz, 2006).

Based on this distribution of emission wavelengths, we say that the emission maximum of this fluorophore is 570 nanometers. The range of wavelengths is represented by the Fluorescence Emission Spectrum, Figure 4.

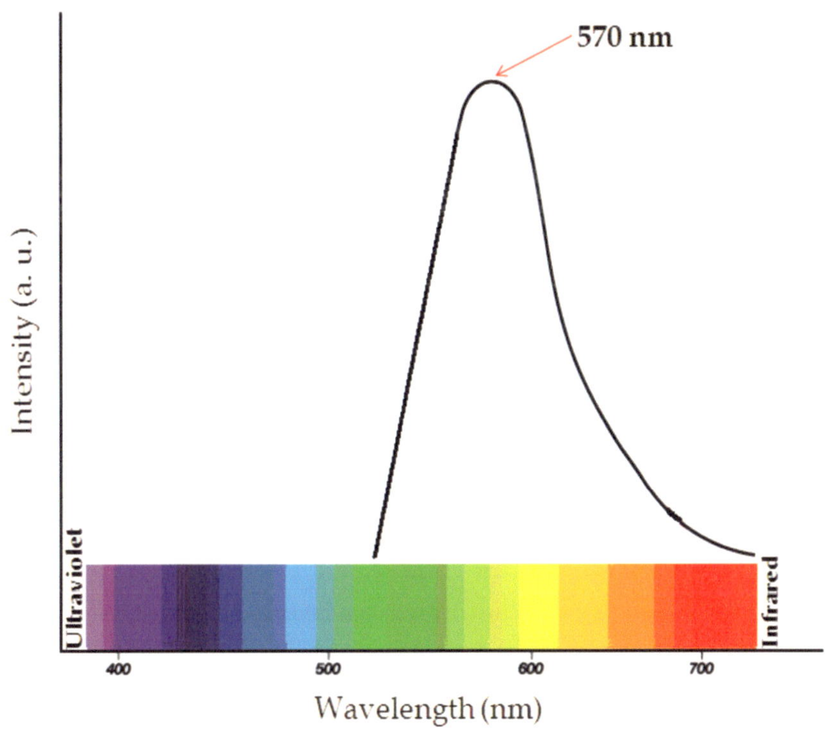

Fig. 4. Fluorescence Emission Spectrum of dye solution monitored at λ_{ex} = 550 nm

The emission intensity is proportional to the amplitude of the fluorescence excitation spectrum at the excitation wavelength. Fluorescence emission intensity depends on the same parameters as absorbance—defined by the Beer-Lambert law as the product of the molar extinction coefficient, optical path-length, and concentration—as well as on the fluorescence quantum yield, the intensity of the excitation source, and the efficiency of the instrument and, in dilute solutions, is linearly proportional to these parameters.

The summary points of this introduction to fluorescence are: 1. Fluorophores are molecules that, upon absorbing light energy, can reach an excited state, and then emit light energy. 2. The three-stage process of excitation, excited lifetime, and emission is called fluorescence. 3. Fluorophores absorb a range of wavelengths of light energy, and also emit a range of wavelengths. Within these ranges are the excitation maximum and the emission maximum. Because the excitation and emission wavelengths are different, the absorbed and emitted lights are detectable as different colors or areas on the visible spectrum.

The purpose of this chapter is to review the articles on the interior cited aspects published since 2000 about various aspects of application of fluorescence spectrophotometry in chemical analysis.

2. Theoretical and instrumental aspects

2.1 Basic theory of fluorescence

This section provides a basic tutorial on specific topic in luminescence, namely fluorescence, and fluorescent instrumentations. To be able to understand the basic theoretical principles of luminescence spectroscopy, which include the electronic transitions, one should have a basic background on quantum mechanics and atomic orbitals, which was developed by Schrödinger in 1926. A tutorial review of Schrödinger's wave equation is out of the scope of this chapter, but briefly the most fruitful outcome of solving Schrödinger's wave equation is a set of wave functions (*called orbitals*), and their corresponding energies. An orbital is described by a set of three quantum numbers (Principal (n), Angular momentum (*l*), and magnetic (m_l) quantum numbers). Later a fourth quantum number, that describes the magnetic field generated due to the spinning of an electron on its own axis, was discovered and named as spin magnetic quantum number (m_s). The spin quantum number has only two allowed values: +1/2 and −1/2. According to Pauli Exclusion Principle, no two electrons in the same atom can have identical sets of four quantum numbers, n, l, m_l, and m_s. Thus any two electrons in same orbital (n, l, m_l) must have different spins either m_s=+1/2 or m_s= -1/2. The total spin is defined by $S = \Sigma m_s$ and the Multiplicity (M) is defined as M = 2S+1(M=1, 2, 3… Singlet, Doublet, and Triplet, respectively,..) as shown in Figure 5.

The photophysical processes that occur from absorption to emission are often shown in a so-called Jabłoński diagram. Of course all possible energy routes cannot be encompassed in single figure, and different forms of the diagram can be found in different contexts. Figure 7 is a simple version of Jabłoński diagram, where absorption ($S_0 \rightarrow S_1$ or $S_0 \rightarrow S_2$), fluorescence (emission not involving spin change, $S_1 \rightarrow S_0$), phosphorescence (emission involving spin change, $T_1 \rightarrow S_0$), intersystem crossing (non-radiative transition with spin change, as an example from singlet to triplet states; $S_1 \rightarrow T_1$), internal conversion (non-radiative transition either to lower state when vibrational energy levels "match" or to lower state by collisional deactivation, $S_2 \rightarrow S_1$)), vibration relaxation (within the vibrational levels in any excited electronic state) as well as intermolecular processes (radiationless relaxations) are shown.

Other intermolecular processes (e.g. quenching, energy transfer, solvent interaction etc.) are omitted (Rendell, 1987; Lakowicz, 2006).

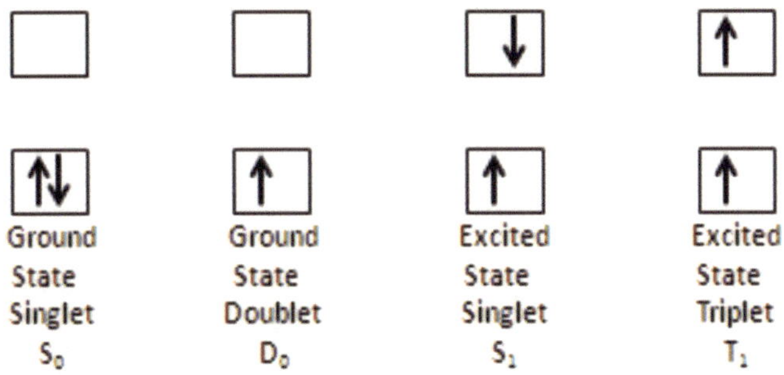

Fig. 5. Possible energy states according to their spin multiplicity

Once a molecule is excited by absorption of light it can return to the ground state with emission of fluorescence, but many other pathways for de-excitation are also possible, these are summarized in Figure 6.

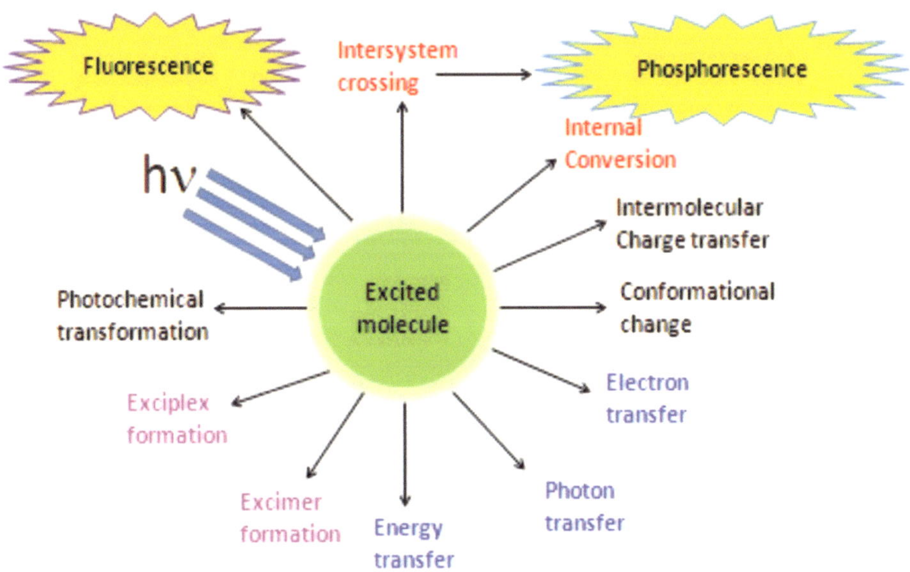

Fig. 6. All possible pathways for de-excitation processes

Jablonski diagram (Fig. 7) explains the mechanism of light emission in most organic and inorganic luminophores. Upon absorption of the light by a molecule, the electron promoted from ground electronic state (S_0) to an excited state that should possess the same spin multiplicity (such as, S_1, S_2,....) this process usually occurs within ~10^{-15} s. This excludes the triplet excited state as the final state of electronic absorption because transitions between states of different multiplicities are improbable "*forbidden*" (e.g. T→S or S→T). According to the quantum mechanical selection rules for electronic transitions, spin state should be maintained upon excitation because it is harder for an electron to go from a singlet state to triplet state since the spin has to be flipped (i.e. *change in spin during the electronic transitions is not allowed*). Therefore, to go from a singlet to a triplet state ($\Delta M = 1$) is so-called forbidden transition and occurs with a small rate constant and typically too weak to be of much importance. At room temperature the higher vibrational energy levels are in general not populated (less than 1% according to Boltzmann statistics). The magnitude of the absorbed energy decides which vibrational level of S_1 (or S_2) becomes populated. This process is very fast and happens within 10^{-15} s. In the next 10^{-12} s the molecule relaxes to the lowest vibrational level of S_1, a process called internal conversion. Since emission typically occurs after 10^{-9} s the molecule is fully relaxed at the time of emission, hence, as a rule, emission occurs from the lowest vibrational level of S_1 (Kasha.s rule) and the fluorescence spectrum is generally independent of the excitation wavelength. After emission the molecule returns to the ground state, possibly after vibrational relaxation. This completes the simplest case of fluorescence: excitation, internal conversion, emission and relaxation. The energy lost to the surroundings, due to vibrational relaxation and internal conversion is the reason why a Stokes' shift (*defined as the energy difference between the emission and absorption peak maxima for the same electronic transition*) is observed (Figure 8). The Stokes shift is due to the fact that some of the energy of the excited fluorophore is lost through molecular vibrations that occur during the brief lifetime of the molecule's excited state. This energy is dissipated as heat to surrounding solvent molecules as they collide with the excited fluorophore. The magnitude of the Stokes shift is determined by the electronic structure of the fluorophore, and is a characteristic of the fluorophore molecule (Rendell, 1987; Lakowicz, 2006).

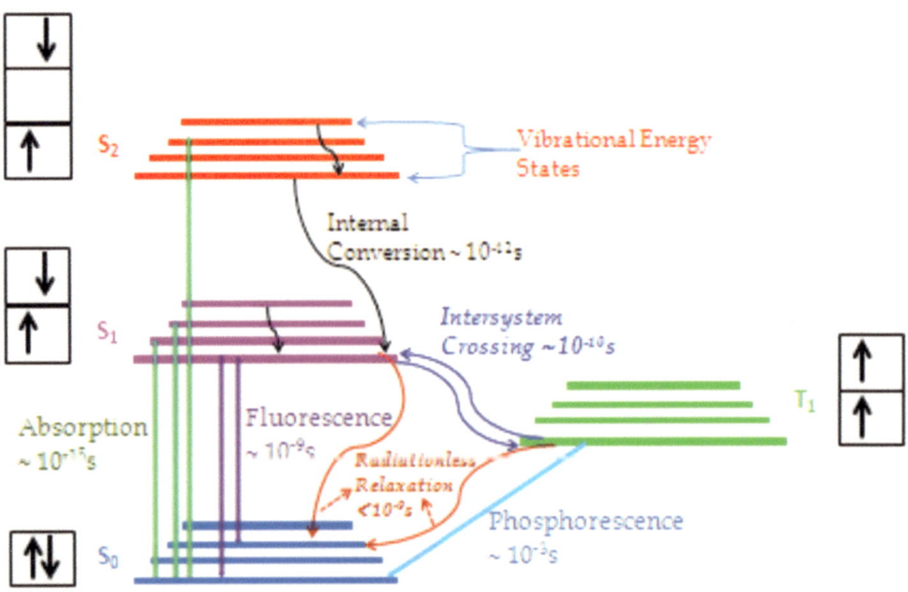

Fig. 7. The Jablonski diagram. Four electronic levels are depicted along with four vibrational energy levels.

Since the energy spacing between the vibrational levels in S_0 or S_1 is of the same size, there often exist mirror image symmetry between the emission spectrum and the $S_0 \rightarrow S_1$ absorption spectrum (not the $S_0 \rightarrow S_2$ absorption) (Fig. 8), needless to say there are plenty of exceptions to the rule. The transitions that will predominate can be justified by Franck-Condon principle. The principle states that since the electronic absorption of light occurs in extremely short time (~10^{-15}s), thus during the time scale of absorption the nuclei are assumed to be frozen, that is that the transitions between various electronic levels are vertical (Rendell, 1987; Lakowicz, 2006).

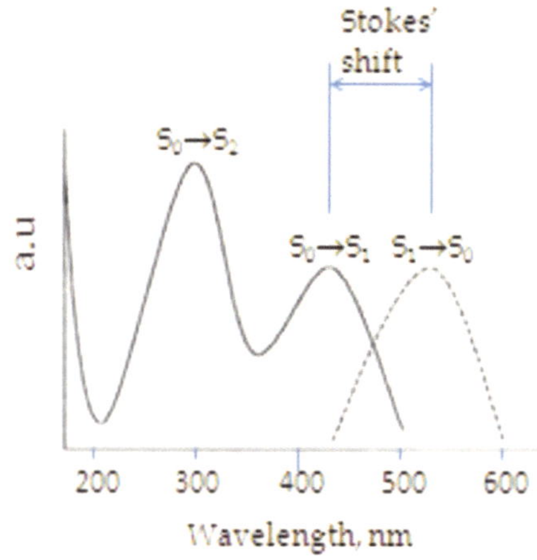

Fig. 8. Typical absorption and emission spectra.

The typical molecular photoluminescence relaxation processes that illustrated by Jablonski diagram (Fig. 7) can be classified to two main type of transition, these are radiative and nonradiative transitions. The non-radiative relaxation processes are vibrational relaxation (a rapid relaxation of excited molecules from higher to lowest vibrational level, occur within ~10^{-14}-10^{-12} s), internal conversion (a rapid relaxation of excited molecules to the lowest energy singlet excited state (S_1) from higher energy excited singlet state (such as S_2 in Fig. 7), occur within a time scale of 10^{-12} s), and intersystem crossing (relaxation between excited states of different spin multiplicity, such as $S_1 \rightarrow T_1$ in Fig. 7, occur within a time scale of 10^{-8} s). Intersystem crossing occurs more slowly than internal conversion since it's a less probable process than internal conversion due to spin multiplicity is not conserved.

The radiative processes are fluorescence and phosphorescence (Figure 9) (Lakowicz, 2006). The fluorescence refers to the emission of light associated with a radiative transition from an excited electronic state that has the same spin multiplicity as the ground electronic state ($S_1 \rightarrow S_0$, Fig. 7). Fluorescence transitions are spin allowed, thus they occur very rapidly and the average lifetimes of the excited states responsible for fluorescence are typically 10^{-9}-10^{-5} s. Phosphorescence refers to the emission of light associated with a radiative transition from an excited electronic state that has a different spin multiplicity from that of the ground electronic state ($T_1 \rightarrow S_0$, Fig. 7). Phosphorescence transitions are spin forbidden, thus they are less probable than spin allowed transitions and the average lifetimes of the excited states responsible for phosphorescence are typically 10^{-3} s. However, spin forbidden transitions become more probable when a significant interaction between the spin angular moment and the orbital angular momentum is observed (spin-orbit coupling increases), this can be observed in the presence of heavy atoms. Furthermore, in solutions, the presences of paramagnetic species such as molecular oxygen increase the probability of intersystem crossing and consequently make the spin forbidden transitions more probable.

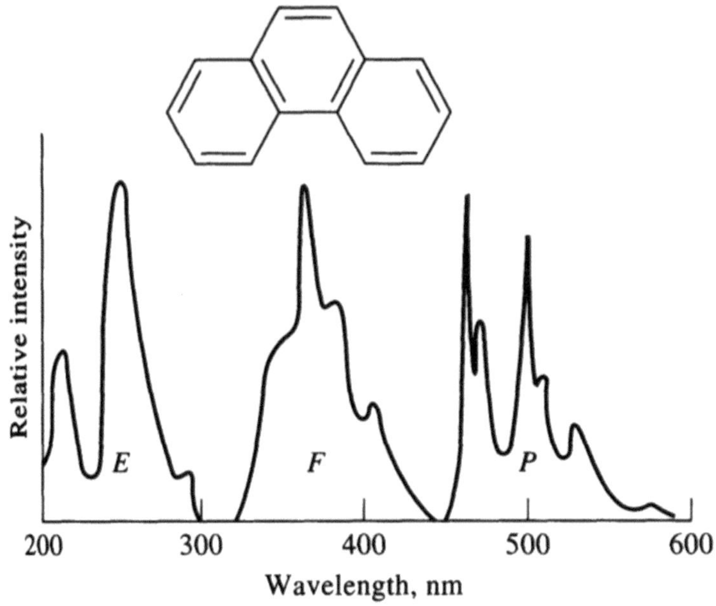

Fig. 9. Typical excitation (E), fluorescence (F), and phosphorescence (P) spectra of phenanthrene (Lakowicz, 2006).

The relative intensity of fluorescence peak is controlled by the Frank-Condon principle, but also the total fluorescence peak intensity (I) is related to the fluorescence quantum yield (Φ_F), which defined as the ratio of number of photons emitted to number of photons absorbed. Furthermore, the fluorescence quantum yield (Φ_F) can be expressed as the rate of photons emitted divided by the total rate of depopulation of the excited state (Equation 1) (Rendell, 1987; Lakowicz, 2006).

$$\Phi_F = \frac{k_F}{k_F + \Sigma k_{nr}} \tag{1}$$

Where k_F and k_{nr} are the rate constant of fluorescence and non-radiative processes, respectively. The fluorescence quantum yield (Φ_F) value in the range of 0.0 to 1.0. If the non-radiative relaxation is fast compared to fluorescence ($k_{nr} > k_r$), Φ will be small, that is the compound will fluoresce very little or not at all. Often different non-radiative events are limited in the solid phase, and long-lived luminescence (e.g. phosphorescence) is often studied in frozen solution or other solid phases. Quenchers make non-radiative relaxation routes more favorable and often there is a simple relation between Φ and the quencher concentration. The best-known quencher is probably O_2, which quench almost all fluorophores; other quenchers only quench a limited range of fluorophores. If a molecule is subject to intramolecular quenching, Φ may yield information about the structure.

The factors that affect the fluorescence quantum yield (Φ_F) are: (1) the excitation wavelength (λ_{ex}), the short wavelengths' break bonds and increase the rate constant of dissociation processes, the most common excitation wavelength are involving the $n \rightarrow \pi^*$ and $\pi \rightarrow \pi^*$

transitions; (2) lifetime of the excited state, the transition probability measured by the molar absorbitivity (ε), large ε implies short lifetime, thus largest fluorescence are observed from short lifetime and high ε state, as an example $\pi^* \rightarrow \pi > \pi^* \rightarrow n$ (10^{-9}-10^{-7} s > 10^{-7}-10^{-5} s); (3) structure of the molecule, emission of light is favored in aromatic molecules (conjugated systems) involving $n \rightarrow \pi^*$ and $\pi \rightarrow \pi^*$ transitions, and fluorescence increased by number of fused rings and substitution on or in the ring, such as pyridine, pyrrole, quinoline and indole; (4) rigidity of the structure, fluorescence quantum yield increases with increasing the rigidity of the molecules specially with chelation, such as fluorine and biphenyl; and (5) the fluorescence quantum yield is highly dependence on the temperature, pH and solvent (Rendell, 1987; Lakowicz, 2006).

The total fluorescence intensity (*I*) is given by Equation 2:

$$I_F = 2.303 I_o \Phi \varepsilon c b \qquad (2)$$

Where I_o is the incident radiant power, the term εcb is deduced from the well known Beers' law expression for the absorption (ε is the molar absorptivity, c is the molar concentration of the fluorescent substance, and b is the path length of the cell). The Beers' law is valid only for diluted solutions ($\varepsilon bc < 0.05$) (Guilbault, 1990)

When an emission spectrum is obtained, data are typically collected for more than 0.1 sec. at each wavelength increment (typically 1nm), but since fluorescence lifetimes typically is measured in nanoseconds, it follows that the obtained spectrum is a time-average of a many events. The time averaging loses much information, and time-resolved experiments are often the more interesting when a system is investigated. The fluorescent lifetime of the excited state, τ_F, is the average time a molecule stays in the excited state before returning to ground state. Thus τ_F can be expressed as the inverse of the total depopulation rate as in Equation 3 (Rendell, 1987; Lakowicz, 2006).

$$\tau_F = \frac{1}{k_F + k_{nr}} \qquad (3)$$

Where k_F and k_{nr} are the rate constant of fluorescence and non-radiative processes, respectively.

Typically fluorescence lifetime values are in the 5-15 ns range. The expression in Eq. 3 is related to the expression for Φ_F, in that way that they have a common denominator. Actually an approximation of τ_F can be obtained by measuring Φ_F in aired and degassed solutions.

In the absence of non-radiative relaxation (k_{nr}= 0), the lifetime becomes the inverse of k_r and is often called the natural lifetime, denoted τ_N. For many compounds the natural lifetime can be calculated from the measured lifetime τ and the measured quantum yield Φ_F, Equation (4) (Rendell, 1987; Lakowicz, 2006).

$$\tau_N = \frac{\tau_F}{\Phi_F} \qquad (4)$$

It is important to notice that the fluorescent lifetime is what is experimentally obtained, and the natural lifetime can be calculated.

The fluorescence lifetime, τ_F, is determined by observing the decay in fluorescence intensity (decay profile) of a fluorophore after excitation. Immediately after a molecule is excited the fluorescence intensity will be at a maximum and then decrease exponentially according to Equation 5 (Rendell, 1987; Lakowicz, 2006).

$$I(t) = I_o e^{-t/\tau_F} \tag{5}$$

Thus after a period of τ_F the intensity has dropped to 37% of I_0, that is 63% of the molecules return to the ground state before τ_F. In many cases the above expression needs to be modified into more complex expressions. First of all it is assumed that the instrument yields an infinite (or very) short light pulse at time zero. In cases where τ_F is small I_0 must be replaced by a function, which describes the lamp profile of the instrument. Also, more than one lifetime parameter is often needed to describe the decay profile, which is I(t) must be expressed as a sum of exponentials. Finally the concept of anisotropy should be mentioned. Anisotropy is based on selectively exciting molecules with their absorption transition moments aligned parallel to the electric vector of polarized light. By looking at the polarization of the emission the orientation of the fluorophore can be measured. The anisotropy of the system is defined as (Equation 6) (Rendell, 1987; Lakowicz, 2006):

$$r = \frac{I_{\parallel} - I_{\perp}}{I_{\parallel} + 2I_{\perp}} = \frac{3\langle \cos^2(\theta) \rangle - 1}{2} \tag{6}$$

The oval in Figure 10 symbolized the absorption transition moment. Vertical polarized excitation light enters along the x-axis and $I_{\perp}$ and $I_{\parallel}$ are measured along the y-axis, setting the emission polarizer perpendicular and parallel to the excitation polarizer respectively. Θ is the angle of the emission to the z-axis (see Figure 10), the squared brackets indicates that it is the average value. If all absorption transition moments are aligned along the z-axis then $I_{\perp}$ = 0 and θ = 0, leading to r = 1, the maximum anisotropy.

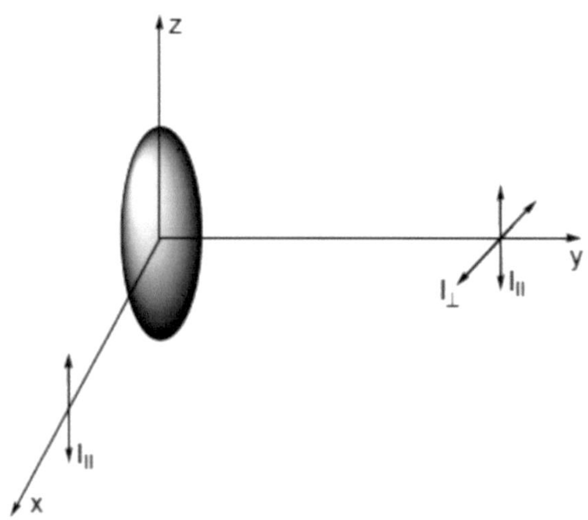

Fig. 10. The absorption dipole is aligned along the z-axis. The excitation light is vertically aligned and follows the x-axis. Emission is measured along the y-axis.

By combining anisotropy with time-resolved measurements it is possible to measure the mobility of a fluorophore. Immediately after excitation all excited molecules will be oriented along a common axis. In the solid phase the system will retain its anisotropy until emission. However, if the fluorophores are free to move, the anisotropy of the system will decrease before emission.

2.2 Instrumentation

The principal sketch of a typical fluorescence spectrophotometer is shown in Figure 11. It consists of a light source, an excitation and emission monochromator (*grooves/mm*), polarizers (*prisms*), sample chamber and a detector (*such as photomultiplier tube*). For steady state measurements the light source usually consists of a 450W xenon arc lamp, and for time resolved measurements it is equipped with nanosecond flash lamp. Most simple spectrometers have a similar geometry, but often extra detectors and/or light sources are fitted resulting in a T- or X-geometry.

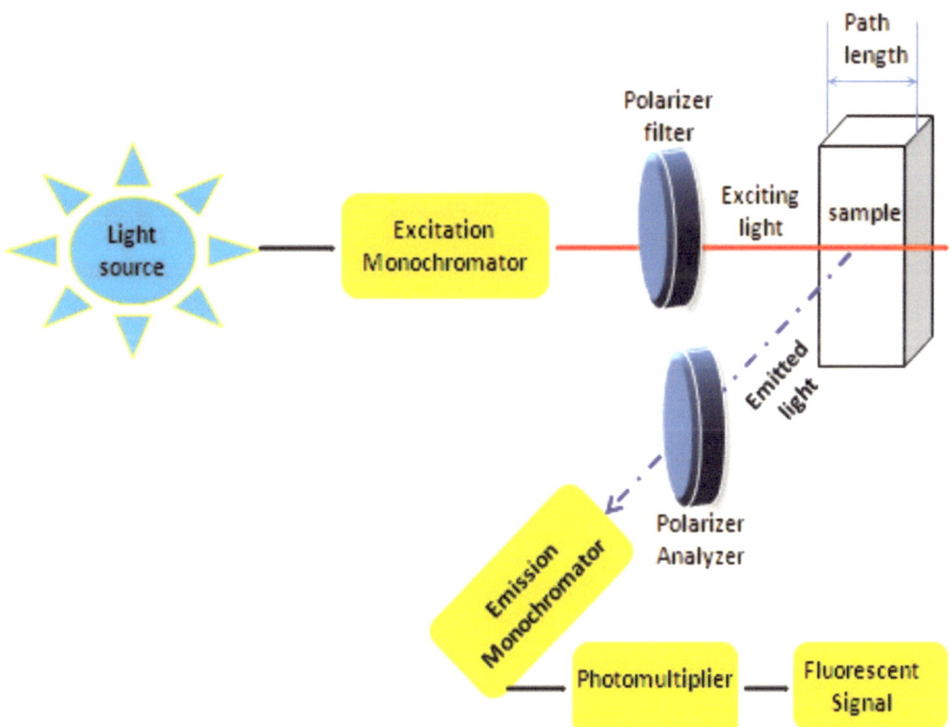

Fig. 11. Schematic representation of a fluorescence spectrophotometer.

The light source produces light photons over a broad energy spectrum, typically ranging from 200 to 900 nm. Photons impinge on the excitation monochromator, which selectively transmits light in a narrow range centered about the specified excitation wavelength. The transmitted light passes through adjustable slits that control magnitude and resolution by further limiting the range of transmitted light. The filtered light passes into the sample cell causing fluorescent emission by fluorphors within the sample. Emitted light enters the emission monochromator, which is positioned at a 90° angle from the excitation light path to eliminate background signal and minimize noise due to stray light. Again, emitted light is transmitted in a narrow range centered about the specified emission wavelength and exits through adjustable slits, finally entering the photomultiplier tube (PMT). The signal is amplified and creates a voltage that is proportional to the measured emitted intensity. Noise in the counting process arises primarily in the PMT. Therefore, spectral resolution and signal to noise is directly related to the selected slit widths (Guilbault, 1990; Rendell, 1987; Lakowicz, 2006).

Not all fluorimeters are configured as described above. Some instruments employ sets of fixed band pass filters rather than variable monochromators. Each filter can transmit only a select range of wavelengths. Units are usually limited to 5 to 8 filters and are therefore less flexible. Fiber optics are also employed for "surface readers", to transmit light from the excitation monochrometers to the sample surface and then transport emitted light to the emission monochrometers. This setup has the advantage of speed, but has the disadvantages of increased signal to noise, due to the inline geometry, and smaller path length which increase the probability of quenching.

Fluorescence requires a source of excitation energy. There are several main types of light sources that are used to excite fluorescent dyes. This section introduces the types of commonly used excitation sources and presents some of the ways that filters can be used to optimize your experimental result. The most popular sources used for exciting fluorescent dyes are broadband sources such as the mercury-arc and tungsten-halogen lamps. These produce white light that has peaks of varying intensity across the spectrum. In contrast, laser excitation sources, which will be described later, offer one or a few well-defined peaks, allowing more selective illumination of your sample. More recently, high-output light emitting diodes, or LEDs, have gained popularity due to their selective wavelengths, low cost and energy consumption, and long lifetime.

When using broadband white light sources it is necessary to filter the desired wavelengths needed for excitation; this is most often done using optical filters. Optical filters can range from simple colored glass to highly engineered interference filters that selectively allow light of certain wavelengths to pass while blocking out undesirable wavelengths. For selective excitation, a filter that transmits a narrow range of wavelengths is typically used. Such a filter is called a band pass excitation filter.

The high intensities and selective wavelengths of lasers make them convenient excitation sources for many dyes. The best performance is achieved when the dye's peak excitation wavelength is close to the wavelength of the laser. Compact violet 405 nm lasers are replacing expensive UV lasers for most biological work. The most commonly used lasers are the 488 nm blue-green argon laser, 543 nm helium-neon green laser and 633 nm helium-neon red laser. Mixed-gas lasers such as the krypton-argon laser can output multiple laser lines and therefore may still require optical filters to achieve selective excitation. While a given dye's excitation maximum may not exactly match the laser's peak wavelength, the high power of the laser can still produce significant fluorescence from the dye when exciting at a suboptimal wavelength. Filters are important for selecting excitation wavelengths. They are also important for isolating the fluorescence emission emanating from the dye of interest. Detecting the fluorescence emission of a sample is complicated by the presence of stray light arising from sources other than the emitting fluorophore—for example, from the excitation source. This stray light must be kept from reaching the light-sensitive detector in order to insure that what the instrument "sees" is due only to the fluorescence of the sample itself. When a single dye is used, a filter that blocks out the excitation light to reduce background noise, but transmits everything else is often a good choice to maximize the signal collected. Such a filter is called a long pass emission filter (Guilbault, 1990; Rendell, 1987; Lakowicz, 2006).

If multiple dyes are used in the sample, a band pass emission filter can be used to isolate the emission from each dye. Careful filter selection helps to ensure that the detector registers only the light you are interested in—the fluorescence emitted from the sample.

LEDs are relatively new light sources for fluorescence excitation. Single-color LEDs are ideal for low-cost instrumentation, where they can be combined with simple long pass filters that block the LED excitation and allow the transmission of the dye signal. However, the range of wavelengths emitted from each LED is still relatively broad. Currently their application may also require the use of a filter to narrow the bandwidth.

There are many options for light sources for fluorescence. Selecting the appropriate light source, and filters for both excitation and emission, can increase the sensitivity of signal detection to an astounding degree. Making fluorescence labeling one of the most sensitive detection technologies available.

With recent advances in sensitive array detectors, fiber optic wave guides, high speed electronics and powerful software, many new generations of spectrofluorometers have been developed. These new spectrofluorometers use charged couple devices (CCDs) or photodiode arrays to replace the photomultipliers and avalanche photodiodes used in conventional spectrometers. They offer excellent performance for a wide range of spectroscopic applications from the UV to the near IR. Because of their unique combination of outstanding sensitivity, high speed, low noise, compactness, instantaneous capture of full spectra, low cost and robustness, these detectors have revolutionized spectroscopic detection. A quick glance at today's instrumentation market indicates the popularity of the CCD as the detector of choice. The overwhelming benefits of array detectors are simultaneous and multi-wavelength data acquisition. On the other hand, the use of fiber optics as light guidance allows a great modularity and flexibility in setting up an optical measurement system. Recent applications and a critical comparison between simple luminescence detectors using a light-emitting diode or a Xe lamp, optical fiber and charge-coupled device, or photomultiplier for determining proteins in capillary electrophoresis are presented by Casado-Terrones et al. (Casado-Terrones, 2007).

In summary, from the preceding discussion, four essential elements of fluorescence detection systems can be identified: (1) an excitation source, (2) a fluorophore, (3) wavelength filters to isolate emission photons from excitation photons and (4) a detector that registers emission photons and produces a recordable output, usually as an electrical signal or a photographic image. Regardless of the application, compatibility of these four elements is essential for optimizing fluorescence detection.

For the sample holders, the majority of fluorescence assays are carried out in solution, the final measurement being made upon the sample contained in a cuvette or in a flowcell.

Cuvettes may be circular, square or rectangular (the latter being uncommon), and must be constructed of a material that will transmit both the incident and emitted light. Square cuvettes or cells will be found to be most precise since the parameters of path length and parallelism are easier to maintain during manufacture. However, round cuvettes are suitable for many more routine applications and have the advantage of being less expensive. The cuvette is placed normal to the incident beam. The resulting fluorescence is given off equally in all directions, and may be collected from either the front surface of the cell, at right angles to the incident beam, or in line with the incident beam. Some instruments will provide the option of choosing which collecting method should be employed, a choice based upon the characteristics of the sample. A very dilute solution will produce fluorescence equally from any point along the path taken by the incident beam through the sample. Under these conditions, the right-angled collection method should be used since it has the benefit of minimizing the effect of light scattering by the solution and cell. This is the usual measuring condition in analytical procedures.

2.3 Sample preparation

Fluorescence is a very sensitive technique. This is the one criterion that makes it a viable replacement to many radioisotope-labeling procedures. However, it is extremely susceptible to interference by contamination of trace levels of organic chemicals. Potential sources of contamination are ubiquitous since any aromatic organic compound can be a possible source of fluorescence signal. For example, the researcher is a possible source of this type of contamination since oils secreted by the skin are fluorescent. Good laboratory procedure is essential in preventing solvents and chemicals from becoming contaminated with high background fluorescence that could hinder low-level measurements. Solvents should be of the highest level purity obtainable commercially. In addition, care must be taken to eliminate all forms of solid interference (suspended particulates such as dust and fibers). These will float in and out of the sampling area of the cuvette via convection currents, and cause false signals due to light scattering while they remain in the instrument's beam.

Fluorescence spectra and quantum yields are generally more dependent on the environment than absorption spectra and extinction coefficients. For example, coupling a single fluorescein label to a protein reduces fluorescein's quantum yields ~60% but only decreases its molar extinction coefficient by ~10%. Interactions either between two adjacent fluorophores or between a fluorophore and other species in the surrounding environment can produce environment-sensitive fluorescence.

Many environmental factors exert influences on fluorescence properties. All fluorophores are subject to intensity variations as a function of temperature, pH of the aqueous medium, and solvents polarity. In general fluorescence intensity decreases with increasing temperature due to increased molecular collisions that occur more frequently at higher temperatures. These collisions bleed energy from the excited state that produces fluorescence. The degree of response of an individual compound to temperature variations is unique to each compound. While many commercially available dyes are selected for their temperature stability, care should be taken to eliminate exposure of samples to drastic temperature changes during measurement. If possible, the temperature of instrument's sample compartment should be regulated via a circulating water bath. At lower assay temperatures, higher fluorescence signal will be generated. It has been found that a 50%

decrease in the fluorescence signal of yellow-green microspheres when exposed to 160°C for 15 minutes (Guilbault, 1990).

Fluorescence variations due to pH changes are caused by the different ionizable chemical species formed by these changes. The results from these pH variations can be quite drastic since new ionization forms of the compound are produced. Fluorescence spectra may be strongly dependent on solvent. This characteristic is most usually observed with fluorophores that have large excited-state dipole moments, resulting in fluorescence spectral shifts to longer wavelengths in polar solvents. As the polarity of environment decreases, the fluorophore shows a shift to longer wavelength with an increase in fluorescence quantum. Also, in polar environments the fluorescence quantum yield decreases with increasing temperature, while in nonpolar environment very little change in the fluorescence quantum yield was observed. Never the less, the environmental sensitivity of a fluorophore can be transformed by structural modifications to achieve desired probe specificity (Guilbault, 1990; Rendell, 1987; Lakowicz, 2006).

In summary, all fluorescence data required for any research project will fall into one of the following categories: (1) the fluorescence emission spectrum, (2) the excitation spectrum of the fluorescence, (3) the quantum yield, and (4) the fluorescence lifetime. In a typical emission spectrum, the excitation wavelength is fixed and the fluorescence intensity versus wavelength is obtained. Early examination of a large number of emission spectra resulted in the formulation of certain general rules: (1) in a pure substance existing in solution in a unique form, the fluorescence spectrum is invariant, remaining the same independent of the excitation wavelength (known as Kasha's rule), (2) the fluorescence spectrum lies at longer wavelengths than the absorption, (3) the fluorescence spectrum is, to a good approximation, a mirror image of the absorption band of least frequency. These general observations follow from consideration of the Jablonski diagram shown earlier. The fluorescence spectrum gives information about processes that happens when the molecules are in the excited stat.

3. Applications

In this section, the applications of fluorescence spectrophotometry as a powerful tool for quantitative analysis, characterization, and quality control in different fields will be reviewed and discussed in details. This section will include the use of fluorescence spectrophotometry as a powerful spectroscopic tool in several fields of science.

3.1 Organic analysis

In the period reviewed so many papers using fluorescence spectrophoometry for analysis, characterization, and as a tool for identification of several compounds are appeared in the literature. In this section, few recent methods will be summarized.

For example, the design and development of artificial molecular systems for sensing anions in biologically relevant conditions is a challenging task in supramolecular chemistry. In particular, sensing fluoride anion has attracted increasing interest in the molecular recognition community because of its pivotal importance in many areas of biological and chemical sciences. In recent years high levels of fluoride in drinking water have caused numerous human diseases, creating a crucial need for artificial sensors to detect fluoride anions in an aqueous environment. Recently, highly sensitive fluorescence "Turn-On" indicators, triisopropylsilyl-protected coumarin derivatives, for fluoride anion with remarkable selectivity in organic and aqueous media have been developed (Sokkalingam & Lee C-H., 2011). This developed method exhibited new fluorescent sensors systems for fluoride anion detection that proved to be simple, inexpensive, and highly selective and, achieve accurate determination with a low detection limit. In addition, the results of this study showed that the system can detect inorganic fluorides as quickly as organic ones by simply introducing chelating agents such as crown ethers. The easily prepared indicator system synthesized here could be an ideal chemodosimeter for detecting and determining fluoride anion in both organic and aqueous solutions and could lead to development of a convenient and reliable detection method for fluoride anion in practical and commercial applications.

Fluorescence is a powerful tool for structural and functional studies of a diversity of molecules. Among the various fluorophores, pyrene derivatives are attractive fluorescent probes. Therefore, pyrene "click" conjugates of 7-deazapurine and 8-aza-7-deazapurine nucleosides and a basic pyrene compound has been synthesized (Ingale et al., 2011). The influence of the nucleobase on fluorescence quenching was studied on nucleoside and oligonucleotide level. This study showed that the favorable photophysical properties of 8-aza-7-deazapurine pyrene conjugates improve the utility of pyrene fluorescence reporters in oligonucleotide sensing as these nucleoside conjugates are not affected by nucleobase induced quenching. This improves the utility of pyrene fluorescence reporters for detection of oligonucleotides.

Selective recognition of Ag^+ ions and amino acids is an important area of research due to their involvement in chemical, biological, and environmental applications. Silver compounds are used as antimicrobial agents, and the activity is closely related with the interaction of Ag^+ with sulfurdryl (–SH) groups. There are several chemosensors reported in the literature for the recognition of Ag^+ ion in nonaqueous and aqueous systems. However, the molecular receptors which can recognize Ag^+ followed by amino acids are indeed limited in the literature. Recently, a new 1,1'-thiobis(2-naphthoxy)-based receptor molecule (L) containing a benzimidazole moiety has been synthesized (Dessingou et al., 2011). The selectivity of L has been explored in aqueous methanol, resulting in selective (7.5 ± 0.5)-fold *switch-on* fluorescence response toward Ag^+ among 14 different transition, alkali, and alkaline earth metal ions studied.

There is a growing interest in the development of molecular sensors that can detect selectively metal ions even in low concentrations. Among the various techniques used for this purpose fluorescence-based methods have gained in importance because of their sensitivity. These methods depend upon the change of fluorescence intensity and/or a shift in the fluorescence band of the sensor upon interaction with the metal ion. Although such a methodology has been successful for diamagnetic metal ions, its application to paramagnetic metal ions is fraught with difficulties in view of the fact that the latter quench fluorescence either via energy or electron transfer and only in some instances a fluorescence enhancement has been observed. In this regard, development of fluorescence-based sensors for Cu(II) has assumed importance in view of the fact that it is an essential trace element and yet at slightly increased concentrations, it is toxic, being implicated in gastrointestinal, liver, and kidney diseases as well as in neurological diseases such as Alzheimer's or Parkinson's. Therefore, Chandrasekhar et al. have demonstrated a simple approach for the design of fluorescence-based sensors (Chandrasekhar et al., 2009). This methodology consists of assembly of phosphorus-supported coordinating platforms whose fluorescence properties are modulated by binding with Cu(II) as well as by the number of coordinating arms that the ligand possesses. We believe that this design is quite general and can be applied for selective detection of other type of metal ions also.

In recent decades, colorimetric and fluorometric sensors have been used in various scientific fields. In biology, for instance, such sensors are useful reagents for living cell imaging. It is important to design novel sensors because they have the potential to overcome many technical limitations in experiments. Imidazo[1,2-a]pyrazin- 3(7H)-one (*imidazopyrazinone*) often is used as a bioluminescent substrate, and it is an attractive core structure for useful sensors. In this regard, a new series of imidazopyrazinones [7-benzylimidazo[1,2 a]pyrazin-3(7H)-one derivatives] have been prepared and their fluorescent properties in the presence of various metal ions (M^{n+}) have been studied (Hirano, et al., 2010).

Copper amine oxidases (CAOs) are a large family of copper-containing quinone-dependent amine oxidases that can be found in all living organisms including bacteria, yeast, plants, and mammals. Human CAOs have been implicated in a number of diseases, including atherosclerosis, cardiovascular diseases, diabetes, Alzheimer's disease, and cancer. To study the kinetic behavior of highly potent inhibitors of the CAO bovine plasma amine oxidase (BPAO), it has been sought a sensitive and real time assay to monitor low levels of enzyme activity. The most sensitive assay for CAOs is the fluorometric coupled assay, which monitors generation of hydrogenperoxide during substrate turnover using horseradish peroxidase (HRP) as a secondary detecting enzyme. In this regard, a novel fluorogenic substrate of bovine plasma amine oxidase (BPAO), namely, (2-(6-(aminomethyl)naphthalen-2-yloxy)ethyl)trimethylammonium (ANETA) has been reported, which displays extremely tight binding to BPAO (K_m 183 (14 nM) and yet is metabolized fairly quickly (k_{cat} 0.690 (0.010 s^{-1}), with the aldehyde turnover product (2-(6-formylnaphthalen-2 yloxy)ethyl)trimethyl-ammonium serving as a real time reporting fluorophore of the enzyme activity (Ling et al., 2009). This allowed for the development of a fluorometric noncoupled assay that is two orders of magnitude more sensitive than the spectrophotometric benzylamine assay. ANETA represents the first highly sensitive, selective, and tight binding fluorogenic substrate of a copper amine oxidase that is able to respond *directly* to the enzyme activity *in real time*.

Organic molecules containing a fluorophoric unit combined with site(s) for guest binding purposes have found application in building up florescent signaling systems for biomedical research and chemical logics. Metal ions can act effectively as guests for these molecules because of their ability to enhance, shift or quench luminescent emissions of these organic ligands by coordination. The changes brought about by metal binding are mechanistically of four types, photoinduced electron transfer (PET), photoinduced charge transfer (PCT), formation of monomer/excimer, energy transfer and proton transfer. Transition metal ions with partially filled d-orbitals are known to induce fluorescence quenching by oxidative or reductive PET and energy transfer processes. In this regard, dioxomolybdenum(VI) complexes [$MoO_2(B^2)H_2O$], [$MoO_2(B^2)$EtOH], [$MoO_2(B^3)$EtOH] and [$MoO_2(B^4)$EtOH] were synthesized using several Schiff base ligands (B^1, B^2, B^3, and B^4) (Gupta et al., 2009). These ligands (B^1, B^2, B^3, and B^4) were prepared by condensation of 1-(2-pyridyl)-5-methyl-3-pyrazole carbohydrazide with salicylaldehyde, o-hydroxy acetophenone, 5-bromo salicylaldehyde and 5-nitro salicylaldehyde, respectively. Due to the presence of a substituted 1-(2-pyridyl) pyrazole unit, these ligands exhibit fluorescent emissions. As the ligands are capable of using different binding modes, according to the demands of the guest metal ions, their emission properties also change accordingly.

3.2 Inorganic analysis

The development of sensors for metal ions in solution has always been of particular importance for cations with biological and environmental interest. The molecular devices converting metal ions recognition in physical recordable signal are continuously growing. In the last few years great attention has been paid to fluorescent chemosensors and many new systems were synthesized. In particular, an effective fluorescent sensor for metal ions is a system able to interact with the metal ion in solution signaling its presence by changing fluorescence properties, as the wavelength or emission intensity, as well as by the appearance of a new fluorescence band different from those of the free sensor. Recently, paper reviews ligand molecules containing fluorophores synthesized and employed in metal ions sensing in solution in the last few years has been published (Formica et al. 2012). The large number of references reported in the review highlights the synthesis of new fluorescent chemosensors able to sense and signal metal ions in solution. This is a prosperous and still emerging field. In this review the authors concluded that the discovery of newer and more efficient emitting units is desirable. In this case, research is now focused on the synthesis of fluorescent units able to shift the emission in an optical range where the biological noise is reduced in the near infrared region (NIR) thus to increase the sensitivity of the chemosensor for in vivo analysis.

Metal-polypyridine complexes are extensively used in photochemical applications, such as solar energy conversion (Akasheh & Al-Rawashdeh, N.A.F, 1990), due to their peculiar excited-state dynamics. These complexes, of which ruthenium tris-bipyridine ($[Ru(bpy)_3]^{2+}$) is considered the prototype, exhibit a visible absorption band due to the singlet metal-to-ligand charge transfer state. The principle of the dye sensitized solar cells is based on the use of such metal-based molecular systems, of which the RuN3 ($[Ru(dcbpyH_2)_2(NCS)_2]$) dye is the most popular, adsorbed onto a semiconductor substrate (usually TiO_2). In this regards, femtosecond-resolved broadband fluorescence studies are recently reported for $[M(bpy)_3]^{2+}$ (M=Fe, Ru), RuN3 and RuN719 ($(Bu_4N)_2[Ru(dcbpyH)_2(NCS)_2]^{2-}$) complexes in solution (Bram et al. 2011). In this study, the pump wavelength dependence of the fluorescence of aqueous $[Fe(bpy)_3]^{2+}$ and the solvent and ligand dependence of the fluorescence of Ru-complexes excited at 400 nm have been investigated. The RuN3 and RuN719 are asymmetric complexes contrary to $[Ru(bpy)_3]^{2+}$, which allows us to explore the effects of molecular geometry on the ultrafast relaxation dynamics of this class of molecules.

Ru(II) polypyridyl complexes have been widely used as DNA, cation, and anion sensors, because their outstanding photophysical and electrochemical properties are quite sensitive to external stimuli (Schmittel, 2007, as cited in Cheng, 2010). Transfer of protons can be regarded as one of the simplest external stimuli and can induce the switching of properties such as fluorescence and UV–Vis absorption for pH sensors, so some Ru(II) polypyridyl complexes containing imidazole fragment have been synthesized. Imidazole-containing ligands are poor π-acceptors and good π-donors and have the appreciable ability to control orbital energies by proton transfer compared with pyridine-, pyrazine-, and pyrimidine-containing ligands, but in most cases these complexes with imidazole rings coordinated to the metal center are nonemissive or weakly emissive and only display deprotonating process [Ayers, 2002, as cited in Cheng, 2010). In this regards, tripodal ligands 1,3,5-tris{4-((1,10-phenanthroline-[5,6-d]imidazol-2-yl)phenoxy)methyl}-2,4,6-trimethylbenzene (L1), 1,1,1 tris{4-((1,10-phenanthroline-[5,6-d]imidazol-2-yl)phenoxy)methyl}propane (L2), 2,2',2''-tris{4-((1,10-phenanthroline-[5,6-d]imidazol-2-yl)phenoxy)ethyl}amine (L3), and corresponding Ru(II) complexes $[(bpy)_6L(Ru(II))_3](PF_6)_6$, have been synthesized (Cheng et al., 2010) . This study showed that the fluorescence spectra of these complexes are strongly dependent on the pH of the buffer solution. These complexes act as pH-induced off–on–off fluorescence switch through protonation and deprotonation of the imidazole-containing ligands. For the same purpose, two novel tetrapodal ligands tetrakis{4-((1,10-phenanthroline-[5,6-d]imidazol-2-yl)phenoxy)methyl}methane (L1), tetrakis {3-((1,10-phenanthroline-[5,6-d]imidazol-2-yl)phenoxy)methyl}methane (L2), and corresponding Ru(II) complexes $[(bpy)_8L(Ru(II))_4](PF_6)_8$ have been synthesized (Cheng et al., 2011). The two complexes act as off–on–off fluorescence pH switches with a maximum on–off ratio of 5. This on–off ratio is moderate compared with those reported for imidazole-containing Ru(II) complexes (Cheng et al., 2010). They have potential utility to detect pH variations of external environment due to their interesting photon-dependent photophysical properties.

Crystal engineering of coordination polymers and supramolecules have attracted lot of attention due to their potential applications as functional materials, as well as their fascinating architectures and topologies (Moulton & Zaworotko, 2010, as cited in Leong, 2009). A successful strategy in the construction of such networks is to utilize appropriate multidentate ligands with flexible backbone that are capable of binding metal ions in different modes. In this regards, a series of metal (Cu(II), Ni(II), Mn(II), Zn(II), Mg(II), Ca(II), and Al(III)) complexes containing the 4-methylumbelliferone-8-methyleneiminodiacetic acid (H_3muia, also named as Calcein Blue) has been synthesized (Leong et al., 2009). In this study, the fluorescence spectroscopy has been used to investigate the hydrogen bonding interactions along with π–π stacking in the synthesized complexes solid-state structures. Solid-state fluorescence studies indicate that complexes of muia have the similar emission properties as in the solution state. Transition metal ions quench the fluorescence of muia while alkali earth and post-transition metal complexes of muia exhibit strong blue emission.

A considerable number of papers have focused on the use of anthracene containing compounds as protein photo cleavers (Hasewage et al., 2006, as cited in Oliveria, 2007), organic light-emitting diodes and materials (Jou, et al., 2006, as cited in Oliveria, 2007), crystal engineering , molecular imprinted polymers, sensors and chemosensors (Magri, et al., 2005, as cited in Oliveria, 2007). Anthracene is one of the most employed chromophores due to its ability to induce PET (photoinduced electron transfer) processes. In this regard, a new scorpionate system (L) containing an emissive anthracene pendant arm, derived of O^1,O^7-bis(2-formylphenyl)-1,4,7-trioxaheptane and tren, has been synthesized (Oliveira et al., 2007). The fluorescence spectroscopy studies conducted suggest that this ligand is an effective complexation molecular device for several divalent metal ions of biological importance as well as for Al(III) and Cr(III), both metals of great relevance in medicine and environmental chemistry. The results of this study could be used as a starting point to develop a more efficiently fluorescence chemosensor based on macrocyclic ligands for these metals.

Since some of lanthanide ions, especially Eu^{3+} and Tb^{3+}, posses good luminescence characteristics (high color purity) based on the 4 f electronic transitions, a variety of rare earth compounds activated by Eu^{3+} and Tb^{3+} have been studied for practical applications (Zhang, et al., 1999, as cited in Xu, et al., 2004). Polydimethylsiloxane (PDMS) materials have broad application in a variety of industrial area so their well-established surface modifying properties. In this regard, Eu(III)-containing polymer complex was synthesized, in which polydimethylsiloxane was used as a polymer ligand (Xu, et al., 2004). The result of this study showed that the fluorescence intensity change of Eu(III)-PDMS complex displays typical fluorescent concentration quenching behavior, in which the emission intensity of the complex was enhanced with increasing content of Eu (III) ion and reaches a maximum at 2 wt.%, and then decreased with further increasing content of Eu(III) ion. These results indicate that the complex contains ionic aggregates in which Eu(III) ions are located close together. Eu(III)-PDMS could transform harmful ultraviolet radiation to blue luminescence that is needed by plants effectively. So Eu(III)-PDMS has a great application foreground.

Schiff base ligands have been extensively studied in coordination chemistry mainly due to their facile syntheses, easily tunable steric, electronic properties and good solubility in common solvents. Transition metal complexes with oxygen and nitrogen donor Schiff bases are of particular interest (You et al., 2004) because of their ability to possess unusual configurations, be structurally labile and their sensitivity to molecular environments. In this regard, the preparation and structures of nickel(II) zinc(II) and cadmium(II) complexes with the related Schiff base ligand N-2-pyridylmethylidene-2- hydroxy-phenylamine have been investigated (Majumder et al. 2006). The results of this study showed that the complexes with zinc(II) and cadmium(II) metals can serve as potential photoactive materials as indicated from their characteristic fluorescence properties. Since the complexes with zinc(II) and cadmium(II) metals possess an intense fluorescence property at room temperature, which is not observed for complexes with Ni(II). It is suggested that complexes with Zn(II) and Cd(II) exhibit potential applications as photoactive materials.

3.3 Pharmaceutical analysis

Over the last three decades, wide ranges of thermodynamic data concerning the guest host complexation have been reported. Cyclodextrins (CDs) are cyclic oligosaccharides composed of glucopyranose units linked together via oxygen bridges at the 1 and 4 positions (a-(l,4)-glycoside bonds) (Bender& Komiyama, 1978). This class of organized media possesses a hydrophilic upper and lower rims lined with hydroxyl groups and a hydrophobic cavity due to C^3H, C^5H, and C^6H hydrogen's and O^4 ether oxygen. This structure gives CDs the ability to extract a variety of organic guest molecules of appropriate size and hydrophobicity from the bulk aqueous solution (Szejtli, 1988). Complexation of various guest compounds with CDs generally results in the improvement of some physical properties of the guest molecules, such as stability, bioavailability, membrane permeability, and solubility. In luminescence studies, CDs have been employed to enhance fluorescence emission of different luminophors (Bender & Komiyama, 1978; Szejtli, 1988) and to induce room temperature phosphorescence under appropriate conditions. In this regards, there is a lot of literature reported in several journals of interest, as an example, the inclusion of the anti-inflammatory drug, Nabumetone (NAB), in γ-cyclodextrin (γ-CD) was studied by fluorescence measurements (Al-Rawashdeh. N.A.F., 2005). Nabumetone is poorly soluble in water and exhibits intrinsic fluorescence. The results of this study showed that the emission fluorescence spectrum, of NAB reveals a maximum whose intensity increases with the different γ-CD's growing concentrations. It is noteworthy mentioning that significant alterations in the physicochemical properties of the included molecule (NAB) are observed upon forming the inclusion complex with γ-cyclodextrin, such as stability and solubility in aqueous media.

Imidazoline-derived drugs are a family of drugs that is structurally distinguished by the existence of the heterocyclic ring of imidazoline that enables these drugs to interact with adrenergic receptors via stimulating presynaptic and postsynaptic a-adrenoceptors (Parini et al., 1996). The majority of the imidazoline-derived drugs are frequently used for their agonist activity, whereas others are used for either antihypertensive, antihistaminic, or agonistic activity (Kaliszan et al., 2006). However, the principle pharmaceutical applicability of these drugs is due to their vascoconstrictive effects. Three typical imidazoline-derived drugs were selected for this study; Naphazoline (NP) (4,5-Dihydro-2-(1-naphthalenylmethyl-1H-imidazole), Antazoline (AN) (4,5-Dihydro-Nphenyl-N-(phenylmethyl)-1H-imidazole-2-methanamine, 2-[(N-benzyl anilino) methyl]-2-imidazole, and Xylometazoline (XM) (2-(4-tert-Butyl-2,6-dimethylbenzyl)-2-imidazoline). While CDs have a wide range of applications, using CDs as additives in various separation and pharmaceutical sciences is still the foremost application of them. Hence, adding the CDs to the separation media can significantly enhance the separation process, whereas employing CDs as additives to drugs formulation can promote the bioavailability through enhancing the stability and solubility of selected drugs. In this regard, the inclusion complexes of selected imidazoline-derived drugs, namely Antazoline (AN), Naphazoline (NP) and Xylometazoline (XM) with β-cyclodextrin (β-CD) were investigated using steady-state fluorescence (Dawoud & Al-Rawashdeh, N. 2008)). Their results confirmed the formation of the inclusion complexes between the three studied drugs and β-cyclodextrin using steady-state fluorescence spectroscopy. Importantly, the results of their study showed that the geometrical size and polarity of various substituents, such as phenyl and naphthyl groups, have dramatically altered the stability and geometrical configuration of the inclusion complexes. These studies present the state-of-the-art of macromolecular binders and provide detailed illustrative examples of recent developments bearing much promise for future pharmaceutical applications.

Propranolol is a beta-adrenergic blocking drug widely prescribed for the treatment of cardiac arrhythmia, sinus tachycardia, angina pectoris and hypertension (Parfitt, 1990). It has also been suggested for use in a number of other conditions including dysfunctional labour and anxiety. When administered over a long period of time it reduces mortality caused by hypertension and lengthens survival in patients with coronary heart disease. It is also used in low activity sports, reducing cardiac frequency, contraction force and coronary flow. Therefore, it has been included in the list of forbidden substances by the International Olympic Committee. The Spanish Olympic Committee has decided that only a qualitative determination of propranolol in urine is necessary. In this regards, a sensitive fluorescence optosensor for the drug propranolol to use in the analysis of pharmaceutical preparations and as a doping test for the qualitative analysis of propranolol in human urine without lengthy preliminary procedures have been developed (Fernandez-Sanchez et al. 2003).The effect of proteins presents in urine samples were evaluated using the developed flow-through fluorescence optosensor. The results of this study showed that the proposed methods for analysis were satisfactorily applied to commercial formulations and urine samples, which offers excellent analytical parameters, such as sensitivity, selectivity, versatility, and ease of use. The development of optosensing techniques has led to a shorter turnaround analysis time and reduced costs for doping controls. As a large part of the samples prove to be non-doped, rapid analytical methods such as doping tests that provide reliable 'yes/no' responses are of increasing interest. These tests can usually be described as systems that 'filter' samples to select those with analyte content levels 'similar to' or 'higher than' a previously established threshold. These 'probably doped' samples must then be examined with more exact instrumental methods. Doping tests can significantly cut costs and save time.

Anthraquinones are known to be present in many different families such as Polygonaceae, Leguminosae, Rubiaceae, Liliaceae and Rhamnaceae. Recently, a number of pharmacological tests revealed that the anthraquinone derivatives present various biological activities including antifungal (S K. Agarwal, 2000, as cited in He, 2009), antimicrobial (Y. W. Wu, as cited in He, 2009), anticancer (J. Koyamma, as cited in He, 2009) , antioxidant (G. C. Yen, as cited in He, 2009), and antihuman cytomegalovirus activity (D. I. Barnard. As cited in He, 2009). In general, fluorescence detection is sensitive and selective. The five anthraquinones are known to possess natural fluorescence, but it is difficult to analyse and determine their contents by conventional fluorimetry due to their similar molecule structures. Therefore, recently a simple, rapid and sensitive reversed-phase high-performance liquid chromatography (RP-HPLC) method, using fluorescence detection, to simultaneously quantify aloe-emodin, emodin, rhein, chrysophanol and physcion in medicinal plants and their pharmaceutical preparations was developed by He et al. (He et al., 2009). Their method was suitable for use as a tool for routine quality assurance and standardization of the anthraquinone from the raw material and commercially available pharmaceutical preparations containing rhubarb.

Fluorimetry from the early fifties has been among the most frequently used techniques for determining both therapeutically and abuse drugs; probably due to its excellent selectivity and the low detection limits. Also, the fluorimetric technique is the recommended choice for quantifying the purity of active principles. The research on analytical fluorescence applications is in continuous expansion to new automated or semi-automate processes like classic or emergent methodologies on the continuous-flow field. In this way, fluorescence-based methods have found a wide range of analytical applications (Calatayud & Zamora, 1995)). In this regard, a new strategic tool to predict the native fluorescence of organic molecules has been proposed (Albert-Garcia et al. 2009). For this purpose, the molecular connectivity indices of different organic substances (pharmaceuticals and pesticides) were calculated. The work presented in this paper was focused to present a new tool for enhancing the research yield on new analytical applications of fluorescence.

3.4 Biological and biomedical analysis

Cancer has overtaken heart disease as the world's top killer by 2011, part of a trend that should be more than double global cancer cases and deaths by 2030. Cisplatin (cis diamminedichloroplatinum(II)) is one of the most effective anticancer drugs in the treatment of a variety of tumors and it has been clinically used widely. However, its limited usefulness in the development of resistance in tumor cells and the significant side effects have generated new areas of research, which mainly focused on searching for new metal-based complexes with low toxicity and improved therapeutic properties. In this regards, 3-Carbaldehyde chromone thiosemicarbazone and its transition metal (Cu(II), Zn(II) and Ni(II)) complexes were synthesized and characterized systematically (Li et al., 2010). The results of this study showed that the Zn(II) complex can emit blue fluorescence under UV light in solid state and may be used as an advanced material for blue light emitting dioxide devices. However, almost the solid fluorescence of Cu(II) and Ni(II) complexes could not be observed. Thus, the fluorescence property of these complexes was used as a helpful tool to understand the interaction mechanism of small molecule compounds binding to DNA. It was believed that the information obtained from the present work would be useful to develop new potential antioxidants and therapeutic agents for some diseases.

The quantitative determination of micro-amounts of nucleic acid has attracted a great deal of attention in the fields of medicine and molecular biology. Many methods have been developed, such as direct determination, including ultraviolet absorption and determination of ribose or deoxyribose in nucleic acid, spectrophotometry, chemiluminescence, electrochemical chromatography, including high-performance liquid chromatography and paper chromatography, capillary electrophoresis, and resonance light scattering. However, low sensitivity and easy interruption by protein and other biomolecules existed in these methods in common. However, the fluorometric methods make predominant concern because of their high sensitivity and selectivity. Generally, the fluorescence intensity of DNA must be enhanced by fluorescent probes because it emits weak fluorescence itself. In this regard, method for the determination of DNA based on the fluorescence intensity of the gatifloxacin-europium(III) (GFLX-Eu^{3+}) complex that could be enhanced by DNA was developed (Wang et al. 2011). The GFLX-Eu^{3+} complex showed an up to 6-fold enhancement of luminescence intensity after adding DNA. On the basis of the above findings, the fluorescence enhancement effect of the GFLX-Eu^{3+} complex by DNA was investigated in this study in detail.

The green fluorescent proteins (GFPs) originated from the bioluminescent jellyfish Aequorea victoria, were discovered by Shimomura in the early 1960s (Shimomura et al., 1962). In the last few years, green fluorescent protein (GFP) has become one of the most widely used tools in molecular and cell biology. As a noninvasive fluorescent marker in living cells, GFP allows for numerous applications where it functions as a probe of gene expression, intercellular tracer or as a measure of protein-protein interactions. The green fluorescent protein (GFP) has emerged as a powerful reporter molecule for monitoring gene expression, protein localization, and protein–protein interaction. However, the detection of low concentrations of GFPs is limited by the weakness of the fluorescent signal and the low photostability. Recently, the proximity of single GFPs to metallic silver nanoparticles increases its fluorescence intensity approximately 6-fold and the decrease in decay time has been observed by Fu et al. (Fu et al., 2008). Furthermore, single protein molecules on the silvered surfaces emitted 10-fold more photons as compared to glass prior to photobleaching. The photostability of single GFP has increased to some extent. Accordingly, longer duration time and suppressed blinking were observed. The single-molecule lifetime histograms indicate the relatively heterogeneous distributions of protein mutants inside the structure.

Detection of DNA in solution is an important problem in a large variety of biochemical assays. The most popular agents for DNA quantitation are fluorescent dyes that strongly interact with nucleic acids and significantly increase their emission intensity in the DNA complex. Fluorescent dyes are used in real-time PCR, DNA-based cell quantitation, gel staining, chromatin and other DNA-based approaches (Glazer and Rye, 1992, Jing et al., 2003, Lakowicz, 2006, Le Pecq and Paoletti, 1967, Lim et al., 1997, Szpechcinski et al., 2008, as cited in Dragan, 2010). In this regard, both a theoretical and experimental analysis of the sensitivity of a DNA quantitation assay using a fluorescent chromophore which non-covalently binds dsDNA were investigated (Dragan et al. 2010). It is well-known that the range of DNA concentrations available for fluorescence quantitation depends on the concentration of the chromophore, its affinity for nucleic acids, the binding site size on DNA and the ratio between the fluorescence intensity of the chromophore when bound to DNA compared to free chromophore in solution. In this study an experimental data obtained for a PicoGreen® (PG)/DNA quantitation assay was presented, which is in complete agreement with the results of our theoretical analysis. Furthermore, it has been shown, both theoretically and experimentally, that DNA assays based on the MEF of PG demonstrate sensitivity to DNA concentration of ≈1 pg/ml, which is several orders of magnitude more sensitive than without the silver nanoparticles, suggesting the broader practical use of this approach (metal-enhanced PicoGreen® fluorescence) for the ultra-sensitive detection of double stranded nucleic acids.

Identification of living organisms (eukaryotes, bacteria, viruses etc.) by means of quantitative analysis of their specific DNA sequences, which represent different genomes, is a challenging aim, which faces many scientists today. It also concerns the search and detection of different microorganism mutations and strains of pathogenic bacteria and causes of severe diseases in humans. In recent short communication, further development of the "Catch and Signal" technology—principles of a 2-color DNA assay for the simultaneous detection/quantification of two genome-specific DNAs in one well was presented (Dragan et al., 2011) In this method a combination of the Metal-Enhanced Fluorescence (MEF) effect and microwave-accelerated DNA hybridization has been utilized. Furthermore, it is shown that fluorescent labels (Alexa 488 and Alexa 594), covalently attached to ssDNA fragments, play the role of biosensor recognition probes, demonstrating strong response upon DNA hybridization, locating fluorophores in close proximity to silver nanoparticles, which is ideal for MEF. The 2-color "Catch and Signal" DNA assay platform can radically expedite quantitative analysis of genome DNA sequences, creating a simple and fast bio-medical platform for nucleic acid analysis. Their results clearly showed that the 2-color DNA assay can effectively be employed as a new "Rapid Catch and Signal" technological platform in the creation of an ultra-sensitive, sequence-specific approach for the fast analysis of genetic material from different organisms, for potential analysis of bacteria and virus pathogens, and search for possible mutations and sequence variations. This technology being fast, ultra-sensitive and inexpensive can effectively compete with the PCR technique, especially for the routine and rapid analysis in Point-of-Care settings and bio-medical laboratories.

3.5 Environmental analysis

For the past 20 years there has been continued growth in the applications of fluorescence spectroscopy in physical and biological sciences. Because of the sensitivity of fluorescence detection, the fluorometeric method has been one of the selected techniques to determine compounds at low concentrations (Lakowicz, 2006). Examples of such compounds are agrochemicals (pesticides, fungicides, insecticides, etc), which have attracted attention worldwide due to their usage in agriculture. Therefore, the potential increase in fluorescence of a benzimidazole-type fungicide (carbendazim) due to complexation with cucurbit[6]uril was reported (Saleh & Al-Rawashdeh, N.A.F., 2006). The fluorescence enhancement of the fungicide carbendazim by cucurbit[6]uril has been observed in solution due to formation of a host–guest inclusion complex. The enhancement of the carbendazim fluorescence (maximum factor of 10) is accompanied by a significant blue shift in the spectrum (11 nm). In general, Saleh and Al-Rawashdeh work (Saleh & Al-Rawashdeh, N.A.F., 2006) demonstrates the potential usefulness of cucurbit[6]uril in fluorometric analysis of fungicide for agricultural and environmental applications.

Recycling in agriculture of organic wastes produced by various animal breeding, such as pig slurry (PS), is a common practice throughout the world, which has recently raised serious environmental concerns. In particular, Cu(II) and Zn(II) ions, which are used abundantly as pig feed additives, may be introduced in relatively large amounts into PS-amended soils, thus representing an actual risk of phytotoxicity and/or leaching downward the soil with potential endangering groundwater quality (L'Herroux et al., 1997, Saviozzi et al., 1997, Giusquiani et al., 1998, Nicholson et al., 1999, Aldrich et al., 2002, Taboada-Castro et al., 2002, as cited in Hernandez, at al. 2006).

Bioavailability, mobility and transport of metal ions in soils are strongly influenced by binding reactions with soil organic matter, and especially its humified fractions, i.e., humic substances (HS), of which humic acid (HA) is the major component (Aldrich et al., 2002, as cited in Hernandez, at al. 2006). For these reasons, the effects of PS application on the compositional, structural and functional properties of native soil HA, and especially on their Cu(II) and Zn(II) binding behavior, need to be accurately evaluated. Therefore, The effect of the consecutive annual additions of pig slurry on the Cu(II) and Zn(II) binding behavior of soil HAs was investigated in a field experiment by a fluorescence titration method (Hernandez et al. 2006). In Hernandez et al. study (Hernandez et al. 2006), a fluorescence titration method was used for determining metal ion complexing capacities and stability constants of metal ion complexes of humic acid isolated from pig slurry and unamended and amended soils. The results of this study is expected to have a large impact on bioavailability, mobilization, and transport of Cu(II) and Zn(II) ions in pig slurry-amended soils (Hernandez et al. 2006).

Moreover, a luminescent sensor utilizing the substrate 2,6pyridine-dicarbox-aldehydebis-(*o*-hydroxyphenylimine) has been developed for low concentration detection of the environmental mercuric ion pollutants (Kanan et. al 2009). The sensor selectively detects mercury in the presence of several ions that are commonly found in aquatic environments. The sensor was found to be highly selective with strongest binding was observed between the mercuric ions and the substrate at a pH range of 6.5-7.5 which makes the substrate a distinctive fluorescent sensor for detecting mercury under normal environmental conditions. No effect was demonstrated with the addition of any other metal ions commonly found in water, making this compound selective to mercuric ions.

Dissolved organic carbon (DOC) refers to the hundreds of dissolved compounds found in water that derive from organic materials, and is composed of 'organic acids', 'organic bases', and 'neutral groups'. The amount of DOC in the hydrosphere (700 gigatons) is almost the same as the amount of carbon in the atmosphere (750 gigatons). However, a large proportion of rainwater DOC is still uncharacterized. A novel method for qualitatively characterizing DOC compounds in rainfall is to use fluorescence. A large proportion of DOC is fluorescent and this fraction can be used to characterize DOC compounds (Baker & Spencer, 2004, as cited in Muller, 2008), yet its full potential for analyzing rainwater DOC compounds has yet to be achieved. Fluorescence can be used to fingerprint fluorophores, identify DOC compounds, and detect small concentration levels which may have otherwise gone undetected, with just 0.04 ml samples. Given the importance of better characterization of rainwater DOC, and the presence of fluorescent DOC compounds, using fluorescence spectrophotometry, primarily sought to examine the fluorescent DOC compounds present in precipitation in Birmingham, UK has been investigated (Catherine et al., 2008). Furthermore, how the fluorescence of the identified DOC compounds varies with meteorological parameters, by assessing the variations with stratiform/convective storm types; examining the variations with air mass type and source area using back-trajectory analysis; and investigating variations with wind speed and wind direction have been investigated (). The results of this study demonstrate the utility of fluorescence analysis for identifying and characterizing rainwater DOC compounds. This research has revealed information regarding fluorescent rainwater DOC compounds in Birmingham, UK and provides evidence in support of using fluorescence spectrophotometry as a means of qualitatively characterizing rainfall DOC.

Organic pollution originating from industrial, agricultural and municipal wastewater discharge has become a common environmental problem in many lakes, rivers, estuaries and coastal waters. The commonly-used methods to study the composition of dissolved organic matter (DOM) (nuclear magnetic resonance, gas chromatography-mass spectrometry, liquid chromatography-mass spectrometry, size exclusion chromatography, etc.) need complicated sample pretreatment procedures and are not suitable for on-line or real-time determination. Excitation-emission matrix (EEM) fluorescence spectroscopy has been suggested as a powerful tool to characterize aquatic DOM which could detect the specific fluorescent fractions in organic matter from different sources with much higher sensitivity (Coble et al., 1996, as cited in Guo, 2010). Therefore, wastewater dissolved organic matter (DOM) from different processing stages of a sewage treatment plant in Xiamen was characterized using fluorescence spectroscopy by Gue at al. (Guo et al., 2010). The results of this study showed that excitation emission matrix fluorescence of wastewater DOM from a sewage treatment plant revealed high concentration of degradable protein-like fluorescence and the occurrence of xenobiotic-like component. This technique provides a fast and sensitive way to monitor the qualitative and quantitative variation of DOM during the whole sewage treatment process.

Perylene-3,4:9,10-tetracarboxylic bisimide, so-called perylene bisimide (PBI) is known as red vat dyes (*fluorescent dyes*), and has been applied to pigments such as automotive finishes due to its light-fastness, thermal stability, and chemical inertness . Perylene bisimide (PBI) is a fluorescent dye which has strong emission and high photostability. Although PBI has been widely used for industrial materials, the application of PBI in analytical fields was limited mainly due to its high hydrophobicity. In recent years, however, unique and useful analytical methods based on PBI platform are being successfully developed by utilizing the characteristic features of this compound including its high hydrophobicity. In this regard, the recent trend of environmental and biological analysis using PBI was reviewed by Soh &

Ueda (Soh & Ueda 2011). Furthermore, the analytical methods presented in this study are classified based on the detection mechanisms.

3.6 Food analysis

Fluorescence spectroscopy is an instrumental technique whose theory and methodology has been widely exploited for studies of molecular structure and function. It is therefore applicable to address molecular problems in food. Since many foods exhibit auto fluorescence it is considered highly relevant for characterization purposes and high-throughput screening. The main advantages of molecular fluorescence spectroscopy are its sensitivity and selectivity, in addition to its ease of use, instrumental versatility, speed of analysis and its non-destructive character.

Recently, it has been shown that excitation-emission matrix (EEM) fluorescence spectroscopy and three-way statistical methods, concretely Parallel Factor Analysis (PARAFAC) can be used for distinguishing between wine samples of different appellations, or between wine samples obtained with different ageing procedure (Airado-Rodrıguez et al., 2011). At first, excitation-emission matrices (EEMs) were obtained and examined with the aim of identifying the main families of fluorescent compounds. Then PARAFAC was applied for exploratory analysis and the scores of the four selected components were obtained. The set of sample score values obtained in the PARAFAC decomposition were plotted against each other, to visualize clustering trends of samples belonging to different appellations. The potential of the autofluorescence of wine, through its Excitation-emission matrices (EEMs), combined with the three-way method PARAFAC for the purpose of discrimination of wine according to the appellation of origin, is shown in this paper. Wine samples were monitored with rapid and non-destructive front face fluorescence spectroscopy, yielding information about the wine throughout its appellation and ageing conditions. PARAFAC revealed four groups of fluorophores that were responsible for the main fluorescence of Spanish wines. Two of them were assigned to benzoic-like phenolic acids and phenolic aldehydes, and to monomeric catequins and polymeric proanthocyanidin dimers, respectively. The exploration of the matrix of score values obtained by PARAFAC reveals some distribution trends of the samples according with their appellation.

Rapid measurements of milk properties and discrimination of milk origin are necessary techniques for quality control of milk products. A study was undertaken to evaluate the potential of using front face fluorescence spectroscopy (FFFS) and synchronous fluorescence spectroscopy (SFS) for monitoring the quality of forty-five ewe's milk samples originating from different feeding systems by Hammamia et al. (Hammamia et al., 2010). Whereas, the physicochemical analyses and fluorescence spectra were conducted on samples during lactation periods (the first 11 weeks). Results obtained showed a good discrimination among milk samples with regard to feeding systems given to the ewes throughout the lactation periods. In addition, a better discrimination was observed with front face fluorescence spectroscopy than with synchronous fluorescence spectroscopy.

Fumonisins (FBs) are worldwide distributed and produced by Fusarium verticillioides and Fusarium proliferatum, mainly in corn and corn-based products (Soriano & Dragacci, 2007, as cited in Silva et al., 2009). Although several other fumonisin analogues have been characterized, fumonisin B1 (FB1) remains the most abundant in naturally contaminated

Corn-based foods followed by fumonisin B2 (FB2).The problems and risks associated with fumonisin contamination have resulted in the development of precise, reliable and sensitive methods for its determination in corn and corn-based foods (Magan & Olsen, 2004, as cited in Silva et al., 2009). Therefore, the quality parameters in the analysis of FB1 and FB2 in corn-based products obtained with LC with fluorescence detector have been investigated (Silva et al., 2009). Furthermore, a comparison study between fluorescence detector (FD), mass spectrometry, and tandem mass spectrometry with a triple quadrupole (QqQ) analyzer using an electrospray ionization interface for the determination of fumonisin B1 and B2 in corn-based products has been performed. A comparative study of the three LC detectors, FD, single quadrupole, QqQ for the analysis of fumonisins in corn samples has been performed. The response achieved by the three detectors was sensitive enough to study the maximum contents established by the EU legislation. These LC detectors would be appropriate for quantification purposes but the acquisition of at least two transitions achieved with QqQ provided a univocal identification.

Low-molecular-weight compounds have been used widely as animal drugs, food additives, and pesticides, to achieve maximum productivity and profits directly or indirectly through food products. However, residues of such low-molecular-weight compounds in food products have been proven to be detrimental to human health. Therefore, development of a microsphere-based competitive fluorescence immunoassay for the determination of hazardous low-molecular-weight compounds in food has been described (Zou et al. 2008). In this method, antigens are covalently bound to carboxy-modified microspheres to compete monoclonal antibody with low-molecular- weight compounds in food samples; mouse IgG/fluorescein isothiocyanate conjugate is used as the fluorescent molecular probe. Thus, the hazardous low-molecular-weight compounds are quantified using a multiparameter flow cytometer. This method has been evaluated using clenbuterol as a model compound. It has a sensitivity of 0.01 ng/mL with dynamic range of 0.01–100 ng/ mL, and the concentration of clenbuterol providing 50% inhibition (IC50) is 1.1 ng/mL. The main advantages of this method are its high efficiency, biocompatibility, and selectivity, as well as ultralow trace sample consumption and low cost.

The aspects of fluorescence spectroscopy that may have value for solving problems in food science and technology have been summarized in a review article by Strasburg & Ludescher (Strasburg et al., 1995). In this review article, the techniques described, which depend on the measurement of the intensity, energy and polarization of fluorescence emission, have been illustrated by examples taken from the food science and related literature.

3.7 Optical sensors

Glucose is considered as a major component of animal and plant carbohydrates in biological systems. It acts not only as a source of energy of the living cells but also as metabolic intermediate in the synthesis of other complex molecules. Furthermore, blood glucose levels are also an indicator of human health conditions: the abnormal amount of glucose provides significant information of many diseases such as diabetes or hypoglycemia. Accurate determination of glucose is very important in clinical diagnosing as well as in food analysis. To date, various sensors for glucose analysis have been reported, and among them, fluorophotometry was used widely owing to its operational simplicity and high sensitivity (Shang et al., 2008, Li et al., 2009, Shiang et al., 2009, as cited in Jin et al., 2011). Recent advances in the noble metal clusters open a promising field toward the development of a satisfying fluorescence probe. In this regard, recently biomolecule-stabilized Au nanoclusters were demonstrated as a novel fluorescence probe for sensitive and selective detection of glucose (Jin et al., 2011). The fluorescence of Au nanoclusters was found to be quenched effectively by the enzymatically generated hydrogen peroxide (H_2O_2). By virtue of the specific response, the present assay allowed for the selective determination of glucose in the range of 1.0×10^{-5} M to 0.5×10^{-3} M with a detection limit of 5.0×10^{-6} M. In addition, it has been demonstrated the application of the present approach in real serum samples, which suggested its great potential for diagnostic purposes. In comparison with previous approaches for glucose detection, this method required no complicated preparation procedure, and used only commercially available materials. It also exhibited environmentally friendly feature and good sensitivity. Furthermore, the present nanosensor possessed red emission and excellent biocompatibility, which presage more opportunities for studying the biological systems in future applications.

Recently, metalloprotein design and semiconductor nanoparticles have been combined to generate a reagent for selective fluorescence imaging of Pb^{2+} ions in the presence red blood cells (Shete et al., 2009). A biosensor system based on semiconductor nanoparticles provides the photonic properties for small molecule measurement in and around red blood cells. Metalloprotein design was used to generate a Pb^{2+} ion selective receptor from a protein that is structurally homologous to a protein used previously in this biosensing system. This designed protein demonstrates a highly sensitive and selective biosensor that can reversibly detect Pb^{2+} ions in aqueous solutions. The modularity of semiconductor nanoparticle-based biosensors has allowed metalloprotein design and different semiconductor materials to be combined to significantly improve the detection of soluble, exchangeable Pb^{2+} ion concentrations to address inefficient Pb^{2+} ion chelation therapy.

The development of artificial receptors for molecular recognition studies of zwitterion amino acids under the physiological conditions is a very important research area since it can help to understand the important roles of free amino acids in biological systems. In this regard, a new fluorescence macrocyclic receptor based on the Zn(II) complex of a C_2-terpyridine and a crown ether has been developed for molecular recognition of zwitterion amino acids in water/DMF solution with remarkable selectivity towards L-aspartate (K = $4.5\text{x}10^4$ M^{-1}) and L-cysteine (K = $2.5\text{x}10^4$ M^{-1}) (Kwong et al., 2009).

Copper is an essential trace element, its deficiency is one of the causes of anemia, but it is toxic at higher concentration levels. The uptake of copper by human beings above a certain level is known to cause gastrointestinal catarrh, Wilson's disease, hypoglycemia, and dyslexia. Increases in copper concentration in water and plants have resulted from industrial and domestic waste discharge, refineries, disposal of mining washings, and the use of copper as a base compound for antifouling paints. Therefore, the trace copper content in water and food must be monitored on a daily basis. In this regard, a highly sensitive and selective optical sensor for the determination of trace amounts of Cu^{2+} based on fluorescence quenching has been developed (Aksuner et al., 2009). The sensing membrane was prepared by immobilization of a novel fluorescent Schiff base ligand 4-(1-phenyl-1-methylcyclobutane 3-yl)-2-(2-hydroxy-5-romobenzylidene) aminothiazole, on polyvinlyl chloride. The accuracy of the proposed sensor was confirmed by analyzing standard reference materials of natural water and peach leaves. The sensor was successfully applied for the determination of copper in tap water and tea samples. This study showed the application of a PCT dye for preparation of a new Cu^{2+} sensitive optical chemical sensor for the first time. The sensor shows a high selectivity and quick response for Cu^{2+} over other common metal ions.

Heavy metal pollution is a global problem and it causes threat to the environment and human beings. Among the different heavy metal ions, mercury has received considerable attention due to its highly toxic and bioaccumulative properties. It is released from coal burning power plants, oceanic and volcanic emissions, gold mining, and solid waste incineration. Mercury vapor lamps, fluorescent lamps, electrical switches, batteries, thermometers and electrodes are the second largest sources of mercury discharge to the environment. In this regard, an ultrasensitive and selective spectrofluorimetric determination of Hg(II) using 2,5-dimercaptothiadiazole (DMT) as a fluorophore was developed (Vasimalai & John, 2011). In this study, the practical application of the present method was demonstrated by determining Hg(II) in tap water, river water and industrial waste water samples. The obtained results have a good agreement with inductively coupled plasma atomic emission spectrometric (ICP-AES) and atomic absorption spectrometric (AAS) methods. According to the literature, this is the first report for the lowest detection with the highest selectivity for Hg(II) in a water medium by the fluorimetric method.

3.8 Nanomaterials

Noble metal nanocrystals exhibit extraordinary plasmonic properties. The excitation of their localized surface Plasmon resonance modes results in the confinement of electromagnetic waves in regions below the diffraction limit near the metal surface. On the other hand, the localized plasmon modes of elongated metal nanocrystals are inherently anisotropic. The optical signal amplification is therefore expected to be strongly dependent on the excitation polarization direction. Recently, the strong polarization dependence of the plasmon-enhanced fluorescence on single gold nanorods has been reported (Ming et al., 2009). In this study, it has been observed that the fluorescence from the organic fluorophores that are embedded in a mesostructured silica shell around individual gold nanorods was enhanced by the longitudinal Plasmon resonance of the nanorods. The polarization dependence is ascribed to the dependence of the averaged electric field intensity enhancement around the individual gold nanorods on the excitation polarization direction. The maximum fluorescence enhancement occurs when the longitudinal Plasmon wavelength of the Au nanorods is nearly equal to the excitation laser wavelength.

Organic dyes, such as fluorescein, rhodamine, and cyanine, and fluorescent proteins, such as green fluorescent protein (GFP) and its variants, are popular probes for cellular organelles and lipid and protein dynamics, due to their small sizes (<5 nm), high specificity, and aqueous solubility. However, because these fluorescent markers rapidly photobleach, have low quantum yield, and exhibit blinking, it is difficult to obtain quantitative spatial and temporal data on the cellular structures they probe (Yao et al., 2005, as cited in Muddana et al., 2009). Recently, the principle photophysical properties of calcium phosphate nanoparticles (CPNPs) using steady-state and time resolved fluorescence spectroscopy, to demonstrate the potential of these particles for biological imaging and drug delivery and to understand the underlying photophysical mechanisms of encapsulation-mediated fluorescence enhancement have been characterized (Muddana et al., 2009). The enhanced photophysical properties together with excellent biocompatibility make CPNPs ideal for bioimaging applications ranging from single-molecule tracking to in vivo tumor detection while pH-dependent dissolvability of calcium phosphate offers the possibility of timed co-delivery of drugs to control cell function.

Fluorescent nanoparticles have attracted increasing research attention due to their promising applications covering electro-optics to bio-nanotechnology. In this regard, monodispersed water-soluble fluorescent carbon nanoparticles (CNPs) were synthesized directly from glucose by a one-step alkali or acid assisted ultrasonic treatment (Li et al., 2011). The results showed that the particle surfaces were rich in hydroxyl groups, giving them high hydrophilicity. The CNPs could emit bright and colorful photoluminescence covering the entire visible-to-near infrared (NIR) spectral range. In this study they conclude that combining free dispersion in water (without any surface modifications) and attractive photoluminescent properties, CNPs should serve as a promising candidate for a new type fluorescence marker, bio-sensors, biomedical imaging, and drug delivery for applications in bioscience and nanobiotechnology.

Recent advances in ultrasensitive protein biosensors have brought significant impacts to proteomics, biomedical diagnostics, and drug discovery (Zhu et al. 2001, as cited in Huang & Chen, 2008). Advanced nanoscale biosensors based on nanoparticles, nanowires, and other nanomaterials have been developed to detect various proteins with improved sensitivity, specificity, and reliability (Fu et al., 2007, as cited in Huang, 2008). Ultrasensitive fluorescence nanosensors can detect the fluorescence signal from a fluorescence tag bound specifically with a single target molecule, but the plain fluorescence intensity measurement can hardly discriminate against a nonspecifically bound tag. Therefore, an electrically modulated fluorescence protein assay that can detect specific fluorescence from a single molecule assembled on an Au nanowire by manipulating the molecule with an electrical potential applied on the nanowire have been developed (Suxian et al., 2008). In their study, they conclude that the simple electrically modulated fluorescence detection method can be generally applied to various bioassays. The essential requirement of the method is to selectively modulate the specific fluorescence from the target molecules by an external reference field, which can be achieved by electrical, optical, magnetic, mechanical, or biochemical interactions etc.

Inorganic nanomaterials have been widely used in biological and environmental fields such as bio-labeling, imaging, drug delivery, separation processes and optical sensing. In these applications, the size and the shape of nanomaterials have very important effects on their properties. Recently, much attention has been given to one- dimensional nanomaterials for building various sensors, due to its high activity, high surface-to-volume ratio, easy assembly in an array for the device and especial suitability for intracellular detection by inserting it into cell (Zhang et al., 2004; Park et al., 2007, as cited in Xu et al., 2011). Therefore, a fluorescence sensor for selective detection of Cu(II) realized by covalently immobilizing derivatives of rhodamine6G (R6G) on the surface of silicon nanowires (SiNWs) has been designed and fabricated (Xu et al., 2011) The fabricated SiNWs-based chemosensor can be electively used for detection of Cu(II) with Cu(II)-special fluorescence enhancement over other metal ions. The Cu(II) sensor exhibits a good selectivity and sensitivity.

Drug delivery systems (DDS) are intensively investigated in the recent years for their great potential to improve the therapeutic index of small molecular drugs. In this regard, recently, new types of fluorescent nanoparticles (FNPs) were prepared through ionic self-assembly of anthracene derivative and chitosan for applications as drug delivery carriers with real-time monitoring of the process of drug release (Wei et al., 2011). In this study, the potential practical applications as drug delivery carriers for real-time detection of the drug release process were demonstrated using Nicardipine as a model drug. Upon loading the drug, the strong blue fluorescence of FNPs was quenched due to electron transfer and fluorescence resonance energy transfer (FRET). With release of drug in vitro, the fluorescence was recovered again. The relationship between the accumulative drug release of FNPs and the recovered fluorescence intensity has been established. Therefore, they conclude that such FNPs may open up new perspectives for designing a new class of detection system for monitoring drug release.

Finally, the current state-of-the-art of gold nanoparticles in biomedical applications targeting cancer has been revised (Cai et al., 2008). In which, gold nanospheres, nanorods, nanoshells, nanocages, and surface enhanced Raman scattering nanoparticles have been discussed in detail regarding their uses in *in vitro* assays, *ex vivo* and *in vivo* imaging, cancer therapy, and drug delivery.

3.9 In vivo fluorescence spectroscopy

Monitoring phytoplankton classes and their abundance is a routine task in marine scientific research. With the frequent occurrence of red tide (a colloquial term used to refer to one of a variety of natural phenomena known as a harmful algal blooms), there is an urgent need for rapid analytical methods that can provide qualitative and quantitative information. Therefore, in vivo synchronous fluorescence spectra (SFS) of phytoplankton samples for determining the relative abundance of specific classes of phytoplankton (plant-like organisms that can form dense, visible patches near the water's surface) was investigated (Li et al., 2008). This study demonstrates the potential for determining phytoplankton class abundance by in vivo SFS. The database could be expanded to include more phytoplankton species grown under different nutrient, temperature, and photon flux densities. This work brings an in situ method for determining major phytoplankton class abundance by fluorescence measurement of seawater and deserves further studies.

Magnetic nanoparticles have been widely used in biomedical research such as magnetic carriers for bioseparation (Dolye et al., 2002, as cited in Yoo at al., 2007), enzyme and protein immobilization (Cao et al., 2003, as cited in Yoo at al., 2007), and contrast-enhancing media (Wu et al., as cited in Yoo at al., 2007). In this regard, in vivo fluorescence imaging method of novel threadlike tissues (Bonghan ducts) inside the lymphatic vessels of rats with fluorescent magnetic nanoparticles has been investigated (Yoo et al., 2007). The results of this study showed new applications of nanoparticles for in vivo imaging of hardly detectable tissues using fluorescence reflectance imaging and magnetophoretic control.

Chlorophyll fluorescence has been used as an accurate and nondestructive probe of photosynthetic efficiency, which can directly or indirectly reflect the impacts of environmental factors and changes in the physiological state of the plants (Baker, 2008, as cited in Falco et al., 2011). Recently, the first in vivo observation of chlorophyll fluorescence quenching induced by gold nanoparticles has been observed (Falco et al., 2011). The results showed that laser-induced fluorescence spectroscopy can be used to investigate the alterations in the physiological response of plants induced by gold nanoparticles, and that both excitation wavelengths, 405 nm and 532 nm, were able to detect the presence of the gold nanoparticles inside the plants. Even though, further investigations must be conducted to clarify the processes of penetration, translocation, and accumulation of nanoparticles in plants.

With the increasing nutrient pollution of freshwater ecosystems, lots of reservoirs, including those used as drinking water resources, suffer from extensive cyanobacterial blooms in the summer. Most of cyanobacteria produce a broad range of compounds with various chemical and toxicological properties. Therefore, their occurrence in recreational or drinking water reservoirs represents high health risk for humans and other organisms. In vivo fluorescence methods have been accepted as a quick, simple, and useful tool for quantification of phytoplankton organisms. In this regard, a case study in which fluorescence methods were employed for the selective detection of potentially toxic cyanobacteria in raw water at the drinking water treatment plant has been presented (Gregor et al., 2007). In this study the author demonstrated that presence of cyanobacteria in raw water can be easily monitored by phycocyanin fluorescence. Measured values were in good correlation with the other parameters of cyanobacterial biomass (chlorophyll a, cell counts). However, the blue light-excited fluorescence of eukaryotic algae should be also monitored to avoid false positive signals.

Photodynamic therapy (PDT) is a treatment modality that may in some cases replace invasive, more harmful or more expensive therapies for certain disorders such as cancer (Dougherty et al., 1998; Ochsner et al., 1997, as cited in Fischer at al., 2002). The PDT treatment process uses photosensitizing compounds that are selectively retained in abnormal tissue some hours to days after administration. At an optimal time after the administration, the photosensitizer is activated with visible light, causing the formation of reactive oxygen species that can initiate a destruction process of the tissue in which they are located. In addition to this photosensitization process, however, most photosensitizers used in PDT show characteristic fluorescence properties (Wagnieres at al., 1998, as cited in Fischer at al., 2002). In this regard, in vivo fluorescence imaging using two excitation and/or emission wavelengths for image contrast enhancement has been described (Fischer et al., 2002). In their study, in vivo measurements in mice where the second image, usually the background signal only, contains new unwanted image data were described. This simple method can successfully resolve the desired image, thus demonstrating the versatility of the image processing procedure.

Currently, the use of 5-aminolevulinic acid (5-ALA) (Zaak, et al., 2001, as cited in Bulgakova et al., 2009) offers the most promising outlook for a fluorescence diagnosis method to reveal superficial bladder cancer. In this regard, a methodological approach combining fluorescence imaging with in vivo local fluorescence spectroscopy (LFS) was clinically tested in order to improve the specificity of photodynamic diagnosis of superficial bladder cancer after intravesical instillation of Alasense (Bulgakova et al., 2009). This preliminary study, which included 62 patients, suggests that in vivo LFS in the course of fluorescence cystoscopy examinations could possibly minimize false positive fluorescence cases and reduce the required number of biopsies from 5-aminolevulinic acid (5-ALA)-based agent induced red fluorescence zones.

4. Comparison of fluorescence spectroscopy with other spectroscopic methods as an analytical technique

Compare to other analysis techniques, it must suffice here to add that fluorescence measurements are rapid, accurate and require only very small quantities of sample (nanomole or less). Fluorescence instrumentation is also relatively inexpensive and easy to use. In general, fluorescence experiments are relatively easy to perform; as in many fields, it is the planning of appropriate experiments, the analysis, and accurate interpretation of the data that require more extensive experience.

Fluorescent methods have three significant advantages over absorption spectroscopy and other typical optical spectroscopy. First, two wavelengths are used in fluorimetry, but only one in absorption spectroscopy. Emitted light from each fluorescent color can be easily separated because each color has unique and narrow excitation spectra. This selectivity can be further enhanced by narrowing the slit width of the emission monochromator so that only emitted light within a narrow spectral range is measured. Multiple fluorescent colors within a single sample can be quantified by sequential measurement of emitted intensity using a set of excitation and emission wavelength pairs specific for each color. The second advantage of fluorescence over absorption spectroscopy is low signal to noise, since emitted light is read at right angles to the exciting light. For absorption spectrophotometry, the excitation source, sample and transmitted light are configured in line, so that the absorption signal is the small difference between the exciting light and the transmitted light, both of which are quite intense. The third advantage is that fluorescent methods have a greater range of linearity. Because of these differences, the sensitivity of fluorescence is approximately 1,000 times greater than absorption spectrophotometric methods (Guilbault, 1990). However, a major disadvantage of fluorescence is the sensitivity of fluorescence intensity to fluctuations in pH and temperature. However, pH effects can be eliminated by using nonaqueous solvents, and normal room temperature fluctuations do not significantly affect the fluorescence intensities of commercial dye solutions.

In addition there are other useful fluorescent techniques that have an advantages over the other typical optical spectroscopy, that not have been mentioned in this chapter, these are: (1) polarization (anisotropy), this technique gives information about the rotation of a fluorophore, and allow to infer protein shape, membrane fluidity (order parameters), and binding analysis; (2) quenching fluorescence, these processes can occur during the excited state lifetime -for example collisional quenching, energy transfer, charge transfer reactions or photochemistry -or they may occur due to formation of complexes in the ground state, thus this technique can give a useful kinetic information of the tested system, and give information about accessibility of fluorophoresis obtained allowing to correlate it with changes in protein or membrane structure; (3) Förster resonance energy transfer (FRET), this technique can be considered as a molecular ruler, which allow to determine distance from 10 to 80 Å; (4) fluorescence microscopy for image analysis, this technique gives magnification (in order to see small parts), resolution (in order to distinguish details of the small parts), and contrast (in order to magnify and resolve details, fluorescence emission provides contrast); (5) multiphoton excitation fluorescence microscopy, as an example for two photon excitation a molecule can be excited by using simultaneous photons and get fluorescence as happens with one photon excitation, this techniques has so many advantages such as sectioning effect without pinholes, low photobleaching and photodamage rate, separation of excitation and emission, no Raman from the solvent, deep penetration in tissues, single excitation wavelength for many dyes, avoid chromatic aberrations, and no expensive UV optics (for UV excited fluorophores) needed. Even though it has some disadvantages such as only is suitable for fluorescence images (reflected light images is not currently available), the technique is not suitable for imaging highly pigmented cells and tissues which absorb near infrared light, and laser source is expensive. It is worth mentioning that all analysis that can be done using fluorescence spectroscopy can be done using a multi-photon excitation fluorescence microscopy. However, multi-photon excitation fluorescence microscopy has a unique analysis applications over the fluorescence spectroscopy such as: analysis of deep tissue imaging (Brain, skin, etc), can prime photochemical reaction within subfemtoliter volumes inside solutions, cells and tissues (photolabile "caged" compounds), can be used for imaging of living specimen for longer period of time, and for live animal imaging (intrinsic fluorophores).

5. Conclusion

It could be concluded at the end of this chapter, that the fluorescence spectroscopy is intensely employed as a powerful spectroscopic tool for quantitative analysis, characterization, and quality control in different fields such as inorganic, organic, pharmaceutical, biological and biomedical, food, and environmental analysis. Furthermore, fluorescence spectroscopy has so many applications in optical sensors, and nanamaterials.

Compare with other analytical techniques, it must suffice here to add that fluorescence measurements are rapid, accurate and require only very small quantities of sample (nanomole or less) with high selectivity. The principal advantage of fluorescence over radioactivity and absorption spectroscopy is the ability to separate compounds on the basis of either their excitation or emission spectra, as opposed to a single spectra. This advantage is further enhanced by commercial fluorescent dyes that have narrow and distinctly separated excitation and emission spectra.

6. References

Airado-Rodrıguez, D., Duran-Meras, I., Galeano-Dıaz, T., Wold, J. P. (2011). Front-face fluorescence spectroscopy: A new tool for control in the wine industry. *J. Food Composition Analysis*, 24, pp.257-264, ISSN: 0889-1575.

Akasheh, T.S., Al-Rawashdeh, N.A. F. (1990). Sodium Lauryl Sulfate-Ruthenium(II) Interactions: Photogalvanic and Photophysical Behavior of Ru(II)-Mimine Complexes. *J. Phys. Chem.*, 94, 23, (Nov. 1990), pp. 8594-8598, ISSN: 0022-3654.

Aksuner, N., Henden, E., Yilmaz, I., Cukurovali, A. (2009). A highly sensitive and selective fluorescent sensor for the determination of copper(II) based on a schiff base. *Dyes and Pigments*. 83, pp. 211-217, ISSN: 0143-7208.

Albert-Garcia, J.R., Antoón-Fos, G.M., Duartb, M.J., Zamoraa, L. L., Calatayuda, J. M. (2009). Theoretical prediction of the native fluorescence of pharmaceuticals. *Talanta*, 79, 2, (July 2009), pp. 412-418, ISSN: 0039-9140.

Al-Rawashdeh. N.A.F. (2005). Interactions of Nabumetone with γ-Cyclodextrin Studied by Fluorescence Measurements. *J. Incl. Phenom. Macrocyclic Chem.*, 51, 1-2, (Feb. 2005), pp. 27-32, ISSN: 0923-0750.

Bender, M. L., Komiyama, M. (1978). *Cyclodextrin Chemistry* (1st Ed.), Springer-Verlag, ISBN: 0387085777, Berlin, Germany.

Bram, O., Messina, F., El-Zohry, A. M., Cannizzo, A., Chergui, M. (available online Dec. 3, 2011). Polychromatic femtosecond fluorescence studies of metal-polypyridine complexes. *Chem. Phys.*, DOI: 10.1016/j.chemphys.2011.11.022, ISSN: 0301-0104.

Bulgakova, N., Ulijanov, R., Vereschagin, K., Sokolov, V., Teplov, A., Rusakov, I., Chissov, V. (2009). In vivo local fluorescence spectroscopy in PDD of superficial bladder cancer. *Med. Laser Appl.*, 24, pp. 247-255, ISSN: 1615-1615.

Cai, W., Gao, T., Hong, H., Sun, J. (2008). Applications of gold nanoparticles in cancer nanotechnology. *Nanotechnology, Science and Applications*, 1 (Sept. 2008), pp. 17-32, ISSN: 1177-8903.

Calatayud, J. M., Zamora, L. L. (1995), in A. Townshend (Ed.), *Encyclopedia of Analytical Science*, Academic Press, ISBN 0-12-226700-1, Oxford.

Catherine L. Muller, C. L., Baker, A., Hutchinson, R., Fairchild, I. J., Kidd, C. (2008). Analysis of rainwater dissolved organic carbon compounds using fluorescence spectrophotometry. *Atmospheric Environment*, 42, pp. 8036-8045, ISSN: 1352-2310

Casado-Terrones, S., Ferna´ndez-Sa´nchez, J. F., Segura-Carretero, A., Ferna´ndez-Gutie´rrez, A. (2007). Simple luminescence detectors using a light-emitting diode or a Xe lamp, optical fiber and charge-coupled device, or photomultiplier for determining proteins in capillary electrophoresis: A critical comparison. *Anal. Biochem.* 365, (February 2007), pp. 82-90, ISSN: 10.1016.

Chandrasekhar, V., Bag, P., Pandey, M. D. (2009). Phosphorus-supported multidentate coumarin-containing fluorescence sensors for Cu^{+2}. *Tetrahedron*, 65, 47, (Nov. 2009), pp. 9876-9883, ISSN: 0040-4020

Cheng, F., Tang, N., Chen , J., Wang, F., Chen, L. (2010). A new family of trinuclear Ru(II) polypyridyl complexes acting as pH-induced fluorescence switch. *Inorg. Chem. Commun.*, 13, pp. 757-761, ISSN: 1387-7003

Cheng, F., Tang, N., Chen , J., Wang, F., Chen, L. (2011). pH-induced fluorescence switch of two novel tetranuclear Ru(II) polypyridyl complexes. *Inorg. Chem. Commun.*, 14, pp. 852-855, ISSN: 1387-7003

Dawoud, A. A., Al-Rawashdeh, N. (2008). Spectrofluorometric, thermal, and molecular mechanics studies of the inclusion complexation of selected imidazoline-derived drugs with β-cyclodextrin in aqueous media. *J. Incl. Phenom. Macrocyclic Chem.*, 60, 3-4, (April 2008), pp. 293-301, ISSN: 0923-0750

Dragan, A.I., Bishop, E.S., Casas-Finet , J.R., Strouse, R.J., Schenerman, M.A., Geddes, C.D. (2010). Metal-enhanced PicoGreen fluorescence: Application to fast and ultra-sensitive pg/ml DNA quantitation. *J. Immunolog. Meth.*, 362, pp. 95-100, ISSN: 0022-1759

Dessingou, J., Mitra, A., Tabbasum, K., Baghel, G. S., Rao, C. P. *J. org. Chem.*, (Nov. 21, 2011), in press, ISSN 10.1021/jo201926q

Dragan, A. I., Golberg, K., Elbaz, A., Marks, R., Zhang, Y., Geddes, C. D. (2011). Two-color, 30 second microwave-accelerated Metal-Enhanced Fluorescence DNA assays: A new Rapid Catch and Signal (RCS) technology. *J. Immunolog. Meth.*, 366, pp. 1-7, ISSN: 0022-1759

Falco, W. F., Falcao, E. A., Santiago, E. F., Bagnato, V. S., Caires, A. R. L. (2011). In vivo observation of chlorophyll fluorescence quenching induced by gold nanoparticles. *J. Photochem Photobio. A: Chem.*, 225, pp. 65-71, ISSN: 1010-6030

Fernandez-Sanchez, J.F., Carretero, A. S., Cruces-Blanco, C., Fernandez-Gutierrez, A. (2003). A sensitive fluorescence optosensor for analysing propranolol in pharmaceutical preparations and a test for its control in urine in sport. *J. Pharm. Biomed. Anal.*, 31, pp.859-865, ISSN: 0731-7085

Fischer, F., Dickson, E. F. G., Pottier, R. H. (2002). In vivo fluorescence imaging using two excitation and/or emission wavelengths for image contrast enhancement. *Vibrational Spectroscopy*, 30, pp. 131-137, ISSN: 0924-2031

Fu, Y., Zhang, J., Lakowicz, J. R. (2008). Metal-enhanced fluorescence of single green fluorescent protein (GFP). *Biochem. Biophys. Res. Commun.*, 376, pp. 712-717.

Formica, M., Fusi, V., Giorgi, L., Micheloni, M. (2012). New fluorescent chemosensors for metal ions in solution. *Coord. Chem. Rev.*, 256, pp. 170-192, ISSN: 0010-8545.

Gregor, J., Marsalek, B., Sipkova, H. (2007). Detection and estimation of potentially toxic cyanobacteria in raw water at the drinking water treatment plant by in vivo fluorescence method. *Water Research*, 41, pp. 228-234, ISSN: 0043-1354.

Gupta, S., Paul, B. K., Barik, A. K., Mandal, T. N., Roy, S., Guchhait, N., Butcher, R. J., Kar, S. K. (2009). Modulation of fluorescence emission of 1-(2-pyridyl) pyrazole derived Schiff base ligands by exploiting their metal ion sensitive binding modes. *Polyhedron*, 28, pp.3577-3585, ISSN: 0277-5387

Guo, W., Xu, J., Wang, J., Wen, Y., Zhuo, J., Yan, Y. (2010). Characterization of dissolved organic matter in urban sewage using excitation emission matrix fluorescence spectroscopy and parallel factor analysis. *J. Environm. Sci.*, 22, 11, (Nov. 2010), pp. 1728-1734, ISSN: 1001-0742

Guilbault, G. G. (1990). *Practical Fluorescence (Modern Monographs in Analytical Chemistry)* (2nd Ed.), Marcel Dekker, INC, ISBN-13: 978-0824783501, New York.

Hammamia, M., Rouissi, H., Salah, N., Selmi, H., Al-Otaibi, M., Blecker, C., Karoui, R. (2010). Fluorescence spectroscopy coupled with factorial discriminant analysis technique to identify sheep milk from different feeding systems. *Food Chemistry*, 122, pp. 1344-1350, ISSN: 0308-8146

He, D., Chena, B., Tiana, Q., Yaoa, S. (2009). Simultaneous determination of five anthraquinones in medicinal plants and pharmaceutical preparations by HPLC with fluorescence detection. *J. Pharm. Biomed. Anal.*, 49, pp. 1123-1127, ISSN: 0731-7085

Hernandez, D., Plaza, C., Senesi, N., Polo, A. (2006). Detection of Cupper(II) and zinc(II) binding to humic acids from pig slurry and amended soils by fluorescence spectroscopy. *Environmental Pollutution*, 143, 2, (Sept. 2006), pp. 212-220, ISSN: 0269-7491

Hirano, T., Sekiguchi, T., Hashizume, D., Ikeda, H., Maki, S., Niwa, H. (2010). Colorimetric and fluorometric sensing of the Lewis acidity of a metal ion by metal-ion complexation of imidazo[1,2-a]pyrazin-3(7H)-ones .*Tetrahedron*, 66, 21, (May 2010), pp. 3842-3848, ISSN 0040-4020.

Ingale, S. A., Pujari, S. S., Sirivolu, V. R., Ding, P., Xiong, H., Mei, H., Seela, F. (2011). 7-Deazapurine and 8-Aza-7-deazapurine Nucleoside and Oligonucleotide Pyrene "Click" Conjugates: Synthesis, Nucleobase Controlled Fluorescence Quenching, and Duplex Stability. *J. Org. Chem.*, (Nov. 30, 2011), in press, ISSN 10.1021/jo202103q

Jin,L., Shang, L., Guo, S., Fang, Y., Wen, D., Wang, L., Yin, J., Dong, S. (2011). Biomolecule-stabilized Au nanoclusters as a fluorescence probe for sensitive detection of glucose. *Biosen. Bioelect.*, 26, pp. 1965-1969, ISSN: 0956-5663.

Kaliszan, W., Petrusewicz, J., Kaliszan, R. (2006). Imidazoline receptors in relaxation of acetylcholine constricted isolated rat jejunum. *Pharm. Rep.* 58, pp. 700.

Kanan, S.M., Abu-Yousef, I.A., Hassouneh, N., Malkawi, A., Abdo, N., Kanan, M.C. (2009). A highly Selective Luminescent Sensor for Detecting Environmental Mercuric Ions. *Aust. J. Chem.*, 62, pp. 1593-1599, ISSN: 0004-942.

Kwong, H.-L., Wonga, W.-L., Lee, Ch-.S., Yeung , Ch.-T., Teng, P.-F. (2009). Zinc(II) complex of terpyridine-crown macrocycle: A new motif in fluorescence sensing of zwitterionic amino acids. *Inorg. Chem. Commun.*, 12, pp. 815-818, ISSN: 1387-7003

Leong, W. L., Vittal. J. J. (2009). Synthesis and characterization of metal complexes of Calcein Blue: Formation of monomeric, ion pair and coordination polymeric structures. *Inorg. Chem. Acta*, 362, pp. 2189-2199, ISSN: 0020-1693

Li, H., Zhang, Q., Zhu, Ch., Wang, X. (2008). Assessment of phytoplankton class abundance using in vivo synchronous fluorescence spectra. *Anal. Biochem.*, 377, pp. 40-45, ISSN: 0003-2697

Li, Y., Yang, Z-Y., Wu, J-C. (2010). Synthesis, crystal structures, biological activities and fluorescence studies oftransition metal complexes with 3-carbaldehyde chromone thiosemicarbazone. *Europ. J. Med. Chem.*, 45, pp. 5692-5701, ISSN: 0223-5234

Li, H., He, X., Liu, Y., Huang, H., Lian, S. Lee, S.-T., Kang, Z. (2011). One-step ultrasonic synthesis of water-soluble carbon nanoparticles with excellent photoluminescent properties. *Carbon*, 49, pp.605-609, ISSN: 0008-6223

Ling, K-Q., Sayre, L. M.. (2009). Discovery of a Sensitive, Selective, and Tightly Binding Fluorogenic Substrate of Bovine Plasma Amine Oxidase. *J. Org. Chem.*, 74, pp. 339-350, ISSN: 10.1021/jo8018945

Lakowicz, J.R. (2006), *Principles of Fluorescence Spectroscopy* (3rd Ed.), Springer, ISBN: 13-978-0387-31278-1, New York.

Majumder, A., Rosair, G. M., Mallick, A., Chattopadhyay, N., Mitra, S. (2006). Synthesis, structures and fluorescence of nickel, zinc and cadmium complexes with the N,N,O-tridentate Schiff base N-2-pyridylmethylidene-2-hydroxy-phenylamine. *Polyhedron*, 25, pp. 1753-1762, ISSN: 0277-5387

Ming, T., Zhao, L., Yang, Z., Chen, H., Sun, L., Wang, J., Yan, C. (2009). Strong Polarization Dependence of Plasmon-Enhanced Fluorescence on Single Gold Nanorods. *Nano lett.*, 9, 11, pp.3896-3903, ISSN: 10.1021

Muddana, H. S., Morgan, T. T., Adair, J. H., Butler. P. J. (2009). Photophysics of Cy3-EncapsulatedCalcium Phosphate Nanoparticles. *Nano let..*, 9, 4, pp. 1559-1566, ISSN: 10.1021

Oliveira, E., Vicente, M., Valencia, L., Macıas, A., Bertolo, E., Bastida, R., Lodeiro, C. (2007). Metal ion interaction with a novel anthracene pendant-armed fluorescent molecular probe. Synthesis, characterization, and fluorescence studies. *Inorg. Chem. Acta*, 360, pp. 2734-2743, ISSN: 0020-1693

Parini, A., Moudanos, C. G., Pizzinat, N., Lanier, S. M. (1996). The elusive family of imidazoline binding sites. *Trends Pharmacol. Sci.*, 17, pp. 13.

Parfitt, K. (Ed.), Martindale: The Complete Drug Reference, Pharmaceutical Press, London, 1990

Rendell, D. (1987), *Fluorescence and Phosphorescence*, Wiley, ISBN-13: 978-0471913801, Chichester, England.

Saleh, N., Al-Rawashdeh, N. A. F. (2006). Fluorescence Enhancement of Carbendazim Fungicide in Cucurbit[6]uril. *J. Fluoresc.*, 16, 4, (July 2006), pp. 487-493, ISSN: 1053-0509

Shimomura, O., Johnson, F.H., Saga, Y. (1962). Extraction, purification and properties of aequorin, a bioluminescent protein from the luminous hydromedusan, Aequorea. *Journal of Cell and Comparative Physiology*, 59pp. 223–229

Shete, V. S., Benson, D. E. (2009). Protein Design Provides Lead(II) Ion Biosensors for Imaging Molecular Fluxes around Red Blood Cells. Biochem., 48, 2, pp. 462-470

Silva, L., Fernndez-Franzon, M., Font, G., Pena, A., Silveira, I., Lino, C., Ma◌es, J. (2009). Analysis of fumonisins in corn-based food by liquid chromatography withfluorescence and mass spectrometry detectors. *Food Chemistry*, 112, pp. 1031-1037, ISSN: : 0308-8146

Sokkalingam, P., Lee C-H. (2011). Highly Sensitive Fluorescence "Turn-On" Indicator for Fluoride Anion with Remarkable Selectivity in Organic and Aqueous Media. *J. Org. Chem.*, 76, 10, (May 2011), pp. 3820-3828, ISSN 10.1021/jo200138t

Soh, N., Ueda, T. (2011). Perylene bisimide as a versatile fluorescent tool for environmental and biological analysis: A review. *Talanta*, 85, 3, (Sept. 2011), pp. 1233-1237, ISSN: 0039-9140

Strasburg, G. M., Ludescher, R. D. (1995). Heory and applications of fluorescence spectroscopy in food research. *Trends in Food Science & Technology*, 6, 3 (March 1995), pp. 69-75, ISSN: 0924-2244

Suxian Huang, s., Chen, Y. (2008). Ultrasensitive Fluorescence Detection of Single Protein Molecules Manipulated Electrically on Au Nanowire. *Nano lett.*, 8, 9, pp. 2829-2833, ISSN: 10.1021

Szejtli, J. (1988). *Cyclodextrins Technology*, Kluwer Academic Publishers, ISBN: 90-277-2314-1 , Dordrecht, Netherlands.

Vasimalai, John, S. A. (2011). Ultrasensitive and selective spectrofluorimetric determination of Hg(II) using a dimercaptothiadiazole fluorophore. *J. Luminesc.*, 131, pp. 2636-2641, ISSN: 0022-2313

Wang, L., Guo, C., Fu,B., Wang, L. (2011). Fluorescence Determination of DNA Using the Gatifloxacin-Europium(III) Complex. *J. Agric. Food Chem.*, 59, 5, (Feb. 2011), pp. 1607-1611, ISSN: 10.1021

Wei Cui, W., Lu, Z., Cui, K., Wu, J., Wei, Y., Lu, Q. (2011). Fluorescent Nanoparticles of Chitosan Complex for Real-Time Monitoring Drug Release., *Langmuir*, 27, 13, (June 2001), pp. 8384-8390, ISSN: 10-1022

Xu, W., Mun, L., Miao, R., Zhang, T., Shi, W. (2011). Fluorescence sensor for Cu(II) based on R6G derivatives modified silicon nanowires. *J. Luminesc.*, 131, pp. 2626-2620, ISSN: 0022-2313

Xu, J., Huang, X.H., Zhou, N.L., Zhang, J.S., Bao, J.Ch., Lu, T.H., Li, C. (2004). Synthesis, XPS and fluorescence properties of Eu^{3+} complex with polydimethylsiloxane. *Mater. Lett.*, 58, pp. 1938-1942, ISSN: 0167-577X

Yoo, J. S., Johng, H.-M., Yoon, T.-J., Shin, H.-S., Lee, B.-Ch., Lee, Ch., Ahn. B. S., Kang, D.-I., Lee, J.-K., Soh, K.-S. (2007). In vivo fluorescence imaging of threadlike tissues (Bonghan ducts) inside lymphatic vessels with nanoparticles. *Curr. Appl. Phys.*, 7, pp. 342-348, ISSN: 1567-1739

You, Z.-L., Zhu, H.-L., Liu, W.-S. (2004). Solvolthermal Syntheses and Crystal Structures of Three Linear Trinuclear Schiff Base Complexes of Zinc(II) and Cadmium(II). *Z. Anorg. Allg. Chem.*, 630, 11, (Sept. 2004), pp. 1617-1622, ISSN: 1521-3749

Zou, M., Gao, H., Li, J., Xu, F., Wang, L., Jiang, J. (2008). Rapid determination of hazardous compounds in food based on a competitive fluorescence microsphere immunoassay. *Anal. Biochem.*, 374, pp. 318-324, ISSN: 0003-2697

Section 5

Other Spectroscopy

13

Basic Principles and Analytical Application of Derivative Spectrophotometry

Joanna Karpinska
Institute of Chemistry, University of Bialystok, Bialystok Poland

1. Introduction

Analytical methods based on measurements of UV or visible light absorption belong to the most popular and most often used in laboratory practice. Commercially available apparatuses are cheap and easy for operation. Spectrophotometric procedures usually are not time- and labour-consuming. The economical aspects of UV-Vis techniques is worth of emphasize too. It is one of the cheapest technique, so spectrophotometers are basic equipment of every laboratory. The main disadvantage and limitation of the spectrophotometry is its low selectivity. The measurement of absorbance is burden by interferences derived from others components of sample. A recorded UV-Vis spectrum is the sum of absorbances of analyte and matrix. Usually, recorded bands are well-defined but more or less distorted by a background. As the background is called absorbance exhibited by matrix (reagents or accompanied compounds). The problem with specific or nonspecific background can be omitted by measurements versus blank. Such procedure can be successfully applied only in the case of simple samples, which composition is stable and well known or when highly selective reagents are used. An isolation of an analyte from matrix is another solution for increasing the selectivity of assay. But every additional operation introduced into sample preparation procedure extents time and costs of single analysis and increases risk of loss or contamination of the analyte.

One of the simplest method for an increasing a selectivity is derivatisation of spectra. This operation allows to remove spectral interferences and as a consequence leads to increase selectivity of assay. Derivatisation of sets of digital data is well known method of separation useful signals from noised data [1]. Historically, the beginning of derivative spectrophotometry is dated on 1953 when the first analogue spectrophotometer was build by Singleton and Cooler [1]. But the fast development of this technique started in 70-s of twentieth century, when new generation of spectrophotometers controlled by computers were constructed. An apogee of its popularity occurred in 80-s of last century. Nowadays, it is only additional technique, rarely used, though it is fully available as a build-in function in software of modern spectrophotometers. I hope that this work gives some light on derivative spectrophotometry and restores it in some way.

2. Basic theory and properties of derivative spectrophotometry

Derivative spectrophotometry is a technique which is based on derivative spectra of a basic, zero-order spectrum. The results of derivatisation of function described a run of absorbance curve is called the derivative spectrum and can be expressed as:

$$^{n}D_{x,\lambda}=d^{n}A/d\lambda^{n}=f(\lambda) \quad \text{or} \quad ^{n}D_{x,v}=d^{n}A/dv^{n}=f(v)$$

where: n - derivative order, $^{n}D_{x,\lambda}$ or $^{n}D_{x,v}$ represents value of n-order derivative of an analyte (x) at analytical wavelength (λ) or at wavelength number (v), A- absorbance.

Derivative spectrophotometry keeps all features of classical spectrophotometry: Lambert-Beer law and law of additivity.

Lambert-Beer low in its differential form is expressed as:

$$^{n}D=\frac{d^{n}A}{d\lambda^{n}}=\frac{d^{n}\varepsilon}{d\lambda^{n}}\cdot c\cdot l$$

Where ε-molar absorption coefficient ($cm^{-1}mol^{-1}l$) , c - concentration of analyte ($mol\ l^{-1}$), l-thickness of solution layer (cm).

Derivative spectrum of **n**-component mixture is a sum of derivative spectra of individual components:

$$^{n}D_{mix}={}^{n}D_{1}+{}^{n}D_{2}+\ldots+{}^{n}D_{n}$$

A new feature of derivative spectrophotometry is a dependence of derivatisation results on geometrical characteristic of starting, zero-order spectrum. A shape and an intensity of the resulted derivative spectrum depend on half- heights width of peak in basic spectrum:

$$^{n}D=P^{n}A_{max}L^{-1}$$

where P^{n}- polynomial described run of n-derivative curve, n- derivative order, L- width of half- heights of peak of zero-order spectrum.

Due to this property broad zero-order spectra are quenched with generation of higher orders of derivatives while narrow undergo amplification. If the zero-order spectrum possess two bands A and B which differ from their half- heights width ($L_B>L_A$), after a generation of n-order derivative a ratio of derivatives intensity can be expressed as:

$$^{n}D_{A}/{}^{n}D_{B}=(L_{B}/L_{A})^{n}$$

This dependence leads to increase in selectivity and/or sensitivity of assay. It allows to use for analytical properties a narrow band, overlapped or completely hooded by a broad ones.

The shape of derivative spectrum is more complicated than its parent one (Fig. 1). New maxima and minima appeared as results of derivatisation. The generation of **n-th** order derivative spectrum produces (n+1) new signals: an intense main signal and weaker bands, so called satellite or wings signals. Position of maxima or minima depend on order of derivative. The main extreme of derivative spectra of even order is situated at the same wavelength as maximum in zero-order spectrum, but for 2, 6 and 10-th order it becomes minimum in derivative spectrum and for 4, 8 and 12-th order it remains as a maximum (Fig. 1). The point of initial maximum converts into the point of inflection in derivative spectra of odd order. A narrowing of new signals is observed during generation of consecutive derivative spectra. This feature leads to narrowing bands and as a consequence to separation of overlapped peaks.

3. Generation of derivative spectra and their properties.

Modern software's controlled spectrophotometers allow not only acquisition and storage of registered spectra. They are equipped in modules enable mathematical operation like addition, subtraction, multiplication as well as derivatisation.

A registered UV-Vis absorption spectrum is a two-dimensional set of points with co-ordinates (λ, A), where λ - wavelength, A- absorbance. Derivative spectra can be obtained by direct calculation of ordinate increment or by fitting a function described a course of spectrum curve and next its derivatisation [1]. Another approach is to find a polynomial representing an absorption curve [1]. A proper form of the polynomial can be found if its all coefficients are known. If long set of n-data is disposed, determination of polynomial coefficients requires to solve **n** equations with **n**-unknowns. This is very hard, laborious job which could be impossible for long sets of data. There are many mathematical approaches simplifying this task [1]. The most popular is Savitzky-Golay algorithm [2] and its modifications [3,4]. Savitzky-Golay algorithm [2] does not analyze a whole set of points but only one exact point and its closest neighbourhood: **m** points from left and **m** points from right of the chosen neighbourhood of central point. A width of analysed set of points is equal **2m+1** and is called a derivatisation window. The coefficients of polynomial are calculated by the least square method for central point and next derivatisation window is moved right by one point and calculations for new central point are repeated. The result of this approach is a set of new points which creates a new - derivative spectrum. Usually the new set of points is shorter by **2m** points in comparison to the parent one. It isn't problem because the recorded spectrum usually is the long set of points and the clipped points are from beginning and end of zero order-spectrum which are useless from analytical point of view. Some improvements of Savitzky-Golay algorithm were done. Originally Savitzky-Golay algorithm was devoted for derivatisation of spectra with uniformly spaced sets of data. Nowadays it can be applied for nonuniformly spaced sequence [3]. There are modification which allow derivatisation without loss of extreme points [4].

The use of Savitzky-Golay algorithm requires optimalisation such parameters as derivative order, polynomial degree, width of derivatisation window and manner how derivative is generated. Analytical usefulness of resulted derivative spectrum depends on proper selection of mentioned parameters. Their selection should be done by taking into account a shape of initial zero-order spectrum and spectral properties of accompanied compounds.

- derivative order

Proper separation of overlapped signals can be achieved if appropriate derivative order is used. Optimal derivative order is a function of signals height, their width at half height and distance between maxima in basic spectrum [5,6]. It is recommended to use low orders if the basic spectrum is a sum of wide bands, while for the spectra consisted of narrow bands - higher orders. Generation of the high order derivative suppress very fast intensity of wide bands and magnify the narrow one[1, 5].

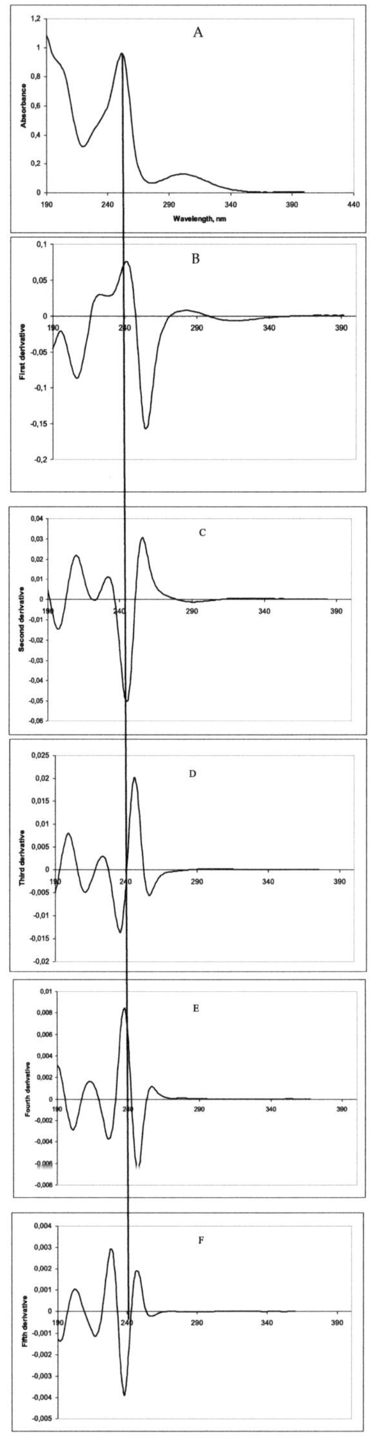

Fig. 1. Zero order (A) and consecutive derivative spectra (B→F) of aqueous solution of promazine hydrochloride (10 ppm); zero order spectrum has been recorded on Hewlett-

Packard HP-8452A diode array spectrophotometer with following working parameters: integration time 1 s, spectral bandwidth 2nm, spectrum scan 0.1 s. Derivative spectra were generated using Savitzky Golay algorithm by PC computer equipped with Excel for Microsoft Windows ($\Delta\lambda$=10 nm, second polynomial degree, derivatives of higher orders were generated by gradual derivatisation of derivative spectra of lower order).

- polynomial degree

The next optimized parameter is the polynomial degree. There is a similar dependence as in the case of derivative order. The high polynomial degrees should be used for spectral curves with sharp and narrow signals. Application of inappropriate polynomial degree gives a distorted derivative spectrum without useful analytical information [5]. In the case of multicomponent analysis, the use of polynomials of different degrees can allow to increase spectral differences of assayed compounds and their selective determination [5].

- width of derivatisation window

A proper selection of this parameter is crucial for quality and quantity of analytical information available in derivative spectrum. Application of the broad derivatisation window gives a smooth averaged derivative spectrum without spectral details. So, the broad derivatisation window is recommended for derivatisation of a zero-order spectra with broad irregular bands with a significant oscillatory constituent [5]. In the case of the basic spectrum with narrow absorption bands the narrow derivatisation window should be used. Otherwise the important analytical information could be lost and resulted maxima of derivative spectrum couldn't correspond to the real one[5].

- manner of generation of derivative spectra

Derivative spectra of higher orders can be obtained using Savitzky-Golay algorithm in two ways: by direct generation of desired derivative spectrum or by gradual generation of first order derivative on consecutive spectra: ${}^{0}A \xrightarrow{1} \frac{dA}{d\lambda} \xrightarrow{1} \frac{d(dA)}{d(d\lambda)} \xrightarrow{\dots} \dots \frac{d^{n}A}{d\lambda^{n}}$

It is very often observed that direct generation of the high-order derivative gives distorted analytically useless spectra (Fig. 2). Selection of derivatisation manner depends on shape of basic spectrum. It is recommended to apply the gradual derivatisation in the case of complicated zero-order spectra. A progressive generation of derivative spectra gives smooth derivative spectrum with advantageous signal-to-noise ratio.

Derivative spectrophotometry can be very useful additional tool which helps to solve some complicated analytical problems. Mathematical processing of spectra is very easy to use as modern spectrophotometers are computer controlled and their software are equipped in derivatisation unit. A proper selection of mathematical parameters gives profits in improved selectivity, sometimes sensitivity and in simplification of analytical procedure.

4. Analytical application of derivative spectrophotometry

Derivative spectrophotometry (DS) has found a wide application in quantitative chemical analysis. As the latest applications have been gathered and described in reviews published previously [8,9], this part is focused on the recent use of DS. Based on scientific literature the following fields of application of derivative spectrophotometry can be distinguished:

a. Multicomponent analysis. This group is the most numerous. The goal of proposed methods is application of DS for determination of one analyte in presence of matrix or for simultaneous assaying of few analytes.
b. Calculation of some physico-chemical constants, e.g. reaction, complexation or binding constants [10].
c. Application for investigation of some processes kinetics[11, 12].

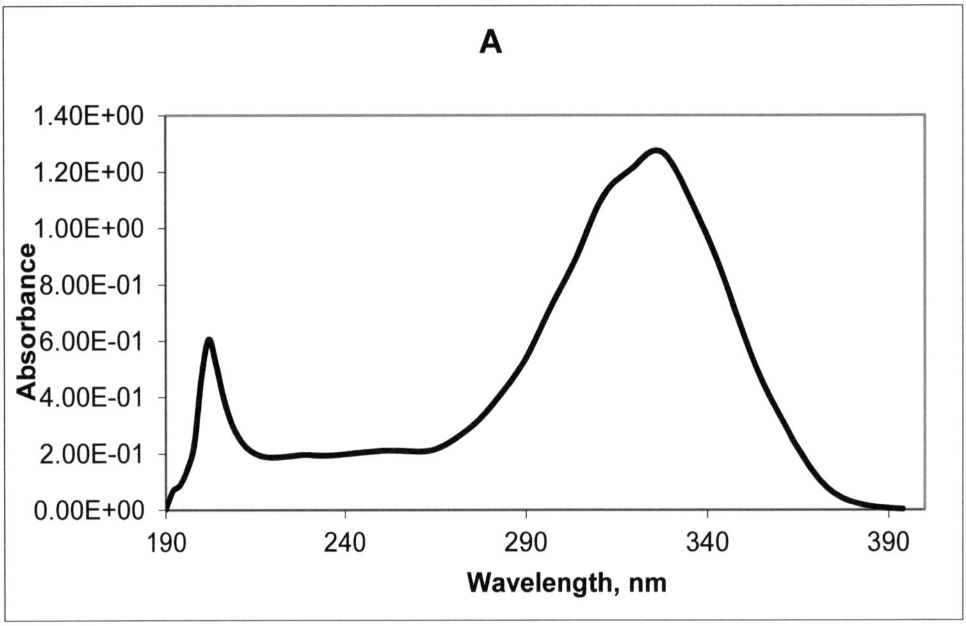

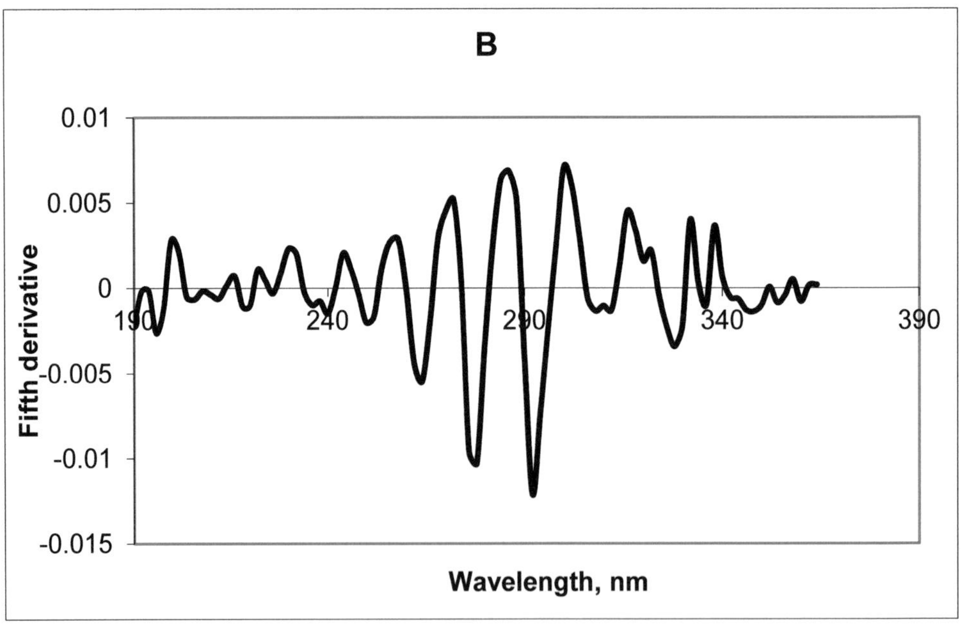

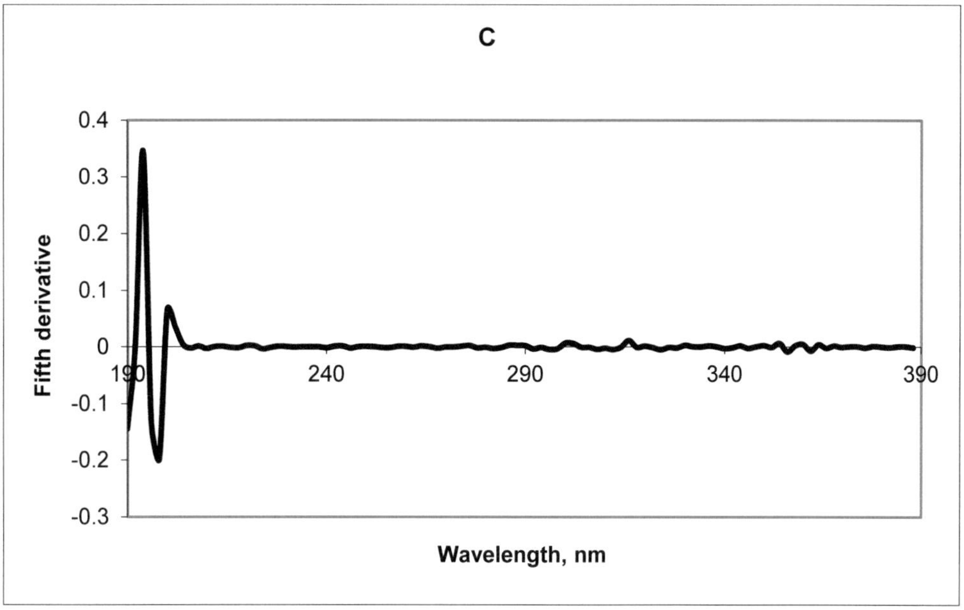

Fig. 2. Zero order (A) and fifth derivative order spectra of ethanolic solution of retinol acetate (10 ppm). Spectrum B has been obtained by gradual derivatisation, spectrum C by direct generation of fifth derivative from zero-order spectrum. Apparatus working conditions: Hewlett-Packard HP-8452A diode array spectrophotometer with following working parameters: integration time 1 s, spectral bandwidth 2nm, spectrum scan 0.1 s. Derivative spectra were generated using Savitzky Golay algorithm by PC computer equipped with Excel for Microsoft Windows ($\Delta\lambda$=14 nm, fourth polynomial degree).

a. Multicomponent analysis

Derivative spectrophotometry (DS) has been mainly used in pharmaceutical analysis for assaying of a main ingredient in a presence of others components or its degradation product. Pharmaceutical samples are characterised by high level of constituents and presence of a relatively simple and stable matrix. The spectral influences of disturbing compounds are easy to remove by derivatisation of spectra. The most numerous procedures based on derivative spectra have been devoted for determination of one components without sample purification. Another field of DS application is the use of it for simultaneous determination of two or more components. As a form of derivative spectrum is more complicated in comparison to its initial zero order, usually derivatives of low orders are employed for analytical purposes. Procedures used DS for pharmaceutical analysis are assembled in Tables 1 and 2.

b. Others applications

Derivative spectrophotometry was applied in different than pharmaceutical analysis areas of analysis. This method was utilised for the determination of amphothericin in various biological samples like plasma, serum, urine and brain tissue [40]. The combination of ratio spectra with their derivatisation allowed to remove spectral interferences caused by a presence of bilirubin in plasma [40].

Compound	Characteristic of the method	Reference
Sertraline	The proposed method is based on reaction of sertraline with chloranilic acid. First derivative spectrophotometry has been evaluated by measuring the derivative signal at 475.72 nm – 588.40 nm (peak to peak amplitude). Calibration graph was established for 5-100 μg mL^{-1} of sertraline	13
Estradiol valerate	The first-order derivative spectra were used for determination of estradiol in tablet. Measurements of derivative were made at 270 nm. The method showed specificity and linearity in the concentration range of 0.20 to 0.40 mg mL^{-1}.	14
Tropicamide	The measurements were carried out at wavelengths of 263.8 and 255.4 nm for third- and fourth-derivative, respectively. The method was found to be linear (r^2> 0.999) in the range of 10-100 μg mL^{-1} for tropicamide in the presence of excipients. The method was applied for analyte determination in eye drops.	15
Nebivolol hydrochloride	Derivative spectrophotometry used for determination of nebivolol in bulk and in preparates.	16
Gemifloxacin mesylate	The proposed methods were based on the reaction of gemifloxacin with chloranilic acid and parachloranil to give highly coloured complexes. The coloured products were quantified spectrophotometrically at 530 nm and 540 nm at zero order, 590 and 610 nm for the first derivative and 630 and 650 nm for second order derivative. Beer's law was obeyed in the concentrations range of 10 to 60 μg mL^{-1}, 5 to25 μg mL^{-1} at zero order, 5 to 25 μg mL^{-1}, 5 to 40 μg mL^{-1} at first order and 2 to 20 μg mL^{-1} and 2 to 14 μg mL^{-1} at second order.	17
Olanzapine	The first derivative values measured at 222 nm and the second derivative values measured at 230 nm (n=6) were used for the quantitative determination of the drug . Calibration graphs were linear in the concentration range of olanzapine using 2-10 μg mL^{-1} for first and second derivative spectrophotometric method.	18
Galanthamine	The 1st derivative zero crossing spectrophotometry was proposed for determination of galanthamine. Absorbance was measured at 277.4 nm. It obeyed Lambert-Beer's law in the range of 30-80 μg mL^{-1}.	19
Doripenon	The first derivative spectrophotometry was used for determination of doripenem in pharmaceuticals in the presence of its degradation products. The Beer low was obeyed in the range (0.42-11.30)x10^{-2} mg L^{-1}.	20
Tropisetron	The first derivative spectra were applied for determination of analyte in the presence of its degradates. The quantification was done by measurement of first-derivative amplitude at 271.9 nm. The obtained results were in a good	21

Compound	Characteristic of the method	Reference
	agreement with those obtained by HPLC and TLC methods.	
Fluphenazine Pernazine Haloperidol Promazine	Derivative spectrophotometry was used for quantification of fluphenazine, pernazine and haloperidol in their preparations. First and second derivative were applied for determination of active ingredients in pharmaceutical preparations.	22
Ezetimibe	First, second and third derivative spectrophotometric methods were proposed and utilised for determination of ezetimibe in pharmaceuticals.	23
Oxybutynin hydrochloride	First derivative of ratio spectra was used for determination of analyte in presence of its degradation product.	24
Ertapenem	First derivative and first derivative of ratio spectra methods were applied for determination of ertapenem in the presence of its degradation product. The analyte was assayed by the first method at 316 nm in the range 4-60 µg mL^{-1}. The second method allowed ertapenem determination at 298 nm and 316 nm in the same concentration range using spectrum of degradant at 28 µg mL^{-1}as a divisor.	25

Table 1. Determination of one component in sample

Compound	Characteristic of the method	Reference
Democlocycline and minocycline	First derivative spectra were used for simultaneous determination of drugs in synthetic mixtures. The linearity in the ranges 10-40 µg mL^{-1} and 10-50 µg mL^{-1} were obeyed for democlocycline and minocycline, respectively. The method was applied for the analysis of these drugs in clinical samples, urine and honey.	26
Rupatadine and montelukast	The quantification was achieved using first-order derivative method. Rupatadine was determined at 273.46 nm, while montelukast at 297.27 nm. The method was applied for determination of both compounds in their combined dosage form.	27
Ambroxol and doxycycline	First derivative of ratio spectra method was applied for simultaneous determination of both analytes in pharmaceutical formulations and in laboratory-made mixtures.	28
Tramadol and ibuprofen	The first-derivative method was proposed for simultaneous determination of both compounds. The measurements of amplitude was done at 230.5 and 280 nm for tramadol (Trama) and ibuprofen (Ibu), respectively. The linearity was obeyed in the rage 5-50 µg mL^{-1} for Trama and 5-100 µg mL^{-1} for Ibu.	29
Sodium rabeprazole and itopride	First derivative of ratio spectra method was applied for simultaneous determination of both analyte. The amplitudes at 231 nm and 260 nm were used for	30

Compound	Characteristic of the method	Reference
hydrochloride	quantification of rabeprazole and itopride, respectively.	
Alprazolam and fluoxetine hydrochloride	Second derivative spectrophotometry (D2) was applied for simultaneous estimation of alprazolam (ALP) and fluoxetine hydrochloride (FXT) in pure powder and formulation. Quantitative determination of the drugs was performed at 232.14 nm and at 225.25 nm for ALP and FXT, respectively. Quantification was achieved over the concentration range 4-14 μg mL^{-1} for both drugs with mean recovery of 99.36 ± 0.84 and 99.60 ± 0.93 % for ALP and FXT, respectively.	31
Drotaverine hydrochloride and mefenamic acid	The second-order derivative spectra were used for simultaneous determination of drotaverine (DRO) and mefenamic acid (MEF). Calibration graphs were constructed over the concentration range of 4-24 μg/mL^{-1} for DRO and MEF. Detection and quantitation limit were 0.4348 and 1.3176 μg/mL^{-1} for DRO and 0.6141and 1.8611 μg/mL^{-1} for MEF. The method was applied for determination of both ingredients in combined dosage forms.	32
Triprolidine hydrochloride and pseudoephedrine hydrochloride	Second derivative spectrophotometric method was proposed for simultaneous determination of pseudoephedrine hydrochloride (PSE) and triprolidine hydrochloride (TRI). The second derivative amplitudes of PSE and TRI were measured at 271 and 321 nm, respectively. The calibration curves were linear in the range of 200 to 1,000 μg mL^{-1} for PSE and 10 to 50 μg $mL^{-1}$for TRI.	33
Clopidogrel bisulphate and aspirin	The method was based on the second-derivative spectra of both ingredients. The amplitude at 254.0 nm was used for clopidogrel bisulphate, while at 216.0 nm for aspirin. The linearity was obeyed in the range 5.0- 30.0 μg mL^{-1} for both compounds.	34
Simvastatin and ezetimibe	The first-order derivative spectrophotometric method was proposed for simultaneous determination of analytes in their mixtures. The measurements were carried out at 219 and 265 nm for simvastatin and ezetimibe respectively. The validation of method was done. The range of application was estimated to be 2-40 μg mL^{-1} for simvastatin in the presence of 10 μg mL^{-1} ezetimibe and 1-20 μg mL^{-1} of ezetimibe in the presence of 20 μg mL^{-1} of simvastatin.	35
Fe(III) and Al(III) ions	The proposed method was based on the first derivative spectra of Al^{3+} and Fe^{3+}complexes with chrome azurol S. The proposed procedure was successfully applied for simultaneous determination of studied ions in standard mixtures, pharmaceuticals and in post-haemodialysis samples.	36

Compound	Characteristic of the method	Reference
Calcium and magnesium ions	The reaction of studied ions with pyrogallol red at presence of Tween 80 was applied. Next the first and second derivative spectra of complexes were applied for quantification of calcium and magnesium in multivitamin preparations, samples of human serum and in drinking water.	37
Copper and palladium	The proposed method utilise the reaction of studied ions with morpholinedithiocarbamate (MDTC). Derivative spectra of generated complexes allowed simultaneous determination of Cu and Pd in pharmaceutical samples, synthetic mixtures, alloys and biological samples.	38
Paracetamol, propiphenazone and caffeine	First to fourth derivative spectra of components were subjected to chemometric analysis (principal component regression, PCR; partial least squares with one dependent variable, PLS-1; three dependent variables, PLS2) and adopted for multicomponent analysis. The third derivative spectra of all ingredients became a basis of quantification method.	39

Table 2. Application of DS for multicomponent analysis

The fourth-derivative spectra of molybdenum complexes of tetramethyldithiocarbamate (tiram) fungicide were used for its quantification in commercial samples and in wheat grains [41]. Atrazine and cyanazine were assayed in food samples by first- derivative spectrophotometry [42]. In order to improve results of assay, the first-derivative spectra of the binary mixture were subjected to chemometric treatment (classical least squares, CLS; principal component regression, PCR and partial least squares, PLS). A combination of first-derivative with PCR and PLS models were applied for determination of both herbicides in biological samples [42]. A first-derivative spectrophotometry was used as a reference method for simultaneous determination Brillant Blue, Sunset Yellow and Tartrazine in food [43].

First derivative of ratio spectra was applied for determination of strontium, magnesium and calcium in Portland cement [44]. The proposed procedure was based on complexation of studied ions with Alizarin Complexone.

As it is mentioned above, derivative spectrophotometry seems to be a very useful tool for physico-chemical studies. It can be applied for investigation of reaction kinetics [11,12], or for determination of chemical reaction constants.

First derivative spectra of levomepromazine (LV) and its sulphoxide were employed for investigation of LV photodegradation [11]. The degradation process of biapenem was monitored by measurement of first-derivative amplitude at 312 nm [12]. The determined rate constants for studied process were in good agreement with those obtained by HPLC method [12]. The second-order derivative spectrophotometric method was used for investigation of solvolytic reaction 2-phenoxypropionate ester of fluocinolone acetonide [45]. The run of process was observed by measurement the second-order amplitude at 274.96 nm corresponded to fluocinolone acetonide. The solvolysis rate constant was calculated using derivative method and compare with those obtained by HPLC methods [45].

An interesting application of derivative spectrophotometry was described by Wu and Zivanovic [46]. They proposed the use of the first derivative spectra for determination of the degree of acetylation of chitin and chitosan. They employed the evaluated procedure for commercial samples.

5. New trends in derivative spectrophotometry

The provided short review shows good and bad sides of derivative spectrophotometry. It has mainly found application in pharmaceutical analysis for control of pharmaceuticals. It gives good results for samples with well defined composition. A main compound usually is present in its commercial forms at a relatively high level, convenient for spectrophotometric determination. An application of derivative spectrophotometry simplified procedure and allows to determine an active compound in presence of matrix (others ingredients, its degradation products) without primary sample preparation.

An analysis of scientific articles shows new trends in the use of derivative spectrophotometry. First direction of development is a combination of derivative spectra with chemometric methods [28, 36, 39, 42]. Procedures based on derivatisation of ratio spectra [24, 25,28, 30, 40, 44, 47] belong to the same group. An interesting modification of derivative spectrophotometric procedure described Eskandari [48]. A fusion of H-point standard addition method with the first derivative of mixture spectra was applied for simultaneous determination palladium and cobalt. The method was applied for their determination in synthetic mixtures and alloys.

The second observed trend is an association of derivatisation with others instrumental methods. Every set of digital data can be subjected derivatisation. So this mathematical approach was applied for data processing with synchronous fluorescence spectroscopy. The second derivative synchronous fluorimetry was used for simultaneous determination of sulpiride and its degradation product [49]. For quantification were used amplitudes of ^{2}D peaks at 295.5 nm and at 342 nm corresponded to main compound and its degradate, respectively. The method was applied for studies of the kinetics of alkaline degradation of drug.

Kang et al. [50] developed the first derivative synchronous fluorescence method for simultaneous assay of traces of some polycyclic aromatic hydrocarbons in human urine. Proposed method was fast, sensitive, selective and reliable. The results were comparable with those obtained by HPLC method.

Derivative spectrophotometry was applied for resolving and quantification of overlapped peaks in capillary electrophoresis [51]. Derivatisation of electropherogram improved separation of compounds. An elaborated procedure was used for determination of eleven derivatives of benzoic acid.

6. Disadvantageous of derivative spectrophotometry

Specific properties of derivative spectrophotometry can be a source of an additional errors. As it is mentioned previously a shape of derivative spectrum is closely connected with the shape of its parent zero-order spectrum. Small changes in a course of curve describing basic spectrum are strongly magnified in derivative spectrum. Application of derivative spectrophotometry requires from analyst knowledge about its specific properties. The main disadvantage of derivative spectrophotometry is its poor reproducibility. It is result of strong dependence of derivative spectrum on recording parameters of used spectrophotometer like scan rate, spectral width of beam, integration time and interpoint distance[1, 5, 7]. Zero-order spectra of the same substance obtained on different spectrophotometers can be identical, but derivatisation of them gives different results. The generated derivative spectra can derived in intensity, shape and positions of maxima and minima. So restoration of given literature method requires to use the same type of apparatus with the same working parameters described in an article or reoptimisation parameters of method on an own spectrophotometer.

Optimisation of used working spectrophotometer parameters should be done when a new derivative-spectrophotometric method is elaborated. A construction of some spectrophotometers does not allow to check influence of whole factors, but if more advanced equipment is available it is worth to do.

As a result of derivatisation is closely connected with geometrical features of a zero-order spectrum, it is obvious that a method of spectrum registration is a key-point. The use of broad beams gives the averaged smoothed zero order spectra. Application of narrow beams results in intensification and narrowing of absorption bands. But from the other hands, the narrowing of monochromator' slit increases an effects connected with beams bending on edges of the slit. The edge phenomenon causes additional noises which are recorded with absorbance. So the absorption spectrum recorded with too narrow monochromator' slit can be distorted by high level of noise.

Interpoint distance of registered spectrum is very important parameter. Absorption spectrum obtained by spectrophotometer possess a digital structure which is the result of construction of a monochromator and a manner of registration. Spectra registered with large interpoint distance are averaged, flat without many spectral details.

A level of noise enclosed in zero order spectrum directly influences a quality of generated derivative spectrum. It was proved that spectra registered with low scan rates and long integration times are less biased by noise. This is advantageous if high order derivative are generated [7].

Taking into account above information, it is obvious that reproducibility of method based on derivative spectrophotometry depends on reproducibility of parameters of registration of zero-order spectra. So, adaptation of elaborated in another laboratory derivative spectrophotometric method, requires application the same working parameters as used by authors. But this problem is completely ignored by scientists. Based on analysis of articles concerned on application of derivative spectrophotometry it could be stated that working parameters of spectra registration are very rarely given [8]. There is noticeable lack of standardisation in description of procedures based on derivative spectra. Very often, authors of scientific articles give only information what model of apparatus they used without any details of its working parameters as well as algorithm for derivatisation of spectra. In this case the published procedure can be used only if our laboratory is equipped with the same model of spectrophotometer supplied with the same software. Otherwise verification of literature' method requires reoptimisation, adaptation to our conditions

A geometrical features of derivative spectra can be a source of analytical errors. A course of derivative curve is different than its initial spectrum. A main band gets narrowing but additional satellite bands appear. If basic spectrum of mixture is subjected to derivatisation a resulted derivative spectrum of mixture is a sum of derivative spectra of each individual components. New peaks in the final spectrum can be the result of addition or subtraction, so their intensity undergo amplification or reduction. Very often their positions are shifted in comparison to their position in derivative spectra of individual components. Some analytical information can be lost during derivatisation or new false peaks can be generated. A careful analysis of course of derivative spectra of components and their mixtures at different compositions should be done to avoid such errors. A selection of optimal derivatisation parameters should be make taking into account influence of others components on intensity of derivative peaks of determined analyte. This procedure seems to be time- and labour-consuming but gives good results. Properly done selection of mathematical parameters of derivatisation and instrumental parameters of spectral analysis allows to elaborate selective method of determination and leads to minimise errors connected with features of derivative spectra.

7. Conclusions

Nowadays, derivative spectrophotometry is fully available with software's controlling modern spectrophotometers. Analysts receive an elegant tool which allows extraction of analytically useful information from spectra. An understanding of specific features of this technique and its proper utilisation leads to simplification of procedure and to increase a selectivity of assay.

8. References

[1] G. Talsky, Derivative Spectrophotometry, 1st ed., VCH, Weinheim, 1994.
[2] A. Savitzky, M. J. E.Golay, Anal. Chem. 36, 1627-1642 (1964).
[3] P. A. Gorry, Anal. Chem. 63, 534-536 (1991).
[4] P. A. Gorry, Anal. Chem. 62, 570-573 (1990).
[5] S. Kuś, Z. Marczenko, N. Obarski, Chem. Anal. (Warsaw) 41, 899-929 (1996).
[6] T. C. O'Haver, G. L. Green, Anal. Chem. 48, 312-318 (1976).
[7] T. C. O'Haver, T. Begely, Anal. Chem. 53, 1876-1878 (1981).
[8] J. Karpinska, Talanta 64, 801-822 (2004).
[9] F. Sanchez Rojas, C. Bosch Ojeda, Anal. Chim. Acta 635, 22-44 (2009).
[10] M. Gumustas, S. Sanli, N. Sanli, S.A.Ozkan, Talanta 82, 1528-1537 (2010).
[11] J. Karpinska, A.Sokol, M. Skoczylas, Spectrochim. Acta Part A 71, 1562-1564 (2008).
[12] J. Cielecka-Piontek, A. Lunzer, A. Jelinska, Cent. Eur. J. Chem. 9, 35-40 (2011).
[13] Y.F. M. Alqahtani, A.A. Alwarthan, S. A. Altamrah, Jordan J. Chem. 4, 399-409 (2009).
[14] A.S. L. Mendez, L.Deconto, C. V. Garcia, Quim. Nova, 33, 981-983 (2010).
[15] E. Souri, M. Amanlou, S. Shahbazi, M. Bayat, Iranian J. Pharm. Sci. 6, 171-178 (2010).
[16] S. M. Malipatil, P.M. Deepthi, S.K.K. Jahan, Int. J. Pharm. Pharmaceut. Sci. 3, 975-1491, (2011).
[17] D. Madhuri , K.B. Chandrasekhar, N. Devanna, G. Somasekhar, Int. J. Pharm. Sci. Res. 1, 222-231 (2010).

[18] V.M. Patel, J. A. Patel, S.S. Havele, S.R. Dhaneshwar, Int. J. Chem. Tech. Res. 2, 756-761, (2010).
[19] K. Mittal, R. Kaushal, R. Mashru, A. Thakkar, J. Biomed. Sci. Eng. 3, 439-441 (2010).
[20] J. Cielecka-Piontek, A. Jelińska, Spectrochim. Acta Part A. 77, (554-557 (2010).
[21] S. L. Abdel-Fattah, A. Z. El-Sherif, K. M. Kilani, D. A. El-Haddad, J. AOAC Int. 93, 1180-1191 (2010).
[22] M. Stolarczyk, A. Apola, J. Krzek, A. Sajdak, Acta Pol. Pharm. 66, 351-356 (2009).
[23] M. Sharma, D.V. Mhaske, M. Mahadik, S.S. Kadam, S. R. Dhaneshwar, Indian J. Pharm. Sci. 70, 258-260 (2008).
[24] N. E. Wagiem, M.A. Pegazy, M. Abdelkawy, E. A. Abdelaleem, Talanta 80, 2007-2015 (2010).
[25] N.Y.Hassan, E. M. Abel-Moety, N.A. Elragey, M.R. Rezk, Spectrochim. Acta Part A 72, 915-921 (2009).
[26] A.R.G.Prasad, V. S. Rao, Sci. World 8, 34-38 (2010).
[27] P.Patel, V. Vaghela, S. Rathi, N. Rajgor, V. Bhaskar, J. Young Pharm. 1, 354-360 (2009).
[28] G. M. Hadad, A. El-Gindy, W. M. M. Mahmoud, Spectrochim. Acta Part A 70, 655-663 (2008).
[29] A. B. Thomas, N. G. Dumbre, R. K. Nanda, L. P. Kothapalli, A.A. Chaudari, A.D. Deshpande, Chromatographia 68, 843-847 (2008).
[30] S. S. Sabnis, N. D. Dhavale, V. Y. Jadhav, S. V. Gandhi, Spectrochim. Acta Part A 69, 849-852 (2008).
[31] R. B. Patel, M. R. Patelb, M. B. Shankara, K. K. Bhattb, Eurasian J. Anal. Chem. 4,76-86 (2009).
[32] J. D. Patel, B. A. Patel, B. P. Raval, V. M. Vaghela, J. Pharm. Res, 3,566-569, (2010).
[33] L. Sriphong, A. Chaidedgumjorn, K. Chaisuroj, World Acad. Sci., Eng. Technol. 55, 573-577 (2009).
[34] M. D.Game, D. M. Sakarkar , Inter. J. Chem. Tech. Res. 2, 1886-1891 (2010).
[35] E. Souri, M. Amanlou, E-Journal of Chemistry, http://www.e-journals.net, 7(S1), S197-S202 (2010).
[36] N. Aguerssif, M. Benamor, M. Kachbi, M.T. Draa, J. Trace Elements Med. Biol. 22, 175-182 (2008).
[37] M. Benamor, N. Aguerssif, Spectrochim. Acta Part A 69, 676-681 (2008).
[38] V. Kaur, A. K. Malik, N. Verma, Anal. Letters 40, 2360-2373 (2007).
[39] M. De Luca, F. Oliviero, G. Ioele, G. Ragno, Chemometr. Intell. Lab. Sys. 96, 14-21 (2009).
[40] J.S. Millership, F.McCaffrey, D. Tierney, J. Pharm. Biomed. Anal. 48, 408-413 (2008).
[41] V. K. Sharma, J. S. Aulakh, A.K. Malik, Talanta 65, 375-379 (2005).
[42] G. Zhang, J. Pan, Spectrochim. Acta Part A 78, 238-242 (2011).
[43] T.M.Coelho, E. C. Vidotti, M. C. Rollemberg, A. N. Medina, M. L. Baesso, N. Celle, A. C. Bento, Talanta 81, 202-207 (2010).
[44] K. A. Idriss, H. Sedaira, S.S. Ahmed, Talanta 78, 81-87 (2009).
[45] B. Markovic, S. Vladimirov, O. Caudina, V. Savic, K. Karljikovic-Rajic, Spectrochim. Acta Part A 75, 930-935 (2010).
[46] T. Wu, S. Zivanovic, Carbohyd. Polym. 73, 248-253 (2008).
[47] F. A. El-Yazbi, H. H. Hammud, S. A. Assi, Spectrochim. Acta Part A 68, 275-278 (2007).
[48] H. Eskandari, Spectrochim. Acta Part A 63, 391-397 (2006).

[49] A. Abdeal, N. El-Enany, F. Belal, Talanta 80, 880-888 (2009).
[50] R-H. Kang, Y-S. Wang, H-M. Yang, G-R. Li, X. Tan, J-H. Xue, J-Q. Anal. Chim. Acta 658, 180-186 (2010).
[51] J. Zhang, Y. Hu, J. Liu, Z.Hu, Microchim. Acta 164, 487-491 (2009).

14

Spectrophotometry as a Tool for Dosage Sugars in Nectar of Crops Pollinated by Honeybees

Vagner de Alencar Arnaut de Toledo[1,*],
Maria Claudia Colla Ruvolo-Takasusuki[2], Arildo José Braz de Oliveira[3],
Emerson Dechechi Chambó[1] and Sheila Mara Sanches Lopes[3]
[1]*Animal Science Department,*
[2]*Cell Biology and Genetics Department,*
[3]*Pharmacy Department, Universidade Estadual de Maringá, Maringá, Paraná*
Universidade Estadual de Maringá, Maringá,
Brazil

1. Introduction

The pollination by honeybees is important to the best performance of several crops. In this interaction plant-insect there is a change of reward between both organisms, and the sugar concentration in the nectar is a keyword. The spectrophotometry allows analyzing the type and the quantities of sugar in the nectar of flowers, and identifying varieties that are more attractive for pollinators.

The nectar is the reward for several pollinators, and the principal is the honeybee *Apis mellifera*. The nectar is produced from sap of phloem by active secretion that results in a solution of sugars like sucrose, fructose and glucose in varied proportion depending on the vegetal.

Besides sugars, other compounds of the nectar has importance for the coevolution between plants and their pollinators like amino acids, proteins, lipids and alkaloids and these may be toxic for visitors and, however, these compounds may have a role of protection against animals that withdraw nectar of flowers without an efficient pollination.

Several researches are carried out to evaluate the effect of crop pollination using honeybees and consequently the increase of productivity in agriculture. The visit and hoarding of nectar and pollen allows rise in grain production, or tasteful fruit with symmetric format.

The study of sugar from floral nectary is important for identification if a rise or decrease in quantity or nectar quality. The plant may secret a little bit of nectar, but with high sugar concentration, or unlike, secret more quantities, but with low sugar concentration. These differences in nectar may vary depend on pollinator visitation. However, the frequency of honeybees that visit flowers may contribute for rising nectar production like change the sugar proportion.

* Corresponding Author

Other factors that must be considered are secreted sugars: sucrose, glucose and fructose. The quantities of them may vary depending on variety and type of vegetal. Honeybees have preference for nectar with more sucrose concentration. Sugars present in nectar are related with honey quality that will be produced by honeybees and, finally will be commercialized.

The association of beekeeping and agriculture provide a rise in profits as for farmer as for beekeeper. However, the quantity and/or quality of the sugars in floral nectar like the pollination of cultivated crops by honeybees have an economic and social role well-established and significant currently.

The proposal of this chapter is perform a review about spectrophotometry in sugar dosages of nectar of the main cultivated crops and show the importance of this tool (spectrophotometry) to improve crop production by honeybee pollination contributing as for agriculture as for beekeeping.

1.1 Pollination and plant reproduction

The pollination process occurs in spermatophitic plants consists in pollen grain transfer to the stigma, which is the receptive part of feminine flowers of superior vegetal. The pollen grains are structures that contain the reproductive male cells, they are produced and storaged in anthers (part of male organ of flowers) until the deiscence, moment wherein are released. The indispensable factor in pollination process and that denotes the success is the need of ovule fertilization and subsequent fruit set and seeds formation.

The pollination of flowers beside the diaspore dispersion is a fundamental process in reproductive success of vegetal species. The involved animals in this process have an important role (Buckmann & Nabhan, 1997), then the efficience in the pollination process means rise in disponibility of food to the human being and animals.

The structure of plant (monoic or dioic) varies to the size and anatomic and physiological characteristics of flower and their position in the plant, may occur the autopollination or cross pollination. This cross pollination provides a rise in gene flow between plants, spreading them and the results are favorable (Malerbo-Souza et al., 2004).

The animals that carried out the pollen transfer of anthers to the stigma flowers are known pollinators, and can be insects like bees, bettles, flies, butterflies, wasps and moth; birds - hummingbirds and parakeets; and small mammals - bats, rodents, and marsupials (Malagodi-Braga, 2005). Among pollinators, animals of Insecta class are the most important, and in the order Hymenoptera you can find the major number of them. Honeybees are the most important pollinators available in the nature.

Unlike of other insects that visit flowers only to collect its own food, honeybees visit a bigger quantity of flowers, besides of food to own survival, they harvesting pollen and nectar to feed their larvae and storaging (Müller et al., 2006). Futhermore, the higher efficiency of honeybees as pollinators is as much as by their number in the nature, as by better adaptation to floral complex structures like mouthparts and adapted body to imbibe the nectar of flowers and harvesting pollen, respectively (Proctor et al., 1996). The bees of Apidae family have higher distinction because their morphological traits are representative, like special structure to load pollen - corbicula, located in tibia of hind leg, similar to basket

- pollen is loaded in this structure in association with nectar or oil, absence of ventral scopa and long tongue (Teixeira & Zampieron, 2008).

Each group among animals that visit flowers is associated to some particular type of floral reward, that is the morphological traits of flowers reflect adaptation to diversified pollinators. The contemporaneous interactions between plants with flowers and insects can be because of long and closer coevolutionary relation (Backer & Hurd, 1968; Prince, 1997). This process of coevolution or the interaction between plants and pollinators is based on a system of mutual dependence. This system was detailed by the first time by Christian Konrad Spengel (1750-1816), in which plants show their rewards like nectar, pollen, oil and resins by floral arrangement, colour, size and odour of flowers, while the pollinators in change of provided resources by plants transfer pollen between flowers increasing the gene flow and promoting the diversification of the species, named by this as key mutualism (Morgado et al., 2002).

Plants, year by year, specialized in attract more efficient pollinators and make transportation of their reproductive cells, and therewith could be benefited with cross pollination. The disposition of flowers is an important factor that can be isolated in the branches or grouped in the same floral axis forming inflorescence, colour, odour, size, nectar, oil, pollen and resins.

The floral rewards provided by Angiosperms are required to attract pollinators, nectar seems the most searched in crop cultivated commercially, however, in searching by this reward many animals, mainly worker honeybees *Apis mellifera* have pollen adhered to their body, and so, later deposit accidentally their loads on the stigma of other flower of the same specie, performing indirectly the cross pollination.

Assays carried out in Marechal Cândido Rondon, Paraná, Brazil, with sunflower make clear worker honeybees that hoarding nectar are more frequent (mean 2.28 honeybees/capitulum) than pollen foragers (0.40 honeybees by capitulum) in anthesis period and schedule of higher visitation in the crop (Chambó et al., 2011). Other experimental results make clear more frequency of nectar foragers than pollen foragers in sunflower crop (Paiva et al., 2002; Teixeira & Zampieron, 2008).

Moreover, it must be considered that quantity of honeybee visitors to different species of superior vegetal can be related to concentration and volume of nectar in the flowers during all day (Pham-Delegue et al., 1990). Experiments with attractiveness sunflower genotypes show significative difference in relation to the number of honeybee visitation, mean of 3.40 (genotypes Helio 360 and Aguará) and 1.60 (genotype Multissol) visits of *A. mellifera* by capitulum in third day of anthesis (Chambó et al., 2011, in press). The researchers did not tested the concentration and volume secreted during all day, but assign to these causes the difference of sunflower genotypes assayed. In hybrid of ornamental coloured sunflower BRS-OASIS had an increase ($p < 0.05$) in four times in number of honeybee visitors using sucrose solution in two concentrations 5% and 7.5% as attractive. These solutions were pulverized on sunflower capitulum in relation to concentration of 2.5% (Martin et al., 2005).

Despite of efficiency of pollination process, it depends on numerous received visits by pollinators (Schirmer, 1985). Vidal et al. (2010) studying the pollination and set fruit in *Cucurbita pepo* by honyebees reported that percentage of set fruit was maximum (100%)

when the flowers received 12 visits of *A. mellifera,* corresponding to load of 1.253 pollen grains deposited on the stigma. In comparision, with two visits of honeybees, 174 pollen grains were deposited on the stigma, and the set fruit only of 50%. Other important factor is the attraction by pollen and nectar of male flower was arised by opening grade of nectary pore of flower (Vidal et al., 2010).

Models of pollen transfer depends on the specific pollinator rate about pollen remotion of anthers and deposition on the stigmas. The pollinators have a high remotion and low deposition (HRLD) of pollen on the stigma of flowers will benefite a plant wherever it is not better available pollinator. In case of pollinators have high remotion of pollen of anthers and high deposition (HRHD) of pollen on the stigmas also visit a plant population, the visits of pollinators HRLD can reduce the total pollen transfer. The HRLDs parasite the plants, displace the pollen grains that would delivery by HRHDs. When two visitors remove equal quantities of pollen, the pollinator more efficient will be that with higher delivery rate. In case of different quantities of remotion, the better do not depends on the deposition rates only, but another variables including schedule of visitation to pollen deposition (Thomson & Goodell, 2002).

The volume and sugar concentration of nectar, important factor in attraction of pollinators alos are known by varing between plant species and affect the answer of pollinator (Lanza et al., 1995). Besides, different varieties of the same species can range to the sugar concentration in the nectar (Free, 1993). In some vegetal species, for example, *Curcubita pepo,* the periodical remotion of nectar of the flower by pollinators do not arise the total volume of produced nectar by plant, then the nectar secretion is not stimulated or inhibited by successed harvesting of this reward. However, the nectar remotion of flowers can reduce or stimulate the process of secretion in several plant species.

The pollinator that drag out the spent time in flowers to nectar hoarding, specially if it is available, increase the probability of pollen deposition and, consequently, the pollination can be well successed. Nevertheless, cultivated commercially crops that do not stimulate the nectar production after several visits of pollinators can have an advantage in pollination process, so the number of honeybee visits to flowers is positively correlated to the nectar secretion in all flower duration (Vidal et al., 2006). The evaluation of nectar secretion rate is an important component in ecological studies related to the pollination process, mainly in that about flower-insect interaction.

2. Use of spectrophotometry in nectar analysis of plants

The quantitative analysis using spectrophotometric methods are widely used because have a good sensitivity, low analysis cost, easily handle, accessible equipments, justifying their application and efficiency for utilization in quantification analysis of several compounds.

The main spectrophotometric techniques used in sugar determination are based on reactions of these carbohydrates with colorimetric reagents, forming a colored complex that can be detected and quantified in a spectrophotometer.

The techniques used for total sugar quantification are anthrone method (Trevelyan & Harrison, 1952 in: Yemm & Wills, 1954), phenol-sulfuric (Dubois et al., 1956) and also can be used the 3-5-dinitrosalicylic acid - DNS modified (Miller, 1959). For reducing sugar determination, also can be used the technique of DNS and the reaction with p-hydroxybenzoic acid hydrazide - *PAHBAH* (Blakeney & Mutton, 1980).

For specific determination of some sugars also are available some other methods. Fructose can be determinated by cysteine/tryptophan/sulphuric acid method (Messineo & Musarra, 1972) and resorcinol method (Roe et al., 1949). Sucrose can be determined by anthrone method after destruction of monosaccharides with KOH (Sala Jr. et al., 2007). Besides these methods, the glucose, fructose and sucrose can be selectively determined by methods using enzymatic reactions (Moernan et al., 2004; Amaral et al., 2007).

3. Total sugar determination

3.1 Phenol-sulphuric method (Dubois et al., 1956)

This method is based on the fact that simple or complex sugar and their derivatives, including methyl esters with free reducing groups or potentially free, when treated with phenol and concentrated sulphuric acid will generate a yellow-orange colour, the reaction is sensible and this colour is stable. The method is simple, quick, sensible, and the results are reproductible.

The changing in colour of solution is measured in visible range and proportional to the quantity of sugar inside the sample.

The carbohydrates are hydrolyzed under heating in strongly acid pH, this reaction produces furan derivative compounds, that when condensed with phenolic compounds produce coloured substances (Figure 1).

The total sugar concentration is determined by spectrophotometry in 490 nm wavelength. Sensitivity of this method range from 10 to 100μg of total sugar and the quantification is made from calibration curve using glucose or pentose as standard and calculation are performed by equation of the linear regression obtained from calibration curve.

3.1.1 Methodology

Reagents: phenol solution 5% (w/v) and concentrate sulphuric acid (95%, p/v).

Get an aliquot of 100μg.mL^{-1} from the sample, add 0.5mL of phenol solution 5%, shake in vortex, add 2.5mL, shake again and keep in water-bath 25°C for 15 minutes. After this period, read the absorbance in spectrophotometer at 490nm.

3.2 Anthrone method (Dreywood, 1946)

The anthrone is the reduction product of anthraquinone and was recognized first as specific reagent for several carbohydrates by Dreywood (1946) because sugar solution in concentrate sulphuric acid form the blue-greenish colour characteristic and since anthrone has been used widely as suitable and specific reagent for sugar colorimetric dosage.

The anthrone reaction is based on hydrolitic and dehydrating action of concentrate sulphuric acid on carbohydrates, in which glicosidic linkages are broken releasing free reducing sugar that are dehydrated and converted to furfural by pentoses and hydroxymethylfurfural by hexoses (Figure 1).

These substances are condensed with anthrone to hydroxyanthracene (9,10-dihydro-9-oxo anthracene) forming a blue-greenish product that have absorption maximum at 620nm.

The sensitivity of this method is from 0 to 100μg.mL^{-1}. Glucose solution is used as standard to build calibration curve and get the straight line equation to quantify samples.

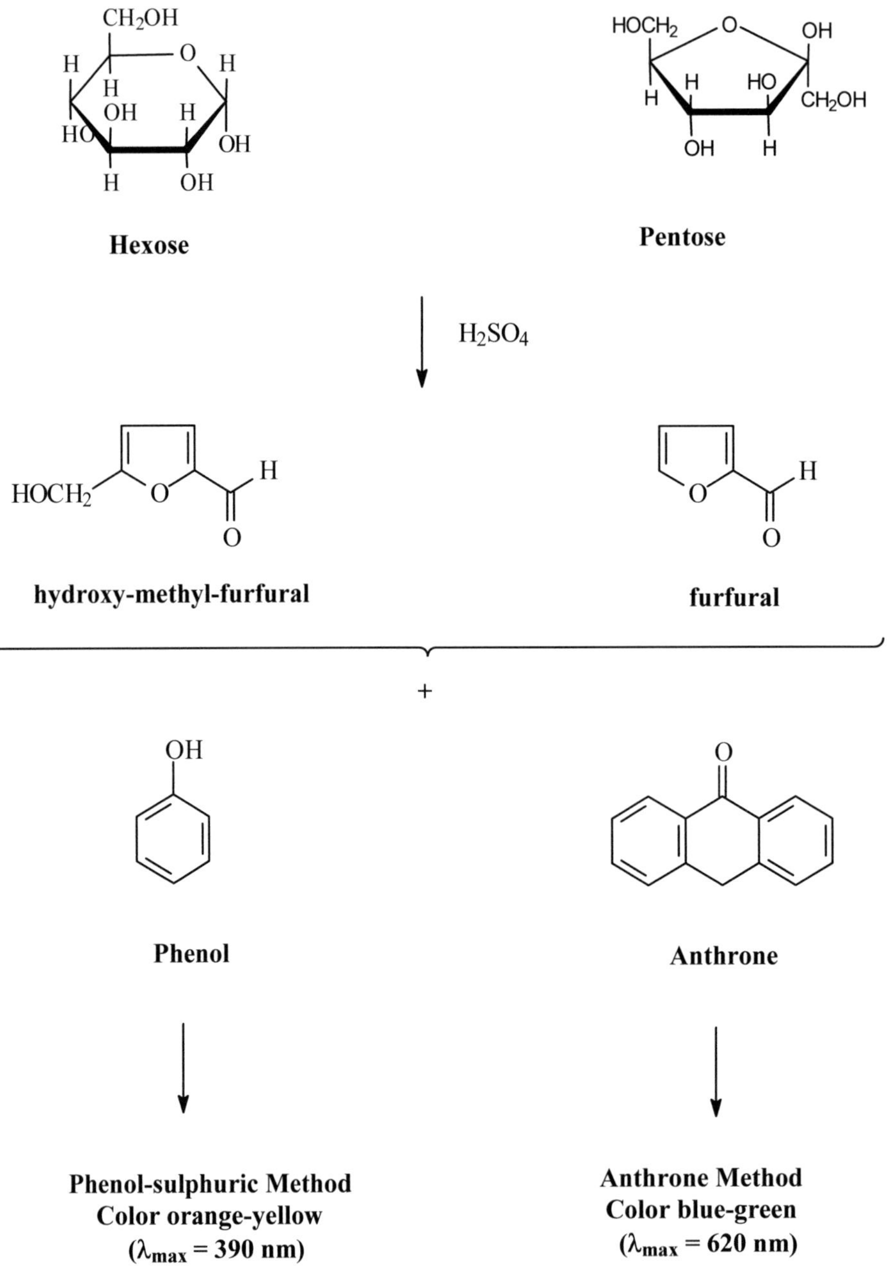

Fig. 1. Reaction of formation of the coloured complex in phenol-sulphuric and anthrone methods.

3.2.1 Methodology

Reagents: anthrone solution 0.2% (0.2g anthrone em q.s. to 100mL of sulphuric acid 95%). This solution must be kept at rest for 30 minutes with sporadic agitation until the solution to be clear, the reagents must be used in 12 hours.

Get an aliquot of 100μg.mL^{-1} from the sample, add 2mL of the anthrone, leave at ice bath and hereupon cool in cooled water and read the absorbance in spectrophotometer at 620nm.

4. Reducing sugar determination

4.1 3,5-dinitrosalicylic acid method (Miller, 1959)

This method was first mentioned by Summer & Sisler (1944) and modified by Miller (1959), with this technique there is a possibility to dosage reducing sugar as total sugar.

The sugar act as a chemical reductor due free aldehyde group or ketone group presence in its molecule. In an alkaline medium, the reducing sugars are able to reduce the 3-5-dinitrosalicylic acid to 3-amino-5-nitrosalicylic acid, wherever, the aldehyde group is oxidized to aldonic acid (Figure 2). The 3-amino-5-nitrosalicylic acid is a orange color product, and the intensity of the color depends on the concentration of the reducing sugar. The sodium hydroxide provides the glucose reaction with 3-5-dinitrosalicylic acid by medium alkalinization.

Besides to 3-5-dinitrosalicylic acid, it is also used in this method the Rochelles salt (Potassium sodium tartrate), phenol, sodium bisulfite, and sodium hydroxide. The phenol optimize the quantity of the colour produced and sodium bisulfite stabilize the colour in the phenol presence (Miller, 1959).

The sensitivity of the method is from 100 to 500μg.mL^{-1} of reducing sugar. A standard glucose or fructose solution is used to build the calibration curve and get the straight line equation to quantify samples.

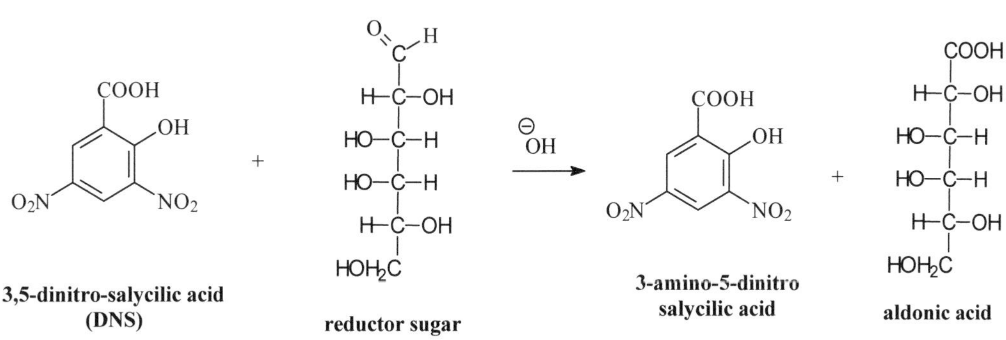

Fig. 2. Reaction of reducing sugar with 3,5-dinitro-salycilic acid reagent.

4.1.1 Methodology

Reagents:

Reagent A – Weight 2.5g DNS (3,5 dinitrosalicylic acid) and add 50mL NaOH 2M.

Reagent B - Weight 75g Potassium sodium tartrate (Rochelle salt), add 125mL of destiled water. Shake under heating until total dissolution.

Subsequently, add reagent A over the B, homogenize under heating until dissolve completely, and after cool the mix, complete the volume of the solution to 250mL.

Get an aliquot 500µg.mL^{-1} of the sample. Add 1mL of DNS, shake in vortex and let in a boiling bath during 5 minutes, put the material in ice bath until cool, add 3.75mL of destiled water, shake again, and read the absorbance in spectrophotometer at 540nm.

An adaptation of the DNS method to determine the total sugar can be getting from a previous hydrolysis before dosage. The hydrolysis is made with 0.5mL HCl concentrate and incubation in water bath at 60°C for 10 minutes. After then, the solution must be neutralized with NaOH 2M and cooled with ice bath until room temperature.

The procedure for total sugar quantification follow the same described for reducing sugar used above.

4.2 Hydrazide method of p-hydroxy-benzoic acid (Blakeney & Mutton, 1980)

This method is based on determination of reducing sugar before and after the digestion by invertase using acid for p-hydroxy benzoic acid hydrazide - PAHBAH. In this method, with some modifications, there is possible to perform the colorimetric determination of glucose, fructose and sucrose, during this analysis procedure with invertase. The major subject of the methodology described in this chapter is for determination of reducing sugar. The p-hydroxy benzoic acid hydrazide also has the advantage that glucose and fructose when reacting produce the same intensity of colour, and then all free monossacharides present in the sample can be determined.

4.2.1 Methodology

Reagents:

Reagent A: 10g p-hydroxy benzoic acid dissolved in 60mL water, add 10mL hydrochloride acid and complete volume to 200mL.

Reagent B: 24.9g trisodium citrate in 500mL water, 2.2g calcium chloride, 40g sodium hydroxide, and to complete the volume to 2.000mL.

In the day of analysis, mix reagents A and B in proportion of 1:10, after keep at 4°C.

Get an aliquot 50µg.mL^{-1} of the reducing sugar sample, add 5mL p-hydroxy benzoic acid hydrazide reagent (mixed in the same day) shake in vortex, incubate in water bath at 100°C by six minutes, cool until room temperature, and read the absorbance in spectrophotometer at 410nm.

5. Fructose determination

5.1 Sulphuric acid-Cysteine-Tryptophan method (Messineo & Edward Musarra,1972)

Fructose is dehydrated in acid medium for formation of furfural derivative, which complexs with cysteine hydrochloride to produce a chromophore at starting of green color that is unstable. Immediately, this first chromophore formed after react with tryptophan hydrochloride forming now a chromophore of pink colour that has greater stability chemical than the first and stability of 48 hours.

The reaction is also sensitive because can detect a low quantities like 1µg of fructose. The reaction also is specific because aldohexoses, aldopentoses, and ketopentoses do not interfere, however these compounds do not react even if its concentrations are higher, as $5mg.mL^{-1}$. The sensitivity of the method is from 1 to 50µg of fructose. A standard solution of fructose is used to build the calibration curve and to get the equation of the linear regression to quantity samples.

5.1.1 Methodology

Reagents: sulphuric acid 75%, cysteine hydrochloride 2.5%, tryptophan solution in hydrochloride acid to formation of tryptophan hydrochloride ($100µg.mL^{-1}$ in HCl 0.1M).

Getting an aliquot $50µg.mL^{-1}$ of the sample, add 2.8mL of sulphuric acid 75%, shake in vortex, add 0.1mL cysteine hydrochloride solution 2.5%, shake in vortex again, let in water-bath 45-50°C for 10 minutes, cool at room temperature and add 1mL tryptophan hydrochloride solution, shake again in vortex.

This sequence must be followed rigorously during assay because the formation of the final chromophore depends on the initial formation of the first formed chromophore by complexation with cysteine hydrochloride. After that, read the absorbance in spectrophotometer at 518nm.

5.2 Resorcinol method (Roe et al., 1949)

This reaction follows the same theoretical principles in which there is formation of furfural from hexoses and hydroxy-methyl-furfural (HMF) and from aldopentoses by acid dehydration (Fig. 1). These two products, singly are colorless, however, it is necessary a phenolic compound addition in the medium to develop a colored compound, in this case redness. This technique is firstly mentioned by Roe (1934), with some posterior modifications by Roe et al. (1949), becoming a quick reaction and with stable color. The reaction uses the hydrochloride acid (HCl) for carbohydrates dehydration and the resorcinol is the phenolic compound that reacts with furfural and HMF.

This test allows distinguish aldoses from ketoses because the reaction with ketoses is faster and more intense than aldoses. Therefore, the formation of the furfural is easier than HMF formation.

The sensitivity of the method ranges from 10 to $80µg.mL^{-1}$ fructose. A standard fructose solution is used to build the calibration curve and to get the equation of the linear regression to quantity samples.

5.2.1 Methodology

Reagents: Resorcinol reagent, 1g resorcinol and 0.25g thiourea in 100mL. This solution must be kept in the dark to keep its stability. Hydrochloride acid is diluted as 5mL HCl and 1mL water.

Getting an aliquot 80μg as maximum of the sample, add 0.5mL resorcinol reagent, shake in vortex, add 3.5mL hydrochloride acid, shake again, let in water bath at 80°C for 10 minutes, after that, read the absorbance in spectrophotometer at 520nm.

6. Enzymatic methods for sugar determination

There are several enzymatic methods for determination of the three principal sugars individually present in nectar - fructose, glucose, and sucrose, and in biological samples like plasma, blood, and urine. A lot of these methods are commercialized in kits and can be used successfully for rapid determination of the sugar from natural products samples. These kits are precise and sensitive, which enable rapid analysis and reliable results.

6.1 Glucose oxidase (Amaral et al., 2007)

The enzyme glucose oxidase is used for quantitative and enzymatic determination of the glucose in food and other materials. The enzyme glucose oxidase test is widely used because it is cheap, stable, and by its specificity well established for glucose.

In this reaction, the glucose is oxidized to gluconic acid and hydrogen peroxide by enzyme glucose oxidase. The hydrogen peroxide reacts with ortho-dianisidine in the presence of peroxidase enzyme to form a colored product (Figure 3). The compound of the ortho-dianisidine oxidation reacts with sulphuric acid to form a coloured product more stable. The intensity of pink colour measured in 540nm is proportional to the glucose concentration in the sample.

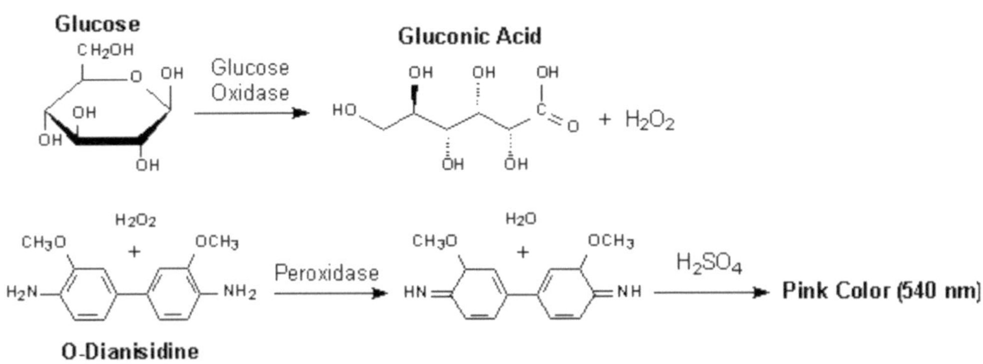

Fig. 3. Reactions of the enzymatic method of glucose oxidase.

6.2 Hexokinase method for simulteneous determination of glucose and fructose (Moerman et al., 2004)

This method is adequate for determination of monosaccharides glucose and fructose.

The principal enzyme of the method is hexokinase. This enzyme catalyzes the fosforilation of the glucose in glucose-6-phosphate, here upon the second enzyme the glucose-6-phosphate dehydrogenase together with cofactor nicotinamide adenine dinucleotide (NAD) oxidized glucose-6-phosphate to gluconate-6-phosphate and the NAD is reduced to NADH, according to stoichiometry of the second reaction (Figure 4), the spectrophotometric quantity of NADH is corresponding to glucose quantity.

Analysis of glucose is according to the following principle: hexokinase, a first enzyme, catalyzes the phosphorylation of glucose to glucose-6-phosphate, with the participation of the enzyme glucose-6-phosphate dehydrogenase and nicotinamide adenine dinucleotide (NAD) is further specifically oxidized to gluconate-6-phosphate. According to the stoichiometry of the last reaction (Figure 4), the photospectrometrically quantified amount of reduced nicotinamide adenine dinucleotide (NADH) is representative for the amount of glucose. Fructose is always determined subsequently to the glucose determination. Fructose undergoes phosphorylation to fructose-6-phosphate, with the same enzyme hexokinase, which is further converted to glucose-6-phosphate with phosphoglucose isomerase. Further oxidation to gluconate-6-phosphate as described above generates a supplementary amount of NADH that is stoichiometric with the amount of fructose.

All methodology of this analysis is performed following instructions of each enzymatic kit. For these enzymatic analysis must carried out the assays criteriously, because the order of addiction of reagents, the time of analysis and reading on spectrophotometer are determinants for an adequate analysis.

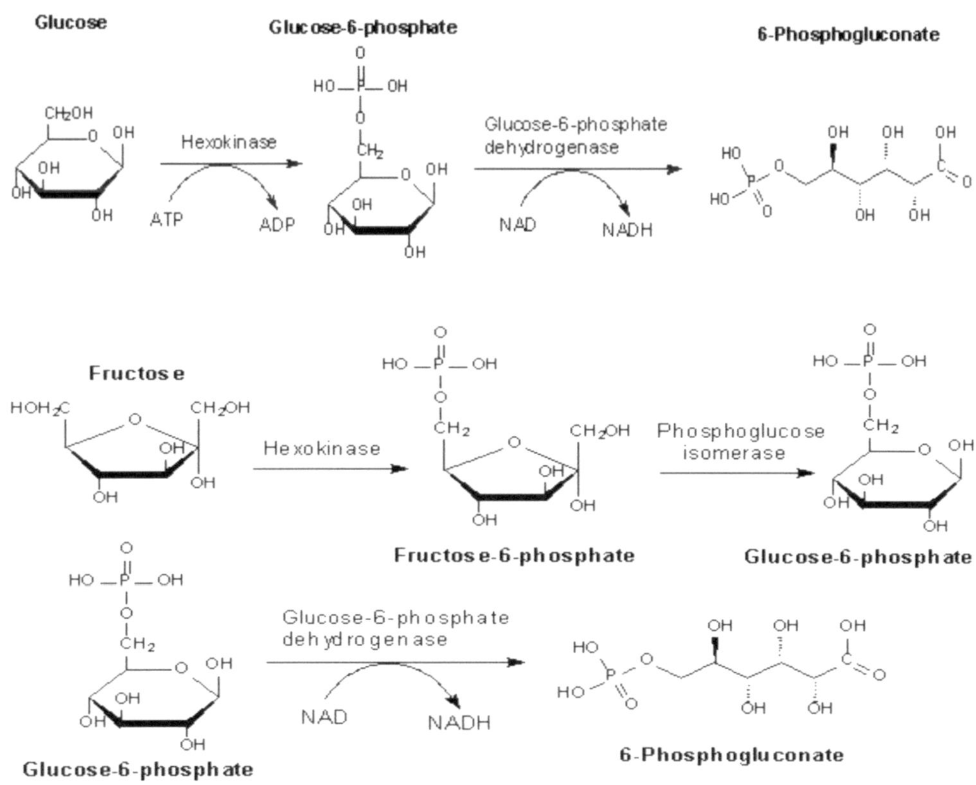

Fig. 4. Enzymatic reactions in enzymatic methods using hexokinase.

7. Sugar concentration in nectar

Nectar is considered the main reward to pollinator (Delaplane & Mayer, 2000) and its sugar concentration is associated to different pollinators, wherever the frequency and duration of visits depend on rate of nectar production (Biernaskie et al., 2002; Shafir et al., 2003; Nicolson & Nepi, 2005).

Toledo et al. (2005) reported total sugar concentration presents variation during the day in *Macroptilium atropurpureum* Urb., and can be related to number of visitor insects, specially, bees that collect nectar and pollen. In *Citrus sinensis*, it was verified that the high sugar concentration is an attractive, in special for *Apis mellifera*, and the availability of concentrated nectar during the day keeps the attractivity to pollinators (Malerbo-Souza et al., 2003). Therefore, the high quantity of nectar can leads to greater pollination rate by increasing in number of visitors bees (Silva & Dean, 2000).

Sugars are principal components of the floral nectar (Baker & Baker, 1973; Baker, 1977). The three most common sugar in nectar are sucrose, glucose and fructose in varying proportions (Freeman et al., 1985; Endress, 1994; Proctor et al., 1996; Baker et al., 1998). The amount of sugar secreted by flowers and consumed by pollinators cause a variation in sugar concentration of the flowers, during their anthesis period. Floral nectar consists of sugar pure solutions, specially glucose, sucrose and fructose (Roberts, 1979), however it can be found traces of oligosacchrides (Harbone, 1998).

Wykes (1952a) reported two oligosaccharides in nectar composition, the trisaccharide raffinose, and disaccharide melibiose in some nectaries. In some varieties of clover the presence of disaccharide maltose also was identified, however, this maltose can be a contaminant from aphids (Furgala et al., 1958). Taufel & Reiss (1952) confirmed that sucrose, glucose, and fructose, sugars promptly accepted by honeybees are current compounds present in nectar but another sugars can be present.

Besides, fructose and glucose, the presence of the monosaccharide D-galactose, in very low quantities (traces) also already was related in honey samples (Goldschmidt & Burket, 1955). However, it is important to emphasize that this monosaccharide when in its free form is considered a toxic compound to the honeybees (Siddiqui, 1970). Moreira & De Maria (2001) reviewed about carbohydrates in honey, and reported several di-, tri-, and oligosaccharides presented in honey and they came from nectar.

High-fructose is commonly used as sugar substitutes in processed foods, especially in soft drinks, mainly for economical reasons (Long, 1991). These products high-fructose corn syrups (HFCS) are obtained by enzymatic isomerization of corn syrups by both acid and enzymatic hydrolysis of cornstarch. Three enzymes are needed to transform cornstarch into the simple sugars glucose and fructose, α-amylase, glucoamylase, and glucose-isomerase. Fructosyl-fructoses were mainly detected in honey from honeybees fed with high-fructose corn syrups but not from those honeys coming from free-flying foragers or workers fed with sugar syrup (Ruiz-Matute et al., 2010).

Percival (1961) examined 889 plant species and found three pattern of carbohydrates to the nectar: a) nectar with high sucrose, nectar with similar quantities of sucrose, glucose and fructose; and nectar with high glucose and fructose. The nectar with sucrose dominant was associated to flowers of long tubes in which the nectar was protected (clovers), wherever the opened flowers had generally only glucose and fructose. These reports confirmed early researches (Wykes, 1953; Bailey et al., 1954) that suggested a relation between three monosaccharides and different species with flower. In another research, the nectar was divided in four different classes in function to sucrose/hexose rate - S/H: sucrose dominant - S/H >0.999, rich in sucrose - 0.5<0.999, rich in hexose - 0.1<0.499 and hexose dominant - S/H<0.1 (Baker & Baker, 1983).

In the research of Alves (2004) and Alves et al. (2010), the means of sucrose.hexose^{-1} (S/H) per flower for all treatments were: 0.91µg.µL^{-1}, for covered area with Africanized honeybee colony – rich in sucrose; 0.74µg.µL^{-1}, semicovered area with free insects visitation – rich in sucrose; 0.86µg.µL^{-1}, uncovered area with free insect visitation – rich in sucrose; and 3.05µg .µL^{-1}, for covered area without Africanized honeybee colony – sucrose dominant. However, Severson e Erickson (1984) reported in several cultivars of soybean values from 1.2:1.0:1.4 to 1.2:1.0:6.7, with sucrose predominance, which sucrose concentration in nectar ranged from 97 to 986µg.µL^{-1}, these means are higher than those reported by Alves et al. (2010) who found 12.06µg.µL^{-1}. This range suggests that sugar concentration in soybean nectar is influenced by other environment factors independently of pollinator action. Robacker et al. (1983) reported that edaphic and climatic factors affect the number of flowers and another floral characteristics during soybean growing. So, the environmental conditions that generate an increase in number and size of flowers, higher anthesis period, colourness more intense, and also greater nectar production are the factors responsible by became flowers more attractive to honeybees (Alves et al., 2010).

Cruden et al. (1983) suggested that the maximum nectar accumulation occurs before or at the beginning of pollination activity. Such fact can be verified in siratro, since the highest sugar concentration was found at 8:30 a.m. (Figure 5) time in which the bee visitation started (Toledo et al., 2005). Variations in the siratro nectar sugar content measured along the day were observed (Figure 6 – Toledo et al., 2005) and probably associated with the intensity of foraging by honeybees, which is directly related to the nectar quantity and quality (Heinrich, 1979; Hagler, 1990) or to its sugar composition (Waller, 1972; Abrol & Kapil, 1991).

In flowers exposed to pollinators, it is possible that nectar secretion ceases if there is not pollinator in the area or can be reabsorbed in old or pollinated flowers (Cruden et al., 1983). A nectar production without reabsorption may be have an impact on reproductive biology (Galleto & Bernardello, 1995). Therefore, plants reabsorb nectar from aging flowers and utilize its carbon in developing seeds and this is a reproductive advantage (Zimmerman, 1988).

Chiari et al. (2005) studying the pollination of Africanized honeybees on soybean flowers (*Glycine max* L. Merrill) var. BRS 133, measured through the manual refractometer the sugar concentration as total solids and concluded that data found presented a big uniformity, different of the results obtained by Sheppard et al. (1978) that observed big variations in these concentrations and attributed these differences to the variation in the soil composition and other environmental conditions, like precipitation. Despite this, the mean values found by Chiari et al. (2005) were 21.33 ± 0.22% in uncovered area and 22.33 ± 0.38% in covered area with honeybees and differed to each other (P=0.0001). Besides, the medium amounts of total sugar and glucose measured in the nectar of the flowers were 14.33 ± 0.96mg/flower and 3.61 ± 0.36mg/flower, respectively, in the same research.

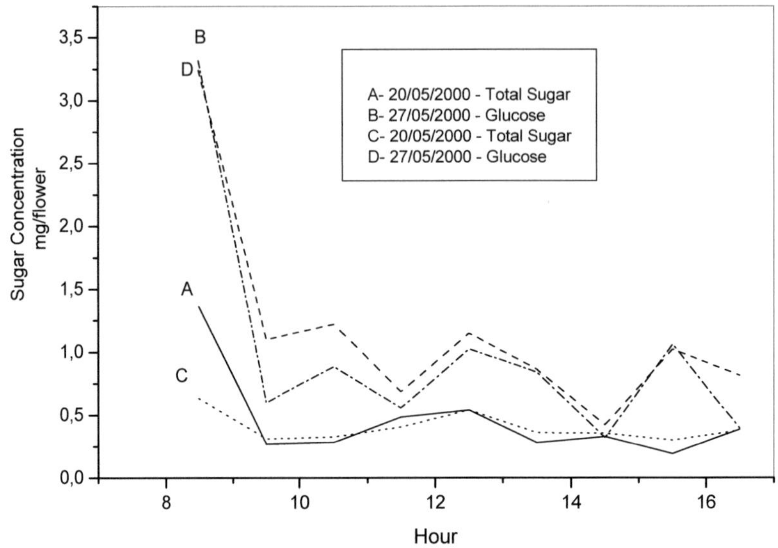

Fig. 5. Total sugar and glucose contents in nectar of siratro flowers along two-day period - reprinted from Toledo et al. (2005) with permission.

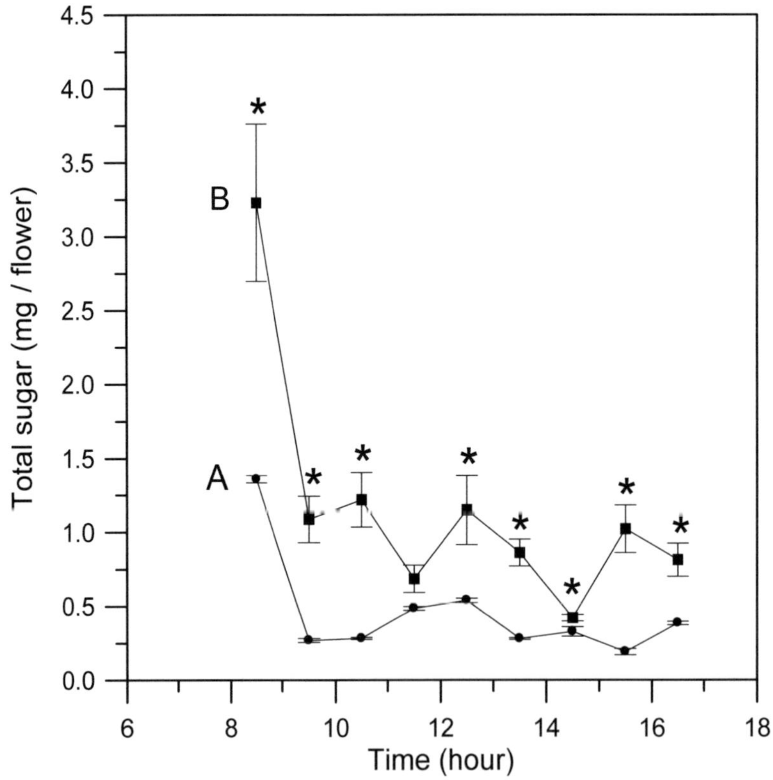

Fig. 6. Total sugar contents in nectar of siratro flowers along two-day period - reprinted from Toledo et al. (2005) with permission.

The study of 25 canola (*Brassica napus x Brassica campestris*) varieties carried out by Kevan et al. (1991) demonstrated that 23 of them had 0.95 or more glucose:fructose rates in their nectar. The same authors reported too that only three varieties had glucose in smaller quantities and none of the samples had detectable quantities of sucrose. Davis et al. (1998) reported higher glucose/fructose rate in lateral chambers than median in Brassicaceae.

Chromatography in paper showed that the nectar from siratro (*Macroptilium atropurpureum* Urb.) is constituted exclusively of glucose, as may be seen in Figure 5 (Toledo et al., 2005). It can be explain the low percentage of honeybee visit on siratro flower - 4%, wherever the another bees were *Trigona spinipes* - 24%, *Bombus morio* - 8%, *Euglossa sp* – 20%, *Megachilidae* - 12%, *Pseudaugochloropsis graminea* – 8% and *Halictidae* - 8% (Vieira et al., 2002). From this, it can be concluded that the main visitor and pollinator in siratro was *Euglossini*, however, *Trigona spinipes* perfurated the external part of the flower avoiding the contact with the pollen grains (Vieira et al., 2002).

Alves (2004) reported a variation of total sugar concentration in soybean nectar var. Codetec 207 from 21.33 to 27.47%, and the higher total sugar concentration was observed in covered area with Africanized honeybee colony. The sucrose concentration ranged from 9.63 to 13.61%, and the higher sucrose concentration was observed in covered area with Africanized honeybee colony. The glucose concentration in this variety was very low. The fructose concentration ranged from 7.93 to 13.75%, so the covered area with Africanized honeybee colony presented higher concentration too. Alves (2004) suggested that Africanized honeybees estimulated the sugar secretion in soybean nectar var. Codetec 207 (Table 1).

Aproximately, 20% of all food-crop production and about 15% of seed crops require the help of pollinators for full pollination (Klein et al., 2007), but Kevan & Phillips (2001) reported that aproximately 73% of cultivated vegetals in the world would be pollinated by some bee species. Gallai et al. (2009) reported a bioeconomic approach, which integrated the production dependence ratio on pollinators, for the 100 crops used directly for human food worldwide as listed by FAO. The total economic value of pollination worldwide amounted to €153 billion, which represented 9.5% of the value of the world agricultural production used for human food in 2005. The honeybee is the most common insect used as agricultural pollinator in many parts of the world like Europe and United States.

In United States, farmers rent more than 2 million honeybee colonies every year for pollination, but the honeybee is being threatened by several problems like Colony Colapse Desorder (James & Pitts-Singer, 2008). The cost of renting bees is up from US$50 per hive in 2003 to US$ 140 per hive in 2006 (Sumner & Boriss, 2006). Some crops require five to seven hives per hectare. In addition, no all crops are well pollinated by honeybees. For example, tomatoes require buzz pollination, which honeybees cannot achieve and alfalfa flowers are not properly worked by honeybees. Fortunately, honeybees are not the only bees that make good pollinators (James & Pitts-Singer, 2008).

Therefore, Greenleaf & Kremen (2006) reported that behavioral interactions between wild and honeybees increase the pollination efficiency of honeybees on hybrid sunflower up to 5-fold, effectively doubling honeybee pollination services on the average field. These indirect contributions caused interspecific interactions between wild and honeybees were more than five times more important than the contributions wild bees make to sunflower pollination directly.

		H=19.88			DF=3	KW=0.0002
Total Sugar	x±se (%)	n*	Rx	T2	T3	T4
covered area with honeybee colony	27.47±1.37	52	137.16	ns	0.0009	0.0008
semi-covered area	26.80±1.81	32	137.81		0.0039	0.0033
uncovered area	21.33±0.86	74	98.24			ns
covered area without honeybee	21.74±1.05	66	96.79			
		H=13.87			DF=3	KW=0.0031
Sucrose	x±se (%)	n	Rx	T2	T3	T4
covered area with honeybee colony	13.01±1.26	51	135.62	ns	0.0004	ns
semi-covered area	12.74±1.49	33	123.61		0.0028	ns
uncovered area	09.63±1.18	78	93.60			0.0366
covered area without honeybee	12.86±1.49	65	116.65			
		H=0.78			DF=3	KW=0.8531
Glucose	x±se (%)	n	Rx	T2	T3	T4
covered area with honeybee colony	0.71±0.05	81	135.62	ns	ns	ns
semi-covered area	0.78±0.09	52	123.61	ns	ns	ns
uncovered area	0.98±0.11	97	93.60	ns	ns	ns
covered area without honeybee	0.95±0.10	83	116.65	ns	ns	ns
		H =12.26			DF=3	KW=0.0065
Fructose	x±se (%)	n	Rx	T2	T3	T4
covered area with honeybee colony	13.75±1.36	51	59.30	ns	ns	0.0016
semi-covered area	13.29±1.38	33	58.42		0.0701	0.0075
uncovered area	10.72±1.05	78	46.33			0.0390
covered area without honeybee	7.93±1.32	65	34.92			

*n- sample size; x-Averages; se-standard error; H - H test; DF-degrees of freedom; KW-Kruskal-Wallis Probability; Rx-Medium position; ns-non significant and probability of the interactions (T1, T2, T3 and T4 vs T2, T3 and T4)

Table 1. Means of total sugar, sucrose, glucose and fructose concentration (%), in soybean nectar (Glycine max L. Merrill) var. Codetec 207 for covered area with honeybee colony (T1); Semi-covered area (T2); uncovered area (T3) and covered area without honeybee colony (T4) – Reprinted from Alves (2004) and Alves et al. (2010) with permission

Analysis of sugar composition in nectar can be used for detecting variation between flowers or nectaries from different taxonomic varieties, and consequently generate differences in type and frequency of visitation of pollinators. Alves et al. (2010) studied the total sugar concentration in soybean nectar (*Glycine max* L. Merrill) var. Codetec 207 by spectrophotometry, using the general method for carbohydrates determination by phenol-sulphuric technique (Dubois et al., 1956).

The research of Alves et al. (2010) was carried out in soybean plants in cages of 24m^2 covered with an Africanized honeybee colony inside in which, semicovered area with free insect visits, uncovered area, covered area without Africanized honeybee colony. Each treatment was five repetitions. In this research, it was emphasized greater total sugar concentration in covered area with Africanized honeybee colony, reduction of sucrose concentration in uncovered area, and lower fructose concentration in covered area without honeybee colony and uncovered area. Besides, none difference among treatments in relation to glucose concentration. So, the high total sugar concentration observed in nectar of soybean var. Codetec 207, in covered area with Africanized honeybee colony suggests that the presence of *Apis mellifera* influenced in this composition, even though a variety with high grade of autopollination. However, the low fructose concentration in uncovered area can be related with low density of honeybees recorded or low presence of another preferential pollinator of fructose. For some researchers, the pollinators can affect the nectar composition (Canto et al., 2008; Herrera et al., 2009). In *Helleborus foetidus*, for example, some species of *Bombus* unchain modifications in sugar composition in nectar reducing the sucrose percentage, and rising the fructose and glucose percentage (Canto et al., 2008).

8. Conclusion

Nectar is an important floral reward for the bee visitation in the flowers. That food is converted in honey into the hive and used as an energy source for the workers. The bee-plant interaction is essential for the maintenance of genetic variability of plants as well as increased production of grains and fruits on a commercial scale.

Honeybees are very important for the pollination of both cultivated and native plants, and then understand the relationship between the collection of floral products (nectar, pollen, resins and oils), biology and behaviour of these insects help making better use for making bee products and agricultural products. Spectrophotometry is a tool to quantify and identify the components of floral products used by bees, especially nectar. These tests allow checking the correlation between the type and intensity of nectar in flower visitation by bees.

The methods employed in this molecular tool began to be developed in the early 50s of last century. Currently, the main analyses performed with the nectar spectrophotometry are: total sugars, reducing sugars, fructose determination (furfural and resorcinol), glucose oxidase (glucose determination in foods), hexokinase (glucose and fructose) and sugar concentration in nectar. As well as, in honey are made several analyses too, like diastase index, hydroxymethylfurfural and others.

Among the tests that can be done is determining sugar concentration in the nectar. The quantification of these sugars allows developing a series of studies associated with the floral visitation by bees collecting nectar and pollination of several species of cultivated plants. These studies contribute to the use of *A. mellifera* to assist beekeepers in increasing the honey production and the farmer in agricultural production. The association between agriculture and beekeeping has been demonstrated by studies presented that showed a positive association between the type of nectar produced by plant and intensity collection by honeybees. Much is still to be undertaken because the number of plant species studied that are pollinated by bees is limited in relation to the number of known species.

Especially in tropical regions, stingless native bees must be preserved and have great potential apicola and sustainable that is not operated, besides to the physicochemical analysis of the sugar composition of honey produced by these bees is not well known and do not have legislation around the world. Floral preferences of native stingless bees are not established yet. Spectrophotometry is an important tool to quantify and identify these components and contribute to knowledge about the interaction of stingless bees and flowers species in neotropical regions, besides to continue to be used in bee products chemical determination.

Finally it is worth mentioning that the spectrophotometry should be more standardized and published as an analysis tool to better use of the bees by the beekeepers in beekeeping industry and in agricultural production.

9. Acknowledgements

To the National Council for Scientific and Technological Development (CNPq), process number 479329/2009-5, and Coordination of Improvement Staff (CAPES) for their financial support.

10. References

Abrol, D.P. & Kapil, R.P. (1991). Foraging strategies of honeybees and solitary bees as determined by nectar sugar components. *Proceedings of the Indian National Academy of Sciences,* Vol. 57-B, p. 127-132, ISSN 0019-5588

Alves, E.M. (2004). *Polinização e composição de açúcares do néctar de soja (Glycine max L. Merrill) variedade Codetec 207.* Master dissertation, Universidade Estadual de Maringá, Maringá, Brasil. 57p.

Alves, E.M.; Toledo, V.A.A.; Oliveira, A.J.B.; Sereia, M.J.; Neves, C.A. & Ruvolo-Takasusuki, M.C.C. (2010). Influência de abelhas africanizadas na concentração de açúcares no nectar de soja (*Glycine max* L. Merrill) var. Codetec 207. *Acta Scientiarum - Animal Science,* Vol. 32, No. 2, pp. 189-195, ISSN 1807-8672

Amaral, L.I.V.; Gaspar, M.; Costa, P.M.F.; Aidar, M.P.M. & Buckeridge, M.S. (2007). Novo método enzimático rápido e sensível de extração e dosagem de amido em materiais vegetais. *Hoehnea,* Vol.34, p. 425-431, ISSN 0073-2877

Bailey, M.E.; Fieger, E.A. & Oertel, E. (1954). Paper chromatographic analyses of some southern nectars. *Gleanings in Bee Culture,* Vol. 82, p. 7-8, ISSN 0017-114X

Baker, H.G. & Hurd, Jr., P.D. (1968). Intrafloral ecology. *Annual Review of Entomology,* Vol.13, No.1, p. 385-414, ISSN 0066-4170

Baker, H.G. (1977). Non-sugar chemical constituents of nectar. *Apidologie,* Vol. 8, p. 349-356, ISSN 12979678

Baker, H.G. & Baker, I. (1973). Some anthecological aspects of the evolution of nectar producing flowers, particularly amino acid production in nectar. In: Heywood, V.H. (Ed.) *Taxonomy and ecology.* Academic Press, ISBN 0-12-346960-0, London, UK

Baker, H.G. & Baker, I. (1983). A brief historical review of the chemistry of floral nectar. In: Bentley, B. & Elias, T.S. (Eds.) *The biology of nectaries.* Columbia University Press, ISBN 978-0-231-04446-2, New York, USA

Baker, H.G.; Baker, I. & Hodges, S.A. (1998). Sugar composition of nectars and fruits consumed by birds and bats in the tropics and subtropics. *Biotropica,* Vol. 30, p. 559-586, ISSN 0006-3606

Biernaskie, J.M.; Cartar, R.V.; Hurly, T.A. (2002). Risk-averse inflorescence departure in hummingbirds and bumble bees: could plants benefit from variable nectar. *Oikos,* Vol. 98, No. 1, p. 98-104, ISSN 1600-0706

Blakeney, A.B. & Mutton, L.L. (1980). A simple colorimetric method for the determination of sugars in fruit and vegetables. *Journal of the Science of Food and Agriculture*, Vol.31, No.9, p. 889-897, ISSN 1097-0010

Buckmann, S.L. & Nabhan, G.P. (1997). *The forgotten pollinators*, Island Press, ISBN 1559633522, Washington, DC, USA

Canto, A.; Herrera, C.M.; Medrano, M.; Perez, R. & Garcia, I.M. (2008). Pollinator foraging modifies nectar sugar composition in *Helleborus foetidus* (Ranunculaceae): an experimental test. *American Journal of Botany*, Vol. 95, No. 3, pp. 315-320, ISSN 1537-2197

Chambó, E.D.; Garcia, R.C.; Oliveira, N.T.E. & Duarte, Jr., J.B. (2011). Honey bee visitation to sunflower: effects on pollination and plant genotype. *Scientia Agricola*, Vol.68, No.6, ISSN 0103-9016

Chiari, W.C.; Toledo, V.A.A.; Ruvolo-Takasusuki, M.C.C.; Attencia, V.M.; Costa, F.M.; Kotaka, C.S.; Sakaguti, E.S. & Magalhaes, H.R. (2005). Floral biology and behavior of Africanized honeybees *Apis mellifera* in soybean (*Glycine max* L. Merrill). *Brazilian Archives of Biology and Technology*, Vol. 48, No. 3, p.367-378, ISSN 1516-8913

Cruden, R.W.; Hermann, S.M. & Peterson, S. (1983). Patterns of nectar production and plant-pollinator coevolution. In: Bentley, B. & Elias, T. (Eds.). *The biology of nectaries.* Columbia, ISBN 023104461, New York, USA, p. 80-125

Delaplane, K.S.; Mayer, D.F. (2000). *Crop pollination by bees.* CABI Publishing, ISBN 9780851994482 London, UK

Dreywood, R. (1946). Qualitative test for carbohydrate materials. *Industrial and Engineering Chemistry Analytical Edition*, Vol.18, p. 499-504, ISSN 0096-4484

Dubois, M.; Giles, K.A. & Hamilton, J.K. (1956). Colorimetric method for determination of sugars and related substances. *Analytical Chemistry*, Vol. 28, p. 350-356, ISSN 0003-2700

Endress, P.K. (1994). *Diversity and evolutionary biology of tropical flowers.* Cambridge, ISBN 0-521-56510-3, Cambridge, USA

Free J.B. (1993). *Insect pollination of crops.* Academic Press, ISBN 0122666518, London, UK

Freeman, C.E.; Worthington, R.D. & Corral, R.D. (1985). Some floral nectar-sugar compositions from Durango and Sinaloa, Mexico. *Biotropica*, Vol. 17, p. 309-313, ISSN 0006-3606

Furgala, B.; Gochnauer, T.A. & Holdaway, F.G. (1958). Constituent sugars of some northern legume nectars. *Bee World, Vol. 39, p.* 203-205, ISSN 0005-772X

Galetto, L. & Bernardello, G. (1995). Characteristics of nectar secretion by *Lycium cestroides, L. Ciliatum* (Solanaceae), and their hybrid. *Plant Species Biology*, Vol. 11, p. 157-163, ISSN 1442-1984

Gallai, N.; Salles, J.-M; Settele, J. & Vaissière, B.E. (2009). Economic valuation of the vulnerability of world agriculture confronted with pollinator decline. *Ecological Economics*, Vol. 68, p. 810-821, ISSN 0921-8009

Goldschmidt, S. & Burkert, H. (1955). Observation of some heretofore unknown sugars in honey. *Zeitschrift für Physiologische Chemie*, Vol. 300, p. 188, ISSN 0018-4888

Greenleaf, S.S. & Kremen, C. (2006). Wild bees enhance honey bees'pollination of hybrid sunflower. *Proceedings of the National Academy of Sciences*, Vol. 103, No. 37, p. 13890-13895, ISSN 1091-6490

Hagler, J.R. (1990). Honeybee (*Apis mellifera* L.) response to simulated onion nectars containg variable sugar and potassium concentrations. *Apidologie,* Vol.21, p. 115-121, ISSN 12979678

Harbone, J.B. (1998). *Phytochemical methods: a guide to modern techniques of plant analysis,* Chapman & Hall, ISSN 0973-1296, New York, USA

Heinrich, B. (1979). Resource heterogeneity and patterns of movement in foraging bumblebees. *Oecologia,* Vol. 40, p. 235-245, ISSN 00298549

Herrera, C.M.; Veja, C.; Canto, A. & Pozo, M.I. (2009). Yeasts in floral nectar: a quantitative survey. *Annals of Botany,* Vol. 103, No. 9, pp. 1425-1423, ISSN 1095-8290

James, R.R. & Pitts-Singer, T.L. (2008). The future of agricultural pollination. In: ___________ *Bee pollination in Agricultural Ecosystems.* Osford University Press, ISBN 978-0-19-531695-7, New York, USA

Kevan, P.; Lee, H. & Shuel, R.W. (1991). Sugar rations in nectars of varieties of canola (*Brassica napus*). *Journal of Apicultural Research,* Vol. 30, No. 2, p. 99-102, ISSN 0021-8839

Kevan, P. & Phillips, T.P. (2001). The economic impacts of pollinator declines: an approach to assessing the consequences. *Conservation Ecology,* Vol. 5, No.1, p. 1-13, available on line URL: http://www.consecol.org/vol5/iss1/art8/ accessed on Ap 14th 2009.

Klein, A.M.; Vaissière, B.E.; Cane, J.H. & Steffan-Dewenter, I. (2007). Importance of pollinators in changing landscapes for world crops. *Proceedings of the Royal Society of London: Series B,* Vol. 274, p. 303-313, ISSN 1471-2954

Lanza, J.S.; Smith G.C. & Sack, S. (1995). Variation in nectar volume and composition of *Impatiens capensis* at the individual, plant, and population levels. *Oecologia,* Vol.102, p. 113-119, ISSN 0029-8549

Long, J.E. (1991). High fructose corn syrup. *Food Science and Technology,* Vol. 48, 247-258, ISSN 1226-7708

Malagodi-Braga, K.S. (2005). Abelhas: por quê manejá-las para a polinização? *Revista Mensagem Doce,* No.80, ISSN 1981-6243

Malerbo-Souza, D.T.; Nogueira-Couto, R.H.; Couto L.A. (2003). Polinização em cultura de laranja (*Citrus sinensis* L. Osbeck, var. Pera-rio). *Brazilian Journal of Veterinary Research and Animal Science,* Vol. 40, No. 4, p. 237-242, ISSN 1413-9596

Malerbo-Souza, D.T.; Nogueira-Couto, R.H. & Couto, L.A. (2004). Honey bee attractants and pollination in sweet Orange, *Citrus sinensis* (L.) Osbeck, var. Pera-Rio. *Journal of Venomous Animals and Toxins including Tropical Diseases,* Vol.10, No.2, p. 144-153, ISSN 1678-9199

Martin, E.A.C.; Machado, R.J.P. & Lopes, J. (2005). Atrativo para abelhas em campos de produção de sementes de girassol colorido híbrido. *Semina: Ciências Agrárias,* Vol.26, No.4, p. 489-494, ISSN 1676-546X

Messineo, L. & Musarra, E. (1972). Sensitive spectrophotometric determination of fructose, sucrose, and inulin without interference from aldohexoses, aldopentoses, and ketopentoses. *International Journal of Biochemistry,* Vol.3, No.18, p. 691-699, ISSN 1357-2725

Moerman, F.T.; Van Leeuwen, M.B. & Delcour, J.A. (2004). Enrichment of higher molecular weight fractions in inulin. *Journal of Agricultural and Food Chemistry,* Vol.52, p. 3780-3783, ISSN 0021-8561

Moreira, R.F.A. & De Maria, C.A.B. (2001). Glicídeos no mel. *Química Nova,* Vol. 24, No. 4, p. 516-525, ISSN 1678-7064

Morgado, L.N.; Carvalho, C.F.; Souza, B. & Santana, M.P. (2002). Fauna de abelhas (Hymenoptera: Apoidea) nas flores de girassol *Helianthus annuus* L., em Lavras - MG. *Revista Ciência e Agrotecnologia*, Vol.26, No.6, p. 1167-1177, ISSN 1413-7054

Miller, G.L. (1959). Use of dinitrosalicylic acid reagent for determination of reducing sugar. *Analytical Chemistry*, Vol.31, No.3, p. 426-428, ISSN 0003-2700

Müller, A.; Diener, S.; Schnyder, S.; Stutz, K.; Sedivy, C.; Dorn, S. (2006). Quantitative pollen requirements of solitary bees: Implications for bee conservation and the evolution of bee-flower relationships. *Biological Conservation*, Vol.130, No.4, p. 604-615, ISSN 0006-3207

Nicolson, S.W.; Nepi, M. (2005). Dilute nectar in dry atmospheres: nectar secretion patterns in *Aloe castanea* (Asphodelaceae). *International Journal of Plant Sciences,* Vol. 166, No. 2, p. 227-233, ISSN 1058-5893

Paiva, G.J.; Terada, Y. & Toledo, V.A.A. (2002). Behavior of *Apis mellifera* L. Africanized honeybees in sunflower (*Helianthus annuus* L.) and evaluation of *Apis mellifera* L. colony inside covered area of sunflower. *Acta Scientiarum*, Vol.24, No.4, p. 851-855, ISSN 1679-9275

Percival, M. (1953). Types of nectar in angiosperms. *New Phytologist,* Vol. 60, p. 235-281, ISSN 1469-8137.

Pham-Delegue, M.H.; Etievant, P.; Ghuichard, E.; Marilleau, R.; Doualt, P.H.; Chauffaille, J. & Masson, C. (1990). Chemicals involved in honeybee sunflower relationship. *Journal of Chemical Ecology*, Vol.16, No.11, p. 3053-3065, ISSN *0098-0331*

Prince, P.W. (1997). *Insect Ecology,* John Wiley & Sons, New York, ISBN 0471161845

Proctor, M.; Yeo, P. & Lack, A. (1996). *The natural history of pollination,* Harper Collins Publishers, London, ISBN 0002199068

Robacker, D.C.; Flottum, P.K. & Sammataro, D. (1983). Effects of climatic and edaphic factors on soybean flowers and of the subsequent attractiveness of the plants to honey bees. *Field Crops Research*, Vol. 6, No. 4, p. 267-278, ISSN 0378-4290

Roberts, R.B. (1979). Spectrophotometric analysis of sugars produced by plants and harvested by insects. *Journal of Apicultural Research,* Vol. 18, p. 191-195, ISSN 2078-6913

Roe, J.H. (1934). A colorimetric method for the determination of fructose in blood and urine. *Journal of Biological Chemistry*, Vol.107, p. 15-22, ISSN 0021-9258

Roe, J.H.; Epstein, J.H. & Goldstein, N.P. (1949). A photometric method for the determination os inulin in plasma and urine. *Journal of Biological Chemistry*, Vol.178, p. 839-845, ISSN 0021-9258

Ruiz-Matute, A.I.; Weiss, M.; Sammataro, D.; Finely, J. & Sanz, M.L. (2010). Carbohydrate composition of high-fructose corn syrups (HFCS) used for bee feeding: effect on honey composition. *Journal of Agricultural and Food Chemistry,* Vol. 58, p.7317-7322, ISSN 0021-8561

Schirmer, L.R. (1985). *Abelhas ecológicas,* Nobel, São Paulo: Nobel, ISBN 9253012536

Severson, D.W. & Erickson, J.E.H. (1984). Quantitative and qualitative variation in floral nectar of soybean cultivars in southeastern Missouri. *Environmental Entomology,* Vol. 13, No. 4, pp. 91-96, ISSN 1938-2936

Shafir, S.; Bechar, A.; Weber, E.U. (2003). Cognition mediated coevolution: context-dependent evaluations and sensitivity of pollinators to variability in nectar rewards. *Plant Systematic and Evolution,* Vol. 238, No. 1-4, p. 195-209, ISSN 1615-6110

Sheppard, W.S.; Jaycox, E.R. & Parise, S.G. (1978). Selection and management of honey bees for pollination of soybean. In: International Symposium of Pollination, 4., University of Maryland College Park, Maryland, USA. *Proceedings*... 1, p. 123-130

Siddiqui, I.R. (1970). The sugars of honey. *Advances in Carbohydrate Chemistry and Biochemistry*, Vol. 7, p. 51-59, ISSN 0065-2318

Silva, E.M.; Dean, B.B. (2000). Effect of nectar composition and nectar concentration on honey bee (Hymenoptera: Apidae) visitations to hybrid onion flowers. *Journal of Economic Entomology*, Vol. 93, No. 4, p. 1216-1221, ISSN 0022-0493

Sumner, D.A. & Boriss, H. (2006). Bee-conomics and the leap in pollination fees. *Giannini Foundation of Agricultural Economics*, Vol. 9, p. 9-11, ISSN 1081-6526

Summer, J.B. & Sisler, E.B. (1944). A simple method for blood sugar. *Archives of Biochemistry and Biophysics*, Vol.4, p. 333-336, ISSN 0003-9861

Täufel, K. & Reiss, R. (1952). Analytische und chromatographische Studien am Bienenhonig. *Zeitschrift für Lebensmittel-Untersuchung und Forschung*, Vol. 94, No. 1, p. 1-10, ISSN 0044-3026

Teixeira, L.M.R. & Zampieron, S.L.M. (2008). Estudo da fenologia, biologia floral do girasol (*Helianthus annuus*, Compositae) e visitantes florais associados, em diferentes estações do ano. *Ciência et Praxis*, Vol.1, No.1, p. 5-14, ISSN 1983192X

Thomson, J.D. & Goodell, K. (2002). Pollen removal and deposition by honeybee and bumblebee visitors to apple and almond flowers. *Journal of Applied Ecology*, Vol.38, No.5, p. 1032-1044, ISSN 365-2664

Toledo, V.A.A.; Oliveira, A.J.B.; Ruvolo-Takasusuki, M.C.C.; Mitsui, M.H.; Vieira, R.E.; Kotaka, C.S.; Chiari, W.C.; Gobbi Filho, L.; Terada, Y. (2005). Sugar content in nectar flowers of siratro (Macroptilium atropurpureum Urb.) *Acta Scientiarum – Animal Science*, Vol. 27, No. 1, p. 105-108, ISSN 1807-8672

Vidal, M.D.G.; De Jong, D.; Wien, H.C. & Morse, A.R. (2006). Nectar and pollen production in pumpkin (*Cucurbita pepo* L.). *Revista Brasileira de Botânica*, Vol.29, No.2, p. 267-273, ISSN 0100-8404

Vidal, M.D.G.; De Jong, D.; Wien, H.C. & Morse, A.R. (2010). Pollination and fruit set in pumpkin (*Cucurbita pepo*) by honey bees. *Revista Brasileira de Botânica*, Vol.33, No.1, p. 106-113, ISSN 0100-8404

Vieira, R.E.; Kotaka, C.S.; Mitsui, M.H.; Taniguchi, A.P.; Toledo, V.A.A.; Ruvolo-Takasusuki, M.C.C.; Terada, Y.; Sofia, S.H.; Costa, F.M. (2002). Biologia floral e polinização por abelhas em siratro (*Macroptilium atropurpureum* Urb.). *Acta Scientiarum*, Vol. 24, No. 4, p. 857-861, ISSN 1806-2636

Waller, G.D. (1972). Evaluating responses of honeybees to sugar solutions using an artificial-flower feeder. *Annals of the Entomological Society of America*, Vol. 65, p. 857-862, ISSN 0013-8746

Wykes, G. R. (1952a). An investigation of the sugars present in the nectar of flowers of various species. *New Phytologist*, Vol. 51, No. 2, p. 210-215, ISSN 1469-8137

Wykes, G.R. (1953). The sugar content of nectars. *Biochemical Journal*, Vol. 53, p. 294-296, ISSN 0264-6021

Yemm, E.W. & Willis, J. (1954). The estimation of carbohydrates in plant extracts by anthrone. *Journal of Biochemical*, Vol.57, No.3, p. 508-514, ISSN 0264-6021

Zimmerman, M. (1988). Nectar production, flowering phenology, and strategies for pollination. In: Doust, J.L. & Doust, L.L. (Eds.). *Plant reproductive ecology, patterns and strategies*. Oxford, ISBN 01-950-51750, New York, USA

15

Multivariate Data Processing in Spectrophotometric Analysis of Complex Chemical Systems

Zoltan Szabadai[2], Vicențiu Vlaia[2], Ioan Țăranu[1], Bogdan-Ovidiu Țăranu[1], Lavinia Vlaia[2] and Iuliana Popa[1]
[1]*National Institute of Research-Development for Electrochemistry and Condensed Matter,*
[2]*University of Medicine and Pharmacy "Victor Babeş",*
Romania

1. Introduction

There are a great variety of processing the analytical spectroscopy data, especially useful in multicomponent systems [Ewing et al., 1953; Garrido et al., 2004; Lykkesfeld, 2001; Oka et al., 1991; Sánchez & Kowalski, 1986]. These methods essentially are based on different strategies of mathematical strategies including specific formalism of mathematical statistics and of matrix algebra [Garrido et al., 2004; Szabadai, 2005]. The matrix-based methods reffer to quantitative analysis [Bosch-Reigh et al., 1991; Garrido et al., 2008; Li et al., 2011; Lozano et al., 2009; Ruckenbusch et al., 2006; Szabadai, 2005], to determination of the number of independent chemical equilibria in multicomponent systems [Szabadai, 2005] and for correction the action of various perturbing factors such as stray light or backgroud absorption [Burnius, 1959; Fox & Mueller, 1950; Melnick, 1952; Morton & Stubbs, 1946, 1947, 1948; Owen, 1995; Page & Berkovitz, 1943; Szabadai, 2005].

In the present chapter original approaches of matrix treatment of the aforementioned items are presented, with special consideration to the simultaneous assay of compounds in a mixter, to backgruond correction procedures and to the standard addition method in a generalized form.

2. Simultaneous assay of nonreacting compunds in a mixture

The issue of the quantitative analysis of a mixture, when the components do not interact chemically, can be approached, in a rigorous and general manner, with the help of matrix computation [Ewing et al., 1953; Garrido et al., 2004, 2008; Lozano et al., 2009; Lykkesfeld, 2001; Oka et al., 1991; Ruckenbusch et al., 2006; Sánchez & Kowalski, 1986; Szabadai, 2005]. In the case of a mixture with M component, the quantitative determination of the components, one has to measure the absorbance at Λ distinct values of wavelength ($\Lambda > M$). Given a set of N standard solutions (N > M and supposing that, as a rule, each standard solution may contain all of M chemical components of interest in known concentrations), absorbances are to be measured at the same set of wavelengths and in identical conditions as done for standard solutions.

The following notations will be used in what follows: $X_m^n(\lambda)$ represents a quantity X referring to the standard mixture of number n (superscript index), at the individual chemical component of number m (subscript index), measured at the wavelength of number λ (between parentheses). Thus,

c_m^n represents the concentration of the component of number m in the standard solution of number "n" ;

c_m represents the concentration of the component of number m in the mixture undergoing the analysis (sample of unknown composition) ;

$A^n(\lambda)$ is the absorbance of the standard mixture of number m measured at the wavelength of number "λ" ;

$A_m(\lambda)$ is the contribution of the pure m-numbered component to the absorbance of the analyzed mixture, registered at the wavelength λ ;

A(λ) is the absorbance of the mixture under analysis, measured at the wavelength of number λ ;

$\varepsilon_m(\lambda)$ is the molar absorptivity of the chemical component of number "m", measured at the wavelength of number λ ;

p^n is the weight percent of the spectrum of the standard solution of number n in the spectrum of the mixture under analysis.

If the components of a mixture do not interact chemically and if the absorbances of each component satisfies the Bouguer-Lambert-Beer relation, then the absorbance of the mixture, at each wavelength taken into account, consists of the sum of contributions of the individual absorbent chemical components. The absorbances of the N standard solutions, measured at Λ distinct values of wavelength, may be arranged in matrix form (1).

$$\begin{bmatrix} A^1(1) & \cdots & A^n(1) & \cdots & A^N(1) \\ \vdots & & \vdots & & \vdots \\ A^1(\lambda) & \cdots & A^n(\lambda) & \cdots & A^N(\lambda) \\ \vdots & & \vdots & & \vdots \\ A^1(\Lambda) & \cdots & A^n(\Lambda) & \cdots & A^N(\Lambda) \end{bmatrix} = d \cdot \begin{bmatrix} \varepsilon_1(1) & \cdots & \varepsilon_m(1) & \cdots & \varepsilon_M(1) \\ \vdots & & \vdots & & \vdots \\ \varepsilon_1(\lambda) & \cdots & \varepsilon_m(\lambda) & \cdots & \varepsilon_M(\lambda) \\ \vdots & & \vdots & & \vdots \\ \varepsilon_1(\Lambda) & \cdots & \varepsilon_m(\Lambda) & \cdots & \varepsilon_M(\Lambda) \end{bmatrix} \cdot \begin{bmatrix} c_1^1 & \cdots & c_1^n & \cdots & c_1^N \\ \vdots & & \vdots & & \vdots \\ c_m^1 & \cdots & c_m^n & \cdots & c_m^N \\ \vdots & & \vdots & & \vdots \\ c_M^1 & \cdots & c_M^n & \cdots & c_M^N \end{bmatrix} \quad (1)$$

The left side of the relation includes the matrix of absorbances of the standard solutions and the optical path the radiation has been covered, „d" (i.e. the width of the cell used).

It may be allowed that the absorbance of the sample, measured at the same set of wavelengths as in the case of standard solutions, consists of the weighted contributions of the standard solutions. The contribution weight of each standard solution to the absorbance of the sample depends on the concentration of the chemical components in the sample under analysis and in the individual standard solutions. This is expressed, in matrix form, according to relation (2).

$$\begin{bmatrix} A(1) \\ \vdots \\ A(\lambda) \\ \vdots \\ A(\Lambda) \end{bmatrix} = \begin{bmatrix} A^1(1) & \cdots & A^n(1) & \cdots & A^N(1) \\ \vdots & & \vdots & & \vdots \\ A^1(\lambda) & \cdots & A^n(\lambda) & \cdots & A^N(\lambda) \\ \vdots & & \vdots & & \vdots \\ A^1(\Lambda) & \cdots & A^n(\Lambda) & \cdots & A^N(\Lambda) \end{bmatrix} \cdot \begin{bmatrix} p^1 \\ \vdots \\ p^n \\ \vdots \\ p^N \end{bmatrix} \quad (2)$$

In what follows, bold characters are used for denoting matrices: the matrix of the absorbances of the sample will be denoted by **A**, the matrix of the absorbances of the standard solutions by $\mathbf{A}_{st}$, the matrix of the concentrations of chemical components in the analysed sample and in the standard solutions by **C** and $\mathbf{C}_{st}$ respectively, the matrix of the molar absorptivities by **E** and the matrix of the contribution weight of the standard solutions, generating the absorbance of the sample, by **P**. In order to comprehend more easily the matrix formalism, the symbol of matrices is followed (between right brackets) by the specification of the number of rows and columns in the respective matrix. Therefore, matrix $\mathbf{A}_{st}$, made up of Λ rows and N columns, is denoted as follows : $\mathbf{A}_{st}[\Lambda,N]$. Relations (1) and (2) are equivalent to matrix expressions (3) and (4).

$$\mathbf{A}_{st}[\Lambda,N] = d \cdot \mathbf{E}[\Lambda,M] \cdot \mathbf{C}_{st}[M,N] \quad (3)$$

$$\mathbf{A}[\Lambda,1] = \mathbf{A}_{st}[\Lambda,N] \cdot \mathbf{P}[N,1] = d \cdot \mathbf{E}[\Lambda,M] \cdot \mathbf{C}[M,1] \quad (4)$$

Relations (1) and (2) may be written in a condensed matrix form (5).

$$\frac{1}{d} \cdot \mathbf{A}[\Lambda,1] = \frac{1}{d} \cdot \mathbf{A}_{st}[\Lambda,N] \cdot \mathbf{P}[N,1] = \mathbf{E}[\Lambda,M] \cdot \mathbf{C}_{st}[M,N] \cdot \mathbf{P}[N,1] \quad (5)$$

The product matrix $\mathbf{A}_{st}[\Lambda,N] \cdot \mathbf{P}[N,1]$ consisting of Λ rows and one column may be presented in the shortened form $(\mathbf{A}_{st} \cdot \mathbf{P})[\Lambda,1]$.

Practically, the aim is to calculate the elements of matrix $\mathbf{C}[M,1]$. In most of the real situations, the molar absorptivities of the chemical components under analysis are not known (especially not for a set of different wavelengths). For this reason, the spectrophotometric analysis is conditioned by the spectrophotometric study of a number of standard solutions, where the concentrations of the chemical components of interest are known. The matrix formalism presented allows for the standard solutions used to contain several chemical components (basically, each of the N standard solutions may contain all the M chemical components at known concentrations). In particular cases, it may happen (but it is not mandatory) that each standard solution contains only one chemical component (different from the other chemical components present in the other standard solutions); in this case the matrix $\mathbf{C}_{st}$ of the concentrations in standard solutions is square (has the same number of rows and columns) and diagonal (i.e. the c_m^n elements are null when m and n are different). In this particular case, the number of standard solutions is identical to the number of chemical components of analytical interest.

After the spectrophotometric measurements are accomplished, the elements of matrices $\mathbf{A}_{st}[\Lambda,N]$, $\mathbf{A}[\Lambda,1]$ and $\mathbf{C}_{st}[M,N]$ are known, and the further aim is to calculate the elements of matrix $\mathbf{C}$[M,1]. These matrices satisfy relations (6) and (7). In what follows, the desired result is to eliminate matrix $\mathbf{E}$[Λ,M] from these two matrix relations and to explicit the resulting relation in relation to matrix $\mathbf{C}$[M,1].

$$\mathbf{A}_{st}[\Lambda,N] = d\cdot\mathbf{E}[\Lambda,M]\cdot\mathbf{C}_{st}[M,N] \quad (6)$$

$$\mathbf{A}[\Lambda,1] = d\cdot\mathbf{E}[\Lambda,M]\cdot\mathbf{C}[M,1] \quad (7)$$

In order to solve the above system of equation in relation to matrix $\mathbf{C}$[M,1], both members of equation (6) are multiplied on the right by the transpose of matrix $\mathbf{C}_{st}$[M,N].

$$\mathbf{A}_{st}[\Lambda,N]\cdot\mathbf{C}^{T}_{st}[N,M] = d\cdot\mathbf{E}[\Lambda,M]\cdot\mathbf{C}_{st}[M,N]\cdot\mathbf{C}^{T}_{st}[N,M] \quad (8)$$

The product $\mathbf{C}_{st}$[M,N] $\mathbf{C}^{T}_{st}$[N,M] is a MxM square matrix represented, according to the adopted notations, as $(\mathbf{C}_{st}\cdot\mathbf{C}^{T}_{st})$ [M,M]. If the determinant of this matrix is not zero (i.e. if the set of wavelengths was selected suitably for relevant absorbance values), then the product matrix has an inverse, represented as $(\mathbf{C}_{st}\cdot\mathbf{C}^{T}_{st})^{-1}$[M,M], with the property expressed by (9).

$$(C_{st}\cdot C^{T}_{st})^{-1}[M,M]\cdot(C_{st}\cdot C^{T}_{st})\,[M,M] = (C_{st}\cdot C^{T}_{st})\,[M,M]\cdot(C_{st}\cdot C^{T}_{st})^{-1}[M,M] = I[M,M] \quad (9)$$

In relation (9) $\mathbf{I}$[M,M] is the unit matrix of order M. The elements of this matrix situated on the main diagonal are equal to the unity, and all its other elements are null. The multiplication operation of any matrix by the unit matrix (of the corresponding order) leaves the matrix unchanged. Consequently, after multiplying the equation (8) on the right by $(\mathbf{C}_{st}\cdot\mathbf{C}^{T}_{st})^{-1}$[M,M], the resulting relation is (10).

$$\mathbf{A}_{st}[\Lambda,N]\cdot\mathbf{C}^{T}_{st}[N,M]\cdot(\mathbf{C}_{st}\cdot\mathbf{C}^{T}_{st})^{-1}[M,M] = d\cdot\mathbf{E}[\Lambda,M]\cdot\mathbf{I}[M,M] = d\cdot\mathbf{E}[\Lambda,M] \quad (10)$$

In what follows, both members of equation (7) are multiplied on the left by the transpose of matrix $\mathbf{E}$[Λ,M], namely by $\mathbf{E}^{T}$[M,Λ]; the result is (11).

$$\mathbf{E}^{T}[M,\Lambda]\cdot\mathbf{A}[\Lambda,1] = d\cdot\mathbf{E}^{T}[M,\Lambda]\cdot\mathbf{E}[\Lambda,M]\cdot\mathbf{C}[M,1] \quad (11)$$

The product $\mathbf{E}^{T}$[M,Λ]·$\mathbf{E}$[Λ,M] = $(\mathbf{E}^{T}\cdot\mathbf{E})$[M,M] in the expression (11) is a square matrix allowing an inverse, $(\mathbf{E}^{T}\cdot\mathbf{E})^{-1}$[M,M], provided that the product matrix is not singular (its determinant is different fron zero). By multiplying equation (11) on the left by matrix $(\mathbf{E}^{T}\cdot\mathbf{E})^{-1}$[M,M], the expression (12) is obtained. This expresses explicitly the seeked column matrix $\mathbf{C}$[M,1] of the concentrations of components in the analysed mixture.

$$(\mathbf{E}^{T}\cdot\mathbf{E})^{-1}[M,M]\cdot\mathbf{E}^{T}[M,\Lambda]\cdot\mathbf{A}[\Lambda,1] = d\cdot\mathbf{C}[M,1] \quad (12)$$

Matrix $\mathbf{E}$[Λ,M], occuring in expression (4.43), can be calculated with relation (10).

In order to express the matrix of concentrations $\mathbf{C}$[M,1] only in relation to quantities resulting directly from spectrophotometric measurements (the elements of matrix $\mathbf{A}$[Λ,1]) and in relation to known quantities (the elements of matrix $\mathbf{C}_{et}$[M,N]), the matrix $\mathbf{E}$[Λ,M] has to be eliminated from relations (10) and (12).

The transpose of matrix $\mathbf{E}[\Lambda,M]$, namely matrix $\mathbf{E}^{T}[M,\Lambda]$, is expressed from relation (10) :

$$d\cdot \mathbf{E}^{T}[M,\Lambda] = (\mathbf{C}_{st}\cdot\mathbf{C}_{st}^{T})^{-1}[M,M]\cdot\mathbf{C}_{st}[M,N]\cdot\mathbf{A}_{st}^{T}[N,\Lambda] \quad (13)$$

whereas the inverse matrix of the product of matrices $\mathbf{E}[\Lambda,M]$ and $\mathbf{E}^{T}[M,\Lambda]$ is expressed from (10) and (13):

$$(\mathbf{E}^{T}\mathbf{E})^{-1}[M,M] = d^{2}\cdot(\mathbf{C}_{st}\cdot\mathbf{C}_{st}^{T})[M,M]\cdot(\mathbf{C}_{st}[\mathbf{M,N}]\cdot\mathbf{A}_{st}^{T}[N,\Lambda]\cdot\mathbf{A}_{st}[\Lambda,N]\cdot \quad (14)$$

$$\cdot\mathbf{C}_{st}^{T}[N,M])^{-1}\cdot(\mathbf{C}_{st}\cdot\mathbf{C}_{st}^{T})^{-1}[M,M]$$

By replacing expressions (13) and (14) in (12), and taking into consideration relation (15),

$$(\mathbf{C}_{st}\cdot\mathbf{C}_{st}^{T})^{-1}[M,M]\cdot(\mathbf{C}_{st}\cdot\mathbf{C}_{st}^{T})[M,M] = \mathbf{I}[M,M] \quad (15)$$

the expression (16) is obtained. This presents, in an explicit form, the matrix of unknown concentrations.

$$\mathbf{C}[M,1] = (\mathbf{C}_{st}\cdot\mathbf{C}_{st}^{T})[M,M]\cdot(\mathbf{C}_{st}[\mathbf{M,N}]\cdot\mathbf{A}_{st}^{T}[N,\Lambda]\cdot\mathbf{A}_{st}[\Lambda,N]\cdot \quad (16)$$

$$\mathbf{C}_{st}^{T}[N,M])^{-1}\cdot\mathbf{C}_{st}[M,N]\cdot\mathbf{A}_{st}^{T}[N,\Lambda]\cdot\mathbf{A}[\Lambda,1]$$

In relation (16) the optical pathway (d) no longer appears if the absorbances of standards $_{st}[\Lambda,N]$ and the absorbances of the sample $\mathbf{A}[\Lambda,1]$ are measured at the same cell thickness.

Relation (10) allows to obtain the elements of matrix $\mathbf{E}[\Lambda,M]$ as well, values which are proportional to the absorbances of the pure components measured at the selected Λ wavelengths. Relation (10) allows thus to obtain the spectrum of the M individual components. This is important if a sufficiently large number of standard solutions are available with known concentrations of components, but individual components are not available for recording their individual spectra.

A particular case of the above reasoning is that with each of standard solutions contain only one dissolved chemical component (other than those present in the other standard solutions), so N = M. In this case notation S refers to their common value (N = M = S). Consequently, matrix $\mathbf{C}_{st}[S,S]$ of the concentrations of components in standard solutions is square and diagonal (only elements on the matrix main diagonal differ from zero) (17)).

$$\mathbf{C}_{st}[S,S] = \begin{bmatrix} c_1^1 & & & \\ & \ddots & 0 & \\ & & c_s^s & \\ & 0 & & \ddots \\ & & & c_S^S \end{bmatrix} \quad (17)$$

If the entry data (the absorbance readings at the selected wavelengths and the concentrations of the standard solutions) do not form sets of relevant data, then singular matrices may be obtained when processing the data (whose determinant is null), namely matrices which do not admit an inverse. In order to avoid this failure, the condition $\Lambda \geq N \geq M$ is imposed. This is the necessary (but no sufficient) condition to avoid the apparition of singular matrices. The necessity of the condition above results after inspecting the relations (10) and (12). In relation (10) the inverse of a matrix $(\mathbf{C_{st}} \cdot \mathbf{C_{st}^T})^{-1}[M,M]$ appears calculated from matrix $\mathbf{C_{st}}[M,N]$. Consequently, the matrix of the concentrations of the standard solutions must have higher - or at least equal - rank to the number M of chemical components in the sample. The necessary (but sufficient) condition for this requirement is $N \geq M$. In relation (12) the inverse of a matrix $(\mathbf{E}^T \cdot \mathbf{E})^{-1}[M,M]$, is calculated from the matrix of molar absorptivities, $\mathbf{E}[\Lambda,M]$. The necessary (but not sufficient) condition of the non-singularity of matrix $(\mathbf{E}^T \cdot \mathbf{E})^{-1}[M,M]$ is the compliance of inegality $\Lambda \geq N$. The two necessary conditions for avoiding matrix singularity are expressed in the united form $\Lambda \geq N \geq M$. Also in order to avoid singularity in relation (10), the appropriate choice of concentrations of standard solutions is imposed, so that in matrix $\mathbf{C_{st}}[M,N]$ both rows and columns should be linearly independent. Otherwise expressed, it is essential that there should not be any significant intercorrelation neither between different columns nor between different rows of the matrix of standard concentrations (in algebraic terms, the concentrations in standard solutions must form a complete basis in the linear M-dimensional field). In other words, the spectra of individual chemical components should differ significantly in the spectral field chosen for analysis (more precisely, for the selected set of wavelengths). The relevance of the choice of the wavelength set, from the point of view of the above-mentioned facts, can be tested by calculating the eigenvalues of the square and symmetric matrix $\mathbf{A}^T_{st}[N,\Lambda] \cdot \mathbf{A}_{st}[\Lambda,N]$. If one eigenvalue of this matrix is null (or very close to the null value), the selection of the wavelength set is not adequate for the intended analysis. The selection of another wavelength set is therefore necessary. The general issue of row (or column) intercorrelation is solved in linear algaebra by taking into consideration the issue of eigenvalues and eigenvectors. However, the complete and rigorous mathematical treatment of the issue of basis vectors in linear algaebra goes beyond the purpose of the present work.

2.1 Example

Let be N = 5 standard solutions containing M = 3 components of known concentrations. The concentrations, expressed in mg/l, are included in matrix $\mathbf{C_{st}}[3,5]$. As illustrated by this matrix, each of the 5 standard solutions contains (in different and known concentrations) all three dissolved chemical components.

$$\mathbf{C_{st}}[3,5] = \begin{bmatrix} 2.50 & 4.25 & 1.25 & 0.85 & 2.22 \\ 3.00 & 1.00 & 1.62 & 1.15 & 3.36 \\ 4.00 & 0.80 & 5.00 & 4.45 & 0.82 \end{bmatrix} ; \quad \mathbf{C}^T_{st}[5,3] = \begin{bmatrix} 2.50 & 3.00 & 4.00 \\ 4.25 & 1.00 & 0.80 \\ 1.25 & 1.62 & 5.00 \\ 0.85 & 1.15 & 4.45 \\ 2.22 & 3.36 & 0.82 \end{bmatrix}$$

The matrix $(\mathbf{C_{st}} \cdot \mathbf{C_{st}^T})[3,3]$ resulting after multiplication and the eigenvalues of the product matrix ($\mathbf{EV}[3,1]$) are illustrated below:

$$(\mathbf{C}_{st} \cdot \mathbf{C}_{st}^{T})[3,3] = \begin{bmatrix} 31.5259 & 22.2117 & 25.2529 \\ 22.2117 & 25.2365 & 28.7727 \\ 25.2529 & 28.7727 & 62.1149 \end{bmatrix} \; ; \; {}_{E}\mathbf{V}[3,1] = \begin{bmatrix} 4.915255 \\ 18.974022 \\ 94.988023 \end{bmatrix}$$

All three eigenvalues are different from zero (taking into account the concentration values and the precision in expressing concentration values), so the rank of the matrix $\mathbf{C}_{st}[3,5]$ is 3. In other words, the set of concentration values allows to determine quantitatively all three chemical components in their mixture (provided that the wavelength set at which the absorbance values are going to be measured is chosen correctly).

The situation would differ if the matrix of concentrations of the standard solutions contained the following values:

$$\mathbf{C}_{st}[3,5] = \begin{bmatrix} 2.50 & 4.25 & 1.25 & 0.85 & 2.22 \\ 3.00 & 1.00 & 1.62 & 1.15 & 3.36 \\ 5.50 & 5.25 & 2.87 & 2.00 & 5.58 \end{bmatrix} \; ; \; \mathbf{C}_{st}^{T}[5,3] = \begin{bmatrix} 2.50 & 3.00 & 5.50 \\ 4.25 & 1.00 & 5.25 \\ 1.25 & 1.62 & 2.87 \\ 0.85 & 1.15 & 2.00 \\ 2.22 & 3.36 & 5.58 \end{bmatrix}$$

In this situation, the product $(\mathbf{C}_{st} \cdot \mathbf{C}_{st}^{T})[3,3]$ of the two matrices has other eigenvalues.

$$(\mathbf{C}_{st} \cdot \mathbf{C}_{st}^{T})[3,3] = \begin{bmatrix} 31.5259 & 22.2117 & 53.7376 \\ 22.2117 & 25.2365 & 47.4482 \\ 53.7376 & 47.4482 & 101.1858 \end{bmatrix}$$

$${}_{E}\mathbf{V}[3,1] = \begin{bmatrix} 5.966037 \\ 2.6393 \cdot 10^{-14} \\ 151.982163 \end{bmatrix}$$

In this case the rank of matrix $\mathbf{C}_{st}[3,5]$ is only two because the second element in the column matrix of eigenvalues ($\mathbf{EV}[3,1]$) is a lot smaller than the elements of the initial matrix and a lot smaller than the estimated accepted errors in expressing the standard concentrations. Consequently, even if a number of N = 5 standard solutions were used (with the considered concentrations), the concentrations of the three components in their mixture cannot be determined (irrespective of the wavelengths set chosen for measuring the absorbances), because the values of the concentrations of the standard solution have not been chosen properly.

2.2 Example

For numeric illustration of the spectrophotometric data processing with matrix formalism, the measurement data obtained analyzing the mixture of salicylic acid, caffeine and acetaminophen will be further presented [Szabadai, 2005]. The number of standard solutions is N = 5 and each standard solution contains all three components (in known concentrations). Table 1 contains absorbance values for the 5 standard solutions (A_{st}) and for the mixture of three substances (A), registered at the same set of 18 wavelengths. Table 1 also presents the known concentrations of the three components in the five standard solutions (elements of matrix $\mathbf{C}_{st}[3,5]$), i.e. M = 3, N = 5, Λ = 18. The matrix of concentrations of the standard solutions $\mathbf{C}_{st}[3,5]$, the matrix of absorbances of the standard solutions $\mathbf{A}_{et}[18,5]$ and the matrix of absorbances of the sample $\mathbf{A}[18,1]$ have the following forms:

Table 1

		Standard solution 1	Standard solution 2	Standard solution 3	Standard solution 4	Standard solution 5	Mixture (sample)
Absorbance values for different wavelengths (cell thickness d = 1 cm)		1.167	0.456	1.179	1.011	0.565	0.581
		1.192	0.435	1.257	1.048	0.513	0.566
		1.169	0.377	1.288	1.123	0.439	0.515
		1.109	0.290	1.265	1.109	0.374	0.443
		1.020	0.244	1.154	1.010	0.362	0.395
		0.932	0.228	1.000	0.867	0.402	0.370
		0.822	0.218	0.799	0.679	0.462	0.350
		0.747	0.217	0.645	0.534	0.524	0.336
		0.714	0.233	0.548	0.440	0.585	0.347
		0.654	0.232	0.487	0.388	0.552	0.329
		0.509	0.209	0.379	0.302	0.422	0.276
		0.295	0.167	0.228	0.183	0.225	0.195
		0.152	0.133	0.115	0.092	0.110	0.130
		0.089	0.100	0.060	0.046	0.069	0.095
		0.049	0.058	0.032	0.025	0.038	0.048
		0.030	0.026	0.023	0.018	0.021	0.030
		0.022	0.013	0.019	0.016	0.015	0.011
		0.020	0.009	0.018	0.015	0.013	0.010
Concenrtations of components (mg/l)	Salicylic acid	2.50	4.25	1.25	0.85	2.22	
	Caffeine	3.00	1.00	1.62	1.15	3.36	
	Acetaminophen	4.00	0.80	5.00	4.45	0.82	

$$\mathbf{C}_{st}[3,5] = \begin{bmatrix} 2.50 & 4.25 & 1.25 & 0.85 & 2.22 \\ 3.00 & 1.00 & 1.62 & 1.15 & 3.36 \\ 4.00 & 0.80 & 5.00 & 4.45 & 0.82 \end{bmatrix}$$

$$\mathbf{A_{st}}[18,5] = \begin{bmatrix} 1.167 & 0.456 & 1.179 & 1.011 & 0.565 \\ 1.192 & 0.435 & 1.257 & 1.087 & 0.513 \\ 1.169 & 0.377 & 1.288 & 1.123 & 0.439 \\ 1.109 & 0.290 & 1.265 & 1.109 & 0.374 \\ 1.020 & 0.244 & 1.154 & 1.010 & 0.362 \\ 0.932 & 0.228 & 1.000 & 0.867 & 0.402 \\ 0.822 & 0.218 & 0.799 & 0.679 & 0.462 \\ 0.747 & 0.217 & 0.645 & 0.534 & 0.524 \\ 0.714 & 0.233 & 0.548 & 0.440 & 0.585 \\ 0.654 & 0.232 & 0.487 & 0.388 & 0.552 \\ 0.509 & 0.209 & 0.379 & 0.302 & 0.422 \\ 0.295 & 0.167 & 0.228 & 0.183 & 0.225 \\ 0.152 & 0.133 & 0.115 & 0.092 & 0.110 \\ 0.089 & 0.100 & 0.060 & 0.046 & 0.069 \\ 0.049 & 0.058 & 0.032 & 0.025 & 0.038 \\ 0.030 & 0.026 & 0.023 & 0.018 & 0.021 \\ 0.022 & 0.013 & 0.019 & 0.016 & 0.015 \\ 0.020 & 0.009 & 0.018 & 0.015 & 0.013 \end{bmatrix} ; \quad \mathbf{A}[18,1] = \begin{bmatrix} 0.581 \\ 0.566 \\ 0.515 \\ 0.443 \\ 0.395 \\ 0.370 \\ 0.350 \\ 0.336 \\ 0.347 \\ 0.329 \\ 0.276 \\ 0.195 \\ 0.130 \\ 0.095 \\ 0.048 \\ 0.030 \\ 0.011 \\ 0.010 \end{bmatrix}$$

After performing the matrix operations in relation (16), the elements of matrix **C**[3,1] are obtained. They represent the concentrations, expressed in mg/l, of the three components of interest (salicylic acid, caffeine and paracetamol) in the analysed sample.

$$\mathbf{C}[3,1] = \begin{bmatrix} 3.538 \\ 1.553 \\ 1.381 \end{bmatrix}$$

3. Generalization of the 3-point method to correct backgroud absorption

Before dealing generally with the issue of foreign components in the sample (components which cannot be found in standard solutions) - which may cause deviations from the hypothesis according to which the sample spectrum is formed by adding (with different weights) the spectra of standard solutions - the quantitative analysis method and the baseline correction algorithm suggested by *Morton* and *Stubbs* [Burnius, 1959; Ewing et al., 1953; Fox & Mueller, 1950; Melnick et al., 1952; Morton & Stubbs, 1946, 1947, 1948; Owen, 1995; Page & Berkovitz, 1943; Szabadai, 2005;] (also known as "3-point method") will be presented.

The *Morton - Stubbs* method takes into account that the sample often contains - besides the chemical substance of interest - other foreign absorbent chemical components. If the chemical removal of these foreign components is difficult, the elimination (or at least the minimisation) of their contribution to the final result of the analysis by correcting the absorbance read could be a confortable solution. Accordind to the original form of the *Morton* and *Stubbs* method [Morton & Stubbs, 1946, 1947, 1948], it is possible to eliminate the disturbing effect of a foreign component only in the case in which the absorption of the disturbing component, manifested in the spectral field taken into consideration, does not present a maximum of absorption, but appears as a baseline absorption, dependent on the wavelength according to a linear function, which overlaps the absorption spectrum of the chemical component of interest.

The absorption spectrum of the component of intereset is deformed because of the background absorption (linearly dependent on the wavelength), and the effect of this deformation is eliminated through the special method of processing the measured absorbance values. According to the original *Morton – Stubbs* formalism, it is essential to determinate the absorbance of the sample at at least three wavelengths [Morton & Stubbs, 1946]. The wavelengths values involved are selected as follows: the wavelength used (λ_{max}) is the one at which the standard solution of the substance of interest (where the disturbing component is not present) presents a local absorbance maximum and another two wavelengths ((λ_1 and λ_2 , λ_{max} being between these wavelengths) at which the substace of interest presents equal molar absorptivities ($\acute{A}(1) = \acute{A}(2)$). Figure 1 represents the spectrum of the standard solution by dotted line whereas the spectrum of the mixture, where the quantitative determination of the substances of interest is intended, is represented by a continuous line. The absorbance values corresponding to the three wavelengths selected (λ_1 , λ_2 and λ_{max}) are denoted as A(1), A(2) and A(max) in the spectrum of the sample and as $\acute{A}(1)$, $\acute{A}(2)$ and $\acute{A}(max)$ in the spectrum of the pure (standard) component. The purpose is to calculate quantity $\acute{A}(max)$ (namely the absorbance associated with the substance of interest but without the backgroud absorbance) from the measured values A(1), A(2) and A(max). The absorbance $\acute{A}(max)$ is obtained by subtracting from the measured value A(max) the value denoted by x + y in Figure 1.

$$\acute{A}(max) = A(max) - (x + y) \tag{18}$$

The value x is expressed from the similarity of two triangles chosen conveniently:

$$\frac{\lambda_2 - \lambda_1}{\lambda_2 - \lambda_{max}} = \frac{A(1) - A(2)}{x}; \quad x = \frac{\lambda_2 - \lambda\max}{\lambda_2 - \lambda_1} \cdot \left[A(1) - A(2)\right] \tag{19}$$

For calculating the value y in expression (18), the ratio of the absorbances $\acute{A}(max)$ and $\acute{A}(2)$ is needed, which can be determined from the spectrum of the standard solution. When elaborating an analytical method in order to determine a certain substance of interest, in a standardized work method, the ratio of the absorbances $\acute{A}(max)$ and $\acute{A}(2)$ once determined, it can be used for subsequent analyses, provided that analyses should be performed strictly in unchanged conditions (in the same solvent, at the same pH, the same temperature, with the same slit program of the spectrophotometer, preferably the same type of spectrophotometer as the one used for determining the above mentioned ratio). Let be denoted the aforementioned ratio as ρ:

$$\rho = \frac{A'(\max)}{A'(2)} \tag{20}$$

In possession of the ratio ρ, the value y is obtained from relation (18) and (21).

$$\acute{A}(2) = A(2) - y \tag{21}$$

After dividing member by member relations (18) and (21), results:

$$\rho = \frac{A'(\max)}{A'(2)} = \frac{A(\max) - (x+y)}{A(2) - y} \; ; \; y = \frac{\rho \cdot A(2) - A(\max) + x}{\rho - 1} \tag{22}$$

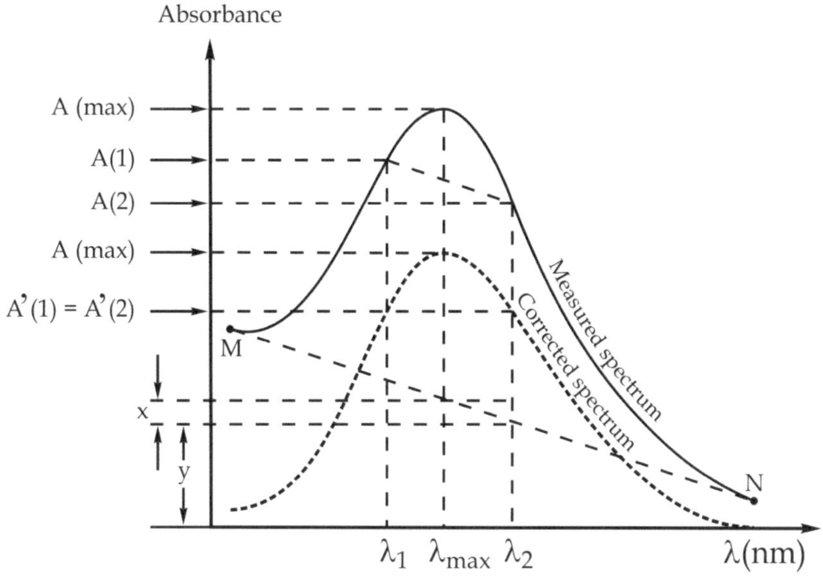

Fig. 1. Illustration of the *Morton – Stubbs* method

After replacing the expressions x and y in the latter relation, relation (23) results. It expresses the absorbance associated to the component of interest $A'(\max)$, which lacks baseline absorption.

$$A'(\max) = A(\max) - \frac{\lambda_2 - \lambda_{\max}}{\lambda_2 - \lambda_1} \cdot [A(1) - A(2)] - \frac{\rho \cdot A(2) - A(\max) + \frac{\lambda_2 - \lambda_{\max}}{\lambda_2 - \lambda_1} \cdot [A(1) - A(2)]}{\rho - 1} \tag{23}$$

The liniarity of the background absorption in a large spectral field is not always satisfied. In the case of wide absorption bands it is recommended to measure the absorbance of the sample at several wavelengths; in these cases however, the processing of the absorbance values measured requires more elaborated mathematical methods.

As it can be noticed, the *Morton – Stubbs* formalism allows the presence in the spectrum of the sample a linear background (a linear foreign spectrum in relation to the wavelength) which cannot be put down to any component of the standard solutions, ensuring corrected results (sample concentrations of the components of interest).

The original algorithm may be extended to ensure the obtention of corrected results in the case in which the sample spectrum contains, besides the chemical components represented in the standard spectra, a G degree polynomial baseline in relation to the wavelength. The spectrum of the sample is thus considered to consist of the spectra of the standard solutions and of the background spectrum, the latter being approximated to an adequate G degree polynomial (relation 24).

$$A(\lambda_i) = \sum_{k=1}^{K} p_k \cdot A_k(\lambda_i) + \sum_{g=0}^{G} q_g \cdot \lambda_i^g \quad ; \quad (i = 1,2,\ldots,N) \tag{24}$$

The purpose is to calculate the contribution weight p_k of each standard solution to the spectrum of the sample, namely the coefficients p_k (k = 1 , 2 , . . . , K). In the ideal case, when the spectrum of the sample does not contain a foreign baseline, but only the components represented in standard solutions, the coefficients q_g (g = 0 , 1 , 2 , . . . , G) are all null. Because of inherent measurement errors these coefficients are not null, but if the polynomial (25) is positive and has small values (for all wavelengths λ_i selected) in relation to the measured absorbances, the approach of the issue is correct and there are still chances to remove, by calculation, the effect of the polynomial backgroud (G degree) from the spectrum of the sample on the results. On the contrary, if the polynomial (25) has a high value or a negative one, even for one wavelength (one i value), the foreign backgroud cannot be approximated to a G degree polynomial form, and forcing the algorithm might lead to an erroneous result.

$$P(i) = \sum_{g=0}^{G} q_g \cdot \lambda_i^g \tag{25}$$

Obviously, the highest the G degree of the polynomial (25) which corrects the foreign backgroud in the spectrum of the sample, The more flexible the correction algorithm of a real backgroud absorption, but the more wavelengths should be selected where the absorbance readings are performed (in other words the inegality $N > G + K + 1$ is imposed in practice in order to obtain, from the measured absorbance values, a supra-determined system of equations).

For the statistical processing of the set of N absorbance values obtained for the sample and N×K absorbance values for the K standard solutions, the function (26) is defined imposing that for the values p_k (k = 1 , 2 , . . . , K) and the values q_g (g = 0 , 1 , . . . , G), which ensure the best global correspondence between the measured absorbances of the sample and the absorbances approximated with the relation (24), the function $F(p_k,q_g)$ should present a local minimum. The condition formulated is equivalent cancel the partial derivatives of the function (26) calculated in relation to p_k (k = 1 , 2 , . . . , . . . , K) and q_g (g = 0 , 1 , . . . , G). The cancellation of partial derivatives in (26) represents the necessary (but sufficient) condition for a local minimum of the function (26).

$$F(p_k,q_g) = \sum_{i=1}^{N} \left[A(\lambda_i) - \sum_{k=1}^{K} p_k \cdot A_k(\lambda_i) - \sum_{g=0}^{G} q_g \cdot \lambda^g \right]^2 = \min$$

$$\frac{\partial F(p_k,q_g)}{\partial q_{g^*}} = 0 \quad ; \quad \frac{\partial F(p_k,q_g)}{\partial p_{k^*}} = 0 \tag{26}$$

$$(k^* = 1,2,\ldots,K \quad ; \quad g^* = 0,1,2,\ldots,G)$$

After derivation and equalization the derivatives to zero, a system of K + G + 1 linear equations is obtained, having the same number of unknowns (27).

$$\sum_{g=0}^{G} q_g \cdot \sum_{i=1}^{N} \lambda_i^g \cdot \lambda_i^{g^*} + \sum_{k=1}^{K} p_k \cdot \sum_{i=1}^{N} A_k(\lambda_i) \cdot \lambda_i^{g^*} = \sum_{i=1}^{N} A(\lambda_i) \cdot \lambda_i^{g^*}$$
$$(g^* = 0,1,2,\ldots,G)$$
$$\sum_{g=0}^{G} q_g \cdot \sum_{i=1}^{N} \lambda_i^g \cdot A_{k^*}(\lambda_i) + \sum_{k=1}^{K} p_k \cdot \sum_{i=1}^{N} A_k(\lambda_i) \cdot A_{k^*}(\lambda_i) = \sum_{i=1}^{N} A(\lambda_i) \cdot A_{k^*}(\lambda_i) \tag{27}$$
$$(k^* = 1,2,\ldots,K)$$

In order to express the wavelength (and its different powers) any unit of measure can be used, provided that the same unit of measure is used in all equations and for all wavelengths.

The generalisation of the *Morton – Stubbs* algorithm for the polynomial correction of the spectrum of the sample can also be presented in a matrix form. The equation system (24), written in a conventional algebraic form, is equivalent to matrix relation (28).

$$\begin{bmatrix} A(\lambda_1) \\ A(\lambda_2) \\ \vdots \\ A(\lambda_i) \\ \vdots \\ A(\lambda_N) \end{bmatrix} = \begin{bmatrix} A_1(\lambda_1) & \cdots & A_k(\lambda_1) & \cdots & A_K(\lambda_1) & 1 & \lambda_1 & \cdots & \lambda_1^g & \cdots & \lambda_1^G \\ A_1(\lambda_2) & \cdots & A_k(\lambda_2) & \cdots & A_K(\lambda_2) & 1 & \lambda_2 & \cdots & \lambda_2^g & \cdots & \lambda_2^G \\ \vdots & & \vdots & & \vdots & \vdots & \vdots & & \vdots & & \vdots \\ A_1(\lambda_i) & \cdots & A_k(\lambda_i) & \cdots & A_K(\lambda_i) & 1 & \lambda_i & \cdots & \lambda_i^g & \cdots & \lambda_i^G \\ \vdots & & \vdots & & \vdots & \vdots & \vdots & & \vdots & & \vdots \\ A_1(\lambda_N) & \cdots & A_k(\lambda_N) & \cdots & A_K(\lambda_N) & 1 & \lambda_N & \cdots & \lambda_N^g & \cdots & \lambda_N^G \end{bmatrix} \cdot \begin{bmatrix} p_1 \\ \vdots \\ p_k \\ \vdots \\ p_K \\ q_0 \\ q_1 \\ \vdots \\ q_g \\ \vdots \\ q_G \end{bmatrix} \tag{28}$$

If the matrix of the absorbance values of the sample (on the left member of the equation (28)) is denoted by **X**[N,1], the first matrix factor on the right member by **Y**[N,K+G+1] and the second matrix factor on the right member by **Z**[K+G+1,1], the equation (28) can have the form (29).

$$\mathbf{X}[N,1] = \mathbf{Y}[N,K+G+1] \cdot \mathbf{Z}[K+G+1,1] \tag{29}$$

The unknowns of interest are found in matrix **Z**[K+G+ 1,1] ; the relative weights are p_k (k = 1, 2 , . . . , K). In order to explain the elements of matrix **Z**, both members of relation (29) are multiplied on the left by the transpose of matrix **Y**.

$$\mathbf{Y}^T[K+G+1,N] \cdot \mathbf{X}[N,1] = \mathbf{Y}^T[K+G+1,N] \cdot \mathbf{Y}[N,K+G+1] \cdot \mathbf{Z}[K+G+1,1] \tag{30}$$

Matrix $(\mathbf{Y}^T \cdot \mathbf{Y}) = \mathbf{Y}^T[K+G+1,N] \cdot \mathbf{Y}[N,K+G+1]$ is square and allows an inverse matrix $(\mathbf{Y}^T \cdot \mathbf{Y})^{-1}[N,N]$ if the associated determinant is not null. By multiplying relation (30) on the left by the inverse matrix, the explicit form of matrix **Z** results.

$$(\mathbf{Y}^T \cdot \mathbf{Y})^{-1}[N,N] \cdot \mathbf{Y}^T\,[K+G+1,N] \cdot \mathbf{X}[N,1] = \mathbf{Z}[K+G+1,1] \quad (31)$$

It is decisively important to determine the correct set of wavelengths at which the absorbance values should be measured in case of a concrete analytical problem. The choice of the optimal wavelength (or wavelengths) is often a difficult issue even in case of a single component of interest. In real samples the component of interest may be accompanied by different other components whithout analytical interest ("the sample ballast"), but which can modify the molar absorptivity of the component of interest and so the sensitivity of the spectral answer of the chemical substance representing the object of the analysis. If one can identify the wavelength value at which the absorption of the sample ballast is negligible and at which the absorption of the component of interest is considerable, the respective wavelength is recommended for the determination. When at this wavelength the component of interest has even a local absorption maximum, this is an additional advantage, because at this wavelength the absorbance value depends in a minimum extent on the possible disorders in setting the wavelengths of the spectrophotometer. In the less fortunate case, where the sample ballast covers the entire spectral field available, more wavelengths are selected in order to determine the component in the sample in order to improve the specificity of the spectral answer in favour of the component of interest.

When the absorption of the component of interest and that of the ballast cannot be separated, a set of wavelengths can often be chosen so that the absorbances measured express the concentration of the component of interest through a multilinear relation (32).

$$c = f(\lambda_1) \cdot A(\lambda_1) + f(\lambda_2) \cdot A(\lambda_2) + \ldots + f(\lambda_i) \cdot A(\lambda_i) + \ldots + f(\lambda_N) \cdot A(\lambda_N) \quad (32)$$

The aim is to determine numerically the coefficients $f(\lambda_i)$; $(i = 1, 2, \ldots, N)$ for each wavelength in the spectral field considered (it is considered that the entire spectrum consists of N absorbance values associated to N discrete wavelength values) to calculate according to (32) the concentration of the chemical substance of interest in different samples, containing a different and unpredictable ballast. This purpose can sometimes be accomplished, sometimes not, according to the ballast variability of the analysed samples.

If there is any chance to determine a set of coefficients in agreement with the requirements mentioned for a component of interest, in presence of a ballast range in different samples, their calculation could be performed through calibration with a number of standard samples (let their number S) containing a ballast range as close as possible to that of real samples (of unknown composition) under analysis. Thus, two different standard samples may have the same concentration of the component of interest if they have a different ballast.

The concentrations of the component of interest in the S standard samples and the absorbances $A_s(\lambda_i)$; $(s = 1, 2, \ldots, S)$ of the standard samples satisfy the equation (32). The equation system obtained with the standard sample data can be rendered in matrix form (33).

If $S = N$, the number of unknowns equals the number of equations, so we dispose of the minimum number of equations necessary to solve the system (33) in relation to the N unknowns. For the reasons discussed above, the creation of a supra-determined system of equations is preferred ($S > N$), as well as the search for a solution with an optimal global fit with least squares method.

$$\begin{bmatrix} c_1^{st} \\ c_2^{st} \\ \vdots \\ c_s^{st} \\ \vdots \\ c_S^{st} \end{bmatrix} = \begin{bmatrix} A_1(\lambda_1) & A_1(\lambda_2) & \cdots & A_1(\lambda_i) & \cdots & A_1(\lambda_N) \\ A_2(\lambda_1) & A_2(\lambda_2) & \cdots & A_2(\lambda_i) & \cdots & A_2(\lambda_N) \\ \vdots & \vdots & & \vdots & & \vdots \\ A_s(\lambda_1) & A_s(\lambda_2) & \cdots & A_s(\lambda_i) & \cdots & A_s(\lambda_N) \\ \vdots & \vdots & & \vdots & & \vdots \\ A_S(\lambda_1) & A_S(\lambda_2) & \cdots & A_S(\lambda_i) & \cdots & A_S(\lambda_N) \end{bmatrix} \cdot \begin{bmatrix} f(\lambda_1) \\ f(\lambda_2) \\ \vdots \\ f(\lambda_i) \\ \vdots \\ f(\lambda_N) \end{bmatrix} \quad (33)$$

Values $f(\lambda_i)$; (i = 1 , 2 , . . . , N), representing the solution to the equation system (33), can be positive and negative numbers, both type being relevant for the analysis. If the absolute value of one of the coefficients $f(\lambda_i)$; (i = 1 , 2 , . . . , N) is small (negligible in relation to the mean of the absolute values of all coefficients), their contribution to the equation (32) is insignificant, they can be considered null, and the respective wavelengths are not relevant for the intended quantitative analysis. Therefore, to each wavelength in the spectrum a coefficient is associated expressing the relevance of that wavelength for the quantitative analysis of the component of interest in the presence of the matrix included in the calibration stage.

By excluding the irrelevant wavelengths, which do not improve the selectivity of the analytical method, one may reduce the number of wavelengths at which the measurement of absorbances is imposed when executing a real sample analysis.

In possession of the coefficients $f(\lambda_i)$; (i = 1 , 2 , . . . , N), the concentrations in the standard samples can be recalculated by relation (4.65) (the concentrations obtained are denoted c_1 , c_2 , . . . , c_s , . . . , c_S). Ideally, concentrations for all standard samples can be found. In reality, the correspondance between the set of existing (and known) concentrations in the S standard samples and the set of concentrations recalculated with relation (33) is not perfect. The success of the calibration operation can be expressed through the value of the linear correlation coefficient between the set of existing concentrations in the standard samples and the recalculated ones. Since the arithmetic mean of the existing (and known) concentrations in the S standard samples and the arithmetic mean of the concentrations recalculated with relation (33) are equal (according to a known theorem of mathematical statistics), their notation with a common symbol is justified:

$$\bar{C} = \frac{1}{S} \cdot \sum_{s=1}^{S} c_s^{st} = \frac{1}{S} \cdot \sum_{s=1}^{S} c_s$$

The linear correlation coefficient between the set of concentrations c_s^{st} and c_s (s = 1 , 2 , ... , S) is calculated with relation (34).

$$r = \frac{\sum_{s=1}^{S} \left(c_s^{st} - \bar{C}\right) \cdot \left(c_s - \bar{C}\right)}{\sqrt{\left[\sum_{s=1}^{S} \left(c_s^{st} - \bar{C}\right)^2\right] \cdot \left[\sum_{s=1}^{S} \left(c_s - \bar{C}\right)^2\right]}} \quad (34)$$

If the correlation coefficient (34) has an acceptable value from a statistical point of view (for example r > 0,95), it is likely that the set of coefficients $f(\lambda_i)$; (i = 1 , 2 , ... , N), obtained by solving the equation system (33) will allow to find the correct concentration of the substance of interest in real samples, provided that the real sample ballast is not completely different from the ballast range covered when calibrating the method (when determining the coefficients $f(\lambda_i)$; (i = 1 , 2 , . . . , N)). This requirement is met to a certain extent in the case of serial analyses, where the nature of individual samples does not differ much, meaning that their ballast is similar.

Presenting a spectrum in a spectral field through pairs of wavelength-absorbance values ($A(\lambda)$ vs. λ , "digitized presentation") implies a large amount of data (for a faithful reprensentation of a spectrum the N number of sampling points is large). It results that, in order to genearate a supra-determinant equation system (33), an even larger number of standard samples is necessary (S > N). This is generally inconvenient to realize in practice because it implies the use of a too large number of standard samples.

If S < N, the equation system (33) allows several sets of wavelengths for which the concentrations in standard samples correlate satisfactorily with the absorbance values, and the remaining problem is to identify at least one of these sets. This method is frequently used in practice, and establishing a profitable set of wavelengths involves the following stages:

(1) The matrix of absorbance values **A**[S,N] turns into a new square matrix **B**[S,S] whose columns are a complete orthogonal basis. The orthogonality of columns in the new matrix **B**[S,S] can be realized, for example, by multiplying the matrix **A**[S,N] on the right by a matrix **Q**[N,S] chosen conveniently (35), so that the elements of matrix **B**[S,S] = **A**[S,N]·**Q**[N,S] satisfy the orthogonality relation of columns (36). The construction of such a matrix **Q**[N,S] is not unique; theoretically, there is an infinite number of such matrices capable of generating orthogonal columns satisfying the requirement (36). In spectrophotometric practice a diagonal-superior form of the matrix **Q**[N,S] is sometimes used (where only elements on the main diagonal and those above this diagonal are different from zero).

$$\begin{bmatrix} A_1(\lambda_1) & \cdots & A_1(\lambda_i) & \cdots & A_1(\lambda_N) \\ \vdots & & \vdots & & \vdots \\ A_s(\lambda_1) & \cdots & A_s(\lambda_i) & \cdots & A_s(\lambda_N) \\ \vdots & & \vdots & & \vdots \\ A_S(\lambda_1) & \cdots & A_S(\lambda_i) & \cdots & A_S(\lambda_N) \end{bmatrix} \cdot \begin{bmatrix} Q_{11} & \cdots & Q_{1j} & \cdots & Q_{1S} \\ \vdots & & \vdots & & \vdots \\ Q_{j1} & \cdots & Q_{jj} & \cdots & Q_{jS} \\ \vdots & & \vdots & & \vdots \\ Q_{N1} & \cdots & Q_{Nj} & \cdots & Q_{NS} \end{bmatrix} = \begin{bmatrix} B_{11} & \cdots & B_{1j} & \cdots & B_{1S} \\ \vdots & & \vdots & & \vdots \\ B_{s1} & \cdots & B_{sj} & \cdots & B_{sS} \\ \vdots & & \vdots & & \vdots \\ B_{S1} & \cdots & B_{Sj} & \cdots & B_{SS} \end{bmatrix} \quad (35)$$

$$\sum_{s=1}^{S} B_s(\lambda_j) \cdot B_s(\lambda_{j^*}) = 0 \quad ; \quad for\ any\ \ j \neq j^* \quad (36)$$

(2) Calculate the correlation coefficient of the elements of matrix $\mathbf{C}_{et}$[S,1] in relation (33) one by one with the columns of matrix **B**[S,S] (for j = 1 , 2 , . . . , S), thus obtaining N correlation coefficient values, in real cases all being smaller than theoretical value 1. The correlation coefficient of the elements of matrix $\mathbf{C}_{st}$[S,1] with the column "j" of matrix **B**[S,S] is calculated by the relation (37).

$$r(\mathbf{C}_{st}[S,1],\mathbf{B}_j[S,1]) = \frac{\sum_{s=1}^{S}\left(c_s^{st}-\bar{C}\right)\cdot\left(B_{sj}-\bar{B}_j\right)}{\sqrt{\left[\sum_{s=1}^{S}\left(c_s^{st}-\bar{C}\right)^2\right]\cdot\left[\sum_{s=1}^{S}\left(B_{sj}-\bar{B}_j\right)^2\right]}} \tag{37}$$

In relation (37) $\mathbf{B}_j[S,1]$ represents the column vector made up of the column of number "j" of matrix $\mathbf{B}[S,S]$, $\bar{C}$ is the mean value of the elements of matrix $\mathbf{C}_{st}[S,1]$ and $\bar{B}_j$ is the mean value of elements in column "j" in matrix $\mathbf{B}[S,S]$.

$$\bar{C} = \frac{1}{S}\cdot\sum_{s=1}^{S} c_s^{st} \quad ; \quad \bar{B}_j = \frac{1}{S}\cdot\sum_{s=1}^{S} B_{sj}$$

The correlation coefficients, calculated with relation (37) for j = 1, 2, . . . , S, in relation to "j" and the value "j" is retained (denoted by $\underline{1}$) for which the correlation coefficient is highest (in case of obtaining equal values of the correlation coefficient for more "j" values, one of these "j" values is retained arbitrarily).

(3) By using the multiple linear regression method, the elements of column matrix $\mathbf{C}_{et}[S,1]$ are correlated with all the pairs of columns of matrix $\mathbf{B}[S,S]$ obtained by combining column $\underline{1}$ with all the other columns of matrix $\mathbf{B}[S,S]$. The values of the multiple correlation coefficient $r(\mathbf{C}_{et}[S,1], (\mathbf{B}_{\underline{1}}\&\mathbf{B}_j)[S,2])$ are calculated in relation to the values taken by "j" and is retained (and denoted by $\underline{2}$) the value " j" for which the multiple correlation coefficient is highest. Two "j" values are thus obtained (denoted by $\underline{1}$ and $\underline{2}$) indicating the pair of columns in matrix $\mathbf{B}[S,S]$ which correlate conveniently with the column matrix $\mathbf{C}_{et}[S,1]$.

(4) By using the multiple linear regression method, the elements of column matrix $\mathbf{C}_{et}[S,1]$ are correlated with all sets of three columns of matrix $\mathbf{B}[S,S]$, obtained by combining columns $\underline{1}$ and $\underline{2}$ with all the other columns of matrix $\mathbf{B}[S,S]$. The values $r(\mathbf{C}_{et}[S,1], (\mathbf{B}_{\underline{1}}\&\mathbf{B}_{\underline{2}}\&\mathbf{B}_j)[S,3])$ are calculated in relation to the values taken by "j" and is retained (and denoted by $\underline{3}$) the value " j" for which the multiple correlation coefficient is highest. Three "j' values result this way (denoted by $\underline{1}$, $\underline{2}$, and $\underline{3}$), indicating the set of three columns of matrix $\mathbf{B}[S,S]$ which correlates conveniently with the column matrix $\mathbf{C}_{et}[S,1]$.

(5) The procedure described above continues by increasing progressively the number of columns of matrix $\mathbf{B}[S,S]$ with which is correlated, by multiple linear regression, the column matrix $\mathbf{C}_{et}[S,1]$. The columns of matrix $\mathbf{B}[S,S]$, involved at this phase, include those retained in the previous phase and a column which hasn not been yet retained. It is obvious that, by increasing the number of columns in $\mathbf{B}[S,S]$, involved in the multiple correlation, the optimal correlation coefficient approaches progressively the ideal value r = 1. Because the columns in matrix $\mathbf{B}[S,S]$ are orthogonal, there is no danger that, at a certain phase, the maximum correlation coefficient will be exceeded by a correlation coefficient corresponding to a combination of columns including a column (therefore a "j" value) which has not been retained in a previous phase. If the columns in matrix $\mathbf{B}[S,S]$ were not orthogonal, the above-mentioned danger would have appeared. This justifies the transformation of matrix $\mathbf{A}[S,N]$ (whose columns are not generally orthogonal) into a matrix $\mathbf{B}[S,S]$ with orthogonal columns. In practice, the procedure continues until obtaining a compromise situation, namely a satisfactory multiple correlation coefficient at a minimum number of involved columns if matrix $\mathbf{B}[S,S]$.

(6) Following the correlations described above, a set of columns of matrix **B**[S,S] results. These have been retained and denoted by $\underline{1}, \underline{2}, \ldots, \underline{J}$. A convenient set is then established, made up of wavelength values or, in other words, a set of $\underline{J}$ columns of the **A**[S,N] matrix. The set of $\underline{J}$ columns of matrix **A**[S,N] (conceived as $\underline{J}$ vectors in an imaginary S-dimensional space) is chosen so that each column of matrix **A**[S,N] presents a maximum covariance with a column of matrix **B**[S,S] retained during the above-mentioned operations. More concretely, if one suppose that the column of order "j" of matrix **B**[S,S] is associated to the column of order "i" of matrix **A**[S,N], it means that for the value "j" the column of order "i" of matrix **A**[S,N] ensures a maximum value of the covariance (of the correlation coefficient) calculated with relation (38).

$$r(\mathbf{A}_i[S,1],\mathbf{B}_j[S,1]) = \frac{\sum_{s=1}^{S}\left(A_s(\lambda_i) - \overline{A}(\lambda_i)\right)\cdot\left(B_{sj} - \overline{B}_j\right)}{\sqrt{\left[\sum_{s=1}^{S}\left(A_s(\lambda_i) - \overline{C}\right)^2\right]\cdot\left[\sum_{s=1}^{S}\left(B_{sj} - \overline{B}_j\right)^2\right]}} \tag{38}$$

In relation (38) $\overline{A}(\lambda_i)$ and $\overline{B}_j$ represent the arithmetic means of the corresponding matrix elements in columns of order "i", and "j" respectively.

$$\overline{A}(\lambda_i) = \frac{1}{S}\cdot\sum_{s=1}^{S} A_s(\lambda_i) \quad ; \quad \overline{B}_j = \frac{1}{S}\cdot\sum_{s=1}^{S} B_{sj}$$

In what follows, the wavelengths selected during phase (6) will be denoted by $\lambda_1^*, \lambda_2^*, \ldots, \lambda_J^*$. By applying relation (38) for i = 1 , 2 , . . . , N and j = 1 , 2 , . . . , $\underline{J}$, the matrix **R**[N,$\underline{J}$] of the correlation coefficients is obtained (39).

$$\begin{bmatrix} r_{11} & \cdots & r_{1j} & \cdots & r_{1J} \\ \vdots & & \vdots & & \vdots \\ r_{i1} & \cdots & r_{ij} & \cdots & r_{iJ} \\ \vdots & & \vdots & & \vdots \\ r_{N1} & \cdots & r_{Nj} & \cdots & r_{NJ} \end{bmatrix} \tag{39}$$

In each column "j" of the matrix (39) an element r_{ij} with maximum absolute value is sought. The set of order numbers "i", which associates an "i" for each column"j", corresponds to the researched set of wavelengths.

(7) The equation system (15) is reconstructed, using only the set of wavelengths $\lambda_1^*, \lambda_2^*, \ldots, \lambda_J^*$ selected in previous phases.

$$\begin{bmatrix} c_1^{et} \\ c_2^{et} \\ \vdots \\ c_s^{et} \\ \vdots \\ c_S^{et} \end{bmatrix} = \begin{bmatrix} A_1(\lambda_1^*) & A_1(\lambda_2^*) & \cdots & A_1(\lambda_J^*) \\ A_2(\lambda_1^*) & A_2(\lambda_2^*) & \cdots & A_2(\lambda_J^*) \\ \vdots & \vdots & & \vdots \\ A_s(\lambda_1^*) & A_s(\lambda_2^*) & \cdots & A_s(\lambda_J^*) \\ \vdots & \vdots & & \vdots \\ A_S(\lambda_1^*) & A_S(\lambda_2^*) & \cdots & A_S(\lambda_J^*) \end{bmatrix} \cdot \begin{bmatrix} f(\lambda_1^*) \\ f(\lambda_2^*) \\ \vdots \\ f(\lambda_J^*) \end{bmatrix} \tag{40}$$

In order for the equation system (40) to be solvable in relation to the unknowns $f(\lambda_1^*)$, $f(\lambda_2^*)$, . . ., $f(\lambda_J^*)$, it is necessary that the number of selected wavelengths (*J*) be smaller than (or equal) to the number of standard samples (S). It is also essential that the determinant of matrix **D**[*J*,*J*], resulting after multiplying the transpose of system matrix $(\mathbf{A}^*)^T[J,S]$ by the system matrix $(\mathbf{A}^*)[S,J]$ be significantly different from zero.

$$(\mathbf{A}^{*T}[J,S]\cdot(\mathbf{A}^*)[S,J] = \mathbf{D}[J,J];\ \det(\mathbf{D}[J,J])\neq 0$$

At the simultaneous determination of several chemical components which do not interact chemically, the equation system (1) and (2) has been constituted, with the help of N standard solutions, measured at Λ distinct wavelength values. In order to correctly solve the analytical problem, it is recommendable that the spectra of the N standard solutions be "as distinct as possible", because in the extreme (and imaginary) case where two standard solutions had identical spectra, the equation system would be undetermined, so impossible to solve. It is necessary to rigorously express the requirement that the spectra be as "different as possible". A method of characterizing the difference between spectra consists in considering the absorbances of a standard solution, measured at the selected set of wavelengths, as components of a vector in the Λ-dimensional space. The N spectra of standard solutions will thus form a set of N vectors.

$$D_{Gramm} = \begin{vmatrix} \sum_{\lambda=1}^{\Lambda} A^1(\lambda)\cdot A^1(\lambda) & \cdots & \sum_{\lambda=1}^{\Lambda} A^1(\lambda)\cdot A^n(\lambda) & \cdots & \sum_{\lambda=1}^{\Lambda} A^1(\lambda)\cdot A^N(\lambda) \\ \vdots & & \vdots & & \vdots \\ \sum_{\lambda=1}^{\Lambda} A^n(\lambda)\cdot A^1(\lambda) & \cdots & \sum_{\lambda=1}^{\Lambda} A^n(\lambda)\cdot A^n(\lambda) & \cdots & \sum_{\lambda=1}^{\Lambda} A^n(\lambda)\cdot A^N(\lambda) \\ \vdots & & \vdots & & \vdots \\ \sum_{\lambda=1}^{\Lambda} A^N(\lambda)\cdot A^1(\lambda) & \cdots & \sum_{\lambda=1}^{\Lambda} A^N(\lambda)\cdot A^n(\lambda) & \cdots & \sum_{\lambda=1}^{\Lambda} A^N(\lambda)\cdot A^N(\lambda) \end{vmatrix} \quad (41)$$

The value of the *Gramm* determinant (41) of the vector set expresses quantitatively the difference between vectors. The higher the value of the determinant (41), the more satisfied the requirement that the standard spectra be "as different as possible". At a higher value of the *Gramm* determinant the absorbance measurement error affects to a smaller extent the precision of the final results.

4. Generalization the standard addition method for several components of interest

In a real sample, subjected to be analyzed, one must take into consideration that the sample contains, besides the substance of interest, various other ingredients. Although it is possible to choose a wavelength at which the absorbance of the substance of interest should be significant and the absorbance of the ingredients negligible, it may happen that the ingredients, through their presence, modify the molar absorptivity of the component of interest, and thus modify the sensitivity of the spectrophotometric response to the component of interest. This possibility is more plausible in real pharmaceutical products, where the ingredients are found, as a rule, in a larger quantity than the active components. In this case, comparing the absorbance of the sample with that of a standard solution (which does not contain any ingredients) could provide erroneous analytical results. In order to realize even in these cases the quantitative determination of the active substance (the component of interest), one may resort to the "standard addition method" [Bosch-Reigh et al., 1991; Lozano et al., 2009 ; Szabadai, 2005; Valderrama & Poppi, 2009].

The reasoning of the addition method in the general case, when aiming to determine several components quantitatively, can be described with the help of the matrix calculation formalism [Szabadai, 2005]. The primary sample, in which the concentrations c_1 , c_2 , . . . , c_j , . . . , c_M of the M chemical components are analysed, is dissolved with an adequate solvent, bringing it to the final known volume V_a . A number of S + 1 equal portions (each having the volume "v") will be drawn from this solution. The portion number "0" is diluted to the final known volume V_b , thus obtaining the final solution of number "0" in which the concentrations of the components of interest are c_{10} , c_{20} , . . . , c_{M0}, and the concentration of ingredients is $c_{b(ing)}$. The portions number 1 , . . . , M are supplemented with known quantities of the M components of interest, so that, after completing to the final volume V_b, "S" solutions with modifications of known concentrations are obtained. In the final solution number "i", which was prepared by adding the masses m_{1i} , m_{2i} , . . . , m_{Mi} of individual components, the concentration modifications of components are Δc_{1i} , Δc_{2i} , . . . , Δc_{Mi} , whereas the concentration of ingredients remains the same in all S solutions, independent of "i". For each final solution the absorbance is measured at the same set of wavelengths λ_1 , λ_2 , . . . , λ_Λ. For the final solution number "0" the values $A_0(\lambda_1)$, $A_0(\lambda_2)$, . . . , $A_0(\lambda_L)$ are obtained. When measuring the absorbances of the final solutions of number 1 , 2 , . . . , S, at the same set of wavelengths and using the same optical path "d", the values $A_i(\lambda_1)$, $A_i(\lambda_2)$, . . . , $A_i(\lambda_L)$, i = 1 , 2 , . . . , S are obtained. The measured absorbances and the concentration modifications, generated by additions, can be arranged in matrix form. If $\varepsilon_j(\lambda)$ denotes the molar absorptivity of the component of order "j" at the wavelength "λ", the absorbances satisfy relations (42) and (43).

$$\frac{1}{d}\cdot\begin{bmatrix} A(\lambda_1) \\ \vdots \\ A(\lambda_\Lambda) \end{bmatrix} = \begin{bmatrix} \varepsilon_1(\lambda_1) & \cdots & \varepsilon_M(\lambda_1) \\ \vdots & & \vdots \\ \varepsilon_1(\lambda_\Lambda) & \cdots & \varepsilon_M(\lambda_\Lambda) \end{bmatrix}\cdot\begin{bmatrix} c_1 \\ \vdots \\ c_M \end{bmatrix} \tag{42}$$

$$\begin{aligned}
&\frac{1}{d}\cdot\begin{bmatrix} A^1(\lambda_1) & \cdots & A^S(\lambda_1) \\ \vdots & & \vdots \\ A^1(\lambda_\Lambda) & \cdots & A^S(\lambda_\Lambda) \end{bmatrix} = \begin{bmatrix} \varepsilon_1(\lambda_1) & \cdots & \varepsilon_M(\lambda_1) \\ \vdots & & \vdots \\ \varepsilon_1(\lambda_\Lambda) & \cdots & \varepsilon_M(\lambda_\Lambda) \end{bmatrix}\cdot\begin{bmatrix} c_1+\Delta c_1^1 & \cdots & c_1+\Delta c_1^S \\ \vdots & & \vdots \\ c_M+\Delta c_M^1 & \cdots & c_M+\Delta c_M^S \end{bmatrix} = \\
&= \begin{bmatrix} \varepsilon_1(\lambda_1) & \cdots & \varepsilon_M(\lambda_1) \\ \vdots & & \vdots \\ \varepsilon_1(\lambda_\Lambda) & \cdots & \varepsilon_M(\lambda_\Lambda) \end{bmatrix}\cdot\left(\begin{bmatrix} c_1 & \cdots & c_1 \\ \vdots & & \vdots \\ c_M & \cdots & c_M \end{bmatrix} + \begin{bmatrix} \Delta c_1^1 & \cdots & \Delta c_1^S \\ \vdots & & \vdots \\ \Delta c_M^1 & \cdots & \Delta c_M^S \end{bmatrix}\right) = \\
&= \frac{1}{d}\cdot\begin{bmatrix} A(\lambda_1) & \cdots & A(\lambda_1) \\ \vdots & & \vdots \\ A(\lambda_\Lambda) & \cdots & A(\lambda_\Lambda) \end{bmatrix} + \begin{bmatrix} \varepsilon_1(\lambda_1) & \cdots & \varepsilon_M(\lambda_1) \\ \vdots & & \vdots \\ \varepsilon_1(\lambda_\Lambda) & \cdots & \varepsilon_M(\lambda_\Lambda) \end{bmatrix}\cdot\begin{bmatrix} \Delta c_1^1 & \cdots & \Delta c_1^S \\ \vdots & & \vdots \\ \Delta c_M^1 & \cdots & \Delta c_M^S \end{bmatrix}
\end{aligned} \tag{43}$$

If the column matrix on the left member of the quation (42) is denoted by **A**[Λ,1], the matrix of molar absorptivities on the right member of the equation (42) by **E**[Λ,M] and the column matrix of the concentrations on the right member of the same equation by **C**[M,1], the equation (42) takes the form (44).

$$(1/d)\cdot\mathbf{A}[\Lambda,1] = \mathbf{E}[\Lambda,M]\cdot\mathbf{C}[M,1] \quad (44)$$

If equation (42) is subtracted, member by member, from equation (43) the result is equation (45).

$$\frac{1}{d}\cdot\left(\begin{bmatrix} A^1(\lambda_1) & \cdots & A^S(\lambda_1) \\ \vdots & & \vdots \\ A^1(\lambda_\Lambda) & \cdots & A^S(\lambda_\Lambda) \end{bmatrix} - \begin{bmatrix} A(\lambda_1) & \cdots & A(\lambda_1) \\ \vdots & & \vdots \\ A(\lambda_\Lambda) & \cdots & A(\lambda_\Lambda) \end{bmatrix}\right) = \\ = \begin{bmatrix} \varepsilon_1(\lambda_1) & \cdots & \varepsilon_M(\lambda_1) \\ \vdots & & \vdots \\ \varepsilon_1(\lambda_\Lambda) & \cdots & \varepsilon_M(\lambda_\Lambda) \end{bmatrix}\cdot\begin{bmatrix} \Delta c_1^1 & \cdots & \Delta c_1^S \\ \vdots & & \vdots \\ \Delta c_M^1 & \cdots & \Delta c_M^S \end{bmatrix} \quad (45)$$

$$\frac{1}{d}\cdot\begin{bmatrix} A^1(\lambda_1)-A(\lambda_1) & \cdots & A^S(\lambda_1)-A(\lambda_1) \\ \vdots & & \vdots \\ A^1(\lambda_\Lambda)-A(\lambda_\Lambda) & \cdots & A^S(\lambda_\Lambda)-A(\lambda_\Lambda) \end{bmatrix} = \\ = \begin{bmatrix} \varepsilon_1(\lambda_1) & \cdots & \varepsilon_M(\lambda_1) \\ \vdots & & \vdots \\ \varepsilon_1(\lambda_\Lambda) & \cdots & \varepsilon_M(\lambda_\Lambda) \end{bmatrix}\cdot\begin{bmatrix} \Delta c_1^1 & \cdots & \Delta c_1^S \\ \vdots & & \vdots \\ \Delta c_M^1 & \cdots & \Delta c_M^S \end{bmatrix}$$

Denoting by **ΔA**[Λ,S] the matrix of differences of absorbances in equation (45) and the matrix of concentration differences in (45), by **ΔC**[M,S], the resulting relation has the form (46).

$$(1/d)\cdot\Delta\mathbf{A}[\Lambda,S] = \mathbf{E}[\Lambda,M]\cdot\Delta\mathbf{C}[M,S] \quad (46)$$

The matrix **E**[Λ,S] is expressed from equation (46), and in its possession the equation (44) may be solved in relation to the column matrix **C**[M,1]. The necessary (but not sufficient) condition for solvency the equations in relation to matrix **C**[M,1] is that Λ should be higher than (or equal) to M or S should be higher than (or equal to) M.

$$\Lambda \geq M \quad sau \quad S \geq M \quad (47)$$

In order to express the matrix **E**[Λ,M], both sides of the relation (46) will be multiplied on the right by the transpose of matrix **ΔC**[M,S].

$$(1/d)\cdot\Delta\mathbf{A}[\Lambda,S]\cdot\Delta\mathbf{C}^{\mathrm{T}}[S,M] = \mathbf{E}[\Lambda,M]\cdot\Delta\mathbf{C}[M,S]\cdot\Delta\mathbf{C}^{\mathrm{T}}[S,M] \quad (48)$$

Both sides of (48) are then multiplied by the inverse of matrix **ΔC**[M,S]·**ΔC** ^{T}[M,S]. Relation (49) is obtained, representing the explicit form of matrix **E**[Λ,M].

$$(1/d)\cdot\mathbf{\Delta A}[\Lambda,S]\cdot\mathbf{\Delta C}^{T}[S,M]\cdot(\mathbf{\Delta C}[M,S]\cdot\mathbf{\Delta C}^{T}[S,M])^{-1} = \mathbf{E}[\Lambda,M] \quad (49)$$

The concentration matrix **C**[M,S] is expressed from relation (44). To this purpose, equation (44) is multiplied on the left by the transpose of matrix **E**[Λ,M].

$$(1/d)\cdot\mathbf{E}^{T}[M,\Lambda]\cdot\mathbf{A}[\Lambda,1] = \mathbf{E}^{T}[M,\Lambda]\cdot\mathbf{E}[\Lambda,M]\cdot\mathbf{C}[M,1] \quad (50)$$

When the above relation is multipled on the left by $(\mathbf{E}^{T}[M,\Lambda]\cdot\mathbf{E}[\Lambda,M])^{-1}$, the explicit form of the concentration matrix results (51).

$$(1/d)\cdot(\mathbf{E}^{T}[M,\Lambda]\cdot\mathbf{E}[\Lambda,M])^{-1}\cdot\mathbf{E}^{T}[M,\Lambda]\cdot\mathbf{A}[\Lambda,1] = \mathbf{C}[M,1] \quad (51)$$

The particular case of standard addition method applied to a system with two components to be determined, is illustrated graphically in Figure 2. In this case, the procedure is reduced to determining the plane π passing through a number of figurative points and to reading the intersection points of this plane with the negative semi-axes of the concentrations.

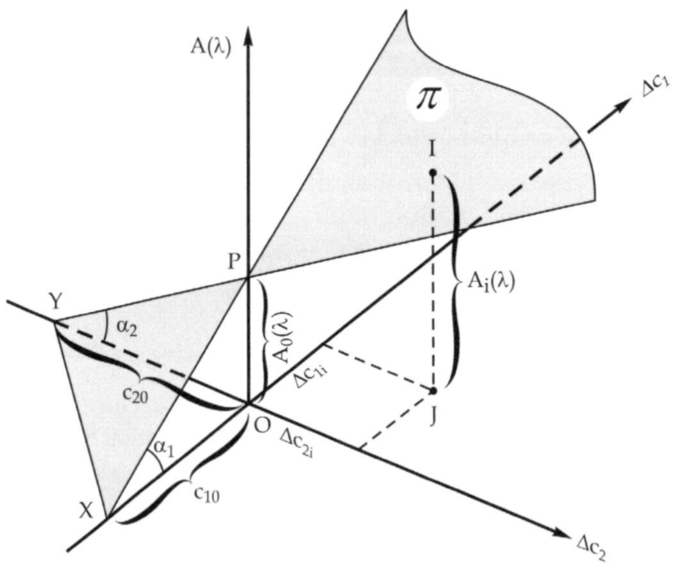

Fig. 2. Graphic representation of absorbances $A_i(\lambda)$ in relation to the modifications of concentrations Δc_{1i} and Δc_{2i} (i = 1 , 2 , . . . , n)

At the graphic representation of absorbances $A_i(\lambda)$ vs. the increase of concentrations Δc_{1i} and Δc_{2i} (i = 1 , 2 , . . . , n), the figurative points are situated theoretically on a plane (denoted by π in Figure 4-20). The axis of absorbances is intersected by plane π in point P, corresponding to the absorbance $A_0(\lambda)$, measured in the case of the solution with i = 0. If at the selected wavelength (λ) the absorbance of the ingredients can be left out, the points X and Y, situated at the intersection of plane π with the negative parts of axes Δc_1 and Δc_2, have the coordinates $-c_{10}$ respectively $-c_{20}$ (in other words, the lengths of the segments OX and OY are proportional to the concentrations c_{10} and c_{20}). From the values c_{10} and c_{20}, and knowing the volumes V_a, v and V_b, one may calculate the concentrations c_1 and c_2 of the components of interest in the first solution, and finally their content in the primary sample.

4.1 Example

In order to illustrate the application of the standard addition method and of the subsequent data processing procedure, let consider the mixture of salicylic acid, caffeine and acetaminophen, discussed in a previous example. The aim is to determine the concentrations of the three chemical components. Table 2 includes the modifications of the component concentrations (5 modifications are performed) and the absorbances both for the original solution (where concentrations have not been modified) and for the five solutions in which the three chemical components have been modified. All absorbance values are read at the same set of 18 wavelengths ($\Lambda = 18$).

The elements of matrix **E** are calculated with relation (49) and are expressed in the tolerated unit of measure l/(mol·cm), employed in spectrophotometric practice, and the elements of matrix **C**, calculated with relation (51) are expressed in μmol/l.

Table 2

		Modified solution 1	Modified solution 2	Modified solution 3	Modified solution 4	Modified solution 5	Original solution
Absorbance values at 18 wavelengths (d = 1 cm)		0.555	0.553	0.579	0.660	0.677	0.453
		0.519	0.525	0.552	0.627	0.637	0.424
		0.454	0.469	0.496	0.564	0.560	0.370
		0.382	0.405	0.430	0.498	0.471	0.311
		0.354	0.374	0.397	0.467	0.431	0.287
		0.368	0.378	0.398	0.476	0.439	0.297
		0.394	0.390	0.408	0.495	0.460	0.317
		0.427	0.410	0.426	0.523	0.490	0.343
		0.466	0.440	0.455	0.560	0.531	0.374
		0.443	0.416	0.430	0.526	0.506	0.356
		0.350	0.329	0.340	0.409	0.405	0.283
		0.207	0.195	0.201	0.229	0.247	0.168
		0.119	0.111	0.115	0.120	0.149	0.098
		0.080	0.074	0.076	0.076	0.101	0.066
		0.045	0.041	0.043	0.042	0.057	0.037
		0.023	0.022	0.022	0.023	0.029	0.019
		0.014	0.014	0.014	0.016	0.018	0.012
		0.012	0.011	0.012	0.013	0.014	0.009
Δc_i μmol/l)	Salicylic acid	3.40	1.50	2.20	0.60	10.00	
	Caffeine	2.60	1.50	1.80	5.00	4.00	
	Acetaminophen	1.20	2.20	2.80	4.00	3.00	

$$\mathbf{E} = \begin{pmatrix} 6721.72 & 17534.60 & 28741.06 \\ 5978.40 & 14156.50 & 32098.95 \\ 4390.01 & 10405.40 & 34854.40 \\ 1832.22 & 8685.62 & 35571.34 \\ 675.67 & 10070.25 & 32288.01 \\ 426.77 & 14575.14 & 26461.11 \\ 463.41 & 20233.58 & 19124.73 \\ 575.88 & 25871.08 & 12545.68 \\ 1045.49 & 29980.13 & 8855.66 \\ 1617.15 & 28153.68 & 7048.81 \\ 2330.58 & 20505.79 & 5443.55 \\ 3248.96 & 8960.11 & 3602.66 \\ 3540.99 & 2713.00 & 1580.26 \\ 2852.64 & 1204.43 & 613.63 \\ 1694.67 & 550.32 & 343.30 \\ 719.08 & 434.12 & 355.28 \\ 292.02 & 443.65 & 350.29 \\ 209.16 & 417.53 & 502.18 \end{pmatrix} \quad ; \quad \mathbf{C} = \begin{pmatrix} 17.819 \\ 10.290 \\ 5.307 \end{pmatrix} \mu mol / l$$

5. Conclusions

The application of matrix algebra to the quantitative spectrophotometry provides a unified formalism for treatment the mathematical issues. Unlike the usual mathematical approaches, the matrix description of the phenomena behind the analytical spectrophotometry promise new dimensions for the automatic processing of results.

6. References

Bosch-Reigh, F., Campins-Falco, P., Sevillano-Cabeza, A., Herraes-Hernandez, R., & Molins-Legua C. (1991). Development of the H-Point Standard Addition Method for Ultraviolet-Visible Spectroscopic Kinetic Analysis of Two-component Systems, *Analytical Chemistry*, Vol.63, No.21, pp. 2424-2429

Burnius, E. (1959). Assay of Vitamin A Oils, *Journal of the American Oil Chemistry's Society*, Vol.35, No.4, pp. 13-14

Ewing, D.T., Sharpe L.H., & Bird O.D. (1953). Determination of Vitamin A in Presence of Tocopherols, *Analytical Chemistry*, Vol.25, No.4, pp. 599-604, ISSN 0003-2700

Fox, S.H. & Mueller, A. (1950). The Influence of Tocopherols on U.S.P. XIV Vitamin Assay, *Journal of the American Pharmaceutical Association*, Vol.39, No.11, pp.621-623

Garrido, M., Lázaro, I., Larrechi, M.S., & Rius F.X. (2004). Multivariate Resolution of Rank-deficient Near-infrared Spectroscopy Data fron the Reaction of Curing Epoxi Resins Using the Rank Augmentation Strategy and Multivariate Curve Resolution Alternating Least Squares Approach, *Analytica Chimica Acta*, Vol.515. No.1, pp.65-73

Garrido, M., Rius, F.X., & Larrechi, M.S. (2008). Multivariate Curve Resolution-alternating Least Squares (MCR-ALS) Applied to Spectroscopic Data from Monitoring Chemical Reaction Processes, *Anal. Bioanal. Chem.*, Vol.390, No.8, pp. 2059-2066

Li, N., Li, X.Y., Zou, Z.X., Lin, L.R., & Li, Y.Q. (2011). A Novel Baseline-correction Method for Standard Addition Based Derivative Spectra and its Applications to Quantitative Analysis of benzo(a)pyrene in Vegetable Oil Samples, *Analyst*, Vol.136, pp. 2802-2810

Lozano, V.A., Ibañez, G.A., & Olivieri, A.C. (2009). A Novel Second-order Standard Addition Analytical Method Based on Data Processing with Multidimensional Partial Least-squares and Residual Bilinearization, *Analytica Chimica Acta*, Vol.651, No.2, pp. 165-172

Lykkesfeld, J. (2001). Determination of Malonaldehyde as Dithiobarbituric Acid Adduct in Biological Samples by HPLC with Fluorescence Detection: Comparison with Ultraviolet-Visible Spectrophotometry, *Clinical Chemistry*, Vol.47, pp.1725-1727

Melnick, D., Luckmann, F.H., & Vahlteich, H.W. (1952). Estimation of Vitamin A in Margarine. I. Collaborative Study of Assay Methods for Estimating the Potencu of the Vitamin A Concentrates, *Journal of the American Oil Chemistry's Society*, Vol.29, No.3, pp. 104-108

Morton, R.A., & Stubbs A.L., (1946). *Photoelectric Spectrophotometry Applied to the Analysis of Mixtures and Vitamin A Oils*, Vol.71, pp. 348-350

Morton, R.A., & Stubbs, A.L. (1947). A Re-examination of Halibut-liver Oil. Relation Between Biological Potency and Ultraviolet Absorption Due to Vitamin A, *Biochem. J.*, Vol.41, pp. 525-529

Morton, R.A., & Stubbs, A.L. (1948). Studies in Vitamin A. 4. Spectrophotometric Determination of Vitamin A in Liver Oils. Correction for Irrelevant Absorption, *Biochem. J.*, Vol.42, No.2, pp. 195-203

Oka, K., Oshima, K., Inamoto, N., & Pishva D. (1991). Chemometrics and Spectroscopy, *Analytical Sciences*, Vol.7, pp.757-760, ISSN 0910-6340

Owen, A.J. (1995). Quantitative UV-Visible Analysis in the Presence of Scattering" Application Notes - Agilent Technologies, *Pharmaceutical Analysis*, (publication number: 5963-3937E)

Page, R.C., & Berkovitz, Z. (1943). The Absorption of Vitamin A in Chronic Ulcerative Colitis, *American Journal of Digestive Diseases*, Vol.10, No.5, pp. 174-177

Ruckenbusch, C., De Juan, A., Duponchel, L., & Huvenne, J.P. (2006). Matrix Augmentation for Breaking Rank-deficiency: A Case Study, *Chemometrics and Intelligent Laboratory Systems*, Vol.80, No.2, pp. 209-214

Sánchez, E.; & Kowalski, B.R. (1986). Generalized Rank Annihilation Factor Analysis, *Analytical Chemistry*, Vol.58, pp. 496-499, ISSN 0003-2700

Szabadai Z. (2005). *Bazele fizico-chimice ale metodelor de control analitic al medicamentelor*, Editura Mirton, Vol. II., pp. 4.38-4.96, ISBN: 973-661-677-0, Timişoara, Romania

Valderrama, P., & Poppi R.J., (2009). Second Order Standard Addition Method and Fluorescence Spectroscopy in the Quantification of Ibuprofen Enantiomers in Biological Fluids, *Chemometrics and Intelligent Laboratory Systems*, Vol.106, No.2, pp. 160-165.

16

Optical and Resonant Non-Linear Optical Properties of J-Aggregates of Pseudoisocyanine Derivatives in Thin Solid Films

Vladimir V. Shelkovnikov[1] and Alexander I. Plekhanov[2]
[1]*Novosibirsk Institute of Organic Chemistry SB RAS, Novosibirsk,*
[2]*Institute of Automation and Electrometry SB RAS, Novosibirsk,*
Russia

1. Introduction

The properties of colligative states of spontaneously aggregated polymethine dyes differ substantially from those of monomeric dye. Excellent examples of such self-organized molecular ensembles are J-aggregates of cyanine dyes. The J-aggregated state is now being considered for a number of non-cyanine dyes, the cyanine dye is still the most known and effective dye for J-aggregate formation (Wurthner, 2011). J-aggregates of cyanine dyes, first discovered by Jelley and Scheibe in 1936 (Jelley, 1936; Scheibe, 1936), have been studied for many years (Kobayashi, 1996). J-aggregates of cyanine dyes attract the attention of the researchers due to their interesting optical properties. J-aggregates are characterized by a strong absorption peak (J-peak) with narrow line widths which are bathochromically shifted relative to the absorption band of the monomeric dye. Their role as photographic sensitizers can hardly be overestimated (Tani, 1996; Trosken et al., 1995; Shapiro, 1994). Aggregates of dye molecules may be used to mimic light harvesting arrays and to prepare artificial photosynthetic systems (McDermott et al., 1995; Blankenship, 1995). Another development is the efficient electroluminescence revealed in single-layer light-emitting diodes based on electron-hole conducting polymers containing nano-crystalline phases of J-aggregates of cyanine dyes (Mal'tsev et al., 1999).

The promise in the property of J-aggregates lies in their high non-linear cubic optical susceptibility, $\chi^{(3)} \sim 10^{-7}$ esu, with a fast response time at the J-peak resonance in solutions and polymer films (Wang, 1991; Bogdanov et al., 1991).

The pseudoisocyanine dye (PIC) is the known dye which forms the J-aggregates in solutions. Of particular interest is the formation and non-linear optical properties of J-aggregates in thin solid films. Films of J-aggregates of organic dyes are promising nanomaterials for non-linear optical switches because they have the unique properties of high non-linear bleaching and non-linear refraction (Markov et al., 2000). As shown in Glaeske et al. (2001), films of J-aggregates with bistable behaviour may be the basis for two-dimensional optical switches, controllable by light. The non-linear optical properties of organic dye J-aggregates have been intensively studied for application in future optical telecommunication and signal processing systems with ultrahigh bit rates (Tbit/s) (Furuki et al., 2000). The observed giant resonant third-order susceptibility in PIC thin solid films $\chi^{[3]}$~10^{-5}-10^{-4} esu (5 orders of magnitude greater than in polyconjugated polymers) and accessible production of optical quality films over a large area gives the possibility of J-aggregates application in telecommunication for terahertz demultiplexing of optical signals.

The methods for obtaining the J-aggregates in solutions do not give stable aggregates, which hampers their application as non-linear optical materials. Besides, a thin film geometry is preferable for applications. Therefore, the preparation of large area thin films of J-aggregates with thermal and photochemical stability, and high optical quality is vital for practical application and a matter of much current interest. The pseudoisocyanine at proper conditions efficiently forms J-aggregates in solid thin films (Shelkovnikov et al., 2002). The aim of this paper is to clarify the influence on the spectral linear and non-linear properties of PIC J-aggregates in thin solid films using a number of factors: structure of pseudoisocyanine dye derivatives, local field factor and character of J-peak broadening.

2. Experiments and discussions

2.1 Experimental part - Materials and methods

We used two methods to stimulate the J-aggregates' formation in the thin solid films. The use of a solution of PIC dyes with long alkyl substituents (Shelkovnikov et al., 2002) and the use of a PIC2-2 solution with the addition of cluster hydroborate anions (Shelkovnikov et al., 2004). Synthesis of the number of the derivatives of PIC with symmetrical and non-symmetrical long alkyl chain substituters for the physical-chemical experiments was carried out and the obtained dyes were isolated, chromatographically refined and characterized using the ^{1}H nuclear magnetic resonance method (Orlova et al., 2002; Orlova et al., 1995). Synthesis and characterization of the PIC2-2 derivative with cluster anion - closo-hexahydrodecaborate ($B_{10}H_{10}^{2-}$) by the infra-red spectroscopy and X-ray spectroscopy methods was carried out [Plekhanov et al., 1998a, 1998c; Cerasimova et al., 2000).

The thin solid films of PIC dye were prepared using spin-coating of the dye solution on the clean glass plates 2,5x2,5 cm^2 with a rate of rotation of 2000-3000 rpm on the custom made spin-coating equipment.

The linear optical spectra of the thin solid samples, depending on the experiments, were measured on the spectrophotometer Hewlett Packard 8453, fast fibre optics spectrophotometer Avantes AVS-SD2000 and fast fibre optics spectrophotometer Calibri VMK Optoelectronics, Novosibirsk (http://www.vmk.ru/). Fast spectrophotometers allow the measurement of 40 spectra per second. Sample heating for determination of the J-aggregates' thermal stability was done in a thermostatic optical chamber with a constant heating rate of 1.5 deg/min.

The thickness, dispersion of refractive index, and absorption of the obtained films were measured on a spectral ellipsometer Ellips developed at the A.V.Rzhanov Institute of Semiconductor Physics SB RAS (http://www.isp.nsc.ru/). The optical parameters of the film according to ellipsometric parameters delta (Δ) and psi (Ψ) were found by approximating the single-phase model of the Si-substrate/absorbing film.

The steady-state luminescence spectra were recorded on the Cary-Eclipse (Varian) and Hitachi 850 spectrofluorimeters. The kinetics of the luminescence decay was measured on the set-up of the Federal Institute for Materials Research and Testing (Berlin, Germany) with the assistance of researcher Ch. Spitz. The excitation was carried out in the cryostat by the pulse irradiation of the dye laser (R6G dye) with pulse duration 80ps synchronously pumped by pulse mode-locking Ar+ laser. The luminescence of the PIC J-aggregates was separated by monochromator and measured in the photon counting regime by the photo-multiplier.

The measurement of non-linear cubic susceptibilities of J-aggregates' PIC in thin solid films was carried out using the Z-scan method on the set-up shown on fig. 1 based on the dye R6G laser pumped using 5 ns pulsed Nd:YAG laser.

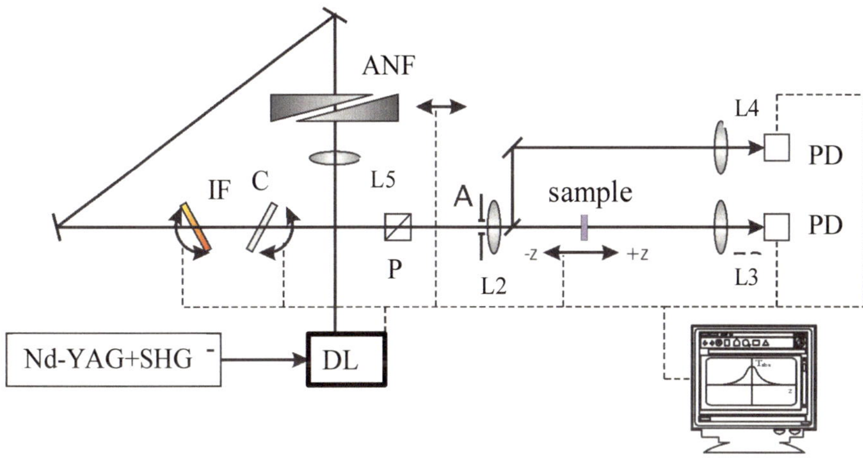

Fig. 1. Z-scan set-up. Nd-YAG+SHG - pulse laser, DL - Dye laser, L1-L4 -lenses, A - Aperture diaphragm, P - Glan prism, IF - interference filter, C- compensator of the beam shift, ANF- adjusted neutral filter, PD-photodiode.

The weak luminescence of the laser dye was cut off by the interference filter IF590 (Carl Zeiss Jena). The light signals were measured by photodiodes and the measured T(z) curves had 200 experimental points with the average value taken from the tens pulses. The calibration was carried out with etalon CS_2 substance (γ=(3.6$\pm$0.3)10^{-14} cm^2/W). The values of the real $\boldsymbol{Re}^{(3)}$ and imaginary $\boldsymbol{Im}^{(3)}$ parts of the cubic susceptibility in the esu units from the measured values' non-linear refraction coefficient γ and absorption coefficient $\boldsymbol{\beta}$ in SI units were calculated by equations:

$$\mathrm{Re}\,\chi^{(3)} = \frac{cn_0^{\;2}}{160\pi^2}\cdot(\gamma - \left(\frac{\lambda}{4\pi}\right)^2 \frac{\alpha_0\beta}{n_0}) \tag{1}$$

$$\mathrm{Im}\,\chi^{(3)} = \frac{\lambda cn_0^{\;2}}{640\pi^3}\cdot(\beta + \frac{\alpha_0\gamma}{n_0}) \tag{2}$$

where, c - light velocity in vacuum, $\boldsymbol{\lambda}$ - irradiation wavelength.

2.2 The formation of J-aggregates' PIC with the long alkyl chain substituters in thin solid films

The derivatives of PIC with symmetrical and non-symmetrical long alkyl chain substituters were synthesized to stimulate the J-aggregates' formation in thin solid films.

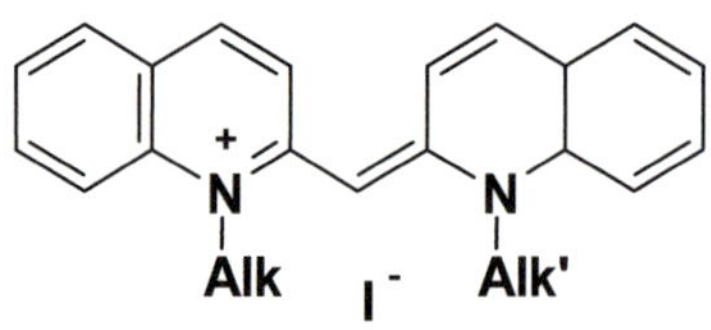

Alk=Alk'=C_2H_5 (PIC2-2),	Alk=C_2H_5, Alk'=C_6H_{13}(PIC2-6),
Alk=Alk'=$C_{10}H_{21}$(PIC10-10),	Alk=C_2H_5, Alk'=$C_{10}H_{21}$(PIC2-10),
Alk=Alk'=$C_{15}H_{31}$(PIC15-15),	Alk=C_2H_5, Alk'=$C_{15}H_{31}$(PIC2-15),
Alk=Alk'=$C_{18}H_{37}$(PIC18-18),	Alk=C_2H_5, Alk'=$C_{18}H_{37}$(PIC2-18).

There is no difference in the absorption spectra of the obtained dye in monomer form in organic solution. Symmetrical dyes do not give J-aggregated form in solid films and have the tendency to form H-aggregates. All non-symmetrical dyes with long alkyl chains PIC2-6, PIC2-10, PIC2-15, PIC2-18 give the J-aggregated form in thin solid films, see fig. 2, with maximum at λ=575-583 nm and FWHM 360-400 cm^{-1}.

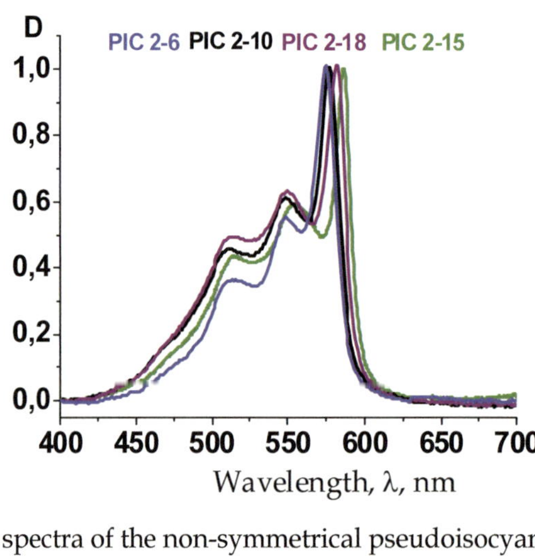

Fig. 2. The absorption spectra of the non-symmetrical pseudoisocyanines in thin solid films.

There are three electron-phonon transitions (526, 492 и 460 nm) shifted relative to each-other to the quant of the double C=C bound oscillation (1400 cm^{-1}) for the monomer form of dye in the solution. The same three transitions shifted to the long wave side are observed in the structural absorption band with the maximum at 556 nm in the spectra of the non-symmetrical pseudoisocyanines in the thin solid films. This band did not change at the thermal decay of the J-aggregate and belongs to the monomer form of the dye. Thus, the spectrum of dye in solid films is the superposition of monomer and J-aggregated forms.

The absorption in the maximum of the J-aggregates' peak exceeds the absorption in the maximum of the monomer form by as much as 1.5-1.8 times. The relation of the optical density in the J-aggregates' peak and in the monomer maximum (J/M) was used to characterize the degree of the conversion dye to aggregate in thin solid films. The length of the alkyl chain has a weak dependence on the degree of the conversion of the monomer to the J-aggregate in the thin solid films. But the alkyl chain length has a strong influence on the rate and degree of the spontaneous recovering of the J-aggregate after thermal destruction. For example, the J-aggregate of PIC2-2 did not restore at all. The recovering begins after 4 day storing in the box at humidity 80% for the J-aggregate of PIC2-6. The J-aggregate of PIC2-10 restored to 80% after a day of storing and PIC2-15 after 12 hours of storing. The J-aggregate of PIC2-18 is completely restored during 4 hours at room temperature. The increase of the alkyl chain length leads to an increasing degree of restoration and shortening of the time of restoring by order of magnitude. To stabilize the J-aggregates on the substrate, a dye with long alkyl substituters is useful. Such substituters lead to an efficient J-aggregation at room temperature, but the aggregates are unstable upon an increase in temperature above 60°C.

2.3 The spectrophotometry of the thermal decay of PIC J-aggregates with long alkyl substituters

The J-aggregates of the PIC with long alkyl substituters are thermally destroyed with transition to the monomer form. The thermal conversion of the J-aggregates of pseudoisocyanine derivatives is studied by measurement of optical absorption spectra at continuous increase of temperature. The spectral change of the J-aggregates of PIC2-18 in the thin film with temperature increased is shown in fig. 3.

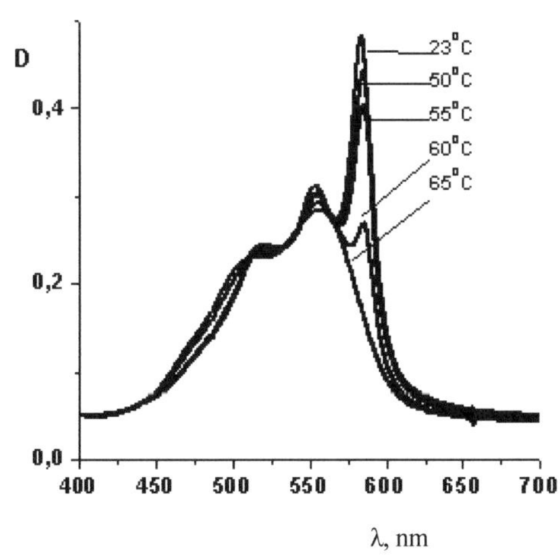

Fig. 3. The spectral change of the J-aggregated film PIC2-18 at sample heating

The J-peak disappears at temperatures above 60°C and the finished spectrum looks like the broadening spectrum of the monomer form of the dye. The J-aggregates' thermal decay curves are plotted (see fig. 4) and from this it is shown that dyes line to the next row of thermal stability determined on the point of inflection of the thermal decay curve: PIC2-2 (73°C) > PIC2-6 (67°C) > PIC2-18 (55°C) > PIC2-10 (47°C) > PIC2-15 (37°C).

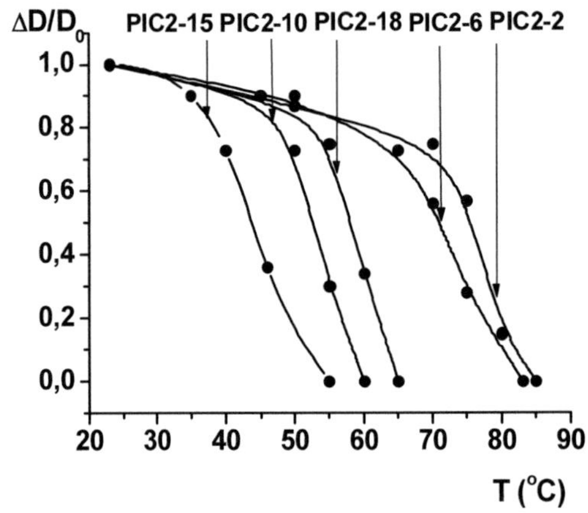

Fig. 4. The thermal decay curves of the J-aggregates of the non-symmetrical dyes in the thin solid films

The effective energies of activation of thermal decay of J-aggregates in thin films were calculated from the non-isothermic curves of the thermal decay. The degree of J-aggregate thermal decay (**α_J**) is $\frac{d\alpha_J}{dt} = k\cdot(1-\alpha_J)$

Where, $k = Z\cdot\exp\left(-\frac{E_a}{RT}\right)$;**Z** - preexponential factor, **E_a** – activation energy, **T** – temperature.

The differential equation for the degree of thermal decay at constant rate of heating is (Wendlandt, 1974)

$$\frac{d\alpha_J}{dT} = \frac{Z}{\beta}\cdot(1-\alpha_J)\cdot\exp\left(-\frac{E_a}{RT}\right) \tag{3}$$

Where, **β** – the rate of the sample heating **(dT/dt)**. After integration we have

$$\alpha_J = 1-\exp\left(-\left(\frac{Z}{\beta}\cdot\frac{RT^2}{E_a}\cdot\left(1-\frac{2RT}{E_a}\right)\cdot\exp\left(-\frac{E_a}{RT}\right)\right)\right) \tag{4}$$

The dependence of the ($\ln\left[-\ln(1-\alpha_J)\right]$) vs. $1/T$ gives the value **E_a** from this equation. The experimental and theoretical curves of **$\alpha_J(T)$** and double-logarithmic line are shown in fig. 5 for the case of PIC-2-15 J-aggregates' thermal decay.

The obtained values of activation energy for the thermal decay of the J-aggregates of PIC with different long alkyl chains are shown in table 1.

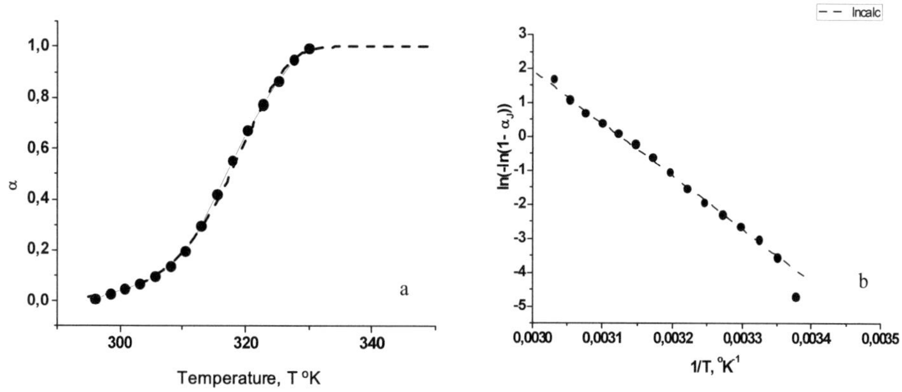

Fig. 5. a,b. Experimental (points) and the model (dash) curves of the degree of the J-aggregates' PIC2-15 thermal decay (a) and its linear approximation vs 1/T

Dye	E_a *kcal/mol*
PIC 2-2	43
PIC 2-6	30
PIC 2-10	39
PIC 2-15	30
PIC 2-18	41

Table 1. The values of the activation energy for the thermal decay of the J-aggregates in the thin solid films

We can conclude that the average value of the effective activation energy for the thermal decay of the J-aggregates in the thin solid film is 36 kcal/mol. Let us compare the value of the effective activation energy for the thermal decay of the J-aggregates with the electrostatic energy of the PIC dimer formation. The main contribution to the dimerization energy gives the Coulomb interaction of the two PIC cations and halogen anions (Krasnov, 1984).

$$E_{ion} = -\frac{q_1 \cdot q_2}{r} + A \cdot \exp\left(-\frac{r}{\rho}\right) \tag{5}$$

Where, q_1 and q_2 are charges of ions, r – the distance between ions, A – the coefficient that characterizes the energy of repulsive exchange interactions of the electron orbitals and ρ is the constant equal 0,34 Å. The A value is calculated from the minimum ($\frac{\partial E}{\partial r} = 0$) of the cation-anion interaction.

$$A = -\frac{q_1 \cdot q_2 \cdot \rho}{r^2_{\min}} \cdot \exp\left(\frac{r_{\min}}{\rho}\right) \tag{6}$$

The value of the energy of the dimer is the minimal at quadruple replacement of the cations and anions as shown in fig. 6. In this case the electrostatic energy of the dimer is

$$E_D = -\frac{4 \cdot q_{PIC} \cdot q_{an}}{r} + 4 \cdot A \cdot \exp\left(-\frac{r}{\rho}\right) + \frac{q^2_{PIC}}{r \cdot \sqrt{2}} + \frac{q^2_{an}}{r \cdot \sqrt{2}} + 2 \cdot A \cdot \exp\left(-\frac{r \cdot \sqrt{2}}{\rho}\right) \tag{7}$$

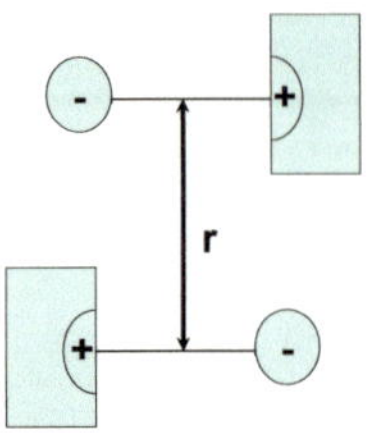

Fig. 6. Quadruple model of the PIC dimer

The effective positive charge of the dye (q_{PIC}~+0,7) and the value of the equilibrium distance cation-anion in the PIC molecule (3 Å) were determined from the quantum chemical calculation using the AM1 method. The map of the calculated electrostatic potential of the PIC iodide is shown in fig. 7.

Fig. 7. Electrostatic potential map of the PIC iodide molecule

The value of the dimerization energy was calculated as the difference between the dimer energy and the two molecules PIC in the monomer form.

$$E_D = -\frac{2 \cdot q_{PIC} \cdot q_{an}}{r} + 2 \cdot A \cdot \exp\left(-\frac{r}{\rho}\right) + \frac{q^2_{PIC}}{r \cdot \sqrt{2}} + \frac{q^2_{an}}{r \cdot \sqrt{2}} + 2 \cdot A \cdot \exp\left(-\frac{r \cdot \sqrt{2}}{\rho}\right) \tag{8}$$

The molecules' distance depending on the dimerization energy is shown in fig. 8.

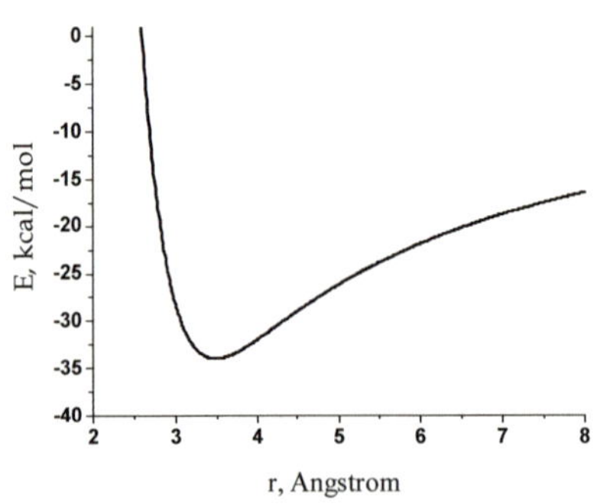

Fig. 8. Distance dependence of the PIC halogen dimerization energy

The minimum of the curve is -34 kcal/mol with the equilibrium distance between PIC molecules 3,5 Å. This is in good agreement with Van der Waals' distance in the dyes' dimers. One can see that the energy of the calculated value of energy of the pair electrostatic interaction of PIC (-34 kcal/mol) is close to the average value of the effective activation energy for the thermal decay of the J-aggregates' PIC (-36 kcal/mol).

The value of the enthalpy of the dyes dimerization in the solution is 6-12 kcal/mol (Ghasemi & Mandoumi, 2008; Coates, 1969; Nygren et al., 1996). It is less than the calculated energy of the dimerization PIC due to the compensatory contribution of the high energy of the dye solvatation. That means that in the solution or in the solvatating polymer the thermal decay of the J-aggregates will happens much more easily than in a non-solvating polymer or in solid films.

2.4 The spectrophotometry of the J-aggregate PIC2-18 iodide to monomer conversion in thin film with the addition of 1-octadecyl-2-methylquinolinium

The number of monomer units in the J-aggregate PIC is an important characteristic in understanding its optical and non-linear optical properties in thin solid films. There are different opinions in the literature about the number of molecular units making up the aggregate: from hundreds (Sundstrom et al., 1988; Tani et al., 1996), tens (Daltrozzo et al., 1974), four (Herz, 1974) or three (Struganova, 2000) molecules in the J-aggregate PIC. Here we describe the difference between the number of molecules which can be in the J-aggregate as in a physical object, for an example in the micelle, and the minimal number of molecules which is enough to give the J-peak in the optical spectrum.

We use another organic cation 1-octadecyl-2-methylquinolinium (MQ18), which has the same charge (+1) and long alkyl tail as PIC2-18, to divide the PIC2-18 molecules in the J-aggregate in a thin solid film. In this way it is possible to determine the number of PIC2-18 molecules in the J-aggregate.

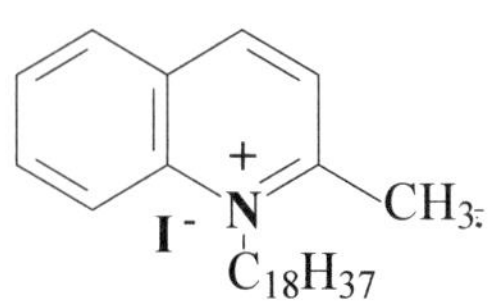

MQ18

The MQ18 is similar to one half of PIC2-18 and statistically replaces the PIC2-18 molecules at the moment of fast J-aggregate assembling during the short time of film formation during spin-coating. In this case the numbers of PIC2-18 molecules in the J-aggregate will depend on the relationship between the concentration of the MQ18 and dye in the film. The probability of PIC2-18 molecule substitution by MQ18 molecule (F_{MQ18}) in the film is equal to the part of the MQ18 molecules in the joint composition PIC18 and MQ18

$$F_{MQ18} = [C_{MQ18}]/([C_{MQ18}]+[C_{PIC18}]) \quad (9)$$

The absorption spectra of the films prepared by spin-coating on the basis of mixture PIC2-18:MQ18 (1:0.05,1:0.1, 1:0.2, 1:0.3, 1:0.5, 1:1) were measured and part of them is shown in fig. 9.

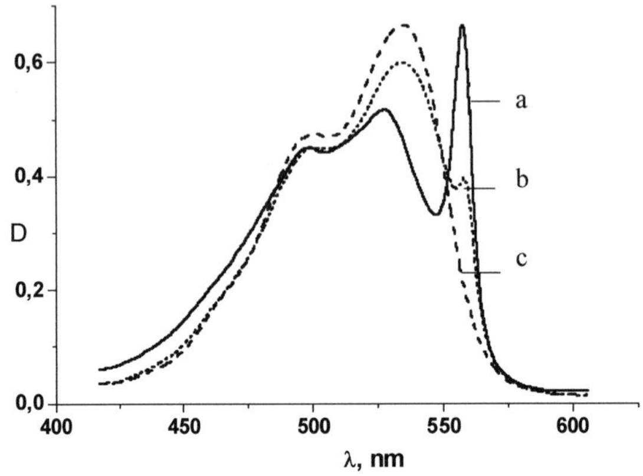

Fig. 9. The film spectra PIC2-18 at MQ18 addition in the mole relation 1:0 (*a*), 1:0.5 (*b*), 1:1(*c*)

One can see the transformation of the J-aggregate spectrum to the monomer dye spectrum with increasing the MQ18 concentration. At the molar relation PIC2-18:MQ18 1:1 we observe only the monomer form of dye. The part of the destroyed J-aggregates (**F_J**), which is determined as the change of the J-peak optical density (**$(D^J_0 - D^J_{MQ18})/D^J_0$**) is equal to the probability of the PIC2-18 molecule substitution multiplied by the number molecules (**N**) needed for J-peak appearance.

$$F_J=(D^J_0 - D^J_{MQ18})/\ D^J_0=N^*\ F_{MQ18} \tag{10}$$

The value **N** was calculated from the tangent of the (**F_J-F_{MQ18}**) dependence shown in fig. 10 is 2.6±0.2. This is between two and three. In accordance with the exciton description of supramolecular assemble (Malyshev, 1993) at the number of molecules in the J-aggregate, at more than two there should be noticeable a hypsochromic shift of J-peak as the aggregate decreases. The J-peak energy depending on the number of molecules in the J-aggregate is:

$$E_k = h\nu + 2V\cos\left(\pi k/\left(N+1\right)\right) \tag{11}$$

where $h\nu$ is the energy of the optical transition of the monomer dye, **V** - the dipole-dipole interaction of the neighbour molecules in the aggregate, **N** - the number of molecules in the J-aggregate. The spectral shift for the Lorenz contour of the J-peak (**Lr_{ex}**) was calculated for the number of PIC molecules in the J-aggregate from 2 to 5 (fig. 11).

$$Lr_{ex} = \frac{\Delta\nu\cdot 1/\sqrt{N}}{4\cdot\left(\frac{10^7}{\lambda}-\left(\nu_0-\cos\left(\frac{\pi}{N+1}\right)\cdot 2V\right)\right)^2+\left(\Delta\nu\cdot 1/\sqrt{N}\right)^2}\cdot N \tag{12}$$

Where, $\Delta\nu$ -the width of J-peak (300 cm^{-1}), ν_0-maximum of the J-peak absorption (18770 cm^{-1}), **V** - 660 cm^{-1}.

The decreasing number of molecules from three to two in the J-aggregate leads to the blue shift of the spectra to 9 nm. In our case the J-peak disappeared, but the blue spectral shift was not observed. It could be explained in the case of the J-aggregate dimer form presence. The dimer form of the dye was determined as the main form for the PIC J-aggregate.

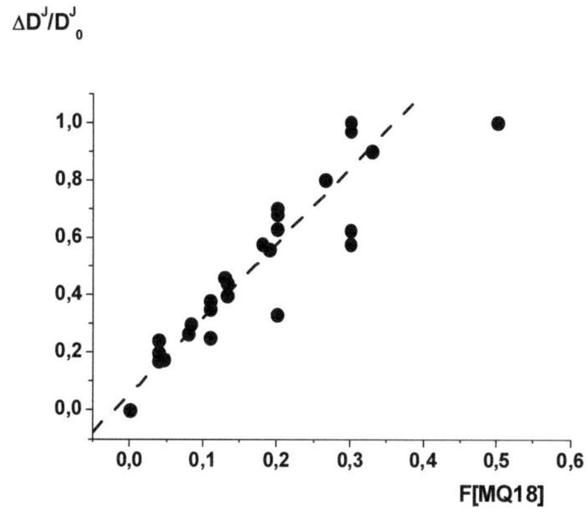

Fig. 10. The relative J-peak optical density change depending on the relative MQ18 content

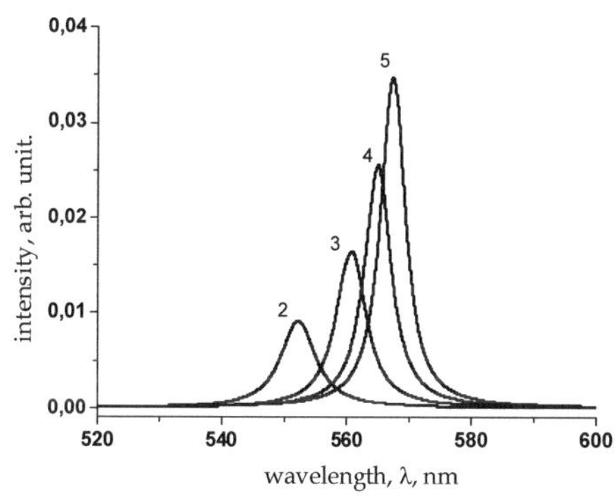

Fig. 11. The absorption spectra of the J-aggregate calculated using eq. 12 for the different number of the molecules in the J-aggregate (the curve number)

2.5 The optical constants of thin solid films of PIC derivatives

The frequency dispersion of the absorption coefficient k and the refractive index n of the films of PIC long chain derivatives in J-aggregated and monomer forms were determined in the wavelength region from 500 to 650 nm. The typical value of measured thin films thickness was 20-30nm. As an illustration, the resulting spectra for two films are shown in Fig. 12. The spectra for other films were similar. One shows anomalous dispersion for n within the region of dye absorption (500-570 nm), with the values n = 3.05 and k = 0.8 at the maximum of the J-peak dispersion curve; at the absorption maximum of the J-peak, n = 2.5 and k = 1.25 for film 21 nm thick.

The value of the index of absorption ($\alpha = 4\pi k / \lambda$) in thin solid films for PIC2-18 is: in the maximum of anomalous dispersion refraction $\alpha = 2{,}5 \cdot 10^5$ cm^{-1} for J-aggregate, $\alpha = 1{,}15 \cdot 10^5$ cm^{-1} in the maximum of absorption for monomer dye (n_{max}=2.1). The electron polarizability for J-aggregated (ρ_J) and monomer forms (ρ_m) of dye were calculated by extrapolation of the refractive index of dispersion curve to infinite wavelength n_∞ using the least squares method and the Zelmeer equation (Verezchagin, 1980).

$$n_\infty = \sqrt{n - \frac{C}{\left(\lambda^2 - \lambda_0^2\right)}} \tag{13}$$

Where, C is constant, λ_0 – wavelength of J-peak maximum.

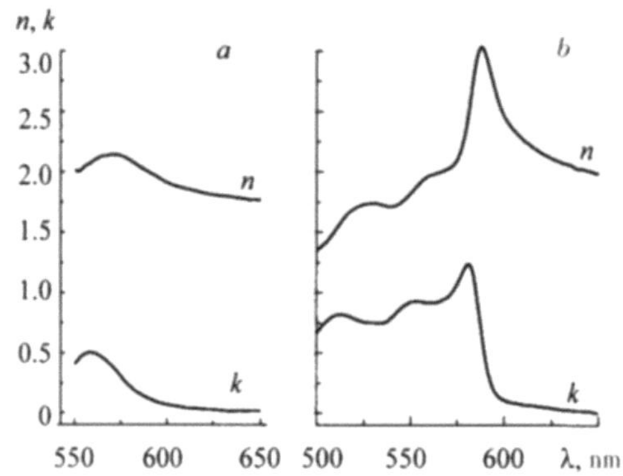

Fig. 12. Refractive indices (n) and absorption (k) of film of PIC 2-18 in the monomeric (a) and J-aggregated (b) form as functions of dispersion measured by spectral ellipsometry

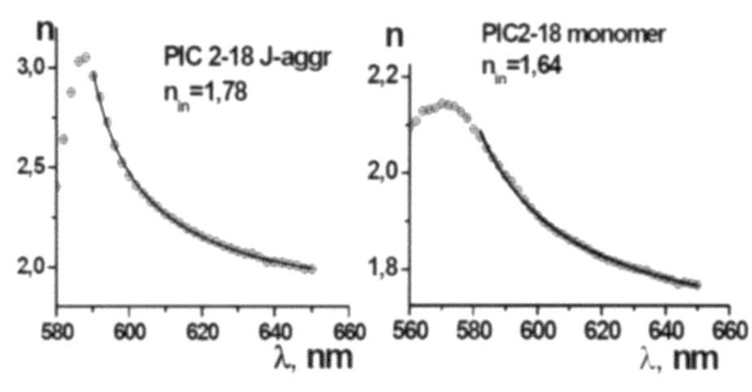

Fig. 13. Dispersion curves for PIC J-aggregate (a) and monomer (b) in the solid film (points) and their least square Zelmeer approximation (line)

The results of the dispersion curves' approximation of J-aggregated and monomer forms of PIC are shown on in fig 13. One can see a good consequence of experimental and fitted curves outside of resonance. The value refractive index for the PIC monomer in film is $n_{\infty m}$=1,64 and for J-aggregate is $n_{\infty J}$=1,78. The values of the electron polarizability (ρ) is obtained by equation

$$\rho = \frac{3}{4\pi N_A} \frac{n_\infty - 1}{n_\infty^2 + 2} \cdot \frac{M}{d} \tag{14}$$

Where M is molecular weight (consider dimer value for J-aggregate), d is density of molecules;

gives the values ρ_J=88 Å^3 for the J-aggregate and ρ_m=66 Å^3 for the monomer form of dye. The electron polarizability of the J-aggregate is 1.33 times more than the polarizability of the monomer. This is not too much to consider the electron delocalization in J-aggregate over the number of dye molecules, but rather it confirms some expansion of electron delocalization due to dye dimerization.

2.6 The luminescent properties of J-aggregate PIC with long alkyl chain

Resonance luminescence takes place for the J-aggregates' PIC in solutions. For example, the PIC forms the J-aggregates in water solution in the presence of phospholipide vesicles with the J-peak of absorption at 580 nm and the J-peak of the luminescence at 582 nm (Sato et al., 1989). The resonance luminescence corresponds to the absence of the molecular coordinate shift between the ground and the excited states' electron energy curves for the $\mathbf{S_{00}}$ and $\mathbf{S^*_{00}}$ oscillatory states. For the number of samples in solutions, the long wavelength luminescent wing with a maximum of 600-630 nm was observed for J-aggregates at a low temperature (Vacha et al., 1998; Katrich et al., 2000). The authors connect the reason for the long wavelength luminescence with the availability of the dimeric nature traps which accept the excitation energy of the J-aggregate.

The measured steady-state spectra of the luminescence of solid films of J-aggregated dyes PIC2-6, PIC2-10, PIC2-15 at room temperature are shown in fig. 14. The determined Stokes shift for the ν_{00}=1/2($\nu_{\max}^{abs} - \nu_{\max}^{fl}$) transition frequency is 90-100 cм$^{-1}$ (see table 2). The measured spectra of the monomer PIC luminescence in water:ethanol (1:1) at concentration 10^{-4} *M/l* is shown in fig. 15. The spectrum has the band with broadening counter reaching 700 nm with the maximum at 565 nm. The Stokes shift for the monomer luminescence is 780 cm^{-1}. In accordance with the Lippert approach, the luminescence Stokes shift of the molecule in the medium appears due to the reaction of the medium which is described by the medium function ($f(\varepsilon,n)$) on the dipole moment change at the molecule excitation. The luminescence Stokes shift depends on the polarization function of medium and the square of change of the dipole moment in ground ($\boldsymbol{M_g}$) and excited state ($\boldsymbol{M_e}$) divided by the radius of the solvated molecule ($\boldsymbol{R_a}$) in cube (Levshin & Salecky, 1989; Lakowics, 1983).

$$\Delta E_{st} = 2\frac{(M_e - M_g)^2}{R_a^{\ 3}} \cdot \left[\frac{(\varepsilon-1)}{(\varepsilon+2)} - \frac{(n^2-1)}{(n^2+2)}\right] \cdot \frac{(2n^2+1)^2}{(n^2+2)^2} \tag{15}$$

The value of the radius of molecule solvate interaction was calculated from

$$R_a = \sqrt[3]{\frac{M}{d} \cdot \frac{3}{4\pi N_A}} \tag{16}$$

Where, ***M, d*** is the corresponding molecular weight and density of the dye. The obtained value $\boldsymbol{R_a}$ for PIC is $\boldsymbol{R_a}$=4.5 Å. The calculation of $\boldsymbol{R_a}$ from the molecular volume of PIC cation (14.52x6.41x4.3=400 Å^3) gives the same value 4.57 Å. The value the radius of molecular cavity for PIC dimer is $\boldsymbol{R_a}$ = 5.66 Å.

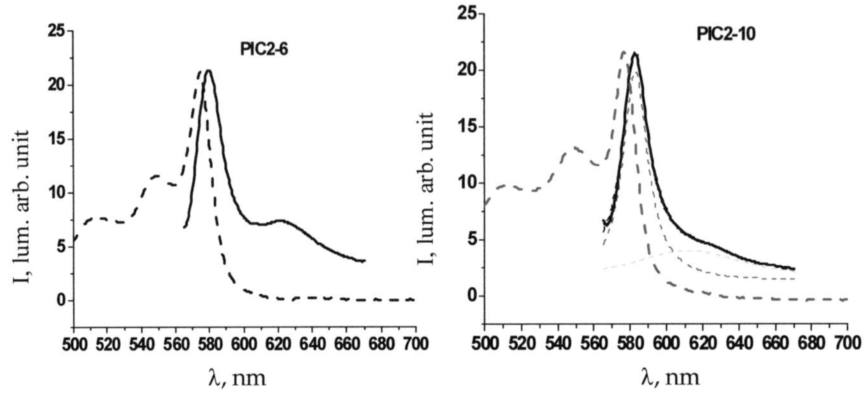

Fig. 14. a,b. The excitation and luminescence spectra in the thin solid films PIC2-6 (a), PIC2-10 (b) with two counter resolution

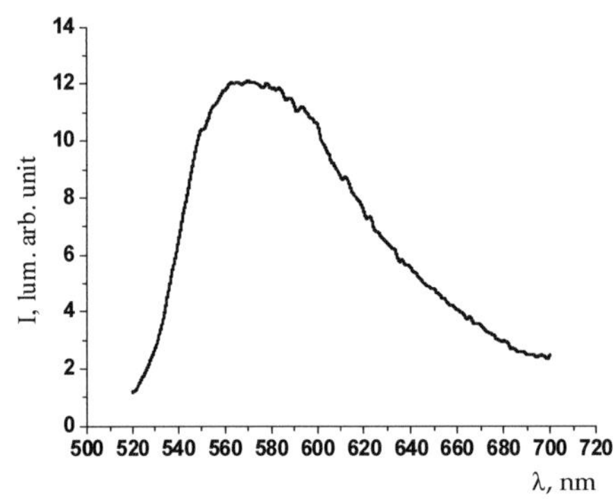

Fig. 15. The luminescence spectrum of the monomer PIC in the water-ethanol solution

Dye,ex. 550 nm	$\lambda_{Макс}$ I lum.,nm	$\nu_{max.abs}$- $\nu_{max.lum}$	Stokes shift, $\Delta\nu_{0-0}$, cm^{-1}	Lorenz counter lum. broadening , cm^{-1}	$\lambda_{Макс}$. 2 lum., nm
PIC2-6	580	150	90	546	624
PIC2-10	583	140	93	502	611
PIC2-15	592	130	102	524	615

Table 2. The luminescent characteristics of the J-aggregates with long alkyl substituters in the thin solid films

The calculation of the dipole moment change for the PIC molecule from the obtained value of the Stokes shift in water-ethanol solution gives the value $\Delta\mu$=2.56D. The analogous calculation for the PIC dimer was carried out in water at Stokes shift 30 cm^{-1} is taken from (Sato et al., 1989) and gives the value $\Delta\mu$=0.7D. One can see from the values of $\Delta\mu$ that the difference of the dipole moments at ground and excited state decreases in 3.66 times with the conversion monomer to J-aggregate form of dye. The calculation of the $\Delta\mu$ change at aggregation PIC at low temperature from the values of the Stokes shift (Renge & Wild 1997) gives the decreasing $\Delta\mu$ at aggregation by 5.5 times. In accordance with the symmetry of the PIC molecule, the dipole moments of the ground and excited state are directed along the short axe of the molecule. It is reasonable to propose that the decreasing of the dipoles moments at the aggregation (dimerization) takes place in the ground and excited state of the aggregate. The decreasing of the dipole moment by 3.7 times is possible at the contra-directed arrangement of the PIC molecules dipoles with the angle between dipoles 43°.

The decreasing of the $\Delta\mu$ apart from the decreasing of the Stokes shift of the luminescence leads to the decreasing of the inhomogeneous broadening of the J-aggregate peak in comparison with the broadening of the dye monomer form absorption band. The square of the inhomogeneous broadening of the molecular spectrum in the medium σ^2_{in} connected with the dipole moment change is proportional to the $\Delta\mu^2/R_a^3$ (Bakhshiev et al., 1989). The decreasing of the $\Delta\mu$ and increasing of the R_a at the dimerization of the dye leads to the decreasing of the inhomogeneous broadening in J-aggregate and the narrowing of the J-peak by 5-7 times in comparison with the spectral width of the monomer form of the molecule. The spectral shift between the maximum of the absorption and luminescence spectra ($\Delta\nu_{a,f}$) depends on the mean-square dispersion of the inhomogeneous broadening of the luminescence spectra of the molecules in the medium σ^2_{in} that corresponds to half width of the inhomogeneous broadening of the luminescence spectrum.

$$\sigma^2_{\phi\lambda} = \frac{kT}{hc} \cdot \Delta\nu_{a,f} \tag{17}$$

The experimental value ($\nu^{abs}_{\max} - \nu^{fl}_{\max}$) for J-aggregate PIC2-10 is 140 cm-1 (see table 2). The calculation of σ^2_{in} for luminescence spectrum of J-aggregates with long alkyl substituters in thin solid films gives a value of full width of inhomogeneous broadening luminescence contour $2\sigma_{in}$=338 cm^{-1}. The comparison of the obtained value $2\sigma_{in}$ (338 cm^{-1}) and experimental value FWHM for luminescence Gauss contour of J-aggregates in thin film 2σ=417 cm^{-1} allows us to conclude that the luminescence spectrum of the J-aggregates has essential contribution of orthe inhomogeneous broadening makes the essential contribution to the luminescence spectrum of the J- aggregates.

The wide long wavelength shoulder of the luminescence is observed on the all measured samples of the J-aggregated dyes in thin films, apart from the main luminescence peak. The maximum of that luminescence is at 610-625 nm depending on the sample and luminescence slump last to 670 nm. The expansion of the PIC2-10 luminescence spectrum on the two Lorentz contours is shown in fig. 14. On the basis of the measured excitation spectra, it was shown that the long wavelength luminescence does not have its own excitation band and its excitation spectra coincide with the absorption and excitation spectra of the J-aggregate. The long wavelength unstructured wide band in J-aggregated films luminescence we attribute to the exciplex luminescence between J-aggregate and monomer dye or between J-aggregates.

The life time of the exciplex luminescence as a rule is longer than the life time of the luminophore luminescence and has the period of luminescence signal growth relates to the stage of the closing in the molecular contact to form the exciplex. The measurement of the kinetics of the luminescence in the maximum of the long wavelength luminescence 625 nm at room temperature shows the signal growing (see fig. 16) with characteristic time 2.1 ns and luminescence decay with life time 3.6 ns. The luminescence life time in the maximum of the J-aggregate has two components: the short component with life time 20-40 ps and the long component with life time 2.0 ns. One can see that the J-aggregate exciplex luminescence has the period of the signal growing and its life time is more than life time measured in the main peak of the J-aggregate luminescence.

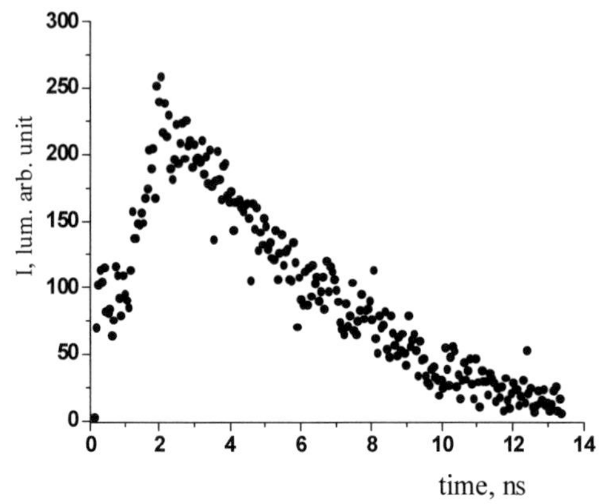

Fig. 16. Kinetics of the luminescence decay of the PIC2-15 at 625 nm in the thin film at the excitation on λ=574 nm

2.7 The formation of J-aggregate PIC in thin solid films. The influence of the cluster anionic derivatives of high boron hydrides

One can see from the map of the electrostatic potential of PIC (fig. 7) that the positive charge of the molecule is distributed in the cavity of electrostatic potential where the anion is situated. From this for the stabilization of the PIC dimer the proper choice is a dianion with dipolar distribution of the negative charge. Cluster dianions of high boron hydrides have the dipolar distribution of the negative charge. The formation of the PIC J-aggregates with the addition of the some cluster high boron hydrides ($K_2B_{10}H_{10}$, $K_2B_{12}H_{12}$, $Cs_2B_{20}H_{18}$, $[Ni^{IV}(1,2\text{-}B_9C_2H_{11})_2]^0$, $Cs[Ni(1,2\text{-}B_9C_2H_{11})_2]$, $NH(CH_3)_3B_9C_2H_{12}$, $NH(C_2H_5)_3[Co(1,2\text{-}B_9C_2H_{11})_2]$, $Cs_2B_{10}H_8I_2$, $[Sn(1,2\text{-}B_9C_2H_{11})]^0$) was studied (Shelkovnikov et al., 2004). It was shown that the addition of anions of high boron hydrides $B_{10}H_{10}^{2-}$ and $B_{10}H_8I_2^{2-}$ to the PIC2-2 spin-coating solution at molar ratio dye:anion salt 1:0.1 leads to the effective formation of J-aggregates of dye in solid films. The absorption spectra of PIC films doped with $[Ni^{(IV)}(1,2\text{-}B_9C_2H_{11})_2]^0$, $[Sn(1,2\text{-}B_9C_2H_{11})]^0$ and anions $B_{10}H_{10}^{2-}$ и $B_{20}H_{18}^{2-}$ are shown in fig. 17. The J-peak is distinctly seen in the absorption spectra of the PIC films doped with $B_{10}H_{10}^{2-}$ anions. This anion leads to an efficient J-aggregation of PIC with the formation of stable J-aggregates in solid films.

Similarly to the $B_{10}H_{10}^{2-}$ anions, the addition of the neutral complex of nickel stimulates the predominant formation of J-aggregates of PIC in a solid film (*J*/*M*=1.52). The addition of the $B_{10}H_8I_2^{2-}$-anion at holding times prior to centrifugation of more than 10 min leads to the formation of stable J-aggregates (*J*/*M* = 1.46). At short holding times prior to centrifugation (less than 5 min), a long wavelength shoulder located at 625 nm and extending up to 800 nm was observed in the absorption spectra (Fig. 18). In this case, the optical density of the J-peak decreases. When a film doped with $B_{10}H_8I_2^{2-}$ anions is applied to a glass surface treated with a silicon-containing surfactant for the surface wettability improvement, a high conversion of the monomer to J-aggregates ($J/M = 2.2$) is observed (Fig. 18, spectrum *3*).

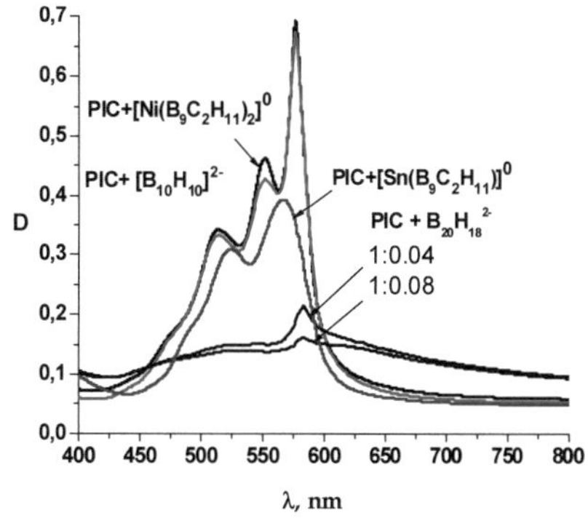

Fig. 17. Absorption spectra of PIC films doped with (1) $[Ni^{IV}(1,2\text{-}B_9C_2H_{11})_2]_0$, (2) $[Sn(1,2\text{-}B_9C_2H_{11})]^0$, (3) $B_{10}H_{10}^{2-}$ and (4, 5) $B_{20}H_{18}^{2-}$. Dye–anion mole ratio: (4) 1 : 0.04; (5) 1 : 0.08

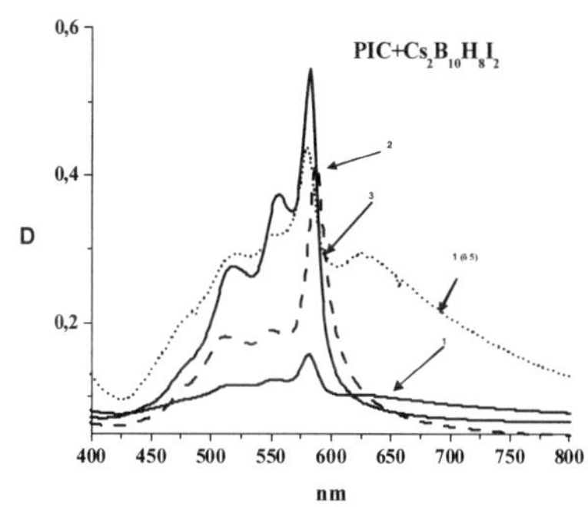

Fig. 18. Absorption spectra of PIC films doped by $B_{10}H_8I_2^{2-}$ anions. The holding time prior to centrifugation: (*1*) 2–3 min; (*2, 3*) longer than 10 min. Sample *2* on a pure glass substrate; sample *3*, a glass substrate treated with a surfactant

A high stability of the J-aggregates of PIC-closo-hydrodecaborate allowed us to perform an additional doping of the dye film with organic cation salts without destroying the J-peak. The following salts of organic cations were used: methylcholine iodide, tetrabutylammonium (TBA) iodide, laurylcholine iodide and dodecylpyridine iodide. The salts of organic cations were introduced into the films at a mole ratio of the dye cation:added cation varying from 1 : 0.5 to 1 : 5. Upon an increase in the concentration of an organic cation to the mole ratio dye–cation 1 : 4, the J-peak increases in intensity and narrows by 1.5–2 times, as is shown in Fig. 19 for the case of TBA. At a higher concentration of organic cations in the film, the J-peak decays. The destruction of the J-aggregates was observed only upon a high dilution of a dye film with organic cations. This demonstrates the high stability of the J-aggregated form of PIC-closo-hydrodecaborate. For example, the addition of octadecyl choline iodide to films of PIC iodide with long alkyl substituters ($C_{18}H_{37}$) leads to the decomposition of the J-aggregates to the monomer at a dye–additive mole ratio even as low as 1 : 0.1. In fact, at a concentration of the additive organic cation at 4 times higher than the concentration of the dye cation, the molecules of the organic salt of the additive cation can be considered as a kind of a matrix for the dye molecules.

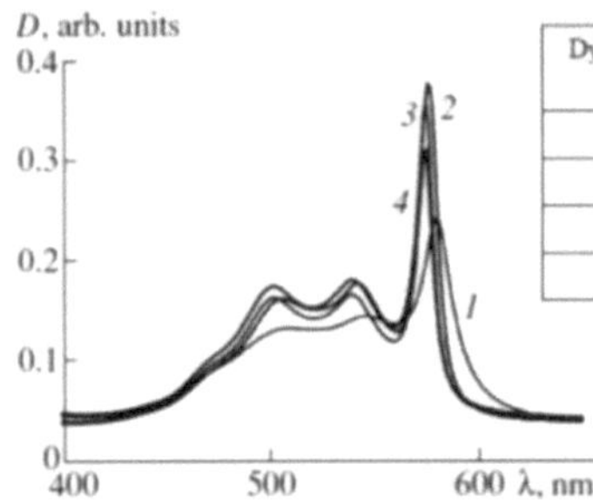

Dye : TBA	λ_{max}, nm	J/M	Full width at half maximum, cm^{-1}
1 : 0	580	1.66	585
1 : 2	576	2.14	326
1 : 4	575	2.15	271
1 : 5	575	1.68	289

Fig. 19. Absorption spectra of films of PIC-closo-hydrodecaborate doped with tetrabutylammonium iodide (the dye–additive mole ratios, maximum wavelengths of the J-peak, *J/M* ratios and the full widths of the peaks at half maximum are given in the inset). Spectra *1–4* correspond to the dye–additive mole ratio varied from 1 : 0 to 1 : 5, respectively

The dilution of a film of PIC-closo-hydrodecaborate with organic cations leads to an increase in the luminescence of J-aggregates. The luminescence spectra of the J-aggregates in a film of PIC-closo-hydrodecaborate and in an analogous film diluted with TBA cations in the ratio dye–TBA 1 : 4 measured under identical excitation conditions are shown in Fig. 20. It is seen from this figure that, upon dilution of the J-aggregated film by TBA cations, the luminescence intensity of the J-aggregates increases by an order of magnitude. This effect can be explained by a decrease in the concentration quenching of luminescence of the J-aggregates.

In our opinion, the changes in the shape of the J-peak in the absorption and luminescence spectra of thin films of PIC-closo-hydrodecaborate observed upon dilution of the films with foreign organic cation can be explained by the fact that this peak arises due to the absorption of strongly coupled PIC dimers. A stable dye dimer may form because of a strong electrostatic interaction of two cationic molecules PIC+ with a doubly charged polyhedral $B_{10}H_{10}^{2-}$ anion. In a dimer PIC_2-$B_{10}H_{10}$, the electrostatic interaction is saturated and, upon dilution with foreign cations, large-size aggregates should decompose.

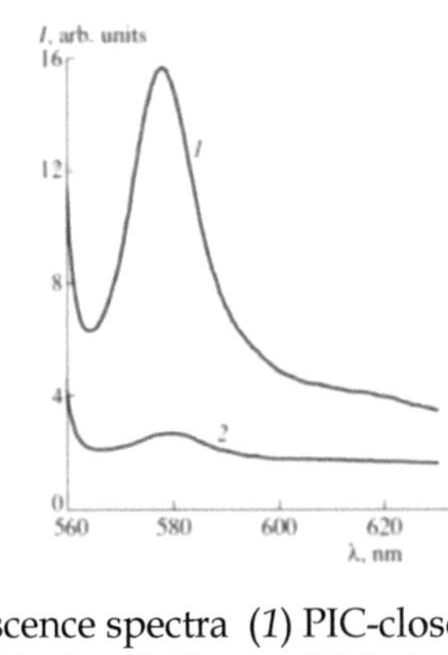

Fig. 20. Thin solid films luminescence spectra (*1*) PIC-closo-hydrodecaborate doped with TBA at a dye–TBA mole ratio of 1 : 4 and (2) pure PIC-closo-hydrodecaborate

The charge distributions in the dye cation and in the anions of the highest boron hydrides were calculated by the AM1 method and are shown on the structures below. Qualitatively, the calculated charge distribution among the atoms of the known anions $B_{10}H_{10}^{2-}$ and $B_{12}H_{12}^{2-}$ agrees with the charge distribution calculated by such methods as PRDDO, 3D Huckel theory and STO-3G (Pilling & Hawthorne, 1964). Quantitatively, the best agreement was obtained for the total charges of the B–H groups.

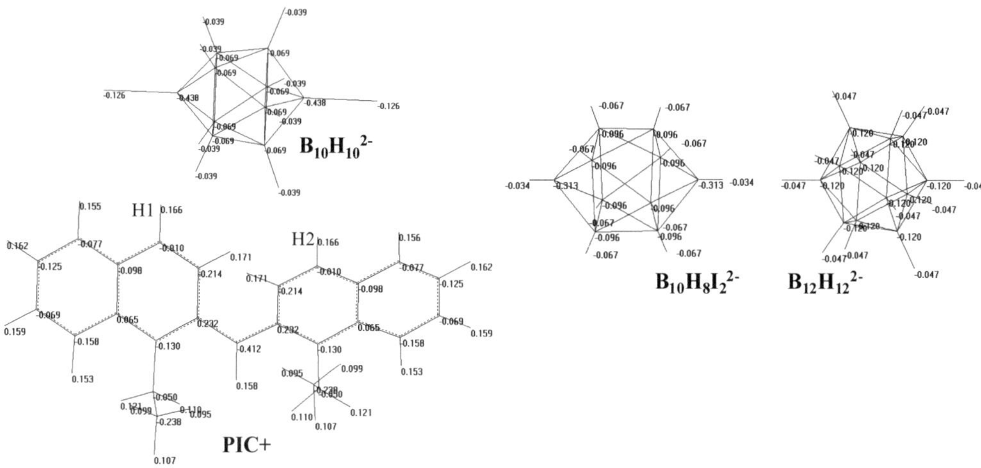

In the $B_{10}H_{10}^{2-}$ anion, a negative charge is distributed non-uniformly. The vertex B–H groups carry a larger negative charge (–0.347) than the equatorial B–H groups (–0.163). In the PIC cation, a highest positive charge is located on two carbon atoms of the heteroaromatic rings involved in the chain of conjugation. The remaining carbon atoms and nitrogen heteroatoms carry negative charges. The remaining positive charge is mainly distributed over the hydrogen atoms of the aromatic rings. The basic atomic centres of the electrostatic attraction of the ions considered are the vertex B–H groups in the $B_{10}H_{10}^{2-}$ anion, the carbon atoms and the hydrogen atoms of the heteroaromatic rings of the dye cation. The geometrical dimensions of the $B_{10}H_{10}^{2-}$ anion (the distance between the polar hydrogen atoms is equal to (5.9 Å)) correspond well to the dimension of the cavity of the electrostatic potential of PIC due to the turn of its aromatic planes, in particular, the distance between the positively charged hydrogen atoms H1 and H2 (5.6 Å). It should be noted that the involvement of the vertex B–H groups of the $B_{10}H_{10}^{2-}$ anion into the formation of a stable PIC-closo-hydrodecaborate complex is confirmed by the data of IR spectroscopy and x-ray diffraction analysis (Cerasimova, 2000). The structure of a dimer of PIC-closo-hydrodecaborate calculated using the MM+ method and the map of the electrostatic potential calculated using the semi-empirical quantum-chemical AM1 method are shown in Fig. 21. It is seen that the aromatic planes of the dye cations interacting via the anion of closo-hydrodecaborate are turned with respect to each other. This turn can be explained by the mutual repulsion of the positively charged aromatic hydrogen atoms in the PIC dimer.

The charge distribution in the $B_{10}H_8I_2^{2-}$ anion is more polarized than in the $B_{10}H_{10}^{2-}$ anion. The negative charge is strongly displaced toward the vertex boron and iodine atoms. If one assumes that the dipole charge distribution in the $B_{10}H_{10}^{2-}$ anion facilitates the formation of J-aggregates of PIC, then, upon addition of $B_{10}H_8I_2^{2-}$ anions, efficient J-aggregation should also be observed. This agrees with the formation of J-aggregates of PIC in the presence of these anions in the case of the solution being held prior to its centrifugation for a time period longer than 10 min (Fig. 18). Upon deposition of a dye solution immediately after the addition of anion, an intensive association of the dye is likely to occur, which is stimulated by a relatively high initial concentration of anions localized near the drop of salt addition (later, the associates gradually dissolve). In this case, the intensive association of the dye leads to the formation of a set of strongly aggregated dimers of PIC, whose absorption extends toward the long wavelength range up to 800 nm. The maximum of this distribution (625 nm) possibly corresponds to the dye tetramers.

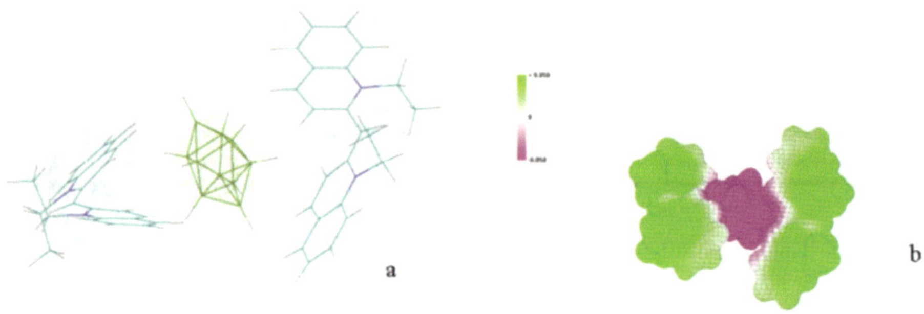

Fig. 21. The calculated structure of the PIC-closo-hydrodecaborate dimer (a) and dimer electrostatic potential map (b).

In the $B_{12}H_{12}^{2-}$ anion, the charge distribution among the boron atoms is completely uniform and the absolute value of the charge of the B–H groups (–0.167) is significantly smaller than that of the vertex B–H groups of the $B_{10}H_{10}^{2-}$ anion. Note that this anion has no poles of concentration of a negative charge and, therefore, no structure-forming effect on the formation of J-aggregates of PIC. Taking into account the charge distribution pattern in the considered cluster anions of boron hydrides, as well as their relative dimensions, we can conclude that a more pronounced bipolar distribution of the negative charge and a smaller volume of an anion facilitate the formation of a J-aggregate of PIC.

In the case of a strongly bound dimeric form of PIC-closo-hydrodecaborate, the dissociation of these dimers to the monomer upon dilution of dye films with a salt of an organic cation does not occur immediately, but rather via the stage of formation of separated dimers of PIC. The weakening of the intermolecular interaction leads to (i) a narrowing of the absorption peak of a J-aggregate, (ii) an increase in the intensity of the J-peak, (iii) a hypsochromic shift of the J-peak and (iv) an increase in the intensity of luminescence of the J-aggregate. We assume that the dimer of a dye is the main structural unit responsible for the absorption of a J-aggregate (J-peak) of PIC.

2.8 The thermal stability of PIC-closo-hexahydrodecaborate in the thin solid films

The thermal stability of the J-aggregates of PIC with long alkyl substituters in the thin solid films (T_{decay} about 60° C) is near the temperature of the film heating at pulse laser excitation. The laser destruction of these films takes place at laser intensity I_0=1-5 MW/cm^2. The film temperature change (ΔT) at the named intensity is 20-90° C from the equation:

$$\Delta T = \frac{I_0(1\text{-}10^{-D})\cdot(1\text{-}R)\cdot\tau}{h\cdot\rho\cdot C_p} \quad (18)$$

where, $\boldsymbol{\rho}$- density 1,4 g/cm^3, C_p – specific heat capacity 3,1 J/g·K^0, $\boldsymbol{D}$ – film optical density 0,5, $\boldsymbol{R}$ – reflection coefficient 0,3, laser pulse duration τ=5ns, film thickness h=30nm.

The J-aggregates of PIC-closo-hexahydrodecaborate in the thin solid films are more suitable for the non-linear laser experiments due to their higher thermal stability compared with that for the J-aggregates of long chain PIC.

The thermal stability of PIC-closo-hexahydrodecaborate with TBA addition in solid films is studied by using the spectrophotometry method. The J-aggregates of PIC-closo-hexahydrodecaborate are stable at sample heating to temperature 90^{0}C and being partially destructed at higher temperature are restored to 100% at sample cooling to room temperature. As an example the thermal decay and restoring of the J-aggregates' PIC-$B_{10}H_{10}$ the spectral change of the J-aggregated film with addition of the TBA for peak narrowing is shown in the fig. 22 a. One of the features of the thermal decay of the J-aggregates' PIC in the thin solid films is the red shift of J-aggregate peak. An example of an evident red shift of absorption peak of J-aggregates is shown in fig. 22 b,c.

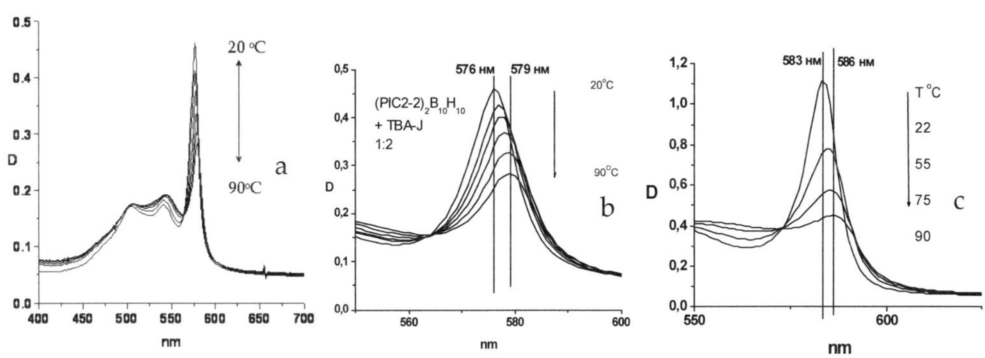

Fig. 22. The absorption spectra change of the J-aggregated films' PIC-closo-hydrodecaborate headed (cooled), doped with TBA at a dye–TBA mole ratio of 1 : 4 (a), expanded scale (b), expanded scale PIC2-2 in polymer anethole (c)

The red shift of the J-aggregate peak in thin solid films and polymer matrix to about 3 nm was observed in the causes of J-aggregates' thermal decay. This effect we connect with the inhomogeneous broadening of the J-aggregate absorption band. The red spectral shift caused by intermolecular interaction follows the depth of the J-aggregate potential in the ground state for the homogeneous absorption profile as it is considered for the positive solvatochromic shifts of the absorption spectra of the molecules (Suppan, 1990). The thermal decomposition takes place for local aggregates having different activation energy. The activation energy has inhomogeneous Gaussian distribution between J-aggregates sites with homogeneous spectral Lorentz profiles. In this case, each homogeneous Lorentz profile has the spectral position with proper own activation energy of thermal decay from aggregate (dimer) to monomer. At the sample heating the aggregates with lower activation energy (ΔE_1) have a higher probability to destroy and have higher excitation energy. The inhomogeneous distribution of J-aggregates leads to spectral-kinetic non-equivalence of the thermal decay. The statistic weight of J-aggregates with high energy of activation increases during thermal decay.The combination of inhomogeneous broadening of initial homogeneous spectral Lorentz profiles of absorbing aggregates gives the Voight function (V) of spectral distribution.

$$V = 2\ln(2)\frac{\gamma}{\Delta w^2}\cdot\int_{-\infty}^{\infty}\frac{e^{-t^2}dt}{\left(2\ln(2)\cdot\frac{w-w_0}{\Delta w}-t\right)^2+\left(\ln(2)\frac{\gamma}{\Delta w}\right)^2} \tag{19}$$

Where, γ – width of Lorentz profile, w - width of Gauss profile and w_o - central position of absorption band. To take into account the individual rate of thermal decomposition of absorbing Lorentz profiles of aggregates, we need to include in the Voight function the equation describing the thermal decay curve, depending on the energy activation distribution. The equation for thermal decomposition as follows from (4) shown in the above curve is

$$\begin{gathered} 1-\alpha_T = \exp(-F_a) \\ F_a = \left[\frac{T^2R}{\Delta E_a}\cdot\left(1-\frac{2TR}{\Delta E_a}\right)\cdot\exp\left(-\frac{\Delta E_a}{TR}\right)\right]\cdot Z \end{gathered} \tag{20}$$

Where, α_T is the degree of J-peak decay, T - temperature ^oK, R - universal gas constant, Z - preexponential factor $Z=10^{18}$ and ΔE_a - activation energy with average value 120 kJ/mol. The values correspond to the thermal decomposition behaviour of J-aggregates' peak for PIC2-15 in thin films. We suppose that the activation energy spectral distribution has the normal distribution profile in spectral region of J-aggregate absorption with deviation ±10 kJ/mol. The Voight function, including the degree of thermal decomposition $V(E_a,T)$, is

$$V(E_a,T) = 2\ln 2\cdot\frac{\gamma}{\Delta w^2}\int_{-\infty}^{\infty}\frac{e^{-t^2}\cdot e^{-F_a}}{(2\ln 2\frac{w-w0}{\Delta w}-t)^2+(\ln(2)\cdot\frac{\gamma}{\Delta w})^2}dt \tag{21}$$

Calculation of the Voight contour of J-peak absorption was carried out. The values of homogeneous and inhomogeneous broadening of the J-peak are approximately equal and make up 235 cm^{-1} at room temperature. The results of modelling of the Voight contour of J-aggregate PIC2-15 absorption at the stage of thermal decomposition was compared with the experimental spectral curves of J-aggregate (see fig. 23).

One can see good agreement between modelling and experimental spectral curves. The results confirm the influence the inhomogeneous broadening of absorption spectra of J-aggregates in thin films on the red shift of absorption band of J-aggregates at thermal decay.

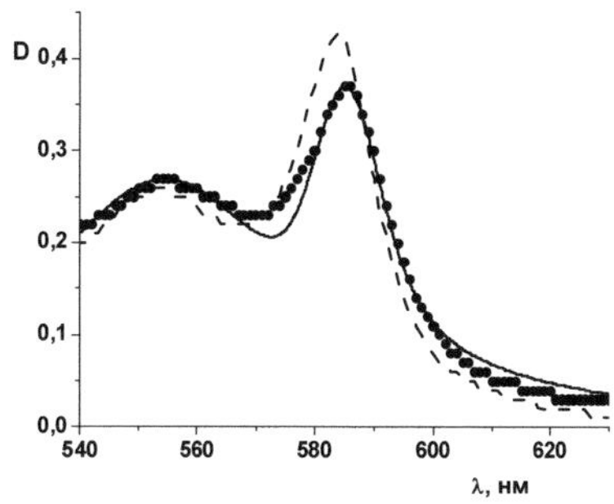

Fig. 23. Calculated Voight contour of J-aggregates at 55°C (solid lines) and experimental absorption curves of thin solid films of J-aggregates of PIC2-15: dash curve - initial spectrum, dot curve - spectrum at 55°C

2.9 The spectrophotometry of J-aggregate thin film formation by spin-coating

Thin film formation of J-aggregated pseudoisocyanine iodide with long alkyl substituents PIC 2-18 and PIC 2-2 with the addition of $B_{10}H_{10}^{2-}$ anion was studied by spectrophotometry directly during spin-coating of the dye solutions. The view of spectra change during spin-coating of a solution of PIC 2-18 at 2000 rpm from the time of placing the solution until formation of the solid J-aggregated film is shown in Fig. 24. The inset shows the change of optical density at the absorption maximum of the monomer form of dye Fig. 24a. The initial spectrum of dye solution on the substrate exhibits optical density noise fluctuations because it exceeds the dynamic range of the spectrophotometer (2.5*D*). After 2.5 sec from the spin-coating stars, the dye absorption in solution stabilizes at optical density D=1.3 and remains unchanged for 0.4 sec. This means that the centrifugal discharge of dye solution from the substrate is finished and the film is held on the substrate by surface-tension forces. The dye concentration increases because of solvent evaporation from the thin film, whereas the number of dye molecules on the substrate remains the same. Fig. 24b shows the spectral changes for a film of PIC 2-2 (+$K_2B_{10}H_{10}$) at rotation rate 1500 rpm. The time for formation of J-aggregates on substrate is 0.45 sec.

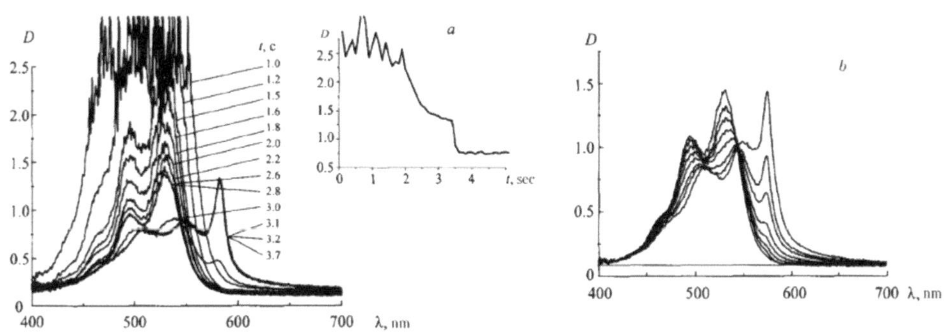

Fig. 24. Spectral changes during spin-coating of solutions of pseudoisocyanine derivatives: PIC 2-18 solution [the time from the moment of placing the solution on the centrifuge is shown; in the inset, optical density at the absorption maximum of the monomeric dye in solution (530 nm) as a function of spin-coating time] (a); PIC 2-2 solution with added $K_2B_{10}H_{10}$ during formation of J-aggregate (spectra recorded every 0.05 sec) (b).

It should be noted that the experimental spectra are not corrected for double passage of light through the sample. Losses of light to double reflection and absorption of dye film must be considered in order to obtain the true value of the optical density in our optical setup. For this, the reflectance spectrum of J-aggregated PIC 2-18 film was measured (Fig. 25). The corrected optical density of J-aggregated film $D(\lambda_{cor})$=log $[I_0/(I_0-A-R)]$ was calculated by solving the quadratic equation for the absorption coefficient obtained from the formula for the measured optical density $D(\lambda)$:

$$10^{D(\lambda)} = \frac{I_0}{I_0 - A - R - R(I_0 - A - R)(I_0 - A - R)},$$
$$A^2 - A(1 - 2R + I_0) - \left(\frac{I_0}{10^{D(\lambda)}} - I_0 + R + RI_0 - R\right) = 0. \tag{22}$$

where A is the absorption coefficient and R the film reflectance coefficient for each wavelength.

A comparison of Fig. 24 and Fig. 25 shows that correction of the spectrum taking into account double absorption and reflection decreases significantly the optical density of the sample at the maximum of the J-peak (to 0.3).

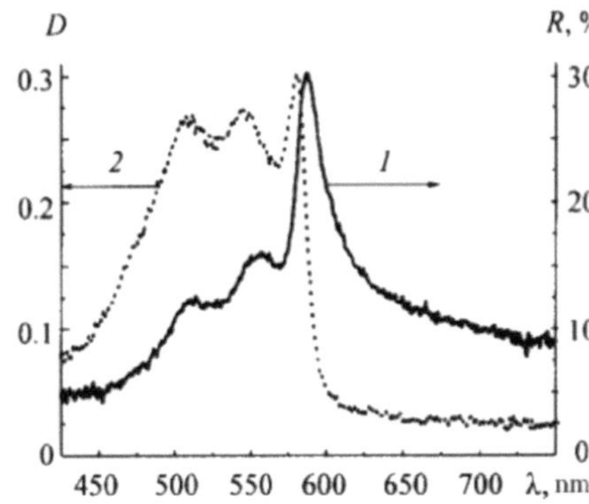

Fig. 25. Reflectance spectrum (1) and absorption spectrum calculated taking into account reflectance of the sample (2), J-aggregated film of PIC 2-18

Spectral changes occurring during spin-coating of J-aggregated PIC dye are related to the increased refractive index of the medium at the time of film formation. The bathochromic shift of the monomer dye absorption maximum is due to solvent evaporation and replacement of the initial medium with refractive index n_1, consisting of solvent molecules, by a medium with refractive index $n2$, consisting of dye molecules themselves. The absorption maximum of the dye for PIC film ($+B_{10}H_{10}^{2-}$) shifts from 529 to 550 nm as the film dries out; for PIC 2-18 film, from 529 to 551 nm. The spectral shift in organic molecules on going from vacuum to a medium with refractive index n_0 is described by the universal interaction function (Bakhshiev et al., 1989)

$$\Delta v = \text{const} \cdot \frac{n_0^2 - 1}{n_0^2 + 2} \tag{23}$$

An empirical expression for the universal interaction function of dye with medium was obtained based on experimental data for absorption of PIC monomer in various media (Renge & Wild 1997)

$$\lambda(n_0) = \frac{10^7}{19716 - 2964\left[\left(n_0^2 - 1\right)/\left(n_0^2 + 2\right)\right]} \tag{24}$$

where $\lambda(n_0)$ is the wavelength (nm) of the absorption maximum for dye in a medium with refractive index n_0.

The refractive index of the medium for dye solution is n_m=1.405. The wavelength of the absorption maximum of monomer dye λmax=526.6 nm was calculated using Eq. (24) and agrees with the experimental value for a dilute dye solution. The refractive index of film after centrifugal discharge of dye solution from the substrate obtained using Eq. (24) was 1.47 for absorption of PIC monomer at 529 nm. From this point the spectral shift of the absorption band of dye monomer during spin-coating from 529 nm to 551 nm corresponds to a transition from a medium with refractive index n=1.47 to a medium with n=2.1. That is in agreement with n value from the dispersion curve for dye in thin film (see fig. 12a).

Thus, the bathochromic shift of the dye absorption spectrum at the time of film formation is due to the increased polarizability of the medium. Therefore, the polarization spectral shift must be considered in determining the exciton splitting of the monomer absorption band upon J-aggregation or the energy of exciton coupling of dye molecules in the J-aggregate (*J*ex). The experimental value *J*ex is determined from the energy difference of the long wavelength transition of PIC in solution (530 nm) and in the J-aggregate using the formula (Kuhn & Kuhn 1996)

$$\Delta E = \sqrt{E_m^2 + 4E_m J_{ex}} - E_m \tag{25}$$

where, *E*m is the energy of the long wavelength transition in the monomer; *J*ex, the exciton coupling energy; and Δ*E*, the energy difference of absorption at the monomer peak and the J-peak. The absorption maximum of the J-peak was observed at λ=570 nm where the absorption of PIC J-aggregates was measured in aqueous ethyleneglycol matrix at liquid helium temperature (Minoshima et al., 1994). In this instance *J*ex=632 cm^{-1}. A similar value of *J*ex is used in theoretical calculations. In particular, *J*ex ≈ 600 cm^{-1} is used to determine the exciton delocalization length (Bakalis & Knoester, 2000).

The change of local field for a dye molecule in the aggregate as a result of the change of medium is not considered for this estimate of *J*ex. The total energy of the J-peak shift of aggregate absorption is the sum of the polarization energy and the exciton coupling energy of neighbouring molecules in the aggregate. Therefore, the bathochromic shift of the dye S_{00}-transition that is due to the medium polarizability with an effective refractive index equal to the refractive index of the dye monomer in the solid phase, i.e., n=2.1, must be taken into account for dye molecules in the aggregate. The exciton coupling energy is *J*ex=293 cm^{-1} for the spectral maximum of monomer absorption shifted to 555 nm and for the absorption maximum of the J-peak located at 574 nm. Thus, the resulting value *J*ex ≈ 300 cm^{-1} is half of that used in the literature (*J*ex ≈ 600 cm^{-1}). The values *J*ex for aggregates of other dyes, known from the literature (Moll, 1995), are also estimated based on the spectral shift of the absorption maximum of J-aggregated dye relative to the monomer in solution. The polarization spectral shift for dyes with highly polarizable π-electrons is significant. This correction must be considered in estimating *Jex*, not only for PIC aggregates, but also for aggregates of any other dyes.

Another aspect of the change the local field factor at the J-aggregate formation is the dramatic increase of the J-peak at the last moment of formation of the solid film without a decrease of the optical density at the absorption maximum of the monomeric dye, i.e., the significant growth of the optical density of the J-peak is not compensated by a decrease of the monomer optical density. An additional increase of the J-peak, for example, for PIC 2-2 ($+K_2B_{10}H_{10}$) by 2.15 times can be seen from the measured absorption spectra.

One reason for the increased absorption as the film dries out is the increased refractive index of the medium. According to Bakhshiev et al. (1989), the ratio of integrals of absorption intensity (Int) in media with different polarizabilities (Int_{max} for n_{max} and Int_{min} for n_{min}) depends on the local field factor

$$\frac{\mathrm{Int}_{max}}{\mathrm{Int}_{min}} = \frac{\left(n_{max}^2 + 2\right)^2}{9n_{max}} \frac{9n_{min}}{\left(n_{min}^2 + 2\right)^2} \qquad (26)$$

The monomer concentration ceases to decrease with a spectral shift of the monomer peak to 539 nm. According to Eq. (24), the refractive index of the medium at this moment n=1.7. The ratio of local field factors is 1.94 for a change of medium refractive index from 1.7 to 2.5. The calculation in the optical density change, taking into account the reflection and absorption growth due to growth of the local field factor, gives the increasing measured optical density of the J-peak by 2.2 times and explains the increase of the J-peak without loss of dye monomer.

The results from the investigation of J-aggregate formation during spin-coating of PIC indicate that the change of local field factor should be considered in interpreting spectral properties of nano-structured aggregates for any type of dye aggregation or for other highly polarizable molecules in both films and solutions.

2.10 The third order non-linear optical properties of pseudoisocyanine J-aggregates in thin solid films

The third order non-linear optical properties of pseudoisocyanine J-aggregates in thin solid films were studied using the Z-scan method as well for PIC iodide anion as for PIC with closo-hexahydrodecaborate anion ($B_{10}H_{10}^{2-}$) (Markov et al., 1998a, 1998b; Plekhanov et al., 1998a, 1998b). The dispersions values of imaginary $\boldsymbol{Im\chi^{(3)}}$ and real $\boldsymbol{Re\chi^{(3)}}$ parts of cubic susceptibility of J-aggregated film are shown in fig. 26. In both causes the non-linear bleaching of the solid film samples in J-peak maximum is observed at resonance laser irradiation excitation ($I>10^5$ W/cm^2) and obvious darkening on the red side of the dispersion curve of imaginary part of $\boldsymbol{\chi^{(3)}}$. J-aggregates of both types of dye have a similar value of the coefficient of non-linear absorption $\beta\approx$-6*10^{-2} cm/W and corresponding value $\boldsymbol{Im\chi^{(3)}}\approx$-1*10^{-5} esu at the bleaching maximum and $\boldsymbol{Im\chi^{(3)}}\approx$0.12*10^{-5} esu at the maximum of darkening. At the intensity of the laser irradiation at more than 3 MW/cm^2, the films become irreversibly burned.

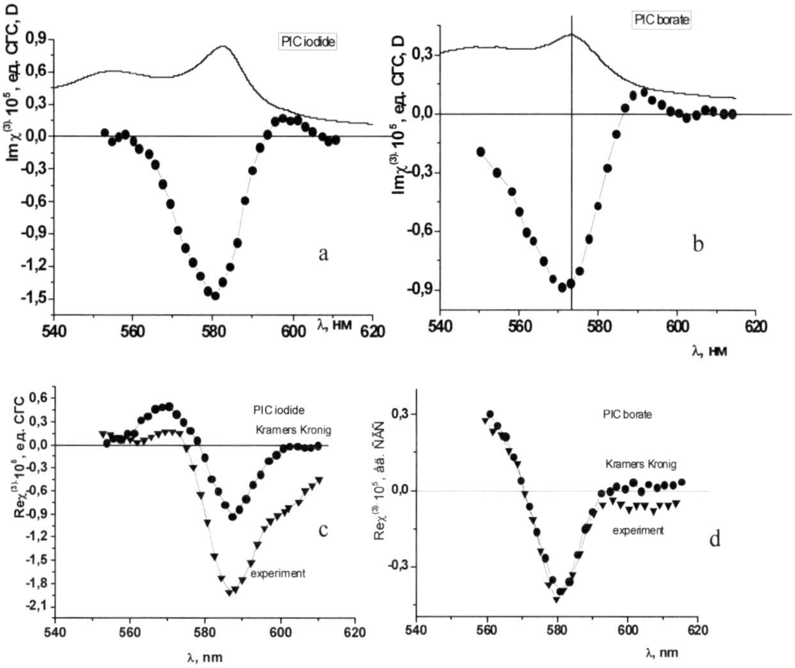

Fig. 26. a-d. The dispersion curves of the imaginary part of the cubic susceptibility of the J-aggregated $\boldsymbol{Im\chi^{(3)}}$ PIC-iodide (a) and closo-hexahydrodecaborate (b) and real $\boldsymbol{Re\chi^{(3)}}$ PIC-iodide (c) and closo-hexahydrodecaborate (d) compared with Kramers-Kronig calculated curves

The maximal value of the real part of the cubic susceptibility for films with J-aggregate of PIC-iodide is $\boldsymbol{Re\chi^{(3)}}$=-1,8*10^{-5} esu and with J-aggregate of PIC-closo-hexahydrodecaborate $\boldsymbol{Re\chi^{(3)}}$=-0,45*10^{-5} esu. This difference is due to the contribution of the refraction thermal change in the higher absorption PIC-iodide film compared with absorption of PIC-($B_{10}H_{10}{}^{2-}$) film as is shown in fig. 27. The imaginary part of the cubic susceptibility does not depend on the thermal contribution. The sample refraction change caused by saturation of the electron transition without the part of thermal contribution was calculated by Kramers-Kronig relation connecting the $\boldsymbol{Im\chi^{(3)}}$ and $\boldsymbol{Re\chi^{(3)}}$ for the non-linear case (Markov et al., 1998c) (see fig. 26 a-d).

$$\operatorname{Re}\chi(\Omega)=\frac{1}{\pi}\int_{-\infty}^{\infty}\frac{\operatorname{Im}\chi(\Omega)}{x-\Omega}dx-2\operatorname{Re}[C^{-1}{}_{(\Omega_p)}(\Omega)] \tag{27}$$

where C^{-1} is the residue in the pole of the $\boldsymbol{\chi(\Omega)/x-\Omega}$, ($\boldsymbol{\Omega=\omega-\omega_J}$) function.

The thermal contribution to the non-linear refraction for the films with the optical density $\boldsymbol{D}$<0.4 is insignificant, but it should be considered for films with optical density $\boldsymbol{D}\approx$1. The maximal values of the real parts of$^{)}$ $\boldsymbol{\chi^{(3)}}$ with and without taking into account the thermal contribution are close to each other and are $\boldsymbol{Re\chi^{(3}}$ ~-(0.5-0,7)*10^{-5} esu. The obtained values of the non-linear response of PIC J-aggregates in the thin solid films ($\chi^{(3)}\approx10^{-5}$ esu) are two orders of magnitude higher than in the water solution (Bogdanov et al., 1991) and polymer matrixes (Plekhanov et al., 1995) ($\chi^{(3)}\approx10^{-7}$ esu). Apart from the much lower values of the cubic susceptibility, there was no observation of non-linear darkening of PIC J-aggregates.

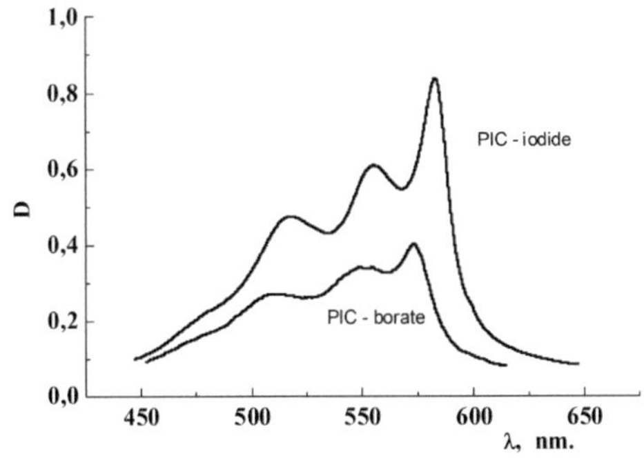

Fig. 27. The absorption spectra of the PIC2-2 iodide and closo-hexahydrodecaborate J-aggregated films

The iodine anion existing in the PIC molecule has the high polarizability and the more the iodine atoms are in the anion molecule, the higher the polarizability. The PIC J-aggregated film with addition of high polarizability tetraethylammonium salt of the closo-tetrahydrohexaiodo-dodecaborate ($B_{10}H_4I_6^{2-}(C_4H_9)_4N^+$) which has six iodine atoms was prepared to raise the J-aggregate surrounding polarizability. The dispersion curve of the imaginary part of the obtained film is shown in fig. 28 compared with the dispersion for the PIC-closo-hexahydrodecaborate film.

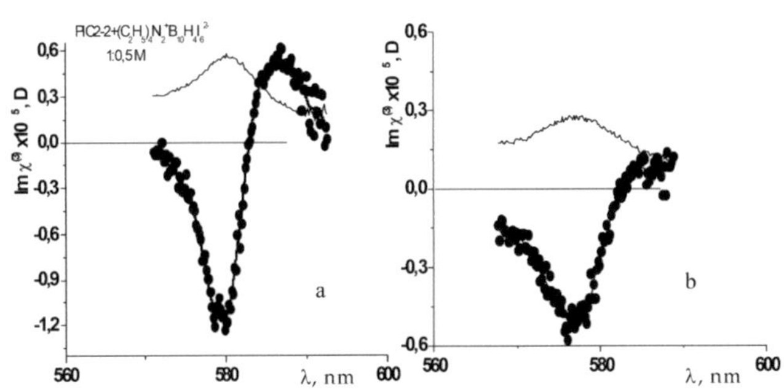

Fig. 28. The cubic susceptibility dispersion curves for the PIC J-aggregated films with - $B_{10}H_4I_6^{2-}$ 1:0,5 (a) and $B_{10}H_{10}^{2-}$ 1:0,5 (b)

As one can see from the adducing figures, the non-linear optical response of the J-aggregated film samples with high polarizable anion ($B_{10}H_4I_6^{2-}$) is higher than for the film with $B_{10}H_{10}^{2-}$ anion. The negative value of the imaginary part of the cubic susceptibility increases 2 times and the red shifted induced non-linear absorption increases in 5 times up to $0.6{\cdot}10^{-5}$ esu. That means it is possible to act upon the non-linear response of the charged aggregated molecules by additions of the proper anions.

The non-linear darkening on the red side of dispersion $Im\chi^{(3)}$ curves is considered on the basis of the four-level model of excitation energy relaxation in PIC J-aggregates shown in fig. 29. The additional level of the relaxation of the first singlet state of J-aggregates in the thin film arises from high polarization of the surrounding J-aggregate medium consisting of the PIC dye molecules. The equations obtained in (Tikhonov & Shpak, 1977) for a four-level model of the excitation relaxation in assumption $k_4 >> k_2$ were used to calculate the induced bleaching and darkening change in the J-aggregated thin film depending on the intensity of the incident laser radiation.

$$\Delta T = \frac{1+\frac{I_0\sigma_{32}}{A_k}\cdot\left(\frac{1}{\tau_2}-\frac{\sigma_{10}}{\tau_3\sigma_{32}}\right)}{1+\frac{1}{A_k}\cdot\left[\sigma_{10}I_0\left(\frac{1}{\tau_2}+\frac{1}{\tau_3}\right)+\sigma_{10}\sigma_{32}I_0^2\right]}-1 \tag{28}$$

$$A_k=\left(\frac{1}{\tau_1}+\frac{1}{\tau_2}\right)\cdot\left(\frac{1}{\tau_3}+\sigma_{32}I_0\right) \qquad \sigma=\frac{2303\cdot\varepsilon}{6.02\cdot 10^{23}}$$

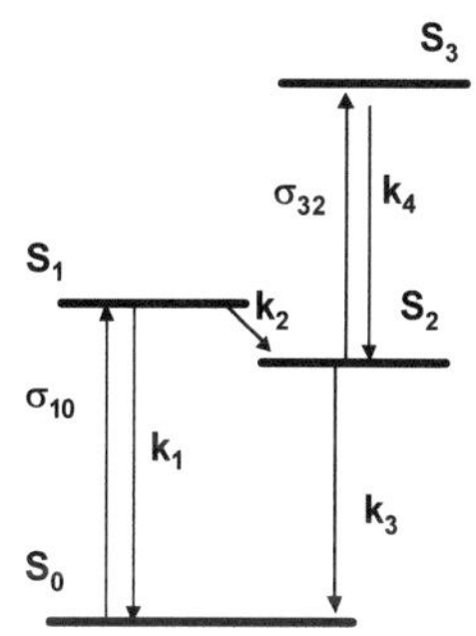

Fig. 29. Four-level model of the J-aggregate excitation relaxation in thin film, where, S_2 excited state is considered here as relaxed S_1 state

The luminescence kinetics decay for J-aggregated film was measured to estimate the values of the characteristic times of excited state relaxation. The luminescence decay of the J-aggregated thin film in the maximum of the J-aggregate luminescence λ=590 nm at the excitation in J-peak maximum λ=574 nm at temperature 94^{0}K is shown in fig. 30.

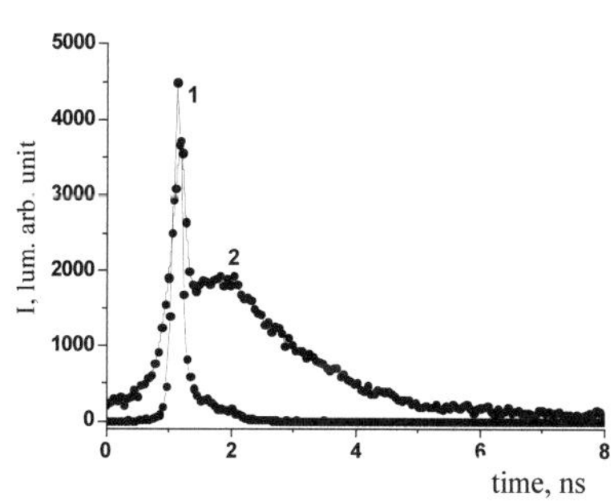

Fig. 30. Signal instrument function (1) and decay kinetics luminescence curve J-aggregates PIC2-15 in the film (2) excitation λ=574 nm measurement at 590 nm,T=96°K

The fast and slow components of the luminescence decay take place on the luminescent decay kinetic curve. The signal instrument function was 80 ps. The expanding of the instrument function due to the appearance of the fast component is 20-40 ps. The life time of the slow component is 2 ns. The period of the luminescence growth of the slow component takes place with characteristic time 0.1-0.2 ns. The existence of the fast and slow components is typical for the dye molecules in polar liquid and solid phases. Some relaxations times from picoseconds to nanoseconds relate to different processes of the solvate shell relaxation depending on the characteristic times of the anionic and cationic parts of the medium relaxation in the case of organic or inorganic salts (Ferrante et al., 1998; Saha et al., 2004; Das et al., 1996; Mandal et al., 2002; Arzhantsev et al., 2003). The luminescence of the J-aggregate in the thin solid film occurs from the two states: the state of the fast matrix relaxation and state of the slow matrix relaxation which is populated in a time of about 100 ps. Different luminescent life times 20-40 ps, 128, 659 пс, 1.7 нс were reported depending on the preparation condition and irradiation intensity (Sundstrom et al., 1988) for the PIC J-aggregates in solvents.

The initial values of the parameters of the model are determined as: J-aggregate extinction coefficient ε_{10}=2 10^5 M^{-1}см$^{-1}$, the same value for S_2-S_3 transition ε_{32}=2 10^5 M^{-1}см$^{-1}$, τ_2=100 ps (from the time of the slow component luminescence increasing), τ_3=2 10^{-9}s (the life time of the luminescence decay slow component), τ_1=10^{-11}s is a typical value of the non-radiation relaxation in organic molecules, the relation of the width values of the transition S_{1-0} and S_{3-2} spectral contours was taken from measurement of dispersion of the non-linear bleaching and darkening as 150 and 120cm^{-1}. The model describing the dispersion of the induced transmission in the non-linear experiment is composed by including into eq. (28) the value of the $\sigma_\lambda(\varepsilon)$ for the transition S_{1-0} and S_{3-2} in the form of the Lorentz contour $\sigma_\lambda = \sigma_0 \cdot \Delta\upsilon^2 / 4 \cdot (\upsilon - \upsilon_0) + \Delta\upsilon^2$.

The model curve (the red curve in fig. 31a) is close to the experimental curve. The moderate change in the relaxation time $\mathbf{S_1}$ level to $\mathbf{S_2}$ level from 100 ps to 150 ps leads to the well fitting of the model (blue curve) and experimental curves. The shift of the maximum of the non-linear absorption (λ=580 nm) from the maximum of the non-linear transmission (λ=575 nm) on the 150 cm^{-1} follows from the calculation. It corresponds to the shift of the luminescence maximum relative to the absorption maximum of the J-aggregates in the thin solid films.

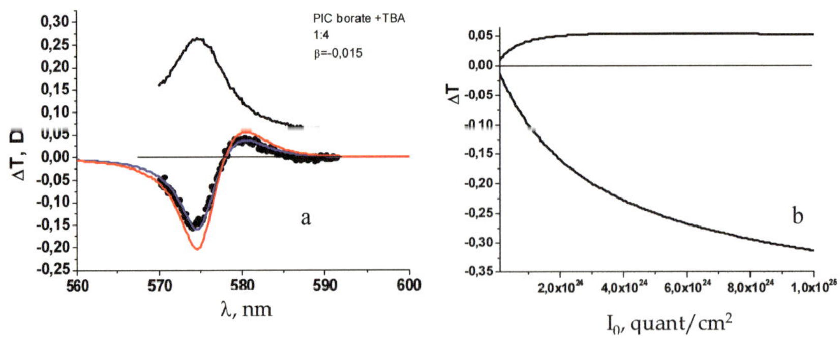

Fig. 31. The calculated (blue and red curves) and experimental (points) non-linear transmission dispersion curves for the J-aggregated PIC ($B_{10}H_{10}^{2-}$)+TBA (1:4) film (a); The calculated induced absorption (upper curve) and transmission (bottom curve) depending on the intensity laser irradiation (b)

The laser radiation intensity dependence on the non-linear transmission of the J-aggregated film sample was calculated for the obtained specified model parameters for the intensity change in the range 10^{23}-10^{25} quant/cm^2 in the maximums of the induced non-linear bleaching and darkening as shown in fig. 31b.

The experimental values of the induced darkening (ΔT=-0.05) and bleaching (ΔT=-0.2) are in good agreement with the calculated values at the experimental achieved intensities 10^6 W/cm^2 ($3 \cdot 10^{24}$ quant/cm^2). The calculated curves also coincide well with the existence of the experimentally observed saturation, as well in induced bleaching as in the induced darkening. At intensity of laser irradiation of more then $2*10^6$ W/cm^2 ($6 \cdot 10^{24}$ quant/cm^2) the irreversible burning of the dye film begins and observation of the non-linear effect becomes difficult.

Using the four-level model for J-aggregates appearing from the polarization relaxation of J-aggregate excites state is closely connected with the inhomogeneous broadening of J-aggregate absorption and luminescence spectrum, and can help us to explain the giant cubic non-linear properties of PIC J-aggregates in thin films. The big resonance values of the cubic susceptibility at the level $\chi^{(3)}=10^{-7}$ esu observed in the water solutions or in the polymer matrixes at high relation polymer:dye (>10:1) are caused by the high oscillator strength of the resonance transition in two level system (Zhuravlev et al., 1992; Shelkovnikov et al., 1993). The J-aggregate in the water is surrounded by weak polarizable water molecules and the excitation-induced medium polarization has a weak influence on the change of the energy of the excited electron level of the aggregate. The non-linear response in this case is the response of the two-level system. In the solid film the J-aggregate is surrounded by highly polarizable molecules of the dye. In this case induced medium polarization leads to the deep relaxation of the energy of the excited $\mathbf{S_1}$ level of J-aggregate to the relaxed state with the energy lowering to 150-170cm^{-1}. This process lasts 150 ps and in the laser pulse duration time 5 ns the effective population of the relaxed level takes place. This effect has the reflection in the linear spectrum as the inhomogeneous broadening of the absorption and luminescence contours of the J-aggregates in films. The decreasing of the population non-relaxed $\mathbf{S_1}$ level leads to the essential non-linear bleaching of the J-aggregates and to the appearance of the giant non-linear susceptibility of the J-aggregates. The excitation of the populated relaxed level leads to the appearance non-linear darkening at the long wavelength slope of the J-peak in solid films. The non-linear response of the J-aggregates in the solid films is the response of the four-level system and this leads to the increasing cubic susceptibility by two orders of magnitude. The additional reason of the non-linear response increasing in solid film is the increasing of the local field factor ($F_a = \frac{\left(n^2+2\right)^2}{9n}$) due to surrounding of the J-aggregate by highly polarizable molecules of the dye. The $\boldsymbol{F_a}$ is increased by 1.65 times at the transition from dye in the solvents to the dye in the film. The cubic susceptibility depends on the four degrees of the local field factor $\chi^{(3)} = \gamma \cdot F_a^{\ 4} \cdot C$ (where C is the concentration, γ is the own polarizability of the molecule). This gives rise to the non-linear response up to seven times more.

2.11 The quantum chemical calculation of monomer and dimer PIC

The calculation of the charge distribution in the ground and excited state in the PIC2-2 molecule was carried out using the semi-empirical quantum chemical AM1 method. The charge distribution in the ground state of the cation PIC is shown in part 8 of this chapter. It coincides with the charge distribution calculated using the theory of the functional density method used in Guo et al. (2002) on the qualitative level. The charge redistribution on the atoms of the PIC molecule takes place at the excitation. Let take the transition dipole moment of the molecule be proportional to the electron density change at the excitation. The value of the electron density change between the ground and excited state was estimated by the change of the values of the dipole moments of the bounds between the neighbour atoms of the PIC π-system. The directions of the charge redistribution in the excited state of the PIC are shown in fig. 32a for one half of the molecule.

The redistribution of the electron density in the excited state of PIC leads to the next induced dipole moments in the molecule: the projection of the induced dipole moment along the X axe M_x=1.38D, along the Y axe M_y=1.44D. That means the total dipole induced by the electromagnetic field of the light wave is oscillated at the angle ~45° in the XY plane of the molecule as shown in fig. 32b.

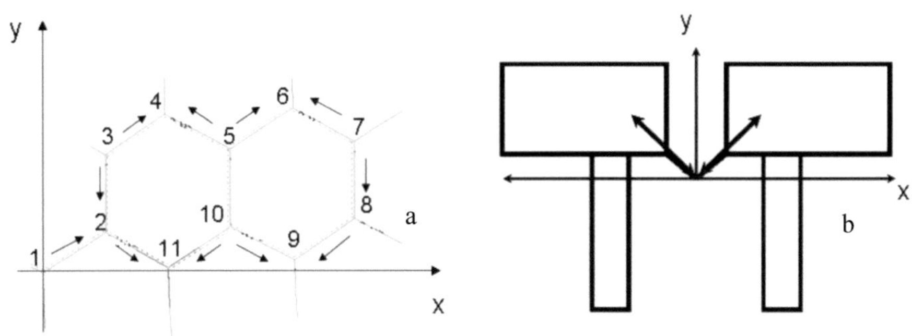

Fig. 32. The charge redistribution (a) and electron density oscillation at the PIC molecule excitation (b)

Since the directions of the projections of the induced dipoles vectors on the X axe are in opposition to each other, then only the induced dipoles vector projection directed along the Y axe is kept safe. It appears from this that the transition dipole moment is polarized along the short axis of PIC molecule.

The approximation of the dipole-dipole interaction between excited molecules in J-aggregate (Kuhn, H., Kuhn, C. 1996) does not consider the orbital overlap of the molecules. On the short distance between molecules in the aggregate, about 3-4 Å orbital overlap is essential. The existence of the orbital overlap between the neighbour dye molecules leads to the existence of the electron exchange between the orbitals and to the splitting of the electron levels of the two combined molecules. The exchange interaction has repulsive character for the system with closed shell as the dye in the ground state. This leads to the increase of energy of the two neighbour molecules in dimer. The excited state is the state with open shell and the dimer excitation takes place as as well to the high as to the low energy excited state. The energy of the splitting has the exponential decay dependence on the distance and at the distance more than 10Å is about some inverse centimetres. But at the distance between molecules in dimer 3-4Å the energy of the splitting is hundreds of the inverse centimetres.

The splitting energy induced by exchanging orbital interaction is calculated via the resonance integral (***β***) distance dependence. The lowering of the energy of the transition is equal ***β*** and the value of the splitting is 2***β***. The excited state of the long wavelength transition of the PIC is π–π* excited state and it is reasonable to estimate the energy of the exchange interaction between molecules in dimer for the π–orbitals. For the estimation in the first approximation of the ***β*** value without application for the calculation of the advanced quantum-chemical methods it is possible to use the distance dependence of the resonance integral $\beta_{res}(r)$ given via calculation of the empiric integrals in the basis of the Slater-type orbitals for the π–orbitals of the conjugated double bounds in molecule that was cited in Warshell (1977).

$$\beta_{res}(r) := \left[2.438 \cdot e^{-2.035 \cdot (r-1.397)}\right] \cdot [1 + 0.405 \cdot (r - 1.397)] \cdot 8065\,(cm^{-1}) \qquad (29)$$

The calculated value β_{res} for the distance 3,4-3,6 Å is 600-420 cm^{-1}. The estimated value β_{res} gives an image of the upper boundary of the lowering of the energy for two interacting π–π orbitals because it was set for the orbitals in molecule. In the some of publication the short-range electron-exchange or electron/hole charge transfer between molecules was included in the calculation for the considering of the excited state of aromatic dimers as the exciton state, (Tretiak, 2000). It was shown that intermolecular electron exchange coherent interaction leads to a crucial red shift in the dimer spectrum and completely invalidates the simple Frenkel exciton model.

The splitting and oscillator strength of the PIC-dimer depend on the configuration of the interacting orbitals and thus the arrangement geometry of the dyes in the dimer. The energetic scheme of the electron levels splitting in the dimer is shown in fig. 33. The energy of singlet transitions in absorption spectrum of PIC in monomer form was calculated using the semi-empirical quantum chemical method ZINDO/S with preliminary geometry optimization utilizing the semi-empirical quantum chemical method AM1. The calculated PIC monomer structure in the model of the overlapping spheres is shown in fig. 34. The calculated long wavelength singlet transition at 509 nm with oscillator strength 1.3 is in accordance with the absorption of molecule PIC in a vacuum (calculation from the equation 24).

The calculations of energy of transitions in PIC dimers on the basis of the supramolecular approach using the semi-empirical quantum chemical method ZINDO/S with dimer geometry optimization utilizing the molecular mechanic method MM+ shows that the two-level splitting of the energy of excited state of dimer and magnitude of splitting depends on the angle between PIC molecules. It is significant that in the previous PIC dimer calculations (Kuhn & Kuhn, 1996; Burshtein et al., 1997) the linear shift of the molecules was considered. Here we consider the model of the molecules rotation in the PIC dimer. The calculation of the energy of the dimer by MM^+ method was accompanied by the optimization anion location at each step of the rotation equal to 5° of the PIC1 molecule in ZX plane relative to the PIC2 molecule as shown in fig. 35. After the whole cycle of the rotation by 90°, the distance L along the Y axe was changed and the cycle was repeated.

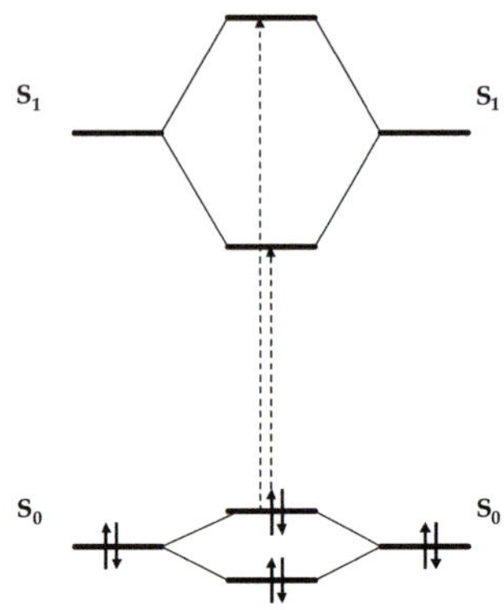

Fig. 33. The energetic scheme of the dimer levels splitting

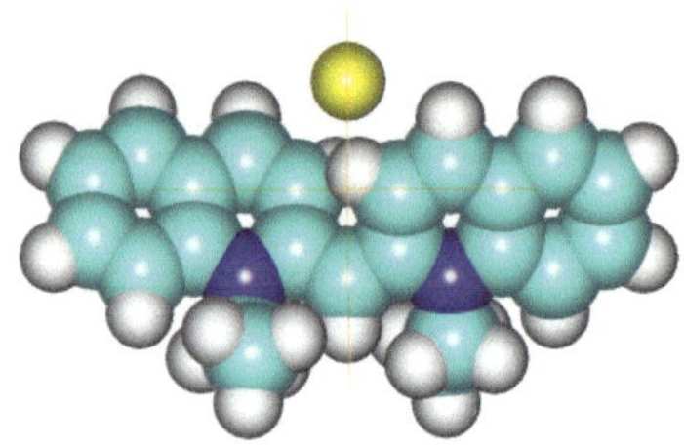

Fig. 34. The overlapping sphere model of the PIC iodide

The dimer energy distance (L) dependence for the three angles PIC molecules rotation perpendicular, parallel and at the angle of rotation 20° is shown in fig. 36. One can see that parallel molecules' orientation shows the weak energy distance dependence at the molecules' rapprochement to the 9.5Å. In further, the high energy growth in dimer takes place. The reciprocal rotation of the molecules on the 20° degree leads to clear energy distance dependence with the minimum at 7.7Å. The close energy minimum at 7.8Å has the dimer with the perpendicular molecular orientation.

The energy and oscillator strength of the PIC dimer at the rotation of the molecules divided on distance 7.8 Å were calculated using the ZINDO/S method, including 3 orbitals in configuration interactions (CI). The results of the calculation for two first allowed singlet transitions are shown in fig. 37.

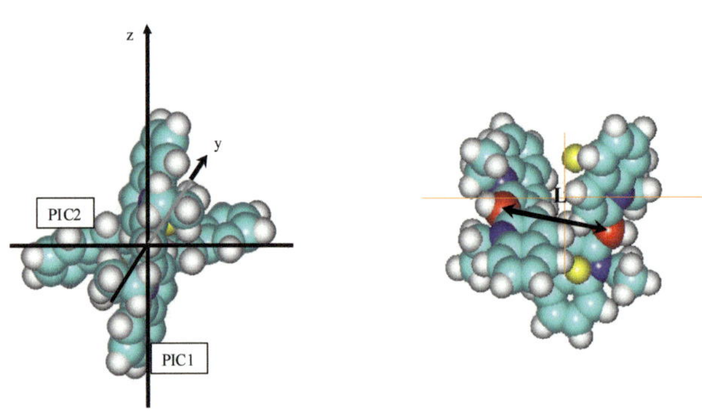

Fig. 35. The two profiles of the PIC molecules orientation in the dimer

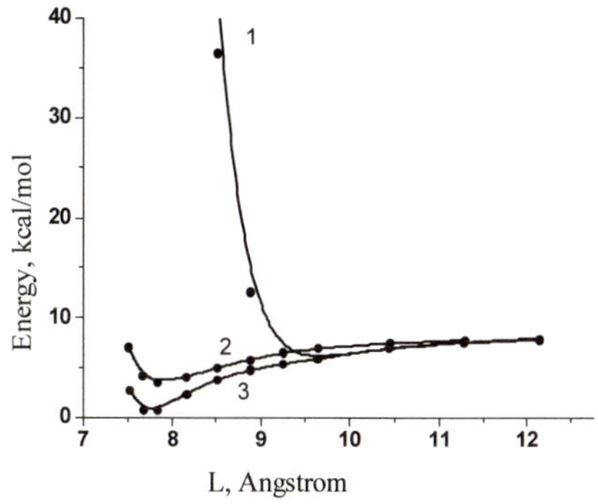

Fig. 36. The energy of the dimer molecular distance dependence for the three angles of the molecules rotation in ZX plane: 0° (1), 90° (2), 20° (3)

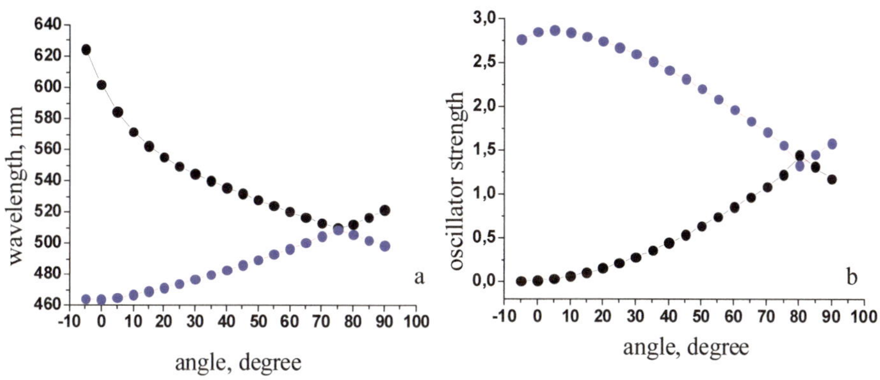

Fig. 37. The values of the wavelengths (a) and ascillator strength (b) for the two allowed singlet transitions in the PIC dimer depending on the rotation angle in ZX plane

The rotation of the dyes in dimer relative to axes connecting the centres of molecules leads to the splitting of the electronic levels giving the rise of the intensity of the short wavelength transition at parallel molecules orientation. That situation corresponds with appearance of the spectral absorption band of H-aggregates of PIC. At the angle between molecules of 75 degrees, which is close to a perpendicular orientation of the dye molecules, the energy level splitting becomes close to zero leading to the degeneration of level splitting with formation of a high intensity absorption batochromic peak with doubled oscillator strength of transition. The red shift of this dimer state in vacuum takes place to 670 cm^{-1} if we include in the CI calculation the enlarged number of orbitals (for an example 15) to consider the more full account of electron exchange in the dimer excited state. From our point of view that situation corresponds to the formation absorption peak of J-aggregates, which has the additional red shift due to the surrounding of a high polarization medium in a thin solid film of dye.

The reason for the rotation at the dimerization of the PIC molecules is the influence of the two forces: the Coulomb interaction of the two cations and anions which try to bring the particles together and repulsive exchange interaction of the molecules' in the ground state . The overlap integral of the external π-electron orbitals of the carbon atoms and σ-electron orbitals of the hydrogen atoms is decreasing at the rotation of the molecules to the perpendicular position of the molecules regarding each other. That leads to the decreasing of the repulsive part of the molecular interaction and the molecules are brought together until the new equilibrium between the forces is established.

3. Conclusion

The medium with the high polarizability compared with the polarizability of the dye in the solutions arises as a result of the dye thin solid film formation. This creates the condition for the appearance, as well the high negative value of the non-linear cubic susceptibility in resonance as the change of the sign on the dispersion curve of the cubic susceptibility and appearance the darkening on the long wavelength slope of the J-peak absorption. In the medium with high polarizability the considering of the four-level system for the J-aggregates excited state description is more appropriate compared with the two-level system typically considered for the J-aggregates. The depopulation of J-aggregate Frank-Condon excited state due to the appearance of lowered relaxed state leads to giant non-linear bleaching of J-aggregate peak and the excitation of the relaxed state leads to the non-linear absorption appearance. The value of non-linear response strongly depends on the relaxation time. The decrease of the relaxation time leads to the growth of non-linear response.

The increase of the medium polarizability enhances the non-linear response and is reflected in the increasing of the inhomogeneous broadening in the absorption and luminescence of the J-aggregates' spectra. The formation of the medium with high polarizability is clear seen from the spectral shift of the dye absorption during spin-coating and helps to explain the J-peak growth at the solid film formation. The inhomogeneous broadening in the J-peak absorption has the connection with aggregate thermal decay energy activation distribution and spectral inequivalent in the kinetics of the thermal decay of the J-aggregate absorption counter. The medium polarization influence on the optical and non-linear optical properties is important for the aggregated or supramolecular state of different dyes and nanostructures in the condensed phase.

The number of facts allow us to consider the properties of the pseudoisocyanine J-aggregates in the thin solid films as the properties of the strongly coupled dimers with inhomogeneous broadened spectral contour: the existence of the isobestic point at the J-aggregates' thermal conversion to monomer dye in the thin solid films or in the polymer films, the absence of the any hypsochromic spectral shift at the transition of the J-aggregate to monomer in the thin films at the addition of the octadecylquinolinium iodide, the high stabilization of the J-aggregate at the addition of the dipole closo-hydrodecaborate anion, the batochromic spectral shift of the J-peak at heating of the dye thin film, the narrowing of the J-peak in the thin solid films in the presence of organic cations and others.

The practical use of the obtained results on the non-linear properties is in the possible application of cyanine dyes' J-aggregates as the elements for teraherz demultiplexing of light signals and in schemes of ultra-short laser pulse stabilization (Plekhanov et al., 2004). The radiation stability of the J-aggregates is at a level of 0.5-1 MW/cm^2. This is useful to increase the radiation stability of the J-aggregates by 5-10 times. For safer J-aggregated films, application is enough to increase the thermal stability to 100-150^oC. However, the search for new highly polarized aggregates of dyes with high absorption coefficients in the visible and infra-red spectral range that possess high thermal and photochemical stability remains a basic and practical task.

4. Acknowledgment

This study was supported by the Russian Foundation for Basic Research (project no. 02-03-33336), the programme "Fundamental Problems of Physics and Chemistry of Nanosystems and Nanomaterials" of the Presidium of the Russian Academy of Sciences (grant no. 8-2) and the programme of interdisciplinary integrated investigations of the Siberian Division of the Russian Academy of Sciences (project no. 84).

5. References

Arzhantsev, S., Ito, N., Heitz, M., Maroncelli, M. (2003). Solvation dynamics of coumarin 153 in several classes of ionic liquids: cation dependence of the ultrafast component. *Chem. Phys. Lett.* Vol. 381. pp. 278-286.

Bakalis, L. D. and Knoester, J. (2000). Linear absorption as a tool to measure the exciton delocalization length in molecular assemblies. *J. Lumin.* Vol. 87-89. pp.67–70.

Bakhshiev, N. G. Libov, V. S. Mazurenko, Yu. T. Amelichev, V. A. Saidov, G. V. and Gorodynskii, V. A. (1989). Solvation Chemistry: Problems and Methods [in Russian], Leningr. Gos. Univ., Leningrad (Rus.)

Blankenship, R.E., Olson, J.M., Miller, M. (1995). Antenna complexes from green photosynthetic bacteria, In: *Anoxygenic photosynthetic bacteria.* R.E. Blankenship, M.T. Madigan, and C.E. Bauer (eds.), Kluwer Academic Publish., Dordrecht, pp. 399-435.

Bogdanov, V.L., Viktorova, E N., Kulya, S V., and Spiro, A.S. (1991). Nonlinear cubic susceptibility and dephasing of exciton transitions in molecular aggregates, *JETP Letters.* Vol. 53, pp.105-108.

Burshtein, K. Ya., Bagaturyanz, A.A., Alfimov M.V. (1997). Computer modeling of the absorption line of the J-aggregates. *Izv. AN. ser. Khim* Vol.1. pp.67-69 (Rus.)

Cerasimova, T.N., Orlova, N. A., Shelkovnikov, V.V., Ivanova, Z. M., Markov, R. V., Plekhanov, A,I., Polyanskaya, T. M., Volkov, V. V. (2000). The Structure of Pseudoisocyanine Decahydro-*closo*-decaborate and Its Nonlinear Optical Properties in Thin Films. *Chemistry for Sustainable Development.* Vol.8, pp.109-114.

Coates, E. (1969). Aggregation of dyes in aqueous solution. *JSDS.* pp. 355-368.

Daltrozzo, E., Scheibe, G., Geschwind, K., Haimerl, F. (1974). On the structure of J-aggregates of pseudoisocyanine *Photogr. Sci. Eng.* Vol. 18. № 4. pp. 441-449.],

Das, K., Sarkar, N., Das, S., Datta, A., Bhattacharya, K. (1996). Solvation dynamics in solid host. Coumarin 480 in zeolite 13X. *Chem. Phys. Lett.* Vol. 249. pp. 323-328.

Ferrante, C., Rau, J., Deeg, F.W., Brauchle, C. (1998). Solvatation dynamics of ionic dyes in the isotropic phase of liquid crystals. *J. Luminesc.* Vol. 76-77. pp. 64-67.

Furuki, M., Tian, M., Sato, Y., Pu L.S. (2000). Terahertz demultiplexing by a single short time-to-space conversion using a film of squaryliun dye J-aggregates. *Appl. Phys. Lett.* Vol.77. pp.472-474.

Ghasemi, J.B., Mandoumi, N.A. (2008). New algorithm for the characterization of thermodynamics of monomer-dimer process of dye stuffs by photometric temperature titration. *Acta Chim. Slov.* Vol. 55. pp. 377-384.

Glaeske H., Malyshev V.A., Feller K.-H. (2001). Mirrorless optical bistability of an ultrathin glassy film built up of oriented J-aggregates: Effects of two-exciton states and exciton-exciton annihilation, *J. Chem. Phys.* Vol.114. pp.1966-1969

Guo, Ch., Aydin, M., Zyu, H-R., Akins, D.L. (2002). Density functional theory used in structure determinations and Raman band assignments for pseudoisocyanine and its aggregate. *J. Phys. Chem. B.* Vol.106. pp. 5447-5454

Herz, A.N. (1974). Dye-Dye interactions of cyanines in solution and at silver bromide surfaces. *Photogr. Sci. Engineering.* Vol. 18. №3. pp. 323-335

Jelley, E., (1936). Spectral absorption and fluorescence of dyes in the molecular state. *Nature,* Vol.138, pp. 1009-1010.,

Katrich, G.S., Kemnitz, K., Malyukin, Yu.V., Ratner, A.M. (2000). Distinctive features of exciton self-trapping in quasi-one-dimensional molecular chains (J-aggregates). *J. Luminesc.* Vol. 90. pp. 55-71

Kobayashi, T., ed. (1996). *J-aggregates,* World Scientific Publish. Co Pte. Ltd., Singapore.

Krasnov, K.S., (1984). *Molecules and chemistry bond,* High school. Moscow. (Rus.)

Kuhn, H., Kuhn, C. (1996). Chromophore coupling effects. In: *J-Aggregates.* T. Kobayashi (ed.)- Singapore: World scientific publishing Co. Pte. Ltd., - 228 p.

Lakowics, J.R. (1983). *Principles of fluorescence spectroscopy.* Plenum press. New York and London

Levshin, L.V., Salecky, A.M. (1989). *Luminescence measurement.* MGY. Moscow. (Rus.)

Mal'tsev E.I., Lypenko D.A., Shapiro B.I., and Brusentseva M.A. (1999). Electroluminescence of polymer/J-aggregate composites. *Appl. Phys. Lett.* Vol. 75, pp.1896-1898.

Malyshev, V.A. (1993). Localization length of one-dimensional exciton and low-temperature behaviour of radiative lifetime of J-aggregated dye solutions. *J. Luminesc.* Vol. 55. pp. 225-230.

Mandal, D., Sen, S., Bhattacharya, K., Tahara, T. (2002). Femtosecond study of solvation dynamics of DCM in micelles. *Chem. Phys. Lett.* Vol. 359. pp. 77-82.

Markov, R. V., Chubakov, P. A., Plekhanov, A. I., Ivanova, Z. M., Orlova, N. A., Gerasimova, T. N., Shelkovnikov, V. V., and Knoester, J. (2000). Optical and nonlinear optical properties of low-dimensional aggregates of amphyphilic cyanine dyes. *Nonlinear Opt.*, Vol.25, pp.365–371

Markov, R.V., Plekhanov, A.I., Rautian, S.G., Orlova, N.A., Shelkovnikov, V.V., Volkov, V.V. (1998). Nonlinear optical properties of two types of PIC J-aggregates in thin films. *Proc. SPIE.* Vol. 3347. pp. 176-183. (a)

Markov, R.V., Plekhanov, A.I., Rautian, S.G., Orlova, N.A., Shelkovnikov, V.V., Volkov, V.V. (1998). Nonlinear optical properties of the two types pseudoisocyanine J-aggregates in thin films. *Jurnal Nauchn. and Prikl. Fotogr.* Vol. 43. pp.41-47. (Rus.) (b)

Markov, R.V., Plekhanov, A.I., Rautian, S.G., Safonov, V.P., Orlova, N.A., Shelkovnikov, V.V., Volkov, V.V. (1998). Dispersion of cubic susceptibility of thin films of pseudoisocyanine J-aggregates as measured by longitudinal scanning, *Optics and Spectroscopy.* Vol. 85. pp.588-594. (c)

McDermott, G., Prince, S.M., Freer, A.A. et al., (1995). Crystal structure of an integral membrane light-harvesting complex from photosynthetic bacteria, *Nature*. Vol.374. pp. 517-521.

Minoshima, K., Taiji, M., Misawa K. and Kobayashi, T. (1994). Femtosecond nonlinear optical dynamics on excitons in J-aggregates. *Chem. Phys. Lett.* Vol.218. pp.67–72

Moll, J., Daehne, S., Durrant, J. R. and Wiersma, D. A. (1995). Optical dynamics of excitons in J-aggregates of carbocyanine dye. *J. Chem. Phys.* Vol.102. No. 16. pp.6362–6370

Nygren, J., Andrade, J. M., Kubista, M. (1996). Characterization of a single sample by combining thermodynamic and spectroscopic information in spectral analysis. *Anal. Chem.* Vol. 68. pp. 1706-1710.

Orlova, N.A., Kolchina, E.F., Zhuravlev, F.A., Shakirov, M.M., Gerasimova, T.N., Shelkovnikov, V.V.(2002). Synthesis of 2,2'-quinocyanines with long alkyl groups. *Chemistry Heterocyclic Compounds*. № 10, pp. 1399-1407. (Rus.)

Orlova, N.A., Zhuravlev, F.A., Shelkovnikov, V.V., Gerasimova, T.N. (1995). Synthesis of pseudoisocyanines with nonsaturated groups in position 1. *Izv. AN, ser, khim.* № 6, pp. 1122-1124 (Rus.)

Pilling, R.L., Hawthorne, M.F. (1964). The boron-11 nuclear magnetic resonance spectrum of $B_{20}H_{18}^{-2}$ at 60 Mc./sec.*J. Amer. Chem. Soc.* p Vol. 86. pp. 3568-3569.,

Plekhanov A.I., Rautian S.G., Safonov V.P., (1995*). Optics and Spectroscopy* Vol.78, 1, p.92. .(Rus.)

Plekhanov, A.I., Kuch'yanov, A.S., Markov, R.V., Simanchuk, A.E., Avdeeva, V.I., Shapiro, B.I., Shelkovnikov, V.V. (2004). Passive mode locking of a Nd3+:YAG laser with a saturable absorber in the form of thin film of J-aggregates. J. *Nonlinear Org. and Polymer Materials.* Vol. 9. № 3. pp. 503–511.

Plekhanov, A.I., Orlova, N.A., Shelkovnikov, V.V., Markov, R.V., Rautian, S.G., Volkov, V.V. (1998). Third-order non-linearity optical properties of the film of the cyanine dye with borate anion. *Proc. SPIE.* Vol. 3473. pp. 100-107. (a)

Plekhanov, A.I., Rautian, S.G., Safonov, V.P., Chubakov, P.A., Orlova, N.A., Shelkovnikov, V.V (1998). Dispersion of the real and imaginary parts of cubic susceptibility in submicron films of pseudoisocyanine J-aggregates. *Proc. SPIE.* Vol. 3485. pp. 418-424.(b)

Plekhanov, A.I., Markov, R.V., Rautian, S.G., Orlova, N.A., Shelkovnikov, V.V., Volkov, V.V (1998). Third-order nonlinearity optical properties of the films of cyanine dye with borate anion. *Proc. SPIE "Third-Order Nonlinear Optical Materials"*. Vol. 3473. pp. 20-31.(c)

Renge, I., Wild, U.P. (1997). Solvent, temperature, and excitonic effects in the optical spectra of pseudoisocyanine monomer and J-aggregates. *J. Phys. Chem. A.* Vol. 01. pp. 7977-7988.

Saha, S., Mandal, P.K., Samanta, A. (2004). Solvation dynamics of Nile Red in a room temperature ionic liquid usingstreak camera. *Chem. Phys.* Vol. 6. pp. 3106-3110.

Sato, T., Yonezawa, Y., Hada, H. (1989). Preparation and luminescence properties of J-aggregates of cyanine dyes at the phospholipid vesicle surface. *J. Phys. Chem.* Vol. 93. pp. 14-16.

Scheibe, G., (1936). Variability of the absorption spectra of some sensitizing dyes and its cause. *Angew. Chem,* Vol.49, p. 563.

Shapiro, B.I. (1994) Aggregates of cyanine dyes: photographic problems, *Russian Chemical Reviews.* Vol.63. pp.231-241.

Shelkovnikov, V. V., Ivanova, Z. M., Orlova, N. A., Volkov, V. V., Drozdova, M. K., Myakishev, K. G., Plekhanov, A. I.(2004). Optical Properties of Solid Pseudoisocyanine Films Doped with Cluster Derivatives of Boron Hydrides. *Optics and Spectroscopy*. Vol. 96, pp.824-833

Shelkovnikov, V.V., Plekhanov, A.I., Safonov, V.P., Zhuravlev, F.A. (1993) Nonlinear optical properties of the assembliesof organic molecules and fractal metal clusters. *Zhurnal struct. chem*. Vol.34. pp.90-105 (Rus.)

Shelkovnikov,V. V., Ivanova, Z. M., Orlova, N. A., Gerasimova, T. N., Plekhanov, A. I. (2002). Formation and properties of long alkyl substituted pseudoisocyanines J-aggregates in thin films. *Optics and Spectroscopy*. Vol.92. pp. 958–966

Struganova, I. (2000) Dynamics of formation of 1,1'-diethyl-2,2'-cyanide iodide J-aggregates in solution. *J. Phys. Chem. A*. Vol. 104. № 43. pp. 9670-9674.

Sundstrom, V., Gillbro, T., Gadonas, R.A., Piskarskas, A. (1988). Annihilation of singlet excitons in J-aggregates of pseudoisocyanine (PIC) studied by pico- and subpicosecond spectroscopy. *J. Chem. Phys.* Vol. 89. № 5. pp. 2754-2762.

Suppan, P. (1990). Solvatochromic shifts - the influence of the medium on the energy of electronic states. *J. Photochem. Photobiol. A: Chemistry*. Vol. 50. pp.293-330.

Tani, I.., Liu-Yi, Sasaki, F., Kobayashi, S., Nakatsuka, H. (1996). Persistent spectral hole-burning of pseudoisocyanine bromide J-aggregates. *J. Luminesc.* № 66-67. pp.157-163.

Tani, T. (1996) J-aggregates in spectral sensitization of photographic materials, In: *J-aggregates*. T. Kobayashi (ed.). World Scientific Publish. Co Pte. Ltd., Singapore. pp. 209-228.,

Tikhonov, E.A., Shpak, M.T. (1977). *Nonlinear optical phenomena in organic compounds*. Naukova dumka. Kiev.

Tretiak, S., Saxena, A., Martin, R. L., and Bishop, A. R. (2000). Interchain Electronic Excitations in Poly(phenylenevinylene) (PPV) Aggregates. *J. Phys. Chem. B*, Vol.104, pp.7029-7037

Trosken, B., Willig, F., Spitle, R. M. (1995) The primary steps in photography: excited J-aggregates on AgBr microcrystals, *Advanced Materials*, Vol. 7, pp. 448-450. ,

Vacha, M., Furuki, M., Tani, T. (1998). Origin of the long wavelength fluorescence band in some preparations of J-aggregates low-temperature fluorescence and hole burning study. *J. Phys. Chem. B*. Vol. 102. pp. 1916-1919.,

Verezchagin A.N. (1980). *Polarisability of molecules*. Nauka. Moskow. (Rus.)

Wang Y. (1991) Resonant third-order optical nonlinearity of molecular aggregates with low-dimensional excitons, *Journal of the Optical Society of America B*. Vol.8, pp.981-990.,

Warshell A. (1977). The self-consistent force field method and quantum chemical generalization, In: *Semiempirical methods of electronic structure calculation*. Segal. G.A. (ed.). Plenum press. New York and London

Wendlandt W.W. (1974). *Thermal Methods of Analysis*. John Wiley & Sons, Inc. New York.

Wurthner, F., Kaiser, Th.E., Saha-Moller, Ch.R. (2011). J-Aggregates: From Serendipitous Discovery to Supramolecular Engineering of Functional Dye Materials. *Angew. Chem. Int. Ed.* Vol.50. pp. 3376 – 3410.

Zhuravlev, F.A., Orlova, N.A., Shelkovnikov, V.V., Plekhanov, A.I., Rautian, S.G., Safonov, V.P. (1992)Giant non-linear susceptibility of the thin films with complexes molecular aggregate – metal cluster. *Pis'ma JETF* . Vol.56. pp.264-267 (Rus.)

17

A Comparative Study of Analytical Methods for Determination of Polyphenols in Wine by HPLC/UV-Vis, Spectrophotometry and Chemiluminometry

Vesna Weingerl*
University of Maribor, Faculty of Agriculture and Life Sciences, Hoče, Slovenia

1. Introduction

Wine, especially red wine, is a very rich source of polyphenols, such as flavanols (catechin, epicatechin, etc.), flavonols (quercetin, rutin, myricetin, etc.), anthocyanins (the most abundant is malvidin-3-o-glucoside), oligomeric and polymeric proanthocyanidins, phenolic acids (gallic acid, caffeic acid, p-coumaric acid, etc.), stilbenes (*trans*-resveratrol) and many others polyphenols. Many of these compounds (e.g. resveratrol, quercetin, rutin, catechin and their oligomers and polymers proanthocyanidins) have been reported to have multiple biological activities, including cardioprotective, anti-inflammatory, anti-carcinogenic, antiviral and antibacterial properties (King et al., 2006; Santos- Buelega & Scalbert, 2000). These biological properties are attributed mainly to their powerful antioxidant and antiradical activity.

Regular, moderate consumption of red wine reduced the incidence of many diseases such as risk of coronary heart disease (CHD), atherosclerosis, cancers, etc. (Cooper et al., 2004; Opie & Lecour, 2007). The most intriguing are the studies which reported the possible association between red wine consumption and decrease in risk, and some suppression and inhibition of cancers (Briviba et al., 2002). Currently, chemoprevention is being used in medicine as a new strategy to prevent cancers. Natural phytochemicals, including red wine polyphenols, appear to be very promising substances to block, reverse, retard or prevent the process of carcinogenesis (Russo, 2007). Many epidemiological studies have found that regular intake of red wine or red wine polyphenols has positive effects on human health. Therefore, determination of the chemical composition, polyphenols content and antioxidant activity of red wine could be very useful for the interpretation of epidemiological studies.

Phenolic antioxidants define total antioxidant potential of wines and have the greatest influence on it. Authors showed that in grape seeds gallic acid, catechins and epicatechins prevailed, whereas in peel ellagic acid, quercetin and trans-resveratrol were most common.

* Corresponding Author

The high antioxidant potential of red wines can be ascribed to the synergistic effect of the mixture of natural phenolic antioxidants (Lopez-Velez et al., 2003).

Production technology is one of the main factors influencing the high antioxidant potential of red wines (Downey et al., 2006; Vršič et al., 2009). During winemaking the grape pulp is fermented, and fruit peel and seeds are very rich in phenolic antioxidants. The concentration of polyphenols in peel is higher than in the flesh (Darias-Martin et al., 2000; Fuhrman et al., 2001). During intensive pressing or during long contact of juice with pulp the content of phenolic compounds increases rapidly (Fuhrman et al., 2001). It was found that in wines fermented with peel the concentration of phenolic antioxidants was 2 times higher than in wines fermented without peel (Darias Martin et al., 2000).

Antioxidant potential and polyphenol composition were assessed in wine of Croatian origin (Katalinic et al., 2004). The concentration of total polyphenols in red wines ranged from 2200 to 3200 mg gallic acid per litter (mg GA/L).

In winemaking, phenolic antioxidants are extracted from berry skins, seeds and stems during crushing and fermentation. Due to the market demand, knowledge of the concentration of phenolic antioxidants in wine and their antioxidant potential is very important.

Wine, especially red wine, is a very rich source of flavonol quercetin and many others polyphenols. Various methods for characterisation of total antioxidant potential are presently in routine use, although some are non-stoichiometric (Alimelli et al., 2007; Campanella et al., 2004; Careri et al., 2003; Carralero Sanz et al., 2005; De Beer et al., 2005; Fernandez-Pachon et al., 2004; Giovanelli, 2005; Gomez-Alonso et al., 2007; Magalhaes et al., 2009; Makris et al., 2003; Malovana et al, 2001; Mozetič et al., 2006; Prior et al., 2005; Prosen et al., 2007; Recamales et al., 2006; Spigno & De Faveri, 2007; Staško et al., 2008; Weingerl et al., 2009; Woraratphoka, 2007). Some of these methods allow for rapid characterization of wines, and allow for evaluation of synergistic effects of various wine components, e.g. transition metals (Strlič et al., 2002), which can have pro-oxidative effects in a mixture with phenolic compounds. The results of such analyses are usually given in equivalents of gallic acid or other reference compounds. Among these methods, determination of total phenolic content using the Folin-Ciocalteu reagent, as described by Singleton and Rossi (Singelton & Rossi, 1965), is very common.

Considering the accumulated knowledge on the effect of phenolic antioxidants on human health and the resulting market requirements it is highly important to have well developed, robust and established methods for their determination (Minussi et al., 2003; Urbano-Cuadrado et al., 2004).

In this study we compared three analytical methods: high pressure liquid chromatography (HPLC) with UV-vis detection, UV-vis spectrophotometry and chemiluminometry.

For separation and determination of phenolic acids and flavonoids, HPLC is the established technique (Nave et al., 2007; Rodriguez-Delgado et al., 2001; Spranger et al., 2004; Vitrac et al., 2002). The chromatographic conditions include the use of, almost exclusively, a reversed phase C18 column; UV-vis diode array detector, and a binary solvent system containing acidified water and a polar organic solvent (Tsao & Deng, 2004).

With use of HPLC we can reach separation of non-stabile and heavy volatile analytes on the base of different chemical interactions of the analytes with mobile phase and stationary phase. We use a non-polar stationary and polar mobile phase (reversed phase chromatography). HPLC with UV-visible detection was used for determination of antioxidant compounds content of gallic acid, (+)-catechin, (-)-epicatechin, *trans*-resveratrol, *cis*-resveratrol and quercetin in numerous wine samples. The selected phenolic compounds are the most important wine antioxidants. Gallic acid, the main hydroxybenzoic acid in red wines, is a very potent antioxidant with three free hydroxyl groups. Because of the relatively slow extraction of gallic acid from grape seeds, higher concentrations are obtained with longer maceration times, which is characteristic for red wines. As the most important flavonol, quercetin was also included in our research.

Spectrophotometric determination of total antioxidant potential (TAP_{SP}) was performed with oxidation of phenolic compounds with Folin-Ciocalteu reagent after spectrophotometric method, described by Singelton in Rossi. Gallic acid was used as modelling solution.

Chemiluminometric determination of polyphenols was another possibility for evaluation of the total antioxidant potential in wine. Chemiluminescence, the emission of light as a consequence of relaxation of kind, which it is evoked between chemical reactions, has become very useful technique for studying oxidation of organic materials (Costin et al., 2003; Garcia-Campana & Baeyens, 2001; Hötzer et al., 2005; Kočar et al., 2008; Kuse et al., 2008). ABEL® (analysis by emitted light) antioxidant test kit, which contains photo protein Pholasin®, was used. Photo protein Pholasin® is the protein-bound luciferin from the bivalve mollusc Pholas dactylus, which reacts with luciferase and molecular oxygen to produce light (Knight, 1997; Michelson, 1978; Roberts et al., 1987). In reaction system substrate-catalyser-oxidant we can therefore inhibit occurrence of chemiluminescence with antioxidants. If there are antioxidants in the sample capable of scavenging superoxide, then these antioxidants will compete with Pholasin® for the superoxide and less light will be detected. Control samples containing no antioxidants were running with each assay. With measuring of decrease of intensity of chemiluminescence we evaluate total antioxidant potential (TAP_{CL}) of the sample (Hipler & Knight, 2001)..

The reactions of Pholasin® have been studied extensively (Dunstan et al., 2000; Müller et al., 1989; Reichl et al., 2000). It was found to have a 50- to 100-fold greater sensitivity towards superoxide than luminol. In addition, the decay of the Pholasin® chemiluminescent product was more rapid than that of the luminol product, leading to a greater accuracy in real-time kinetic studies. For these reasons, Pholasin® offers several advantages over luminol (Roberts et al., 1987). The luminescence of Pholasin® elicited with luciferase has a maximum at 490 nm, and that with Fe^{2+} shows a maximum at 484 nm (Shimomura, 2006).

In addition, we compared the results of the chromatographic method with those obtained using the spectrophotometric and chemiluminometric method.

In order to evaluate their potential individual contributions to TAP, phenolic antioxidants were analysed as pure solutions in the same concentration. The order of contributions of individual phenolic antioxidants to TAP determined according to spectrophotometric method was different than those determined with chemiluminometric method. Generalised, *cis*-resveratrol has the biggest contribution to TAP, following by *trans*-resveratrol, (-)-

epicatechin, (+)-catechin and quercetin. Interesting, gallic acid, as modelling solution by spectrophotometric determination of TAP_{SP}, shows the lowest contribution to TAP.

2. Experimental

137 wine samples (73 red, 54 white and 10 rosé wines) were purchased from wineries and directly analysed. Most of the red wine samples were from grape varieties Blue Frankish, Merlot, Cabernet Sauvignon, Pinot Noir, Refošk and Barbera. Cviček was used in the study as a typical Slovenian mixture of red and white wines, whereas from among the white wines, most samples were from grape varieties Welsh Riesling, Chardonnay, Traminer, Rhine Riesling, Yellow Muscat and Sauvignon. Vintages ranged from 1997 to 2006.

2.1 HPLC- UV/VIS method

The HPLC system Waters 600E was composed of isocratic pump W600, autosampler and Waters 996 photodiode array detector. The HPLC column Synergi Hydro RP 150 × 4.6 mm, 4 µm (Phenomenex, Torrance, California, USA) was used at 35 °C. Wavelengths of detection: (+)-catechin and (-)-epicatechin 210 nm, quercetin 253 nm, gallic acid 278 nm and *trans*-resveratrol 303 nm. Gallic acid, (+)-catechin hydrate, (-)-epicatechin, *trans*-resveratrol and quercetin dihydrate were purchased from Sigma-Aldrich (St. Louis, USA). All reagents and standards were prepared using Milli Q deionized water (Millipore, Bedford, USA).

The experimental conditions were: mobile phase A: 0.1% H_3PO_4; mobile phase B: MeOH; gradient elution: 0 min 90% A, 10% B; 15 min 78% A, 22% B; 25 min 50% A, 50% B; 34 min 34% A, 66% B; 35 min 90% A, 10% B for reconditioning of the system (8min); flow rate: 1.0 mL/min; injection volume: 50 µL; MeOH and ortophosphoric acid were HPLC-grade (Fluka, St. Gallen, Suisse) and were filtered and degassed before their use. The wine samples were diluted ten times with the respective mobile phases described above.

Stock solutions of standards were diluted in the mobile phase to obtain working standard solutions. Concentrations of the analytes were calculated from chromatogram peak areas on the basis of calibration curves. The method linearity was assessed by means of linear regression of the mass of analyte injected vs. its peak area. The repeatability was expressed as standard deviation (SD) of three separate determinations.

Typical standard deviations for determinations of the sum of phenolic antioxidants determined using HPLC/UV-vis were 0.015 mmol/L for red wines, 0.004 mmol/L for rosé and 0.050 mmol/L for white wines.

2.2 Spectrophotometry

Spectrophotometric determination of total antioxidant potential (TAP_{SP}) was performed according to the Singleton-Rossi procedure (Singelton & Rossi, 1965). TAP_{SP} of an antioxidant sample was estimated by measuring its reducing capacity with the Folin-Ciocalteu reagent using a spectrophotometer. The Folin-Ciocalteu reagent is a mixture of phosphowolframic acid ($H_3PW_{12}O_{40}$) and phosphomolybdenic acid ($H_3PMo_{12}O_{40}$), the absorbance of which was measured after the reaction at 765 nm using a Cary 1E spectrophotometer (Varian, California, USA). The Folin-Ciocalteu reagent was purchased from Merck (Darmstadt, Germany). It contains sodium tungstate, sodium molybdate, ortophosphoric acid, hydrochloric acid, lithium sulphate, bromine, hydrogen peroxide (Folin & Ciocalteu, 1927).

Briefly, 25 µL of a red and rosé wine sample or 250 µL of a white wine sample, 15 mL of distilled water, 1.25 mL of the diluted (1:2) Folin–Ciocalteu reagent, 3.75 mL of a sodium carbonate solution (20%) were mixed and distilled water was added to make up the total volume of 25 mL. The solution was agitated and left to stand for 120 min at room temperature for the reaction to take place. The calibration curve was prepared with gallic acid solutions in the concentration range from 0 to 1000 mg/L. The results are expressed as mmol gallic acid per litter (gallic acid equivalents - GAE).

The results for standards were highly reproducible (calibration curve squared regression coefficient >0.9993). All determinations were performed in triplicates. Typical standard deviation for spectrophotometric determinations of total phenolic content (TAP_{SP}) was 0.10 mmol/L for red wines, 0.09 mmol/L for rosé and 0.02 mmol/L for white wines.

2.3 Chemiluminometry

The Abel®-21 M2 antioxidant test kit (Knight Scientific Limited, Plymouth, UK) was used for chemiluminometric determination of total antioxidant potential (TAP_{CL}). Superoxide, generated in a tube containing Pholasin® with and without a sample of wine with unknown antioxidant potential leads to appearance of chemiluminescence, which was measured using a micro plate luminometer model Lucy (Anthos Labtec Instruments, Wals, Austria). Pholasin® is a bioluminescent photo protein of *Pholas dactylus*, which is a marine, rock-boring, bivalve mollusc. Antioxidant test kit used contains assay buffer (pH 7.2).

The Antioxidant Test procedure for Superoxide is provided by the supplier of the test kit (Hipler and Knight, 2001). The amount of sample was optimised to obtain not more than 90% and typically 50% signal inhibition. This signal was then corrected for sample dilution: 10 µL of sample was used, however, red wines and rosé wines were first diluted with water (1:10) while white wines were not.

The results are calculated as TAP_{CL}, expressed as % signal inhibition. Typical measurement uncertainty was 0.024 mmol/L for red wines, 0.016 mmol/L for rosé and 0.002 mmol/L for white wines.

2.4 Statistical analysis

Measurements are expressed as means ± standard deviations (SD) for three replicate determinations. Multivariate analysis was performed using SPSS 17.0 for Windows (SPSS Inc., Chicago, USA). As a typical data reduction and visualisation technique, principal component analysis (PCA) was used.

3. Results and discussion

Comparison of determinations of total antioxidant potential in different wines was performed with spectrophotometric and chemiluminometric method, while comparison of antioxidant compounds content in the same samples was performed using HPLC with UV/VIS detection.

Sum of phenolic compounds, determined using HPLC with UV-vis detection, summarized six manly phenolic compounds (gallic acid, (+)-catechin, (-)-epicatechin, *trans*-resveratrol, *cis*-resveratrol and quercetin). Regarding the knowledge about the influence of phenolic antioxidants on human health, selected phenolic compounds are the most important antioxidants of red wines. Gallic acid, the main hydroxybenzoic acid in red wines, is with three free hydroxyl groups a very strong antioxidant. Because of relative slow extraction of gallic acid from grape seeds, we obtain higher concentrations with longer maceration time, which is characteristic for red wines. In the group of non-flavonoid phenols we analysed also the main representative of stilbenes - resveratrol. Resveratrol (3,5,4'-trihydroxystilbene) is present in wine in four forms; we ware determining *trans*-resveratrol and *cis*-resveratrol. In the group of flavonoid phenols we choose isomers (+)-catechin and (-)-epicatechin, more specific, flavan-3-ols. As most important flavonol we include in our research quercetin. Chemically are flavonols 3-glycosides.

Relative good correlation is result between sum of phenolic compounds determined with HPLC/UV-vis and spectrophotometric determinations of total antioxidant potential TAP_{SP} ($r^2 = 0.91$) (Weingerl et al., 2009).

Light emitted by chemiluminescent substrate was expressed in relative light units (RLU). The time dependence of light intensity was measured and the peak intensity was converted into percent of inhibition relative to control. We expressed the results as total antioxidant potential, expressed as percent inhibition: $TAP_{CL} = ((\max RLU_{control}) - (\max RLU_{sample}))\ 100\ /\ (\max RLU_{control})$. Figure 1 shows the correlation between sum of phenolic compounds

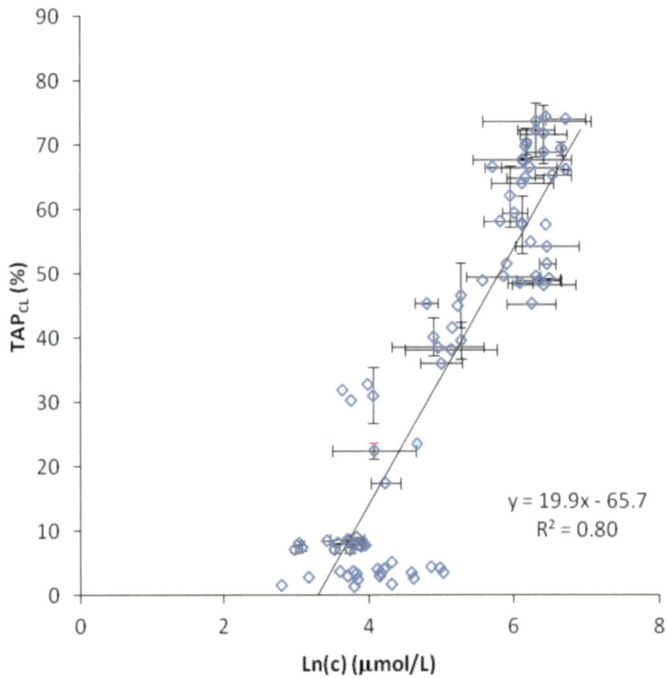

Fig. 1. Comparison between natural logarithm of sum of gallic acid, (+)-catechin, (-)-epicatechin, *trans*-resveratrol, *cis*-resveratrol and quercetin, determined using HPLC and TAP_{CL}. All wine samples included.

determined with HPLC/UV-vis and total antioxidant potential TAP_{CL} determined with chemiluminometry.

Sum of phenolic compounds, determined with HPLC method, correlate well with total antioxidant potential TAP_{CL}, determined with chemiluminometric method ($r^2 = 0.80$) (Fig 1).

With use of HPLC method we have the information about single phenolic antioxidants content, which doesn't consider synergistic influences between phenolic compounds in wine. Those effects considered only methods for determining total antioxidant potential. On the other hand, matrix effects may lead to different results obtained using other methods. In order to obtain a better insight into the extent of these effects, we compared all used methods.

To estimate the quantification limits, weighted tolerance intervals were used. The limit of quantification (in response units) is defined as 10 times the standard deviation at the lowest detectable signal (*LC*) plus the weighted intercept. The corresponding concentration LQ can be obtained by $LQ = (yQ - a)/b$, where a is the intercept, b is the slope of correlation curve (Zorn et al. 1997). The quantification limit for TAP_{SP} was calculated from data represented in Fig. 2, and amounts to 844 µmol/L gallic acid (Weingerl et al., 2011).

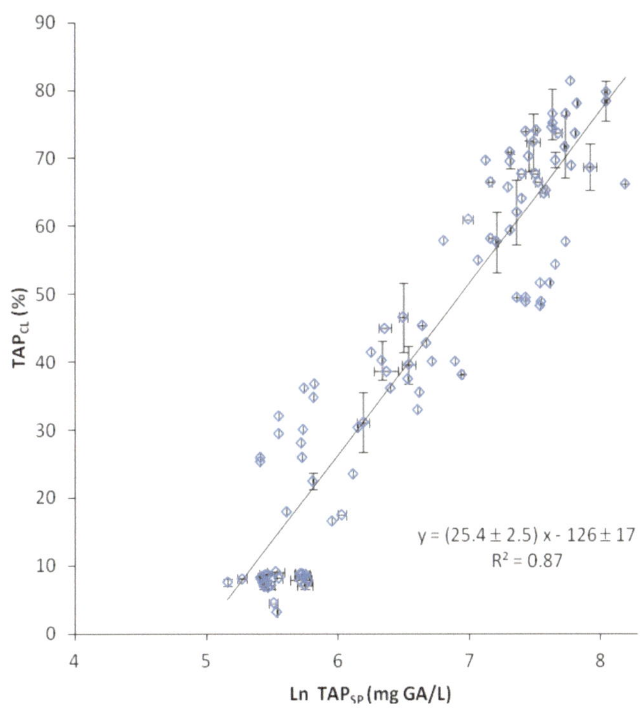

Fig. 2. Comparison of spectrophotometric determinations of TAP_{SP} and chemiluminometric results (TAP_{CL}).

For comparison, Table 1 shows further limits of quantification for determination of sum of HPLC determined phenolic antioxidants, TAP_{CL} and TAP_{SP} in wine.

Method	LOQ	Unit
Chemiluminometric method	32	μmol/L ascorbic acid
Sum HPLC/UV-vis	27	μmol/L
Spectrophotometric method	844	μmol/L gallic acid

Table 1. Limits of quantification (LOQ) for determination of polyphenols in wine.

As it is evident from Table 1, the chemiluminometric method exhibits similar quantification limits to the HPLC method for the determination of the sum of gallic acid, (+)-catechin, (-)-epicatechin, *trans*-resveratrol, *cis*-resveratrol and quercetin in wines (32 μmol/L). This is much lower quantification limit comparable with the spectrophotometric method, for the determination of total polyphenols, expressed in μmol/L of gallic acid.

Like total antioxidant potential determined with spectrophotometric or chemiluminometric method, like sum of six determined phenolic compounds are higher by red wines than white wines, rosé wines are giving intermediate results.

Evaluation of individual contributions of selected phenolic antioxidants to total antioxidant potential shows different results for spectrophotometric and chemiluminometric method (Fig.3 and Fig.4).

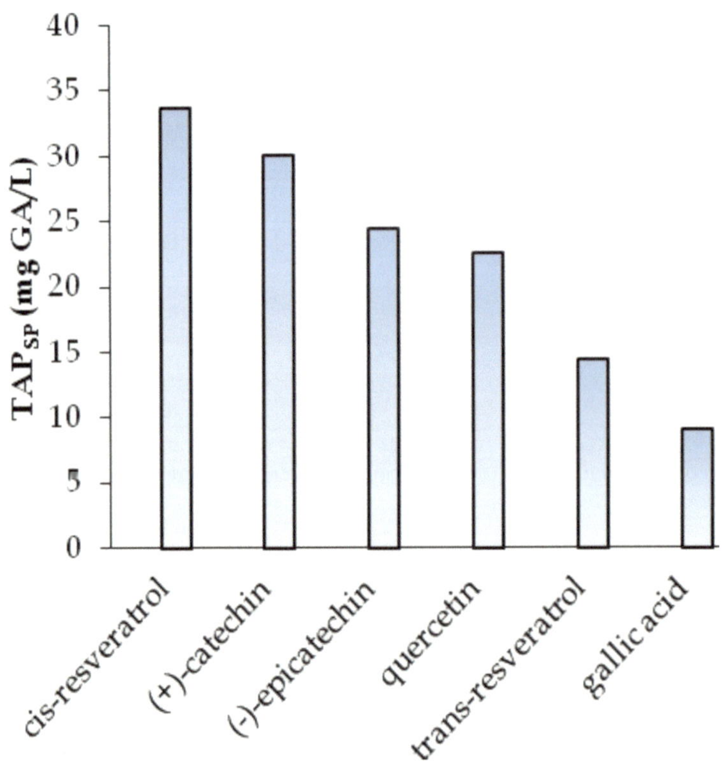

Fig. 3. TAP_{SP} of individual phenolic antioxidant solutions (50 μmol/L).

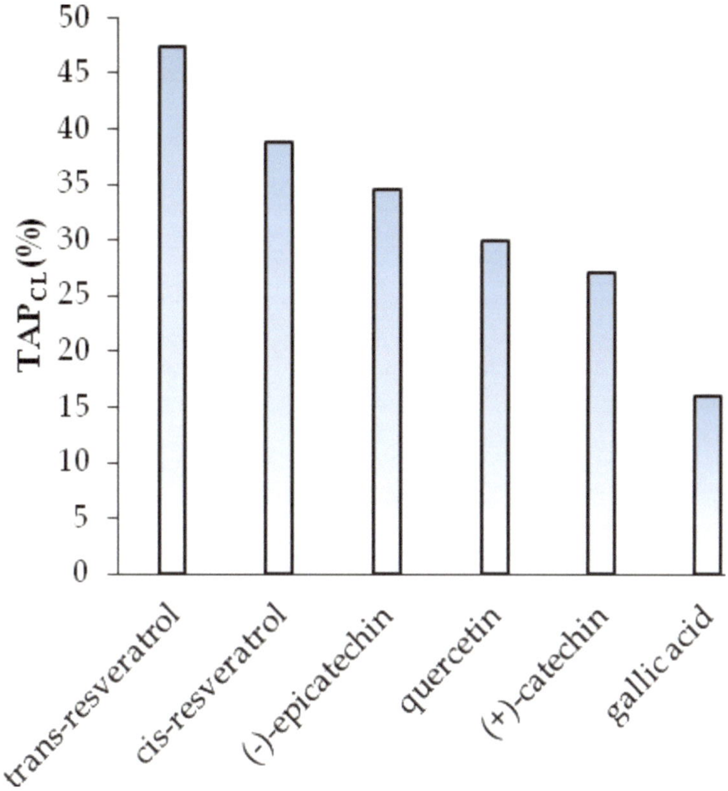

Fig. 4. TAP$_{CL}$ of individual phenolic antioxidant solutions (50 μmol/L).

Regarding their antioxidant potential, *cis*-resveratrol has been found to be the most powerful scavenger among the analysed phenolic antioxidants. *cis*-resveratrol, (+)-catechin, (-)-epicatechin, quercetin, *trans*-resveratrol and gallic acid was the order of contributions to TAP$_{SP}$, determined according the spectrophotometric method. *trans*-resveratrol, *cis*-resveratrol, (-)-epicatechin, quercetin, (+)-catechin and gallic acid was the order of contributions to TAP$_{CL}$, determined according to chemiluminometric method. Gallic acid has, interesting, the lowest contribution to TAP. Generalised, *cis*-resveratrol has the biggest contribution to TAP, following by *trans*-resveratrol, (-)-epicatechin, (+)-catechin and quercetin.

To reduce the number of variables and to investigate the extent of correlation between the six individual phenolic antioxidants and total antioxidant potentials, determined with spectrophotometric and chemiluminometric method, principal component analysis (PCA) was performed. While 57.5% of the variation is explained by PC1 and another 11 % by PC2, we compared the loading factors in Figure 4 to investigate how the different variables might be co-correlated. TAP$_{SP}$ and amount of gallic acid, determined with HPLC/UV-vis method are very strongly co-correlated and most strongly affected by quercetin content.

As is evident from Figure 4, there is a very strong co-correlation between TAP$_{CL}$ and content of *cis*-resveratrol. Referred to this co-correlation TAP$_{CL}$ is also strongly correlated to further antioxidants present especially in red grape varieties: (+)-catechin, (-)-epicatechin and *trans*-resveratrol.

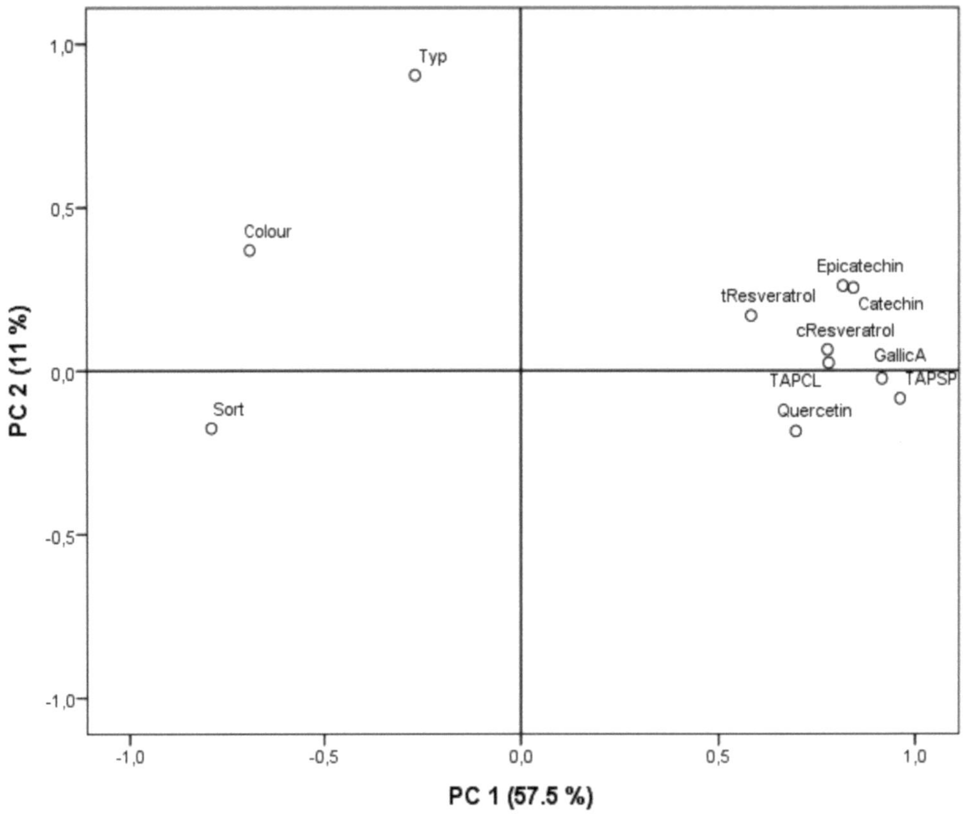

Fig. 5. Loading plot for PCA performed with all measured variables: single phenolic antioxidants, TAP_{SP} and TAP_{CL}.

4. Conclusions

In this work, we compared two routinely used methods of determination of the following wine phenolic antioxidants: gallic acid, (+)-catechin, (-)-epicatechin, *trans*-resveratrol, cis-resveratrol and quercetin. We used liquid chromatographic method with UV-vis detection, the conventional Folin-Ciocalteu spectrophotometric method and chemiluminometric method. The comparisons led us to the following conclusions:

- Determinations of the total antioxidant potential determined with chemiluminometric method correlate well with the sum of individual contents of gallic acid, (+)-catechin, (-)-epicatechin, *trans*-resveratrol, *cis*-resveratrol and quercetin as obtained using HPLC method in all wine samples.
- *cis*-resveratrol has been found to be the most powerful scavenger among the analysed phenolic antioxidants.

Information about single phenolic antioxidants content doesn't consider synergistic influences between phenolic compounds in wine, we ware limiting only on components, which ware analysed individually. Information about content of antioxidant components also doesn't consider prooxidative influence of transition metals content. Those effects considered both described methods for determining total antioxidant potential (TAP_{SP} and TAP_{CL}), on which indicates quality of correlation determining according to these two methods.

- PCA was applied to investigate the correlations between TAP_{SP}, TAP_{CL} and individual phenolic antioxidants and especially good co-correlation was found with TAP_{CL} and *cis*-resveratrol. TAP_{SP} and amount of gallic acid, determined with HPLC method are very strongly co-correlated and most strongly affected by quercetin content.

If we consider content of individually determined phenolic antioxidants and not only antioxidative capacity of wine, we may use only HPLC method with UV-vis detection.

5. Acknowledgements

I offer my regards and blessings to all of those precious people who supported me in any respect during my academic growth.

6. References

Alimelli, A., D. Filippini, R. Paolesse, S. Moretti, G. Ciolfi, A. D'Amico, I. Lundström, C. Di Natale. (2007). Direct quantitative evaluation of complex substances using computer screen photo-assisted technology: the case of red wine. *Anal. Chim. Acta,* 597, 103-112.

Briviba, K., L. Pan and G. Rechkemmer. (2002). Red wine polyphenols inhibit the growth of colon carcinoma cells and modulate the activation pattern of mitogen-activated protein kinases. *The Journal of Nutrition.* 132, 2814-2818.

Campanella, L., A. Bonanni, E. Finotti, M. Tomassetti. (2004). Biosensors for determination of total antioxidant capacity of phytotherapeutic integrators: comparison with other spectrophotometric, fluorimetric and voltammetric methods. *Biosens. Bioelectr.* 19, 641-651.

Careri, M., C. Corradini, L. Elviri, I. Nicoletti, and I. Zagnoni. (2003). Direct HPLC Analysis of Quercetin and trans-Resveratrol in Red Wine, Grape, and Winemaking Byproducts, *J. Agric. Food Chem.* 51, (18), 5226-5231.

Carralero Sanz, V., M. Luz Mena, A. González-Cortés, P. Yáñez-Sedeño, J. M. Pingarrón. (2005). Development of a tyrosinase biosensor based on gold nanoparticles-modified glassy carbon electrodes. Application to the measurement of a bioelectrochemical polyphenols index in wines, *Anal. Chim. Acta,* 528, 1-8.

Cooper, K. A., M. Chopra and D.I. Thurnham. (2004). Wine polyphenols and promotion of cardiac health. *Nutrition Research Reviews,* 17, 111-129.

Costin, J. W., N. W. Barnett, S. W., Lewis, D. J., McGillivery. (2003). Monitoring the total phenolic/antioxidant levels in wine using flow injection analysis with acidic potassium permanganate chemiluminescence detection. *Anal. Chim. Acta.* 499, 47-56.

Darias-Martin, J.J., O. Rodrigues, E. Diaz, R.M. Lamuela-Raventos. (2000). Effect of skin contact on the antioxidant phenolics in white wine. *Food Chemistry,* 71, 483-487.

De Beer, D., E. Joubert, W.C.A. Gelderblom, M. Manley. (2005). Changes in the Phenolic Composition and Antioxidant Activity of Pinotage, Cabernet Sauvignon, Chardonnay and Chenin blanc Wines During Bottle Ageing *Food Chem.* 90, 569-577.

Downey, M.O., N.K. Dokoozlian, M.P. Krstic. (2006). Cultural practice and environmental impacts on the flavonoid composition of grapes and wine: A review of recent research. *Am. J. Enol. Vitic.* 57, 257-268.

Dunstan, S. L., G. B. Sala-Newby, A. B. Fajardo, K. M. Taylor, and A. K. Campbell. (2000). Cloning and expression of the bioluminescent photoprotein pholasin from the bivalve mollusc Pholas Dactylus. *J. Biol. Chem.* 275, 9403.

Fernandez-Pachon, M. S., D. Villano, M.C. Garcia-Parrilla & A.M. Troncoso. (2004). Antioxidant activity of wines and relation with their polyphenolic composition. *Analytica Chimica Acta*, 513, 113-118.

Folin, O., V. Ciocalteu. (1927). On tyrosine and tryptophane determinations in proteins. *J. Biol. Chem.* 73, 627-650.

Fuhrman, B., N. Volkova, A. Suraski, M. Aviram. (2001). White wine with red winelike properties: increased extraction of grape skin polyphenols improves the antioxidant capacity of the derived white wine. *Journal of Agricultural and Food Chemistry* 49 (7), 3164-3168.

Garcia-Campana, A. M., and W. R. G. E. Baeyens. (2001). *Chemiluminescence in Analytical Chemistry*, Marcel Dekker, New york.

Giovanelli, G. (2005). Evaluation of the antioxidant activity of red wines in relationship to their phenolic content. *Italian Journal of Food Science*, 17, 381-393.

Gómez-Alonso, S., E. García-Romero, I. Hermosín-Gutiérrez. (2007). HPLC analysis of diverse grape and wine phenolics using direct injection and multidetection by DAD and fluorescence. *J. Food Comp. Anal.* 20, 618-626.

Hipler, B., and J. Knight. (2001). ABEL® Antioxidant test kit with Pholasin® for Vitamin C type antioxidants. Knight Scientific Ltd., Application note 108.

Hötzer, A. K., C. Henriquez, E. Po, S. Miranda-Rottmann, A. Aspillaga, F. Leighton, E. Lissi. (2005). Antioxidant and prooxidant effects of red wine and its fractions on Cu(II) induced LDL oxidation evaluated by absorbance and Chemiluminescence measurements. *Free radical research.* 39, 2, 175-183.

Katalinic, V., M. Milos, D. Modun, I. Music, M. Boban. (2004). Antioxidant effectiveness of selected wines in comparison with (+)-catechin. *Food Chemistry.* 86, 593-600.

King, R. E., J.A. Bomser and D.B. Min. (2006). Bioactivity of resveratrol. Comprehensive *Reviews in Food Science and Food Safety.* 5, 65-70.

Knight, J. (1997). The piddock and the immunologist. *Immunol. News.* 4, 26.

Kočar, D., M. Strlič, J. Kolar, V. S. Šelih, and B. Pihlar. (2008). Peroxide-related chemiluminescence of cellulose and its auto-absorption. *Polymer Degradation and Stability.* 93, 263-267.

Kuse, M., E. Tanaka, and T. Nishikawa. (2008). Pholasin luminescence is enhanced by addition of dehydrocoelenterazine. *Bioorganic and Medicinal Chemistry Letters.* 18, 5657-5659.

Lopez-Velez, M., F. Martinem-Martinez, C. Dell Valle-Ribes. (2003). The study of phenolic compounds as natural antioxidants in wine. *Critical Reviews in Food Science and Nutrition.* 43, 3, 233-244.

Magalhães, L.M., M. Santos, M. A. Segundo, S. Reis, J.L.F.C. Lima. (2009). Flow injection based methods for fast screening of antioxidant capacity. *Talanta.* 77, 5, 1559-1566.

Makris,D. P., E. Psarra, S. Kallithraka, P. Kefalas. (2003). The effect of polyphenolic composition as related to antioxidant capacity in white wines. *Food Res. Int.* 36, 805-814.

Malovana, S., F. J. Montelongo Garcia, J. P. Perez, and M. A. Rodriguez-Delgado. (2001). Optimisation of sample preparation for the determination of trans-resveratrol and other polyphenolic compounds in wines by high performance liquid chromatography. *Anal. Chim. Acta.* 428, 245-253.

Michelson, A. M. (1978). Purification and properties of Pholas dactylus luciferin and luciferase. In: DeLuca MA ed) *Methods in Enzymology*, Academic Press, London 57, 385.

Minussi, R.C., M. Rossi, L. Bologna, L. Cordi, D. Rotilio, G. M. Pastore. (2003). Phenolic compounds and total antioxidant potential of commercial wines. *Food Chem.* 82, 409-416.

Mozetič, B., I. Tomažič, A. Škvarč, P. Trebše. (2006). Determination of polyphenols in white grape berries cv. Rebula. *Acta chim. slov.* 53, 58-64.

Müller, T., E. V. Davies, and A. K. Campbell. (1989). Pholasin chemiluminescence detects mostly superoxide anion released from activated human neutrophils. *J. Biolumin. Chemilumin.* 3, 105.

Nave, F., M. João Cabrita, C. T. Da Costa. (2007). Use of solid-supported liquid-liquid extraction in the analysis of polyphenols in wine. *J. Chromatogr. A.* 1169, 23-30.

Opie, L. H. and S. Lecour. (2007). The red wine hypothesis: from concepts to protective signalling molecules. *Eur. Heart J.* 28, 14, 1683-1693.

Paixao, N., R. Perestrelo, J.C. Marques & J.S. Camara. (2007). Relationship between antioxidant capacity and total phenolic content of red, rose and white wines. *Food Chemistry*, 105, 204-214.

Prior, R. L., X. Wu, and K. Schaich. (2005). Standardized Methods for the Determination of Antioxidant Capacity and Phenolics in Foods and Dietary Supplements. *J. Agric. Food Chem.* 53, 4290-4302.

Prosen, H., D. Kočar, M. Strlič, and D. Rusjan. (2007). In vino veritas: LC-MS in wine analysis. *LC-GC Eur.* 20, 617-621.

Rastija, V., G. Srečnik & M. Medić-Šarić. (2009). Polyphenolic composition of Croatian wines with different geographical origins. *Food Chemistry*, 115, 54-60.

Recamales, A. F., A. Sayago, M. L. González-Miret, D. Hernanz. (2006). The effect of time and storage conditions on the phenolic composition and colour of white wine. *Food Res. Int.* 39, 220-229.

Reichl, S., J. Arnhold, J. Knight, J. Schiller, and K. Arnold. (2000). Reactions of Pholasin with peroxidases and hypochlorous acid. *Free Radical Biol. Med.* 28, 1555.

Roberts, P. A., J. Knight, A. K. Campbell. (1987). Pholasin: a bioluminescent indicator for detecting activation of single neutrophils. *Anal. Biochem.* 160, 139-148.

Rodríguez-Delgado, M.A., S. Malovaná, J. P. Pérez, T. Borges, F. J. García Montelongo. (2001). HPLC-analysis of polyphenolic compounds in spanish red Wines and determination of their antioxidant activity by Radical scavenging assay. *J. Chromatogr. A* 912, 249-25.

Russo, G. L. (2007). Ins and outs of dietary phytochemicals in cancer chemoprevention. *Biochemical Pharmacology*. 74, 533-544.

Santos-Buelga, C. And A. Scalbert. (2000). Proanthocyanidins and tannin-like compounds-nature, occurrence, dietary intake and effects on nutrition and health. *Journal of the Science of Food and Agriculture.* 80, 1094-1117.

Shimomura, O. (2006). *Bioluminescence: chemical principles and methods.* World Scientific Publishing, New Jersey, USA.

Singleton, V. L., and J. A. Rossi. (1965). Colorimetry of total phenolics with phosphomolybdic-phosphotungstic acid reagents. *Am. J. Enol. Vitic.* 16, 144.

Spigno, G., D. M. De Faveri, (2007). Assessment of process conditions on the extraction of antioxidants from grape marc. *J. Food Eng.* 78, 793-801.

Spranger, I.M., C. M. Clímaco, B. Sun, N. Eiriz, C. Fortunato, A. Nunes, C. M. Leandro, M. L. Avelar, P. A. Belchior. (2004). Total polyphenolic compounds contents (TPC), total antioxidant activities (TAA) and HPLC determination of individual polyphenolic compounds in selected Moravian and Austrian wines. *Anal. Chim. Acta.* 513, 151-161.

Staško, A., V. Brezova, M. Mazur, M. Čertik, M. Kalinak & G. Gescheidt. (2008). A comparative study on the antioxidant properties of Slovakian and Austrian wines. *LWT- Food Science and Technology*, 41, 2126-2135.

Strlič, M., T. Radovič, J. Kolar, and B. Pihlar. (2002). Anti- and pro-oxidative properties of gallic acid in Fenton-like systems. *J. Agric. Food Chem.* 50, 6313-6317.

Tsao, R., Z. Deng. (2004). Separation procedures for naturally occurring antioxidant phytochemicals. *J. Chrom. B* 812, 85-99.

Urbano-Cuadrado, M., M.D. Luque de Castro, P.M. Pérez-Juan, J. García-Olmo, M. A. Gómez-Nieto. (2004). Near infrared reflectance spectroscopy and multivariate analysis in enology Determination or screening of fifteen parameters in different types of wines. *Anal. Chim. Acta.* 527, 81-88.

Vitrac, X., J. P. Monti, J. Vercauteren, G. Deffieux, J. M. Mérillon. (2002). Direct liquid chromatographic analysis of resveratrol derivatives and flavanonols in wines with absorbance and fluorescence detection. *Anal. Chim. Acta.* 458, 103-110.

Vršič, S., B. Pulko, J. Valdhuber. (2009). Influence of defoliation on carbohydrate reserves of young grapevines in the nursery. *European journal of horticultural science.* 74, 5, 218-222.

Weingerl V, M. Strlič, D. Kočar. (2009). Comparison of methods for determination of polyphenols in wine by HPLC-UV/VIS, LC/MS/MS and spectrophotometry. *Acta Chim. Slov.* 56, 3, 698-703.

Weingerl, V., M. Strlič, D. Kočar. (2011). Evaluation of the chemiluminometric method for determination of polyphenols in wine. *Analytical Letters.* 44, 1310-1322.

Woraratphoka, J., K. O. Intarapichet, K. Indrapichate. (2007). Phenolic compounds and antioxidative properties of selected wines from the northeast of Thailand. *Food Chem.* 104, 1485-1490.

Zorn, M. E., R. D. Gibbons, and W. C. Sonzogni. 1997. Weighted least squares approach to calculating limits of detection and quantification by modeling variability as a function of concentration. *Anal. Chem.* 69 (15), 3069–3075.

18

A Review of Spectrophotometric and Chromatographic Methods and Sample Preparation Procedures for Determination of Iodine in Miscellaneous Matrices

Anna Błażewicz
Department of Analytical Chemistry, Medical University of Lublin, Poland

1. Introduction

Why nearly 200 years after the accidental discovery of natural iodine by Bernard Courtois (Dijon, France, 1811) are researchers still intrigued by this element? Why do they constantly search for more sensitive and reliable methods of its determination? During the last 200 years the status of research into the role of iodine in living organisms and the environment has progressed through many phases, from rapture over its interesting properties and healing powers to even some kind of "iodophobia". According to the World Health Organization (WHO), an estimated 2 billion people, including 285 million school-age children, are iodine deficient. And among them, iodine deficiency disorders (IDD) affect some 740 million -- with almost 50 million of them suffering from some form of brain damage resulting from iodine deficiency (United Nations Administrative Committee on Coordination/ Sub-Committee on Nutrition, 2000). On the other hand, it is known that large amounts of iodine are able to block the thyroid's ability to make hormones and worsen infiltration of the thyroid by lymphocytes. According to Teng's studies (Teng et al., 2009) giving iodine to people who had adequate or excessive iodine intake increases the incidence of autoimmune thyroiditis.

It became evident that increasing familiarity with the role of iodine would translate into development and discovery of more and more powerful analytical methods and techniques. Present-day analytical techniques are capable of detecting extremely small quantities. Some of them have become routine ultra-trace measurement tools in analytical and clinical laboratories.

The aim of this review is to explore available information regarding iodine determination in various samples (mainly of biological and environmental origin) focusing on spectrophotometry and chromatography as sensitive and reliable analytical methods of its measurement.

2. Iodine species in nature

Iodine plays an integral role in a diverse array of processes. As such, it exists in a variety of forms reflecting either the environment in which it is found or its biological function. Water, air, soil and food constitute the most common group of analyzed samples derived from our external environment.

2.1 Water

In water iodine is predominantly found in the iodide (I^-) or iodate (IO_3^-) form (Gilfedder et al., 2007; Schwehr & Santschi, 2003). Other forms of iodine species in water include: IO_4^- (periodate), IO^-(hypoiodite), CH_3I (methyl iodide), CH_2I_2 (methyl diiodide), C_2H_5I (ethyl iodide), C_3H_7I (propyl iodide), C_4H_9I (butyl iodide), and CH_2BrI (methyl iodide bromide) (Hou, 1999, Hou et al., 2009). Organic iodine concentrations may be especially high in fresh water (from rivers, lakes and rain).

2.2 Air

Generally, the air contains iodine in particulate form, as inorganic gaseous iodine (I_2, HIO) and as some forms of organic gaseous iodine (CH_3I, CH_2I_2). High iodine concentrations are found in urban areas due to the combustion of oil and coal. Coastal areas also have high iodine concentration due to the emission of gaseous I_2 from algae, seawater and sea spray, which varies with location, season and climate (Hou, 2009; Yoshida & Muramatsu, 1995). There are numerous pathways that may be responsible for transferring I_2 from the sea to air. Photochemical oxidation of I^- from seawater to elemental iodine has been recreated in the laboratory (Miyake & Tsunogai, 1963). Another means of obtaining elemental I_2 is through the reaction of I^- with ozone (O_3)(Garland & Curtis, 1981). The greatest source of atmospheric I_2 is thought to be from microbial activity within the oceans, through the transformation of I^- and IO_3^- into organic CH_3I, which has a residence time between 1.1-8 days (Cicerone, 1981). Lovelock et al. (Lovelock et al., 1973) measured the mean atmospheric CH_3I concentration above the Atlantic as 1.2 ppt (6.8 ng/m^3). Rasmussen et al. (Rasmussen et al., 1982) found the background level of CH_3I to vary between 1 and 3 ppt (5.7-17 ng/m^3) with measurements near oceans with high biomass productivity to be around 7-22 ppt (40-125 ng/m^3).

2.3 Soil

Strong evidence suggests that atmospheric transport from the oceans is responsible for the deposition of iodine in soil. Iodine exists in various forms in soil and varies largely with respect to its concentration. I^- (more mobile form) is believed to be the dominant species in acidic soils whilst IO_3^- (less mobile) will occur in alkaline soils. In low pH oxidizers, e.g. Fe^{3+} and Mn^{4+} convert I^- into molecular I_2. The activity of reducing bacteria impacts the form of iodine in the soil as well. CH_2I_2 and other volatile organic complexes of iodine are generated by microbial activity (Johnson, C.C. 2003). Generally, organic bound iodine is more abundant in soil samples. The secondary environment (soil) has high iodine content compared to the primary environment (parent rocks) from which it is derived as a result of weathering. Weathered rocks and soils are richer in iodine than the unweathered bedrocks (Fuge & Johnson, 1986). On average, the igneous rocks contain an average of 0.25 mg/kg of iodine, the sedimentary rocks have 2.3 mg /kg iodine and the metamorphic rocks have 0.81 mg/kg of iodine. Organic matter is the major concentrator of iodine in sedimentary basins (Mani et al., 2007). The highest values were found in soil samples from areas close to the coast, where there is high rainfall, and from areas with high soil organic matter. In one study, the iodine concentration in Japanese soils was found to range from 0.2 mg/kg to 150 mg/kg (Muramatsu, 2004). Retention of iodine in the soil is influenced by a number of factors, including the soil pH, moisture content, porosity and composition of the organic and inorganic components (Sheppard et al., 1995; Whitehead, 1984).

2.4 Food and plants

Geographic differences in topsoil iodine content and irrigation procedures determine food iodine levels. In most diets the mainstay sources of iodine are fish, shellfish, milk and iodinated salt. Food supplements constitute an alternative means of obtaining dietary iodine. Fish contain iodine in similar forms to those found in humans. Perhaps the widest assortment of forms is encountered in different species of seaweed. Brown seaweeds contain mostly I^-, however green seaweeds play host to a wide array of organic molecules to which iodine is bound, including numerous proteins and polyphenols. Dietary iodine is obtained from a variety of sources and individual dietary habits contribute to the wide disparity in iodine intake among populations. In 1993, the World Health Organization [WHO] published the first version of the WHO Global Database on Iodine Deficiency with global estimates on the prevalence of iodine deficiency based on total goitre prevalence (TGP), using data from 121 countries. Since the international community and the authorities in most countries where IDD was identified as a public health problem have taken measures to control iodine deficiency, in particular through salt iodization programmes - the WHO recommended strategy to prevent and control IDD (World Health Organization [WHO], 2004). Salt iodization programmes are carried out in more than seventy countries, including the United States of America and Canada. There is a wide variation in the scope of iodine supplement; almost 90% of households in North and South America utilize iodized salt while in Europe and the East Mediterranean regions this figure is less than 50%, with a worldwide figure about 70% (WHO, 2007). WHO, ICCIDD (International Council for the Control of Iodine Deficiency Disorders) and UNICEF (United Nations International Children's Emergency Fund) recommend that the term "iodized" be used to designate the addition of iodine to any substance, regardless of the form. Iodine is commonly added as the I^- or IO_3^- of potassium, calcium or sodium. 70% of salt sold for household use in the U.S.A. is iodized with 100 ppm KI (400 µg iodine per teaspoon) (U.S. Salt Institute, 2007). In Canada all salt must be iodized with 77 ppm KI. Mexico requires 20 ppm levels of iodization. Recommendations for the maximum and minimum levels of iodization of salt are calculated as iodine and determined by local national health authorities in accordance with regional variations in iodine deficiency. In Poland iodine deficiency prophylaxis was first started in 1935. Near the end of the 1940's and 80's, the practice of iodizing table salt was abandoned. A nonobligatory recommendation of iodizing salt took place in 1989. In 1991 the Polish Council for Control of Iodine Deficiency Disorders (PCCIDD) was established and an epidemiological survey performed in 1992-1993, defined Poland as an area with moderated - the seaside region as light - severity of iodine deficiency. In 1996 the production of table salt without the addition of KI was made illegal. An obligatory law was passed, mandating the addition of 30(+/-10) mg of KI per kilogram of salt (Szybiński, 2009). Iodization of salt in Turkey has been mandatory since 1998 and the recommended iodine concentration is 50-70 mg KI/kg or 25-40 mg KIO_3/kg (Gurkan et al., 2004). In Switzerland, iodization of salt was altered three times. Iodization was first introduced in 1922 at 3.75 mg/kg. In 1962 the concentration was doubled and in 1980 it was doubled again giving a present level of 15 mg/kg. Although it's use is voluntary, by 1988, 92% of retail salt and 76% of all salt for human consumption (including food industry) was iodized. Most other countries add from 10 to 40 μg iodine per gram of salt (10-40 ppm) (Bürgi et al., 1990).

Bread (0.14 mg/kg), milk (0.32 mg/kg), eggs (0.48 mg/kg), meat (0.13 mg/kg), and poultry (0.1 mg/kg) constitute other important sources of iodine (figures in parentheses represent the average iodine content per fresh weight) (Food Standards Agency [FSA], 2000). In certain individuals, medications may contribute to the ingested daily iodine. Examples include amiodarone, an antiarrythmic agent (Fang et al, 2004), iodized intravenous radiographic contrast agents and certain topical antiseptics (Aiba et al., 1999).

When considering multivitamins and mineral supplements as a source of iodine, one can find that the majority of iodine they contain is in the KI or NaI forms. According to Zimmerman's research, iodine concentrations in plant matter can range from as little as 10 μg/kg to 1 mg/kg dry weight (Zimmermann, 2009). This variability is relevant because plant matter affects the iodine content of meat and animal products (Pennington et al., 1995). Iodine content of different seaweed species varies greatly (Teas et al., 2004). Japanese iodine intake from edible seaweeds is relatively high compared to the rest of the world. Having taken into consideration many factors, such as information from dietary records, food surveys, urinalysis and seaweed iodine content, Zava and Zava estimated that the daily iodine intake in Japan averages approximately 1,000 to 3,000 μg/day (Zava & Zava, 2011). In certain diets, seafood is a large source of iodine, containing 2 to 10 times more iodine than meat (Hemken, 1979). Saltwater seafood usually contains significantly more iodine than freshwater food, some edible seaweeds may contain up to 2500 μg iodine per gram (Teas et al., 2004).

Simon et al. (Simon et al., 2002) presented an example of the value of the determination of iodine compounds in fish. The authors analyzed whole-body homogenates of zebrafish (Danio rerio) and tadpoles of the African clawed frog (Xenopus laevis). They detected five previously unknown iodinated compounds and measured the concentrations of I^-, MIT, DIT, T4, T3 and rT3 in these species.

2.5 Human body

In relation to iodine determinations, in clinical practice the most frequent analytical samples include urine, serum, blood, and a variety of tissues. Therefore, some examples of research studies related to iodine determinations in the mentioned matrices are presented below. The bioavailability of organic iodine, especially associated with macromolecules, is low (Hou et al., 2009), whereas I^- and IO_3^- have high bioavailability. According to recent estimates, KI is almost completely absorbed in humans (96.4%) (U.S. Food and Drug Administration [FDA], 2009).

2.5.1 Thyroid

Iodine plays a key structural role in the thyroid hormones of humans and other mammals, primarily in the form of T3 (triiodothyronine) and T4 (thyroxine). In such samples precursor forms such as MIT (monoiodotyrosine) and DIT (diiodotyrosine) or isomer forms such as rT3 (reverse triiodothyronine) may also be measured. Iodine accounts for 65% of the molecular weight of T4 and 59% of the T3. 15–20 mg of iodine is concentrated in the thyroid and hormones with 70% distributed in other tissues. In the cells of these tissues, iodide enters via the sodium-iodide symporter (NIS).

According to Hou (Hou et al. 1997), the contents of iodine expressed as ng/g wet weight tissue±1SD) in five tissues, plus hair, averaged over 9–11 individuals were: the heart (46.6±14.9), liver (170±34), spleen (26±8.6), lung (33.3±10.6), muscle (23.5±14.3), and hair (927±528). In the U.S. population, Okerlund found a mean value of 10 mg iodine per thyroid, with a range of 4-19 mg. In 56 patients suffering from autoimmune thyroiditis but with normal thyroid function, a mean value of 4.8 mg/thyroid was reported. In 13 patients with autoimmune thyroiditis and hypothyroidism, the mean value was 2.3 mg/thyroid (Okerlund, 1997).

Zaichick and Zaichick (Zaichick & Zaichick, 1997) used instrumental neutron activation and X-ray fluorescent analyses to determine the concentration and total iodine content of iodine within thyroids. They obtained 90 samples (at autopsy) from subjects of a broad age spectrum, from 2 to 87 years old and calculated correlations between iodine concentration and age. All their thyroid samples were weighed, lyophilised and homogenised. Iodine was analyzed in approximately 50-mg samples. The mean intrathyroidal iodine concentration (mean +/- S.E.) of a normal subject aged 26-65 averaged 345 +/- 21 µg/g dry tissue in non-endemic goitre region with no obligatory salt iodination. Maximum iodine concentration was found to be 494 +/- 65 µg/g ($P < 0.05$) for the age of 16-25. For the elderly aged over 65 an increase in iodine of 668 +/- 60 µg/g was shown ($P < 0.001$). When comparing the right and left lobes, the authors found no variation in weight, iodine concentration or the total content. An inverse correlation was found between the thyroid weight and intrathyroidal iodine concentration (-0.32, $P < 0.01$).

Tadros et al. (Tadros et al., 1981) determined iodine in 48 normal thyroids obtained at autopsy. According to the authors' findings, the iodine concentration ranged from 0.02 to 3.12 mg/g of tissue with a mean value of 1.03 +/- 0.67 mg/g. In 91 surgical thyroid specimens with a variety of abnormalities they found that iodine concentration was much lower. The samples of thyroids with cancer had the lowest values. Sixteen (76%) of 21 analyzed malignant thyroid specimens had undetectable iodine (less than 0.02 mg/g), whereas 22 (96%) of 23 benign nodules had measurable iodine concentrations. Błażewicz et al. (Błażewicz et al., 2011) examined correlations between the content of iodides in 66 nodular goitres and 100 healthy human thyroid tissues. The authors presented an accurate assessment of the iodine content in the thyroids of patients with a nodular goitre (mean concentration was 77. 1 +/- 14.02 µg/g) and in the thyroids obtained at autopsy - considered as a control group (mean concentration 622.62 +/- 187.11 µg/g -for frozen samples and 601.49 +/- 192.11 µg/g- for formalin fixed samples). Statistical analysis showed approx. 8-fold reduction of iodine concentration in the pathological tissues in comparison with the control group.

Interesting research into iodine content in human thyroids was also conducted by Zabala et al. (Zabala et al., 2009). Their study focuses on the determination of iodine content in healthy thyroid samples on male population from Caracas in Venezuela. The authors aimed at establishing a baseline of iodine content in thyroid glands and hence to compare the iodine thyroid concentration of the Venezuelan population with other countries. Male post-mortem individual samples were analyzed using a spectrophotometric flow injection method, based on the Sandell-Kolthoff reaction. The median intrathyroidal iodine concentration was 1443+/-677 µg/g (wet weight), ranging from 419 to 3430 µg/g, which corresponds to a median of total iodine content of 15+/-8 mg (ranging from 4 to 37). These results were higher than those values which were found in the literature. No correlation of iodine content with the age or weight of the healthy gland was observed.

2.5.2 Plasma

The inorganic form of iodine represents about 0.5 % of the total plasma iodine. The rest occurs in bound form with specific plasma protein (protein - bound iodine, PBI) which has gained wide use as an indicator of thyroid activity in humans. It has been reported that the total plasma iodine concentration in healthy subjects is between 40 and 80 μg/l. According to Allain's studies when plasma iodine concentrations are below 40 μg/l, hypothyroidism is highly likely, when they are between 80 and 250 μg/l, hyperthyroidism, particularly Graves' disease is probable. Above 250 μg/l – iodine overload is almost certainly indicated (Allain et al., 1993).

2.5.3 Brain

Despite the fact that iodine is one of the most important essential elements, the quantitative data on its concentration in the human brain is really scarce. The nature and site of iodine binding in the human brain is still unknown. The results of Andrasi et al. (Andrási et al., 2004) investigations on iodine distribution between the lipid fraction and in the brain tissue without lipid have indicated that its mean contents vary between 910 ± 147 ng/g dry weight and 281 ± 68 ng/g dry weight depending on the brain region (the highest content was found in susbstantia nigra and the lowest in vermis crebelli).

2.5.4 Hair

Levine et al. (Levine et al., 2007) presented a study on determining iodine concentration in tiny (less than 25 mg) human hair samples. Iodine concentrations from the blinded hair autism study samples ranged from 0.483 to 15.9 μg/g. In Adams' et al. studies (Adams et al., 2006) the mean concentration of iodine in the hair of autistic children has been reported to be lower than in the hair from the control group children. The low level of iodine in the hair of children with autism suggests that iodine could be important in the aetiology of autism, presumably due to its effect on thyroid function.

2.5.5 Human milk

Approximately 80% of iodine in human milk is present as I^-, while, mainly, another six high molecular weight iodine containing molecules (Braetter et al., 1998) account for the remaining 20%. European breast milk samples have been determined to contain 95 ± 60 μg/l total iodine. The total iodine content varied depending on the lactation state, and iodine was associated with fat at approximately 30% and 70% of the low molecular weight fraction (Michalke, 2006). A study of iodine species in milk samples obtained from humans from several different European countries and in infant formulas from different manufacturers was carried out by Fernández-Sanchez and Szpunar (Fernández-Sanchez & Szpunar, 1999). The authors also developed a method to determine iodine in human milk and infant formulas using ICP-MS. In the human milk the values found were between 144 ± 93.2 μg/kg, whereas in the infant formulas the values were 53.3 ± 19.5 (Fernández-Sánchez et al., 2007).

2.5.6 Urine

In urine, iodine occurs as I^-, but some organic species can also be found. Urinary iodine concentration is the prime indicator of nutritional iodine status and is used to evaluate population-based iodine supplementation. In 1994, WHO, UNICEF and ICCIDD recommended median urinary iodine concentrations for populations of 100- 200 μg/l, assuming the 100 μg/l threshold would limit concentrations <50 μg/l to </=20% of people (Delange et al., 2002). During the period between the years 1994-2002, the urinary iodine concentration was determined in 29,612 samples at the Institute of Endocrinology in the Czech Republic. The mean basal urinary iodine concentrations +/-SD were 115+/-69 μg/l. Out of all the samples, 0.7% were in severe (<20 μg/l), 9.6% in moderate (20-49 μg/l), 40.1% in mild (50-99 μg/l), 35.6% in adequate (100-200 μg/l), and 14.0% in more than adequate (>200 μg/l) subsets of iodine nutrition. A statistically significant (p<0.00001) difference was found between the mean male (127 μg/l) and female (112 μg/l) urinary iodine, and an inversely proportional trend also existed in the age-related data (Bílek et al., 2005). It is also known that patients with iodine induced hyperthyrosis have 10- to 100-fold more urinary iodide than healthy patients (Mura et al., 1995).

Delange et al. (Delange et al., 2002) determined the frequency distribution of urinary iodine in iodine-replete populations (schoolchildren and adults) and the proportion of concentrations <50 μg/l. The findings were as follows: nineteen groups reported data from 48 populations with median urinary iodine concentrations >100 μg/l. The total population was 55 892, including 35 661 (64%) schoolchildren. Median urinary iodine concentrations were 111-540 (median 201) μg/l for all populations, 100-199 μg/l in 23 (48%) populations and >/=200 μg/l in 25 (52%). The frequencies of values <50 μg/l were 0-20.8 (mean 4.8%) overall and 7.2% and 2.5% in populations with medians of 100-199 μg/l and >200 μg/l, respectively. The frequency reached 20% only in two places where iodine had been supplemented for <2 years. According to the authors' conclusions the frequency of urinary iodine concentrations <50 μg/l in populations with median urinary iodine concentrations >/=100 μg/l has been overestimated, and the threshold of 100 μg/l does not need to be increased. The main conclusion of the cited work was that in populations, median urinary iodine concentrations of 100-200 μg/l indicate adequate iodine intake and optimal iodine nutrition.

According to Verheesen and Schweitzer (Verheesen and Schweitzer, 2011) the threshold of 100 μg/l is only to make sure that severe iodine deficiency (beneath 50 μg/l) is not present in more than 20% of the population. Although the WHO is concerned about the negative effects of even mild iodine deficiency, the 100 μg/l threshold was never intended to prevent mild iodine deficiency. In order to combat mild deficiency the threshold should be reconsidered. The authors also emphasized the need to test for other biomarkers in individual cases in order to be able to adequately establish iodine deficiency. Since there is a lack of trusted biomarkers, thus far statistics have been used to estimate the percentage of the population being deficient, instead of showing prevalence figures. Furthermore, population figures are typically described only by a median; variables such as % being deficient, % being pregnant, % women, % men and age related figures should be thoroughly investigated.

3. Problems with analysis of iodine in biological matrices

Despite a wide choice of available analytical methodologies, determination of iodide in biological matrices remains a difficult problem. Biological samples belong to so-called complex samples (with complex matrices). In such samples, the analyte content is usually much scarcer when compared with the accompanying macrocomponents. Aside from the necessity of choosing the appropriately sensitive method, it is equally important to comply with the sample preservation, pretreatment, and preparation conditions.

Historically, published values of the I_2/I^- concentration of both tissue and body fluids from healthy subjects have varied greatly. These great differences were attributed to numerous variables, such as age, sex, dietary habits, physiological conditions, environmental factors and numerous other X-factors. Given the delicate nature and the instability of biological samples, it has been concluded that improper sample collection methods and manipulation drastically affects the iodine content of biological matrices.

Analytical methods are often versatile in nature. Thus, in order to achieve successful and satisfactory results, the process of analysis needs to be carefully tailored to its needs. Before applying the appropriate method for a particular application, many factors have to be considered and some of them are discussed below.

It is well known that sources of errors that affect the final error of an analytical result are connected, among other things, with incorrect obtaining of the samples, their improper transport, storage and transformation, wrong methodology, wrong measurement (instruments, parameters) or human errors (Konieczka & Namieśnik, 2007). When it comes to quantitative evaluations of iodine concentrations (in all chemical forms), the proper storage of biological samples is of paramount importance. The tissues must be preserved in such a way that a potential loss of the analyte (i.e. iodine) is minimized. The choice of the tissue-fixing agents is quite wide. Formalin is a routinely used tissue-fixing agent after surgical procedures. Other recommended agents for tissue preservation include, e.g. a mixture of 50% glutaraldehyde, 16 % paraformaldehyde, and 0.2M sodium phosphate buffer solutions (i.e. original composition of Karnovsky fixative) or its modification. The other possibility to preserve the tissues is sample freezing.

Hansson at al. (Hansson et al., 2008) used X-ray fluorescence analysis (XRF) and secondary ion mass spectrometry (SIMS) for evaluation of a freezing technique for preserving samples (XRF analysis) and for evaluation of the efficacy of using aldehyde fixatives to prepare samples (SIMS analysis). There were no significant changes in the iodine content due to freezing. Freezing for 4 weeks produced no more than a 10% change in the iodine content. For all the samples fixed in an aldehyde, there was a loss of iodine. The decrease in iodine content from baseline was significant for samples fixed in aldehyde ($p < 0.05$). Karnovsky was the best fixative in this regard, yielding a mean 14% loss compared to 20% and 30% for glutaraldehyde and formaldehyde, respectively. For SIMS method, Fragu et al. (Fragu et al., 1992) recommended chemical fixation with a mixture of Karnovsky fixative, followed by embedding in methacrylate. This method, which was evaluated for iodine loss by Rognoni et al. (Rognoni et al., 1974), has proven suitable for preservation of substances bound to macromolecules (like iodine bound to thyroglobulin [Tg]).

The effect of sample preservation on determination of I^- in healthy and pathological human thyroids has also been studied (Błażewicz et al., 2011). It was pointed out that the way of tissue preservation (either in formalin or by freezing) had no significant effect on the iodine determination result ($\alpha = 0.1$) by ion chromatography combined with the pulsed amperometic detection method (IC-PAD). Sample decomposition is a critical step in iodides' analysis as well. All reported methods have a digestion or ashing step prior to the final determination of iodine. The procedure usually requires the use of alkaline media and a high temperature (Moxon & Dixon, 1980). A catalytic spectrophotometric method based on the Sandell-Kolthoff reaction together with many modifications and improvements of this method is a low-cost assay of iodine, however, it is not free of possible interferences (especially for foodstuffs with low range of iodine levels). Most iodine in biological media is covalently bonded and there are some substances that interfere with the determination reaction (eg. SCN^-, NO_3^-, or Fe^{2+}). A high lipid content in the sample (eg. milk) can cause problems in spectrophotometric readings, so a mineralization step is absolutely necessary before analysis.

When preparing a sample for analysis, it is necessary to take into consideration also the loss of analyte due to erroneous application of decomposition procedures. Some of them, e.g. "Schoeniger combustion", require a highly homogenous sample, which is sometimes difficult to obtain (Knapp et al., 1998). There are many options for biological sample preparations, among which alkaline digestion using tetramethylammonium hydroxide (TMAH) is the most common before the analysis of I^- (Fecher et al., 1998; Fecher & Nagengast, 1994; Schramel & Hasse, 1994). Alkaline conditions during the extraction procedure have some advantages in comparison with acidic media, where I^- may be oxidized into volatile forms (I_2 or HI). A destruction of the organic matrix using a typical digestion reagent, like HNO_3, is not possible because of the losses due to volatile iodine formation (therefore, no stable sample solution can be achieved). However, acid digestion procedures (by the use of mainly H_2SO_4, HNO_3, and $HClO_4$) have also been applied (Fischer et al., 1986). What is more, despite its obvious weak point (loss of analyte), the US Food and Drug Administration (FDA, 2009) still recommends such procedures.

Błażewicz et al. (Błażewicz et al., 2011) used the alkaline digestion with 25 % TMAH water solution for the thyroid glands` preparation before the IC method of analysis. A diluted TMAH solution has also been used for serum samples by Schramel and Hasse (Schramel & Hasse, 1994) to analyse iodine in the serum, milk, plants and tissues by using the ICP-MS method.

Unfortunately, despite the abovementioned advantages, the use of TMAH solution for the digestion of samples has some disadvantages as well. Since for all modern sample pretreatment methods time is a very important factor, the long procedure time still remains a huge problem. As reported in the literature, digestion of biological materials with TMAH usually requires up to 6 hours (Gamallo-Lorenzo et al., 2005). However, the assistance of microwaves significantly shortens the time of the sample preparation step (less than 20 minutes). Such microwave-assisted alkaline digestion has been developed before the IC analysis of thyroids' samples (Błażewicz et al., 2011). It is important to monitor all conditions of the digestion procedure, especially temperature, in order to avoid the decomposition of TMAH and bursting of the closed vessels (therefore a temperature lower than 100 ^{0}C is recommended). Each time the vessels must be thoroughly cleaned with a digestion mixture in order to avoid memory effects (adsorption by the walls of containers) and consequently the loss of analyte.

4. Spectrophotometry and chromatography as tools for iodine assessment in miscellaneous matrices

It is known that the choice of the proper analytical method depends on the intended application, the number of samples, the cost of analysis and the technical capability.

Currently, there are multiple distinctive analytical methods for determining concentrations of iodine species. The methods vary in principle, reliability, accuracy, precision, availability, detection limit, sample throughput, time and reagent consumption, ease of performance and cost of analysis. These factors all play a role in the choice of the most suitable method but ultimately the purpose of the analysis determines the method, e.g., whether the analysis is routine or if an analysis of a reference material is necessary. For any given purpose, one of the first factors taken into account is whether the method's detection limit is adequately low. Several methods of iodine determination have been proposed, including catalytic methods (with LOD=0.1 µg/ l) (Kamavisdar & Patel, 2002), chromatography in various modes (eg., IC with LOD = 0.1-0.8 µg/l (Hu et al., 1999; Bichsel & Von-Gunten, 1999), (chromatographic methods are especially useful for iodine speciation when coupled with ICP-MS or elecrochemical detection), GC-EC: gas chromatography with electron capture detection (0.11µg/l) (Maros et al., 1989), GC–MS: gas chromatography–mass spectrometry(0.010 µg/l) (Das et al., 2004), FAAS: flame atomic absorption spectrometry (2.75 µg/l) (Yebra & Bollaín, 2010), NAA (0.1-0.2 µg/l) (Hou et al., 1999), ETAAS: electrothermal atomic absorption spectrometry (1.2 -3.7 µg/l) (Bermejo-Barrera et al., 1999), inductively coupled plasma mass spectrometry ICP-MS (1.0-9.0 µg/l) (Fernandez-Sanchez & Szpunar 1999), ICP-AES (40.0-470.0 µg/l) (Anderson & Markowski, 2000), inductively coupled plasma optical emission spectrometry (ICP- OES) (2 µg/l) (Naozuka et al., 2003), ion selective electrodes (1.96 µg/l) (Kandhro et al., 2009), X-rayfluorescence (XRF) (180 µg/L) (Varga, 2007), VG-ICP-OES: vapour generation inductively coupled plasma optical emission spektrometry (20 µg/l) (Niedobová at al., 2005). The iodine content can also be measured by the use of titrimetric methods usually combined with potentiometric measurements. They are also used for verifying other methods (Gottardi, 1998). The titrimetric method is mainly used for samples without complex matrices (i.e. water or salt). Generally such methods involve acidification of the sample solution and adding an excess of KI solution to determine the liberated iodine by titration with sodium thiosulphate. Despite numerous advantages of the above-mentioned methods, very few of them are widely used due to very high costs of instrumentation, software, and maintenance. Spectrophotometric and chromatographic methods are used very frequently for the analysis of iodine and its various chemical forms. Chosen examples of applications of iodine determinations are presented below.

4.1 Spectrophotometric methods

4.1.1 Water samples

Spectrophotometric analysis continues to be one of the most widely used analytical techniques available. Kinetic spectrophotometric methods, which are based on the reaction, found by Sandell and Kolthoff (1934) set the foundation for the development of different methods for the determination of iodine in environmental samples (mostly water). The said reaction proceeds according to the following equation (1):

$$2\,Ce^{4+} + As^{3+} \rightarrow 2\,Ce^{3+} + As^{5+} \qquad (1)$$

By adding an arsenious acid (H_3AsO_3) solution and an ammonium cerium sulfate ($(NH_4)_2Ce(SO_4)_3$) solution as reagents to I^- in a specimen, yellow Ce^{4+} is reduced to produce colorless Ce^{3+} ((2) and (3)).

$$2Ce^{4+} + 2I^- \rightarrow 2Ce^{3+} + I_2 \qquad (2)$$

$$I_2 + As^{3+} \rightarrow 2I^- + As^{5+} \quad (3)$$

Iodine has a catalytic effect upon the course of reaction (1), i.e., the more iodine is present in the preparation to be analyzed, the more rapidly proceeds the reaction (1). The speed of reaction is proportional to the iodine concentration. In this manner it is possible to determine iodine even in the nanogram range. Sandell and Kolthoff found that Os and Ru catalyse this reaction in the same manner as I_2, while Mn and MnO_4^- do so in the presence of Br^-. Among substances reducing Ce^{4+} they listed NO_2^-, SCN^-, Fe^{2+}, while BrO_3^-, MnO_4 were classed by them as oxidising As^{3+}. These authors also pointed out that certain substances, such as F^-, form compounds with Ce^{4+} giving a stable complex. Ag^+, CN^-, and Hg^+ react with I^- as well. The effect of various concentrations of NaCl, NaF, KH_2PO_4, $ZnSO_4$, KCl, $MgSO_4$, KBr and of $CuSO_4$ on the described reaction was studied by Stolc (Stolc, 1961). According to the author the substances studied may be grouped into two categories, i.e. reaction inhibiting (NaF, KH_2PO_4, $ZnSO_4$, KCl) and reaction stimulating agents (NaCl, $MgSO_4$, KBr, $CuSO_4$).

The method is achieved in the following manner: a measured amount of an arsenous oxide (As_2O_3) solution in concentrated H_2SO_4 is combined with the test solution. This mixture is then adjusted to its reaction temperature, usually between 20 and 60 degree C. Cerium (IV) sulfate in sulfuric acid is then added, after which the solution is able to react for a limited time at the set temperature. The reaction time ranges from 10 to 40 minutes, and subsequently the content of the test solution of cerium (IV) ions is photometrically determined. The lower the determined concentration of cerium (IV) ions, the faster the reaction, thus a larger amount of catalyzing agent, i.e., iodine. By these means it is possible to directly and quantitatively measure the iodine concentration of the test solution, though execution of such processes is complicated and demands extensive measuring times (Sandell & Kolthoff, 1934, 1937). The above-described method was modified in various ways, for example by replacing H_2SO_4 with HNO_3 (used for acidifying the reaction mixture). It was found that the catalytic activity of iodine in HNO_3 solution is 20 times that in H_2SO_4 and is also far less sensitive towards accompanying ions, making the system far more useful for the determination of traces of iodine (Knapp & Spitzy, 1969). The reaction mixture's change in composition multiplies the sensitivity of the reaction by twenty. Consequently, test solutions of an iodine content that, according to the conventional catalytic reaction method utilizing sulfuric acid, required a reaction time of approximately 20 minutes in order to display a notable decrease in cerium (IV) ion concentration need only 1 minute to produce the same result. These results were achieved while operating at the same reaction temperature. Rodriguez and Pardue (Rodriguez & Pardue, 1969) studied the effect of H_2SO_4, $HClO_4$, Ag(I), Hg(II), Cl^- and temperature on the aforemetioned kinetic reaction. Their studies utilized the catalytic action of iodide on the decomposition of the $FeSCN^{2+}$complex ion. This indicator reaction is characterized by an induction period, the length of which depends on the reagent concentration, pH and temperature. The mentioned method was adopted as a standard method for iodide determination in natural and waste waters as well as in food and biological samples. However, high inter-laboratory relative standard deviations have frequently been reported for this method. Some authors have suggested that this might be partly attributed to the limitations of the method to quantitatively detect or tolerate IO_3^- that are found in natural waters (Heckwan, 1979).

An alternative flow injection spectrophotometric method for the determination of I^- in the ground and surface water was reported by Kamavisdar and Patel (Kamavisdar & Patel, 2002). The method was based on the catalytic destruction of the colour of the Fe(III)–SCN^-–CP^+–$_nBP_y$ quarternary complex. The detection limit of the method was reported to be 0.1 ng ml^{-1} of iodide. Another redox reaction between chloramine-T and N,N'-tetramethyldiaminodiphenylmethane (Feigl's Catalytic Reaction) was applied for the determination of traces of iodine in drinking water (Jungreis & Gedalia, 1960).

An alternative to the Sandell-Kolthoff method was developed by Gurkan et al., (Gurkan et al., 2004). Iodides were determined in waters by inhibition kinetic spectrophotometric method based on the inhibitory effect of I^- on the Pd(II)-catalyzed reduction of Co(III)-EDTA by the hypophosphite ion in a weak acid medium. The main advantage of this method was related to the pretreatment step of the analysis which would be omitted (a time-consuming alkaline ashing preparative procedure is necessary in order to apply the standard method). The sensitivity of the method allowed determinations in the range of 2-35 ng/ml of I− (LOD=1.2 ng/ml). Koh et al. (Koh et al., 1988) separated I^- from other chemical species by its oxidation and subsequent extraction into carbon tetrachloride. The proposed spectrophotometric method was based on the extraction of the back-extracted iodide into 1,2-dichloroethane as an ion pair with methylene blue. The authors applied that method to determine various amounts of iodide in natural water samples (at the 10^{-6} mol l^{-1} level). Spectrophotometric determination of the total dissolved sulfide in natural waters allowed also simultaneous determination of other UV-absorbing ions, including I^- (Guenther et al., 2001).

4.1.2 Soil and plant samples

Different analytical techniques have been developed to extract and measure iodine concentration from the soil. The reduction of Ce (IV) by As (III) catalyzed by iodine can be used to determine the low concentration of iodine in plant and soil samples. The sample preparation requires a specialized combustion apparatus and trapping systems for iodine. For plant samples and biological materials, halogen extraction using TMAH under mild conditions has proved to be effective (Knapp et al., 1998).

Kesari et al. (Kesari et al., 1998) developed a simple and sensitive spectrophotometric method for the determination of iodine in tap water, sea water, soil, iodized salt and pharmaceuticals samples. The said method was based on oxidation of I^- to IO_3^- with bromine water and liberation of free I_2 from IO_3^- by addition of KI in acidic medium. I_2 is then reacted with leuco crystal violet and the crystal violet dye liberated shows maximum absorbance at 591 nm. Beer's law is obeyed over the concentration range from 0.04 to 0.36 mg/l of iodine in a final solution volume of 25 ml. The method is free of interference of other major toxicants.

Lu et al. (Lu et al. 2005) applied the arsenic-cerium redox method for assessment of iodine content in soils and waters. Mean iodine concentration in soil samples was found to be 1.32 ± 0.14 mg/kg, and its content correlated positively with the water iodine content. In association with the photometric analytical technique, the alkaline dry ashing method (adding KOH and $ZnSO_4$), along with digestion via the calorimetric bomb and the utilization of the Schoniger digestion arc, provide a means for obtaining reliable results. For this investigation, the influence of iodine fertilisation on the iodine concentration of cress

(Lepidium sativum) was determined by an experiment in which different amounts of iodine were added to the potted plants. The iodine fertiliser used was natural caliche. The results show a very close correlation between the iodine supply and iodine concentration in the cress which increased to more than 30 mg/kg dry matter (Jopke et al., 1996).

4.1.3 Foodstuffs

A semi-automated method for determination of the total iodine in milk was described by Aumont (Aumont, 1982). The method involved destruction of organic matter by alkaline incineration and automated spectrophometric determination of iodide based on the Sandell and Kolthoff's reaction. The recoveries of the added iodide before calcination were between 90.05 +/- 7.36% and 97.14 +/- 4.56% (mean +/- S.D.). The coefficient of variation ranged from 2.15 to 7.21% depending on the iodine content in the milk. The limit of detection was estimated to be around 2 μg/kg.

The iodide-catalyzed reaction between As(III) and Ce(IV) stopped by the addition of diphenylamine-4-sulfonic acid was used for the development of a sensitive kinetic procedure for determining iodides with a detection limit of 2 ng/mL. The developed procedure was suitable for the determination of the total iodine in foodstuffs (Trokhimenko & Zaitsev, 2004).

Another modification of the catalytic kinetic spectrophotometric method has been established for the determination of iodine using the principle that potassium periodate oxidize rhodamine B (RhB) to discolor and I^- has a catalytic effect on the reaction. The absorbance difference (ΔA) is linearly related with the concentration of iodine in the range of 0 - 2.6 μg/mL and fits the equation $\Delta A = 0.1578$ C(C: μg/mL) + 0.0052, with a regression coefficient of 0.9965. The detection limit of the method is 7.10 ng/mL. The method was used to determine iodine in kelp, potato, tap water, and rain water samples. The relative standard deviation of 13 replicate determinations was 1.81–2.10%. The recovery of the standard addition of the method was 96.2–99.2% (Zhai et al., 2010).

Some researchers reported that the spectrophotometric methods for the determination of IO_3^- are based on its reaction with the excess I^- to liberate I_2 which forms tri iodide (Afkhami & Zarei 2001; Ensafi & Dehaghi, 2000).

Balasubramanian and Nagaraja (Balasubramanian & Nagaraja, 2008) described a sensitive spectrophotometric method for the determination of multiple iodine species such as I^-, I_2, IO_3^- and IO_4^-. The method involved oxidation of iodide to ICl_2^- in the presence of iodate and chloride in an acidic medium. The formed ICl_2^- bleaches the dye methyl red. The decrease in the intensity of the colour of the dye is measured at 520 nm. Beer's law is obeyed in the concentration range 0-3.5 μg of iodide in an overall volume of 10 ml. The relative standard deviation was 3.6% (n=10) at 2 μg of iodide. The developed method can be applied to the samples containing iodine, iodate and periodate by prereduction to iodide using Zn/H($^+$) or NH_2NH_2/H($^+$). The effect of interfering ions on the determination was pointed out. The described method was successfully applied to determine iodide and iodate in salt samples and iodine in pharmaceutical preparations.

Silva *et al.* (Silva et al., 1998) outlined a new method for the determination of iodate in table salt. KIO_3, after being converted to I_3 by reacting with iodide in the presence of phosphoric acid, was spectrophotometrically determined at two well defined UV absorption maxima of 352 and 288 nm. The results were comparable with a standard, ranging from 37.39 (±0.15) to 63.67 (±0.16) mg KIO_3 per kg of salt with samples of 0.15-0.21 g.

A flow injection method based on the catalytic action of iodide on the colour-fading reaction of the $FeSCN^{2+}$ complex was proposed and applied in order to determine iodine in milk. At pH 5.0, temperature 32°C and measurements at 460 nm, the decrease in absorbancy of Fe^{3+}-SCN (0.10 and 0.0020 mol /l) in the presence of NO_2^- (0.3 mol/ l) is proportional to the concentration of iodide, with a linear response up to 100.0 µg/l. The detection limit was determined as 0.99 µg/l and the system handles 48 samples per hour. Organic matter was destroyed by means of a dry procedure carried out under alkaline conditions. Alternatively, the use of a Schöninger combustion after the milk dehydration was evaluated. The residue was taken up in 0.12 mol/l KOH solubilization. For typical samples, recoveries varied from 94.5 to 105%, based on the amounts of both organic matter destroyed. The accuracy of the method was established by using a certified reference material (IAEA A-11, milk powder) and a manual method. The proposed flow injection method is now applied as an indicator of milk quality on the Brazilian market (de Araujo Nogueira et al., 1998).

Another spectrophotometric flow injection method for the determination of I^- and based on the catalytic effect of this ion on the oxidation of pyrocatechol violet by potassium persulphate has been developed. The method allows the determination of 0.5–5 mg/l I^- at a rate of 60 samples per hour and is subject to very little interference. It was successfully applied to the determination of iodide in table salt (Cerda et al., 1993).

4.1.4 Biological fluids and tissues

Due to the noninvasive way of sampling, urine is the most commonly analysed biological fluid. Efficient management of national salt iodization programmes depends on quality data on iodine concentrations in the urine and salt samples. These data are crucial in the evaluation of iodine interventions. Most of the analytical methods for urinary iodine concentration are based on the manual spectrophotometric measurement of Sandell-Kolthoff reduction reaction catalyzed by iodine using different oxidising reagents in the initial digestion step (Jooste & Strydom, 2010). Bilek et al. (Bilek et al., 2005) used a method which was based on alkaline ashing of urine specimens preceding the Sandell-Kolthoff reaction using brucine as a colorimetric marker. The detection limit was 2.6 µg/l and the limit of quantification was 11.7 µg/l, with intra-assay precision of 4% and inter-assay precision of 4.9%.

Another study described simple photometric determination of the iodine concentration in the thyroid tissue of small animals. Again, the method was based on the well-known catalytic Sandell-Kolthoff reaction. Prior to the analysis, the tissue was digested in a mixture of sodium chlorate and perchloric acid at 100 degrees C. Using this manner of digestion between 94 and 110% of iodine in the sample was recovered. Comparison with the neutron activation analysis showed excellent agreement of the obtained values (Tiran et al., 1991).

4.2 Chromatographic methods

While the main advantage of catalytic spectrophotometric methods is low cost of the needed equipment, chromatography is arguably the most widely used separation technique in the modern analytical laboratory. Fast, simple, reliable and sensitive chromatographic systems coupled with various detectors became the basic tool in many analytical laboratories. Routine analysis of iodine compounds can be carried out by means of gas chromatography (GC) and high performance liquid chromatography (HPLC). Analysis of inorganic iodine species in waters is mainly carried out with the use of ion chromatography (IC) or IC inductively coupled-mass spectrometry (IC-MS). Separation methods enable direct determination of various species of iodine in the presence of various kinds of complex components with the detection limit in the range of sub µg/l or µg/l (Hu et al., 1999; Schwer & Santschi, 2003). The IC method can separate I^- directly by using anion-exchange column, while HPLC method usually uses the reverse phase column modified by an ionpairing reagent in the mobile phase. Both spectrophotometric and electrochemical detectors are commonly used. Pulsed amperometric detection (PAD) typically utilizes gold, silver, platinum and glass carbon electrodes.

It has to be emphasized that both spectrophotometric and chromatographic methods are not applicable to a wide range of matrices. As far as complex matrices are concerned (e.g. seawater with high content of Cl^- and Br^- and relatively small of I^-) IC and HPLC are useful tools for iodine determination. Unfortunately, none of these methods are flexible enough to measure all iodine species, including organo-iodine in biological samples. The oxidative pretreatment of biological materials limits the application of the described methods (IO_3^- and I^- can be converted to I_2 under acidic conditions). In order to analyse IO_3^-, I^-, and organic forms of iodine in the same sample, IC is often coupled with ICP-MS. What is more, the coupling of highly efficient IC to multi-dimensional detectors such as MS or ICP/MS significantly increases sensitivity, while simultaneously reducing possible matrix interference to the absolute minimum.

4.2.1 Water samples

Liang et al. (Liang et al., 2005) applied the disposable electrode for the determination of iodide in soil and seawater samples with the spiked recovery ranging from 96-104% and the detection limit of 0.5 µg/L. Rong et al. (Rong et al., 2005) performed a direct determination of iodide and thiocyanate ions in seawater collected from the coasts of Japan. No sample pretreatment was needed. Liquid chromatography (LC) with a UV detection of 220 nm was applied. The separation was achieved on a C_{30} column of conventional size (150 mm × 4.6 mm i.d.) modified with poly(ethylene glycol). Such stationary phase enables the determination of I^- in seawater without any interference. Anions such as NO_3^-, NO_2^-, Br^- which absorb in the UV region do not interfere because the I^- peak is well resolved from the others. An aqueous solution of 300 mM sodium sulfate and 50 mM sodium chloride was used as the mobile phase. Detection limits (S/N=3) were obtained by injecting a 20-µL sample with 0.5 and 6 ng / ml for iodide and thiocyanate, respectively.

Buchberger (Buchberger, 1988) determined I^- (among other ions) in water samples using an anion-exchange stationary phase (Vydac 302-IC) and methanesulphonic acid solution as the mobile phase. A post-column reaction detector was developed based on the reaction between iodide or bromide, chloramine-T and 4,4′ bis (dimethylamino)diphenylmethane. The detection limit was *ca.* 20 pg iodide injected.

A non-suppressed ion chromatography (IC) with inductively coupled plasma mass spectrometry (ICP-MS) was developed for simultaneous determination of trace IO_3^- and iodide in seawater. An anion-exchange column (G3154A/101, Agilent) was used for the separation of IO_3^- and I^- with an eluent containing 20 mM NH_4NO_3 at pH 5.6. NH_4NO_3 used in mobile phase minimizes salt deposition on the sampler and skimmer cones of mass spectrometer. Linear plots were obtained in a concentration range of 5.0–500 µg/l and the detection limit was 1.5 µg/L for IO_3^- and 2.0 µg/l for I^-. The proposed method was used to determine IO_3^- and I^- in seawaters without sample pre-treatment (with exception of dilution) (Chen et al., 2007).

Using IC-ICP-MS, Tagami and Uchida (Tagami & Uchida, 2006) measured concentrations of halogens (Cl, Br and I) in 30 Japanese rivers. Cesium was used as an internal standard during I counting. The typical detection limit was calculated as three times the standard deviation of the blank, between 0.01–0.04 µg/l. The ranges of geometric means of I in each river were 0.18-8.34 µg/l.

Bruggink et al. developed an anion-exchange chromatography method in combination with the pulsed amperometric detection (PAD) for the analysis of dissolved I^- in surface water and in absorption solutions obtained from adsorbable organic iodide (AOI) determination. The development of the amperometric waveform for a selective detection using a silver-working electrode together with the optimization of the injection volume and digital signal smoothing was performed. This method exhibited a detection limit of 0.02 µg/L, without any need of sample treatment other than micro-filtration. The results of AOI determination of the method described in this article were compared with results obtained with a different ion chromatography approach utilizing UV detection (Bruggink et al., 2007).

4.2.2 Seaweed

A gas chromatography (GC) method was reported for the trace analysis of I^- in processed seaweed by Lin et al. (Lin et al., 2003). The method is based on the derivatization of aqueous iodide extracted from seaweed with 2-(pentafluorophenoxy)ethyl 2 (piperidino) ethanesulfonate in toluene using tetra-*n*-hexylammonium bromide as a phase-transfer catalyst.

4.2.3 Food

GC method has been developed for determination of total iodine in food, based on the reaction of iodine with 3-pentanone. Organic matter of a sample was destroyed by an alkaline ashing technique. Iodide in a water extract of the ash residues was oxidized in order to free I_2 by adding $Cr_2O_7^{2-}$ in the presence of H_2SO_4. Liberated iodine reacted with 3-pentanone to form 2-iodo-3-pentanone, extracted into n-hexane, and then determined by gas chromatography with an electron-capture detector. Recoveries of I^- from spiked food samples ranged from 91.4 to 99.6%. Detection limit for iodine was 0.05 µg/g (Mitsuhashi & Kaneda Y, 1990).

Two methods were described for the preparation of samples for total iodine measurement in milk and oyster tissue. In the first method, the samples were combusted in a stream of oxygen to release iodine that, subsequently, was trapped in a solution as iodide. The second method used a new approach in which the samples were oxidized in a basic solution of peroxydisulfate. In this case, iodine was retained in the solution as an iodate. Total iodine was measured by means of the GC analysis of the 2-iodopentan-3-one derivative. The methods were tested using Standard Reference Materials (SRMs) 1549 Non-Fat Milk Powder, and 1566a and 1566 Oyster Tissue. Also, KI and KIO_3 were used for testing the procedures. The results obtained for the SRMs, given as average +/- standard deviation in µg/l, were: 3.39 +/- 0.14 and 3.40 +/- 0.23 for SRM 1549; 4.60 +/- 0.42 and 4.51 +/- 0.45 for SRM 1566a; and 2.84 +/- 0.16 and 2.76 +/- 0.06 for SRM 1566; values corresponding to combustion and wet oxidation, respectively. Overall, the absolute recoveries varied between 91 and 103% (Gu et al., 1997).

Cataldi and Ciriello (Cataldi & Ciriello, 2005) described a sensitive method based on anion-exchange chromatographic separation coupled with amperometric detection at a modified platinum electrode under constant applied potential (+0.85 V vs. Ag AgCl). An experimental setup with an in-line and very effective method of electrode modification was proposed using an amperometric thin-layer cross-flow detector and a flowing 300 mg/l solution of iodide. The working electrode was polarized to the limiting current for oxidation of iodide to iodine in acidic solutions with the consequent formation of an iodine-based film. The results confirmed that the modified electrode exhibits high analytical response for iodide electro-oxidation with a good stability and long-life. The detection limit of iodide was estimated to be 0.5 µg/l (S/N=3) with an injection volume of 50 µL. This method was applied successfully to quantify the iodide content of milk samples, wastewaters, common vegetables and solutions containing high chloride levels. The iodide peak was always observed without interferences from the excess of coexisting anions (e.g. Cl^-, $SO_4{}^{2-}$ or Br^-). Chloride (the main component of marine samples) exhibited no effect upon the separation and detection of iodide. The same method (RP ion pair HPLC with an electrochemical detector and a silver working electrode) was considered by the International Organization for Standardization (the determination of iodide content of pasteurized whole milk and dried skimmed milk when present at levels from 0.03 µg/g to 1 µg/g and from 0.3 µg/g to 10 µg/g) (International Standard ISO, 2009).

Xu et al. (Xu et al., 2004) described a method for determination of iodate developed by RP-HPLC with UV detection. Iodate was converted to iodine, which was separated from the matrix using a reversed-phase Ultrasphere C18 column (250×4.6 mm, 5 µm) with methanol (1M) H_3PO_4 (1:4) as the mobile phase at 1.00 ml/min and UV detection at 224 nm. The calibration graph was linear from 0.05 µg/ml to 5.00 µg/ml for iodine with a correlation coefficient of 0.9994 (n=7). The detection limit was 0.01 µg/ml. The recovery was from 96% to 101% and the relative standard deviation was in the range of 1.5% to 2.9%.

A method based on the coupling of size-exclusion chromatography (SEC) with on-line selective detection of iodine by ICP MS was developed allowing determination of iodine species in milk and infant formulas. Iodine species were quantitatively eluted with 30 mM Tris buffer which was prepared by dissolving 30 mM of tris [tris(hydroxymethyl)-aminomethane] in water and adjusting the pH to 7.0 by the addition of hydrochloric acid (1 : 10, v/v) within 40 min and detected by ICP MS with a detection limit of 1 µg l^{-1} (as I). A systematic study of iodine speciation in milk samples of different animals (cow, goat) and humans, of different geographic origin (several European countries) and in infant formulas from different manufacturers was carried out. When obtained after centrifugation of fresh milk or reconstituted , milk powders contained more than 95% of the iodine initially present in the milk of all the investigated samples with the exception of the infant formulas in which only 15-50% of the total iodine was found in the milk whey. Adding sodium dodecyl sulfonate (SDS) improved considerably the recovery of iodine from these samples (in case of the natural milk samples, this increase was ca. 10±20% but for infant formula samples the amount of iodine recovered in the supernatant was more than twice that in the samples not incubated with SDS). Iodine was found to be principally present as iodide in all the samples except infant formulas. In the latter, more than half of iodine was bound to a high molecular (>1000 kDa) species. The sum of all the species recovered from a size-exclusion column accounted for more than 95% of the iodine present in a milk sample. For the determination of total iodine in milk, a rapid method based on microwave-assisted digestion of milk with ammonia followed by ICP MS was optimized and validated using CRM 151 Skim Milk Powder (Fernandez-Sanchez & Szpunar, 1999).

4.2.4 Biological fluids and tissues

Odink et al. (Odink et al., 1988) presented a simple method for the routine analysis of iodide in urine. Iodide was separated by means of ion-pair reversed phase chromatography (RP-HPLC) and detected electrochemically with a silver electrode after a one-step sample clean-up. The coefficient of variation of a single analysis of iodide in a pooled urine sample (530 nmol/l) was 7.6%. The detection limit was 3 pmol (S/N 3), corresponding to 0.06 µmol/l. The recovery of iodide added to urine was 96±7 %.

There are also studies that compare spectrophotometric and RP-HPLC determinations of iodine concentrations in urine (Bier et al., 1998). In the first one ammonium persulfate was used as an oxidant in the modified ceric arsenite method. With the use of this sensitive method iodine concentrations can be determined in very small specimens (50 µL). A Technicon Autoanalyzer II and a paired-ion-RP HPLC were the basic analytical equipment. The authors found that the precision of this optimized ammonium persulfate method yielded inter assay CVs of <10% for urinary iodine concentrations >10 µg/dL. The detection limit was 0.0029 µg iodine. There was a high correlation between all three methods (r > 0.94 in any case) and the interpretation of the results was consistent. The authors suggested that the manual ammonium persulfate method could be performed in any routine clinical laboratory for urinary iodine analysis. Another benefit of the described methods is a possibility to process a large number of samples with high accuracy and minimal technician`s time.

When using the HPLC assay method, contaminations from the protein bound iodine do not interfere with the determination of the serum inorganic iodide (SII), making it the method of choice for detection in the serum. Although the clinical relevance of the measurement of SII is limited, it allows calculation of the absolute iodine uptake , which has a great value in certain pathophysiological studies (Rendl et al., 1998).

Błażewicz et al. (Błażewicz et al., 2011) examined correlations between the content of iodides in 66 nodular goitres and 100 healthy human thyroid tissues (50 - frozen and 50 formalin - fixed). A fast, accurate and precise ion chromatography method on the IonPac AS11 chromatographic column (Dionex, USA) with a pulsed amperometric detection (IC-PAD) followed by alkaline digestion with tetramethylammonium hydroxide (TMAH) in a closed system and with the assistance of microwaves was developed and used for the comparative analysis of two types of human thyroid samples (healthy and pathological). A good correspondence (for 10 additional determinations) between the certified (3.38 ± 0.02 ppm with variation coefficient /V.C. / of 0.59 % for Standard Reference Material (SRM) NIST 1549- non-fat milk powder) and the measured iodine concentrations (3.52 ± 0.29 ppm; V.C. = 10 %) was achieved. Suitability of the developed IC method was supported by validation results.

Ion chromatography coupled with electrospray ionization tandem mass spectrometry was applied for quantifying iodide, as well as perchlorate and other sodium-iodide symporter (NIS) inhibitors in the human amniotic fluid. The use of selective chromatography and tandem mass spectrometry decreased the need to clean up samples, leading to a quick and rugged method that is capable of the routine analysis of 75 samples per day. Along the physiologically relevant concentration range for the analytes, the analytical response was linear. The analysis of a set of 48 samples of amniotic fluid identified the range and median levels for iodide as: 1.7–170, 8.1μg/l (Blount & Valentin-Blasini, 2006).

5. Conclusion

There are many analytical methods available for detecting, and/or measuring iodine and its various species in complex matrices. Unfortunately, there is no perfect method which would be accurate, sensitive, cheap, fast, simple, and free of interferences at the same time. This review has been focused mainly on applications of spectophotometric and chromatographic methods of iodine analysis because they are widely used in practice, and relatively cheap. What is more, to achieve lower detection limits, they can also be coupled with other more sophisticated techniques (eg. ICP-MS). Although, these two methods have their own limitations, connected mainly with sample preatretment step (often timeconsuming), the literature data show continuous progress in the search for the best spectrophotometric and chromatographic conditions in iodine determinations. Reduction of time necessary for sample preparation still remains a challenge for analysts. Summarizing, future directions of iodine analysis lie rather in the simplification of methodologies and their extensive accessibility rather than in the tendency to decrease the limit of detection. Some recently published papers on the determination of iodine include: the evaluation of urinary iodide by the use of micro-photometric method compared to ICP-MS results (Grimm et al., 2011); determination of iodine and its species in plant samples using IC-ICP/MS (Lin et al., 2011); spectrophotometric determination of I^-, IO_3^-, IO_4^- in table salt, pharmaceutical preparations and sea water (George et al., 2011); investigation of the concentration-dependent mobility, retardation, and speciation of iodine in surface sediment from the river (Zhang et al., 2011); comparison of Sandell-Kolthoff reaction with potentiometric measurements of urinary iodide in female thyroid patients (Kandhro et al., 2011). One of the newest studies concerns the analysis of food samples by ICP-MS after alkaline digestion with TMAH (Tinggi et al., 2012). As it turns out, the newest published works utilize the most common already existing methods.

6. Acknowledgment

My appreciation and thanks are given to Prof. Ryszard Maciejewski, Vice Rector for Research at the Medical University of Lublin, for financial support of the research.

7. References

Adams, J.B., Holloway, C.E., George, F. & Quig, D. (2006). Analyses Of Toxic Metals And Essential Minerals In The Hair Of Arizona Children With Autism And Associated

Conditions And Their Mothers. *Biol Trace Elem Res,* Vol.110, pp 193–209, ISSN 1559-0720

Afkhami, A. & Zarei, A. R. (2001). Spectrophotometric Determination Of Periodate And Iodate By A Differential Kinetic Method. Talanta Vol.53, pp 815–820, ISSN *0039-9140*

Aiba, M., Ninomiya, J., Furuya, K., Arai, H., Ishikawa, H., Asaumi, S., Takagi, A., Ohwada, S. & Morishita, Y. (1999). Induction Of A Critical Elevation Of Povidone-Iodine Absorption In The Treatment Of A Burn Patient: Report Of A Case. *Surg Today* Vol.29, pp 157-159, ISSN 1436-2813

Allain, P., Berre, S., Krari, N., Laine-Cessac, P.,Le Bouil, A., Barbot, N., Rohmer,V.,Bigorgne, J. C. (1993). Use of plasma iodine assay for diagnosing thyroid Disorders. *J Clin Pathol.,* Vol.46, pp453-455, ISSN 0021-9746

Anderson, K. A. & Markowski, P. (2000). Speciation Of Iodide, Iodine, And Iodate In Environmental Matrixes By Inductively Coupled Plasma Atomic Emission Spectrometry Using In Situ Chemical Manipulation. *J. Aoac. Int.*, Vol.83, No.1, pp 225-230, ISSN 1944-7922

Andrási, E., Bélavári, C., Stibilj, V., Dermelj, M. & Gawlik, D. (2004). Iodine Concentration In Different Human Brain Parts. *Anal Bioanal Chem.*, Vol.378, No.1, pp 129-33, ISSN 1618-2642

Aumont, G. (1982). A Semi-Automated Method For The Determination Of Total Iodine In Milk. *Ann Rech Vet.* Vol.13, No.2, pp 205-210, ISSN 0003-4193

Balasubramanian, M. G. & Nagaraja K.S. (2008). Spectrophotometric Determination Of Iodine Species In Table Salt And Pharmaceutical Preparations. *Chem Pharm Bull (Tokyo),* Vol.56, No.7, pp 888-893, ISSN 1347-5223

Bermejo-Barrera, P., Anllo-Sendín, R.Ma., Aboal-Somoza, M. & Bermejo-Barrera, A. (1999). Contribution To The Development Of Indirect Atomic Absorption Methods: Application Of The Ion Pair 1,10-Phenanthroline-Mercury(II)-Iodide To Iodide Determination In Water And Infant Formulae Samples. *Mikrochimica Acta,* Vol.131, No.3-4, pp 145-151, ISSN: 00263672

Bichsel, Y. & Von-Gunten, U. (1999). Determination of Iodide and Iodate by Ion Chromatography with Postcolumn Reaction and UV/Visible Detection. *Anal. Chem,* Vo.71, No.1, pp 34–38, ISSN 0003-2700

Bier, D., Rendl, J., Ziemann, M., Freystadt, D. & Reiners, Ch. (1998). Methodological And Analytical Aspects Of Simple Methods For Measuring Iodine In Urine. Comparison With HPLC And Technicon Autoanalyzer II. *Exp Clin Endocrinol Diabetes,* Vol.106, pp 27-31, ISSN 0947-7349

Bílek, R., Bednár, J. & Zamrazil, V. (2005). Spectrophotometric determination of urinary iodine by the Sandell-Kolthoff reaction subsequent to dry alkaline ashing. Results from the Czech Republic in the period 1994-2002. *Clin Chem Lab Med.* Vol.43, No.6, pp 573-580, ISNN 1434-6621

Błażewicz, A., Orlicz-Szczęsna, G., Szczęsny, P., Prystupa, A., Grzywa-Celińska & A., Trojnar, M. (2011). A Comparative Analytical Assessment Of Iodides In Healthy And Pathological Human Thyroids Based On IC-PAD Method Preceded By Microwave Digestion. *J Chromatogr B Analyt Technol Biomed Life Sci.*, Vol.15, No.879 (9-10), pp 573-578, ISSN1570-0232

Blount, B. C. & Valentin-Blasini, L. (2006). Analysis Of Perchlorate, Thiocyanate, Nitrate And Iodide In Human Amniotic Fluid Using Ion Chromatography And Electrospray Tandem Mass Spectrometry. *Analytica Chimica Acta,* Vol.567, No.1, pp 87-93, ISSN 0003-2670

Braetter, P., Blasco, I.N., Negretti de Braetter, V. E. & Raab A. (1998). Speciation As An Analytical Aid In Trace Element Research In Infant Nutrition. *Analyst,* Vol.123, pp 821-826, ISSN 0003-2654

Bruggink, C., van Rossum, W. J. M., Spijkerman, E. & van Beelen, E. S. E. (2007). Iodide analysis by anion-exchange chromatography and pulsed amperometric detection in surface water and adsorbable organic iodide. *Journal of Chromatography A,* Vol. 1144, No.2, pp 170-174, ISSN 0021-9673

Buchberger, W. (1988). Determination Of Iodide And Bromide By Ion Chromatography With Post-Column Reaction Detection. *Journal of Chromatography A,* Vol.439, No.1, pp 129- 135, ISSN 0021-9673

Bürgi, H., Supersaxo, Z. & Selz, B.(1990). Iodine Deficiency Diseases In Switzerland One Hundred Years After Theodor Kocher's Survey: A Historical Review With Some New Goitre Prevalence Data. *Acta Endocrinol (Copenh),* Vol.123, No.6, pp 577-590, ISSN 0001-5598

Cataldi, T. R. I., Rubino, A. & Ciriello, R. (2005). Sensitive Quantification Of Iodide By Ion-Exchange Chromatography With Electrochemical Detection At A Modified Platinum electrode. *Anal Bioanal Chem,* Vol.382, pp 134–141, ISSN 1618-2642

Cerda, A., Forteza, R. & Cerda, V.(1993). Determination Of Iodide In Table Salt By Flow Injection Analysis Using Pyrocatechol Violet. *Food Chemistry,* Vol.46, No.1, pp 95 - 99, ISSN 0308-8146

Chen, Z. L., Megharaj, M. & Naidu, R. (2007). Speciation Of Iodate And Iodide In Seawater By Non-Suppressed Ion Chromatography With Inductively Coupled Plasma Mass Spectrometry. *Talanta,* Vol.72, No.5, pp 1842-1846, ISSN *0039-9140*

Cicerone, R. J. (1981). Halogens in the Atmosphere. *Rev. Geophys. Space Phys.,* Vol.19, pp 123-139, ISSN 0034-6853

Das, P., Gupta, M., Jain, A. & Verma, K. K. (2004). Single Drop Microextraction Or Solid Phase Microextraction-Gas Chromatography-Mass Spectrometry For The Determination Of Iodine In Pharmaceuticals, Iodized Salt, Milk Powder And Vegetables Involving Conversion Into 4-Iodo-N,N-Dimethylaniline. *J. Chromatogr. A,* Vol.1023, pp 33–39, ISSN: 00219673

De Araujo Nogueira, A. R., Mockiuti, F., De Souza, G. B. & Primavesi, O. (1998). Flow Injection Spectrophotometric Catalytic Determination of Iodine in Milk. *Anal. Sci.,* Vol.14, pp 559-564, ISSN 1348-2246

Delange, F., de Benoist, B., Burgi, H. (2002). Determining median urinary iodine concentration that indicates adequate iodine intake at population level. *Bull World Health Organ.,* Vol.80, No.8, pp633-666, ISSN 0042-9686

Ensafi, A. & Dehaghi, G.B. (2000). Flow-Injection Simultaneous Determination Of Iodate And Periodate By Spectrophotometric And Spectrofluorometric Detection. *Anal. Sci.* Vol.16, pp 61–64, ISSN 09106340

Fang, M.C., Stafford, R.S., Ruskin, J.N., & Singer, D.E. (2004). National Trends In Antiarrhythmic And Antithrombotic Medication Use In Atrial Fibrillation. *Arch Intern Med,* Vol.164, pp 55-60, ISSN 1538-3679

Fecher P.A. & A. Nagengast. (1994). Trace analysis in high matrix aqueous-solutions using helium microwave-induced plasma-mass spectrometry. *J. Anal. At. Spectrom.*, Vol.9, pp1021-1027, ISSN 1364-5544

Fecher P.A., Goldman, I. & A. Nagengast. (1998). Determination Of Iodine In Food Samples By Inductively Coupled Plasma Mass Spectrometry After Alkaline Extraction. *Journal of Analytical Atomic Spectometry*, Vol.13, pp 977-982, ISSN 1364-5544

Fernández-Sánchez, L. & Szpunar, J. (1999). Speciation Analysis For Iodine In Milk By Size - Exclusion Chromatography With Inductively Coupled Plasma Mass Spectrometric Detection (SEC-ICP MS). *J.Anal. At. Spectrom.*, Vol.14, pp 1697-1702, ISSN 1364-5544

Fernández-Sánchez, L., Bermejo-Barreraa, P., Fraga-Bermudez, Szpunar, J. & Łobiński R. (2007). Determination Of Iodine In Human Milk And Infant Formulas. *Journal of Trace Elements in Medicine and Biology*, Vol.21, Supp.1, pp 10-13, ISSN 0946-672X

Fischer, P.W.F., Labbe, M. R. & Giroux, A. (1986). Colorimetric determination of total iodine in foods by iodide-catalyzed reduction of Ce+4. *J. Assoc. Off. Anal. Chem.* , Vol.69, No.4, pp 687 – 689, ISSN 0004-5756

FOOD STANDARDS AGENCY (FSA) UK - 2000. Total Diet Study - Fluorine, Bromine and Iodine, Available from
http://www.food.gov.uk/science/surveillance/fsis2000/5tds

Fragu, P., Briançon, C., Fourré, C., Clerc, J., Casiraghi, O., Jeusset, J., Omri, F. & Halpern, S. (1992). SIMS Microscopy In The Biomedical Field. *Biol Cell.*, Vol.74, No.1, pp 5-18, ISSN 1768-322X

Fuge, R., Johnson, C. C. (1986).The geochemistry of iodine - a review. *Environmental Geochemistry and Health*,Vol.8, No. 2, pp 31-54, ISSN 0269-4042

Gamallo-Lorenzo,D., Barciela-Alonso, M. C., Moreda-Pineiro, A., Bermejo-Barrera, A. & Bermejo-Barrera, P. (2005). Microwave-assisted alkaline digestion combined with microwave-assisted distillation for the determination of iodide and total iodine in edible seaweed by catalytic spectrophotometry. *Analitica Chimica Acta, Vol.* 542, pp 287-295, ISSN 0003-2670

Garland, J. A. & Curtis, H. J. (1981). Emission of iodine from the sea surface in the presence of ozone. *J. Geophys. Res.*, Vol.86 (C4), pp 3183–3186, ISSN 0148-0227

George, M., Nagaraja, K. S., Natesan Balasubramanian, N. (2011). Spectrophotometric Determination of Iodine Species in Table Salt,Pharmaceutical Preparations and Sea Water., *Eurasian J Anal Chem.*, Vol.6, No.2, pp129-139, ISSN 1306-3057

Gilfedder, B.S., Althoff, F., Petri, M. & Biester, H. (2007). A Thermo Extraction-UV/Vis Spectrophotometric Method For Total Iodine Quantification In Soils And Sediments. *Analytical And Bioanalytical Chemistry*, Vol.389, No(7-8), pp. 2323-2329, ISSN 1618-2642

Gottardi, W. (1998). Redox-potentiometric/titrimetric analysis of aqueous iodine solutions. *Fresenius J Anal Chem*, Vol.362, pp263–269, ISSN 0937-0633

Grimm, G., Lindorfer, H., Kieweg, H., Marculescu, R., Hoffmann, M., Gessl, A., Sager, M.,Bieglmayer, C. (2011). A simple micro-photometric method for urinary iodine determination. *Clin Chem Lab Med.*, Vol.49, pp1749-51, ISSN 0008-4212

Gu, F., Marchetti, A.A. & Straume, T. (1997). Determination Of Iodine In Milk And Oyster Tissue Samples Using Combustion And Peroxydisulfate Oxidation. *Analyst.* Vol.122, No.6, pp 535-537, ISSN 0003-2654

Guenther, E. A., Johnson, K. S. & Coale, K. H. (2001). Direct Ultraviolet Spectrophotometric Determination of Total Sulfide and Iodide in Natural Waters. *Anal. Chem.* Vol.73, | pp 3481-3487, ISSN 0003-2700

Gurkan, R., Bicer, N., Ozkan, M. H., & Akcay, M. (2004). Determination Of Trace Amounts Of Iodide By An Inhibition Kinetic Spectrophotometric Method. *Turk J Chem,* Vol.28, pp 181-191, ISSN 1303-6130

Hansson, M., Isaksson, M. & Berg, G. (2008). Sample Preparation For In Vitro Analysis Of Iodine In Thyroid Tissue Using X-Ray Fluorescence. *Cancer Inform.* Vol.6, pp 51-57, ISSN 1176-9351

Heckwan, M.M. (1979). Analysis of foods for iodine: interlaboratory study. *J Assoc Off Anal Chem,* Vol.62, pp 1045- 1049, ISSN 0004-5756

Hemken, R.W. (1979). Factors That Influence The Iodine Content Of Milk And Meat: A Review. *J Anim Sci.* Vol.48, pp 981-985, ISSN1525-3163

Hou, X. (2009). Iodine Speciation in Foodstuffs, Tissues, and Environmental Samples, In: Comprehensive Handbook of Iodine, Preedy, (Ed.), 139-150, Elsevier, ISBN 978- 0-12-374135-6, Burlington, USA

Hou, X., Chai, C. & Qian, Q. (1997). Determination Of Bromine And Iodine In Biological And Environmental Materials Using Epithermal Neutron Activation Analysis. Fresenius Anal Chem, Vol.357, pp 1106–1110, ISSN 1432-1130

Hou, X., Dahlgaard, H., Rietz, B., Jacobsen U., Nielsen, S. P. & Aarkrog, A. (1999). Determination of Chemical Species of Iodine in Seawater by Radiochemical Neutron Activation Analysis Combined with Ion-Exchange Preseparation. Anal. Chem., Vol.71, No. 14, pp 2745–2750, ISSN 0003-2700

Hu, W., Haddad, P. R., Hasebe, K., Tanaka, K., Tong, P. & Khoo, C. (1999). Direct Determination of Bromide, Nitrate, and Iodide in Saline Matrixes Using Electrostatic Ion Chromatography with an Electrolyte as Eluent. *Anal. Chem.,* Vol.71, pp 1617-1620, ISSN 0003-2700

International Standard ISO 14378, IDF 167, 2009 -10-01. Milk And Dried Milk - Determination Of Iodide Content - Method Using High-Performance Liquid Chromatography, Available from http://www.iso.org/iso/iso_catalogue/catalogue_tc/catalogue_detail.htm?csnumber=53832

Johnson, C. C. (2003).The geochemistry of iodine and its application to environmental strategies for reducing the risks from iodine deficiency disorders. *British Geological Survey Commissioned Report,* CR/03/057N. pp 54

Jooste, P.L. & Strydom, E. (2010). Methods For Determination Of Iodine In Urine And Salt. *Best Pract Res Clin Endocrinol Metab.,* Vol.24, No.1, pp 77-88, ISSN 1521-690X

Jopke, P., Bahadir, M., Fleckenstein, J. & Schnug, E. (1996). Iodine Determination In Plant Materials. *Communications In Soil Science And Plant Analysis,* Vol.27, No.3-4, pp 741-751, ISSN 1532-2416

Jungreis, E. & Gedalia, I. (1960). Ultramicro Determination Of Iodine In Drinking Water On The Basis Of Feigl's Catalytic Reaction. *Microchimica Acta,* Vol.48, No.1, pp 145-149, ISSN 1436-5073

Kamavisdar, A. & Patel, R. M. (2002). Flow Injection Spectrophotometric Determination of Iodide in Environmental Samples, *Microchimica Acta,* Vol.140, No.1-2; pp. 119-124, ISSN 1436-5073

Kandhro, G. A., Kazi, T. G., Kazi, S. N., Afridi, H. I., Arain, M. B., Baig, J. A., Shah, A. Q. & Syed, N. (2009). Evaluation of the Iodine Concentration in Serum and Urine of Hypothyroid Males Using an Inexpensive and Rapid Method. *Pak. J. Anal. Environ. Chem.*, Vol. 10, No. 1 & 2, pp 67-75, ISSN 1996-918X

Kandhro, G. A., Gul Kazi,T. G., Sirajuddin, Kazi, N., Afridi, H. I., Arainc, M. B., Baig, J. A., Shah, A. Q., Wadhwa, *S. K.*, Shah, *F.* (2011). Comparison of Urinary Iodide Determination in Female Thyroid Patients by two Techniques. *Russian Journal of Electrochemistry., Vol.47, No.12, pp 1355–1362, ISSN 1023-1935*

Kesari, R., Rastogi, R. & Gupta, V. K. (1998). A Simple And Sensitive Spectrophotometric Method For The Determination Of Iodine In Environmental Samples. *Chemia Analityczna* , Vol.43, No.2, pp 201-207, ISSN 0009-2223

Knapp, G. & H. Spitzy, H. (1969). Untersuchungen Zur Optimierung Der Reaktionsbedingungen Für Die Katalytische Jodwirkung Auf Das System CE(IV)-Arsenige Säure (Eine Modifizierte Sandell-Kolthoffreaktion). *Talanta,* Vol.16, No.10, pp 1353-1360, ISSN 0039-9140

Knapp, G., Maichin, B., Fecher, P., Hasse, S. & Schramel, P. (1998). Iodine Determination In Biological Materials Options For Sample Preparation And Final Determination. *Fresenius J Anal Chem,* Vol. 362, pp 508 - 513, ISSN 1432-1130

Koh, T., Ono, M. & Makino, I. (1988). Spectrophotometric determination of iodide at the 10–6 mol l–1 level by solvent extraction with methylene blue. *Analyst,* Vol.113, pp 945-948, ISSN 0003-2654

Konieczka, P. & Namieśnik J. (Ed.), Ocena i kontrola jakości wyników pomiarów analitycznych. Wydawnictwa Naukowo-Techniczne, ISBN 978-83-204-3255-8, Warszawa, 2007

Levine, K.E., Essader, A.S., Weber, F.X., Perlmutter, J.M., Milstein, L.S., Fernando, R.A., Levine, M.A., Collins, B.J., Adams, J.B. & Grohse, P.M. (2007).Determination Of Iodine In Low Mass Human Hair Samples By Inductively Coupled Plasma Mass Spectrometry.*Bull Environ Contam Toxicol,* Vol.79, No.4, pp 401-404, ISSN 0007-4861

Liang, L., Cai, Y., Mou, S. & Cheng, J. (2005). Comparisons Of Disposable And Conventional Silver Working Electrode For The Determination Of Iodide Using High Performance Anion-Exchange Chromatography With Pulsed Amperometric Detection. *Journal of Chromatography A,* Vol.1085, No.1, pp 37-41, ISSN 0021-9673

Lin, F.-M., Wu, H.-L., Kou, H.-S. & Lin, S.-J. (2003). Highly Sensitive Analysis of Iodide Anion in Seaweed as Pentafluorophenoxyethyl Derivative by Capillary Gas Chromatography. J. Agric. Food Chem., Vol.51, No.4, pp 867–870 , ISSN 0021-8561

Lin, L., Chen, G., Chen, Y. (2011). Determination of iodine and its species in plant samplesusing ion chromatography-inductively coupled plasma mass spectrometry. *Chinese Journal of Chromatography,* Vol.29, pp662-666, ISSN 1872-2059

Lovelock, J.E., Maggs, R.J. & Wade, R.J. (1973). Halogenated Hydrocarbons In And Over The Atlantic. *Nature,* Vol.241, No. 5386, pp 194-196, ISSN 0028-0836

Lu, Y.-L., Wang, N.-J., Zhu, L., Wang, G.-X., Wu, H., Kuang, L. & Zhu W.-M. (2005). Investigation Of Iodine Concentration In Salt, Water &Soil Along The Coast Of Zhejiang, China. *J Zhejiang Univ Sci B.*, Vol.6, No.12, pp 1200–1205, ISSN 1862-1783

Mani, D., Gnaneshwar Rao T., Balaram, V., Dayal, A.M. & Kumar, B. (2007). Rapid Determination Of Iodine In Soils Using Inductively Coupled Plasma Mass Spectrometry. *Current Science,* Vol.93, No. 9, 10, pp 1219-1221, ISSN 0011-3891

Maros, L., Kaldy, M. & Igaz, S. (1989). Simultaneous Determination Of Bromide And Iodide As Acetone Derivatives By Gas Chromatography And Electron Capture Detection In Natural Waters And Biological Fluids. *Anal. Chem.*, Vol. 61, 733–735, ISSN 0003-2700

Michalke, B. (2006). Trace Element Speciation In Human Milk. *Pure Appl. Chem.*, Vol.78, No.1, pp. 79-90, ISSN 0033-4545

Mitsuhashi, T. & Kaneda, Y. (1990). Gas Chromatographic Determination Of Total Iodine In Foods. *J Assoc Off Anal Chem.*, Vol.73, No.5, pp 790-792, ISSN 0004-5756

Miyake, Y. & Tsunogai, S. (1963). Evaporation of iodine from the ocean. *J. Geophys. Res.*, Vol.68, pp 3989-3993, ISSN 0148-0227

Moxon, R.E. & Dixon, E.J. (1980). Semi-Automatic Method For The Determination Of Total Iodine In Food. *Analyst*, Vol.105, pp 344–352, ISSN 0003-2654

Mura, P., Papet, Y., Sanchez, A. & Piriou, A. (1995). Rapid And Specific High-Performance Liquid Chromatographic Method For The Determination Of Iodide In Urine, *J. Chromatogr. B*, Vol.664, pp 440–443, ISSN 1570-0232

Muramatsu, Y., Uchida, S., Fehn, U., Amachi, S. & Ohmomo, Y. (2004). Studies With Natural And Anthropogenic Iodine Isotopes: Iodine Distribution And Cycling In The Global Environment. *J. Environ. Radioact.*, Vol.74, pp 221–223, ISSN 0265-931X

Naozuka J., Mesquita Silva da Veiga M.A., Oliveira P.V. & de Oliveira E. (2003). Determination Of Chlorine, Bromine And Iodine In Milk Samples By ICP-OES. *J. Anal. Atomic Spectrom.* Vol.18, pp 917–921, ISSN 1364-5544

Niedobová, E., Machát, J., Otruba, V. & Kanicky, V. (2005). Vapour Generation Inductively Coupled Plasma Optical Emission Spectrometry In Determination Of Total Iodine In Milk. *J.Anal.Atom. Spectrom.*, Vol.20, pp 945–949, ISSN 1364-5544

Odink, J., Bogaards, J.J.P., Sandman, H., Egger, R.J., Arkesteyn, G.A. & de Jong, P. (1988). Excretion Of Iodide In 24-H Urine As Determined By Ion-Pair Reversed-Phase Liquid Chromatography With Electrochemical Detection. *Journal Of Chromatography B: Biomedical Sciences And Applications* Vol.431, pp 309-316, ISSN 1570-0232

Okerlund, M.D. (1997). The Clinical Utility of Fluorescent Scanning of the Thyroid. In: Medical Applications of Fluorescent Excitation Analysis, Kaufman & Price (Ed.), 149-160, CRC Press, ISBN 0849355079, Boca Raton Florida, USA

Pearce, E.N., Gerber, A.R., Gootnick, D.B., Khan, L., Li, R., Pino, S. & Braverman, L. E.(2002). Effects Of Chronic Iodine Excess In A Cohort Of Long-Term American Workers In West Africa. *J ClinEndocrinol Metab*, Vol.87, pp 5499 -5502, ISSN 1945-7197

Pennington, J.A.T., Schoen, S.A., Salmon, G.D., Young, B., Johnson, R.D. & Marts, R.W. (1995). Composition of Core Foods of the U.S. Food Supply, 1982-1991. III. Copper, Manganese, Selenium, and Iodine. *J Food Comp Anal.*, Vol.8, No.2, pp 171-217, ISSN 1096-0481

Rasmussen, R. A., Khalil, M.A.K., Gunawardena, R. & Hoyt, S. D. (1982). Atmospheric Methyl Iodide (CH_3I). *J. Geophys. Res.*, Vol.87, pp 3086-3090, ISSN 0148-0227

Rendl, J., Bier, D. & Reiners, C. (1998). Methods For Measuring Iodine In Urine And Serum. *Exp Clin Endocrinol Diabetes*,Vol.106, Suppl 4, pp 34-41, ISSN 09477349

Rodriguez, P. A. & Pardue, H. L. (1969). Kinetics of the Iodide-Catalyzed Reaction between Cerium(IV) and Arsenic(III) in Sulfuric Acid Medium. *Anal. Chem.,* Vol.41, pp 1369-1376, ISSN 0003-2700

Rognoni, J. & Simon, C. (1974). Critical Analysis Of The Glutaraldehyde Fixation Of The Thyroid Gland: A Double-Labelling Experiment. *J. Microscopie., Vol.* 21, pp 119–128, ISSN 1365-2818

Rong, L., Lim, L. W. & Takeuchi, T. (2005). Determination of Iodide and Thiocyanate in Seawater by Liquid Chromatography with Poly(ethylene glycol) Stationary Phase. *Chromatographia,* Vol.61, pp 371-374, ISSN 1612-1112

Sandell, E. B. & Kolthoff, I. M. (1934). Chronometric catalytic method for the determination of micro quantities of iodine. *J. Am. Soc.*, 56, pp 1426-1435 ISSN: 0002-7863

Sandell, E. B. & Kolthoff, I. M. (1937). Micro Determination of Iodine by a Catalytic Method. *Microchim. Acta,* Vol.1, pp 9-25, ISSN 00263672

Schramel, P. & Hasse, S. (1994). Iodine Determination In Biological Materials By ICP-MS. *Microchimica Acta,* Vol.116, No.4, pp 205-209, ISSN 1436-5073

Schwehr, K. A. & Santschi, P. H. (2003). Sensitive Determination Of Iodine Species, Including Organo-Iodine, For Freshwater And Seawater Samples Using High Performance Liquid Chromatography And Spectrophotometric Detection. Anal. Chim. Acta, Vol.482, pp 59-71, ISSN 00032670

Sheppard, M.I., Thibault, D.H., Mcmurray, J. & Smith, P. A. (1995). Factors Affecting The Soil Sorption Of Iodine. Water Air Soil Pollut., Vol.83, No.1-2, pp 51-67, ISSN *0049-6979*

Silva, R. L., De Oliveira, A. F. & Neves, E. A. (1998). Spectrophotometric Determination Of Iodate In Table Salt. *J. Braz. Chem. Soc.,*Vol.9, No.2, pp 171-174, ISSN 0103-5053

Simon, R., Tietge, J. E., Michalke, B. Degitz, S. & Schramm K.-W. (2002). Iodine Species And The Endocrine System: Thyroid Hormone Levels In Adult Danio Rerio And Developing Xenopus Laevis. *Anal Bioanal Chem, Vol.* 372, pp 481–448, ISSN 1618-2642

Stolc, V. (1961). Interference of Certain Ions with the Catalytic Action of Iodine in the Sandell-Kolthoff Reaction *Fresenius J Anal Chem,* Vol.183, No.4, pp262–267, ISSN 0937-0633

Szybiński, Z. (2009). Iodine prophylaxis in Poland in the lights of the WHO recommendation on reduction of the daily salt intake. *Pediatric Endocrinology, Diabetology and Metabolism.*, Vol.15, No.2, pp103-107, ISSN 2083-8441

Tadros, T.G., Maisey, M.N., Ng Tang Fui S.C. & Turner, P,C.(1981). The Iodine Concentration In Benign And Malignant Thyroid Nodules Measured By X-Ray Fluorescence, *Br J Radiol.*, Vol.54, No.643, pp 626-629, ISSN 1748-880X

Tagami, K. & Uchida, S. (2006). Concentrations Of Chlorine, Bromine And Iodine In Japanese Rivers. *Chemosphere,* Vol.65, No.11, pp 2358-2365, ISSN *0045-6535*

Teas, J., Pino, S., Critchley, A. & Braverman, L.E. (2004). Variability Of Iodine Content In Common Commercially Available Edible Seaweeds. *Thyroid, Vol.*14, No.10, pp 836-841, ISSN 1050-7256

Teng, W.; Shan, Z. & Teng, X. (2009). Effect of an Increased Iodine Intake on Thyroid, In: Comprehensive Handbook of Iodine, Preedy, (Ed.), 1213-1220, Elsevier, ISBN 978-0-12-374135-6, Burlington, USA

Tinggi, U., Schoendorfer, N. Davies, P. S. W., Scheelings, P., Olszowy, H. *(2012).* Determination of iodine in selected foods and diets by inductively coupled plasma-mass spectrometry., *Pure Appl. Chem.,* Vol.84, No.2, pp 291-299, ISSN1365-3075

Tiran, B., Wawschinek, O., Eber, O., Beham, A., Lax, S. & Dermelj, M. (1991). Simple Determination Of Iodine In Small Specimens Of Thyroid Tissue. *Exp Clin Endocrinol.*, Vol.98, No.1, pp 32-36, ISSN 0232-7384

Trokhimenko, O.M. & Zaitsev, V.N. (2004). Kinetic Determination of Iodide by the Sandell-Kolthoff Reaction Using Diphenylamine-4-Sulfonic Acid. *J Anal Chem.*, Vol.59,pp 491–494, ISSN 1608-3199

United Nations Administrative Committee on Coordination/ Sub-Committee on Nutrition (2000) 4th Report on World Nutrition Situation - Nutrition Throughout the Life Cycle. Available from http://www.unsystem.org/scn/archives/rwns04/index.htm (accessed June 2007)

U.S. Food and Drug Administration, Code of Federal Regulations, CFR 21, Sections 184.1634 and 184.1265. Revised April 1, 2009. Available from http://ods.od.nih.gov/factsheets/Iodine-HealthProfessional

U.S. Salt Institute, accessed July 9, 2007, Available from http://www.saltinstitute.org/html

Varga, I. (2007). Iodine Determination In Dietary Supplement Products By TXRF And ICP-AES Spectrometry. *Microchemical Journal*, Vol.85, pp 127-131, ISSN 0026265X

Verheesen, R. H. & Schweitzer, C. M. (2011). Iodine and Brain Metabolism. In: Handbook of Behavior, Food and Nuirtion, Preedy et al., (Ed.), 2411-2425, Springer, ISBN 978- 0-387-92270-6, New York Dordrecht Heidelberg London

Whitehead, D.C. (1984). The Distribution And Transformations Of Iodine In The Environment. Environ. Int., Vol.10, pp 321–339, ISSN 0160-4120

World Health Organization. United Nations Children's Fund & International Council for the Control of Iodine Deficiency Disorders. (2007). Assessment Of Iodine Deficiency Disorders And Monitoring Their Elimination. In: A Guide For Programme Managers, 3rd Ed. Geneva, Switzerland, pp 1-108, Available from

Xu, X. R., Li, H. B., Gu, J. D. & Paeng, K. J. (2004). Determination of Iodate in Iodized Salt by Reversed-Phase High-Performance Liquid Chromatography with UV Detection. *Chromatographia*, Vol.60, No.11-12, pp 721-723, ISSN 1612-1112

Yebra, M.C. & Bollaín, M.H. (2010). A Simple Indirect Automatic Method To Determine Total Iodine In Milk Products By Flame Atomic Absorption Spectrometry. *Talanta*, Vol.82, No.2, pp 828-33, ISSN *0039-9140*

Yoshida, S. & Muramatsu, Y. (1995). Determination Of Organic, Inorganic And Particulate Iodine In The Coastal Atmosphere Of Japan. *Journal Of Radioanalytical And Nuclear Chemistry*, Vol.196, No. 2, pp 295-302, ISSN 1588-2780

Zabala, J., Carrión, N., Murillo, M., Quintana, M., Chirinos, J., Seijas, N., Duarte, L. & Brätter, P. (2009). Determination Of Normal Human Intrathyroidal Iodine In Caracas Population. *Trace Elem Med Biol.*, Vol.23, No.1, pp 9-14, ISNN 0946-672X

Zaichick, V. & Zaichick, S. (1997). Normal Human Intrathyroidal Iodine. *Sci Total Environ.* Vol.27, No.206(1), pp 39-56, ISSN 0048-9697

Zhang, S., Du, J., Xu, C., Schwehr, K. A., Ho, Y.-F., Li, H.-P., Roberts, K. A., Kaplan, D. I., Brinkmeyer, R., Yeager, C. M., Chang, Hyun-shik, P. H. Santschi, P. H. (2011). Concentration-Dependent Mobility, Retardation, and Speciation of Iodine in Surface Sediment from the Savannah River Site. Environ. Sci. Technol., Vol.45, No.13, pp 5543–5549, ISSN 0013-936X

Zava, T. T. & Zava, D. T. (2011). Assessment of Japanese iodine intake based on seaweed consumption in Japan: A literature-based analysis. *Thyroid Research*, Vol.4, p 14, ISSN 1756-6614

Zhai, Q.-Z., Zhang, X.-X. & Goupages, X.-L. (2010). Catalytic Kinetic Spectrophotometric Determination Of Trace Amounts Of Iodine. *Instrumentation Science & Technology*, Vol.38, No.2, pp 125-134, ISSN 1073-9149

Zimmermann, M. (2009). Iodine Deficiency. *Endocr Rev.*, Vol.30, No.4, pp 376-408, ISSN 0163-769X

19

Quality Control of Herbal Medicines with Spectrophotometry and Chemometric Techniques – Application to *Baccharis* L. Species Belonging to Sect – Caulopterae DC. (Asteraceae)

María Victoria Rodriguez[1,2], María Laura Martínez[1], Adriana Cortadi[1], María Noel Campagna[1], Osvaldo Di Sapio[1], Marcos Derita[2], Susana Zacchino[2] and Martha Gattuso[1]

[1]*Farmacobotánica, Área Biología Vegetal, Departamento Cs. Biológicas*
[2]*Farmacognosia, Área Farmacognosia, Departamento de Química Orgánica*
Facultad de Ciencias Bioquímicas y Farmacéuticas
Universidad Nacional de Rosario
Argentina

1. Introduction

Medicinal plants constitute a rich cultural and biological heritage in many countries, which could be very useful in meeting the therapeutic needs of the population (Rodriguez, 2010a). Traditional herbal medicines have been widely used for many years in many eastern countries (Liang et al., 2004). However, little work has been done to validate and standardize these products properly in order to match phytotherapy to chemotherapy which currently receives almost unconditional support from formal systems of health care. For several years now activities have been undertaken to systematize the identification, validation, production and use of medicinal plants, for both primary health care as well as a semi-industrial or industrial process, which implies their transformation into safe, reliable and stable phytopharmaceutical products. Therefore it is suggested that medicinal plants and their derived products would be a viable option for national development as an agricultural and therapeutic alternative, but standardization and industrialization, involving sustained yields, a quality control system and honest and reliable marketing would be needed for widespread implementation and official support. Consequently, on account of the above, education and research should be in agreement if any advance is to be made in this area (Rodriguez, 2010a). However, the necessary criteria for data quality, safety and efficacy of traditional medicine that would support its use in the world do not exist. Appropriate, accepted research methodology for evaluating traditional medicine is also lacking (Liang et al., 2004).

Most countries have developed organisms for controlling the quality of herbal remedies destined for the internal market or for export. The officially recognized drugs are subject to testing and specifications of identity, purity and contents of the principal active ingredients or markers for each plant drug, in order to guarantee the conservation of the species and the specifications of microbiological purity. Currently the quality requisites of the different national pharmacopoeias vary between countries, however the International Federation of Pharmaceutical Manufacturer Associations, have taken a first step towards the global implementation of organized criteria (International Federation of Pharmaceutical Manufacturer Associations [IFPMA], 1997). The World Health Organization has conducted a review of the legal status of traditional medicine and complementary or "alternative" therapies in 123 countries (World Health Organization [WHO], 2001). Moreover, some countries and organizations have developed, or are developing, their own monographs, for example, the Commission E of the German Ministry of Health, WHO and European Scientific Cooperative on Phytotherapy, China, Brazil and Argentina, among others. These monographs recognise the quality standards applicable to drugs and herbal remedies in the pharmaceutical market. (Blumenthal, 1998; European Scientific Cooperative on Phytotherapy [ESCOP], 2003; Keller, 1991; WHO, 1991, 1999, 2002).

In Argentina there is great interest in controlling the Quality, Security and Efficacy of herbal medicines. In order to achieve this ambitious objective the following steps must be taken: a- Registration of the products, b- Verification of good practices of production and quality control, of the crude drug and its products, and c- Pharmacovigilance. ANMAT (National Administration of Drug, Food and Medical Technology) is the organism responsible for authorizing all activities related to medicines through its technical organism, the INAME (National Institute of Drug). The products based on herbal medicine and aromatics are considered as Phytotherapeutic Drugs in Resolution N° 144/98 and the Supplementary Provisions (**Provision 2671/99; Provision 2672/99; Provision 2673/99)** that are now in effect. Provision 1788/2000 contains a list of 109 herbal medicines that can not be authorised for the production of herbal medicine on account of the contradictions or toxic effects reported in their traditional use.

It is very important that plant material is of the highest quality as it is used as a medicine. The aforementioned decrees aim to guarantee the quality of the raw material, the intermediate and finished products and a series of tests have been established among which, as a minimum requisite, are botanical identity, purity (physical-organoleptic and microbiological - for health) and their activity or composition (methodological analysis), taking into account the Good Practices of Agricultural Production and the Good Practices of Manufacturing (Rodriguez, 2010a).

1.1 Spectrophotometry and chemical fingerprints of herbal medicines

In general one or two markers or pharmacologically active compounds of herbal components or herbal blends were used to evaluate the quality and authenticity of herbal medicines in the identification of a single herb or preparation of herbal medicine and to evaluate the quantitative composition of a herbal product. However, this type of determination does not give a complete picture of the herbal products because multiple ingredients are usually responsible for its therapeutic effects. These therapeutic effects may work synergistically and change their activity on being separated into their active parts. So various analytical techniques can be applied for this type of registration and the complete herbal product can be considered as the "active compound" (Liang et al., 2004).

The concept of phytoequivalence was developed in Germany to ensure the consistency of herbal products (Tyler, 1999). According to this concept, a chemical profile, such as a spectrophotometric fingerprint for a herbal product should be constructed and compared with the profile of a clinically tested reference product. Therefore, a spectrophotometric fingerprint of a herbal medicine is a spectrophotometric pattern of an extract of some chemical components that are pharmalogically active and/or with chemical characteristics. It is suggested that with the help of spectrophotometric fingerprints obtained, the authentication and identification of herbal medicines would be appropriate, even when the amount and/or concentration of some typical chemical ingredients are not exactly the same for different samples of these herbal medicines. In this way, we are broadly considering multiple ingredients in the extracts of herbal medicines rather than one or two marker ingredients for evaluating the quality of the herbal products.

In general various analytical techniques can be used to obtain fingerprints of herbal medicines of which chromatography is most often used on account of its great efficiency in separating the different components of an extract. For the above mentioned reasons these chromatographic techniques have a high cost, especially if they are to be used for routine quality control. A simple alternative technique such as UV/Visible spectrophotometry, coupled to chemometric methods was carried out by Lonni et al. (2005) for taxonomic purposes. Authors were able to separate populations of three species of *Baccharis* according to their UV / Visible spectra. Here, we propose to apply the same methodology as part of quality control of herbal medicines.

Figure 1 shows the spectrophotometric profiles of ethanol extracts of the aerial parts of *Baccharis gaudichaudiana* DC. obtained from five areas in Misiones province, Argentina (MI1, MI2, MI3, MI4, MI5) (Rodriguez, 2010b). Here, of course, some differences can be observed in the profiles; however phytoequivalence for *B. gaudichaudiana* can be seen.

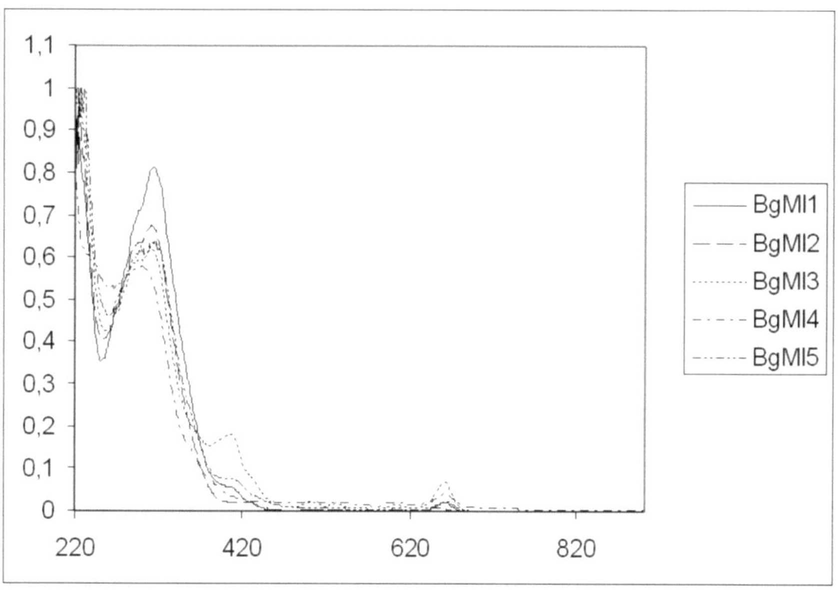

Fig. 1. Absorption UV/Visible spectra of *Baccharis gaudichaudiana* populations (Bg). MI, Misiones. Numbers indicate when there is more than one population of the same species.

1.2 Chemometric methods and quality control

The chemometric methods consist of a number of statistical, mathematical and graphic techniques that analyze many variables simultaneously (Lonni, 2005). The method used in this study is as follows:

Principal components analysis (PCA).

This method is based on the transformation of a group of original quatitative variables into another group of unrelated independent variables, known as principal components. The components have to be interpreted independently of one another, as they contain part of the variance that is not expressed in any other principal component (Pla, 1986; López & Hidalgo, 1994).

The criterion of Cliff in 1987 was adopted for selecting the number of components that should be used for the analysis, which states that the eigenvalues of acceptable components should explain 70% of the total variance (López & Hidalgo, 1994).

Proportion of variance explained: in general statistical programs provide information of the eigenvectors and, in some cases, the correlation between the original variables and the principal components. However, these correlations can be calculated from the eigenvectors in the following formula (Pla, 1986):

$$r(jk) = l(jk) \times (\lambda(\kappa))\ 1/2/s(ij) \quad (1)$$

where,

$r(jk)$= correlation between the original variable $x(j)$ and the k-esim component.

$l(jk)$ = j-esim element of the k-esim eigenvector.

$\lambda(\kappa)$= k-esim eigenvalue.

$s(ij)$ = variances of the correlation matrix.

In most studies it is important to determine the degree of discrimination of the variables so that those with the most and least variation can be identified.

Using PCA it is possible to determine the degree of discrimination, quantifying the proportion of variance explained by each original variable of the selected components; to do this it is necessary to add the squares of the correlation formed by each original variable with the selected components. This is possible as the components are not correlated (Pla, 1986). In the case of a variable in series: $rx12 + rx22 + rx32$ = proportion of the variance explained, having selected three components. It should to be taken into account that the variables that explain a larger proportion of variance are the most discrimatory and therefore they are more important.

The UV/Visible spectra of populations of two species, *Baccharis articulata* (Lam.) Pers. and *B. trimera* can be seen in Figure 2. In this figure it is difficult to distinguish which spectrum corresponds to which species, but the resolution is greater when PCA is applied (Figure 3) and different coordinates can be seen for the corresponding populations of one species or another.

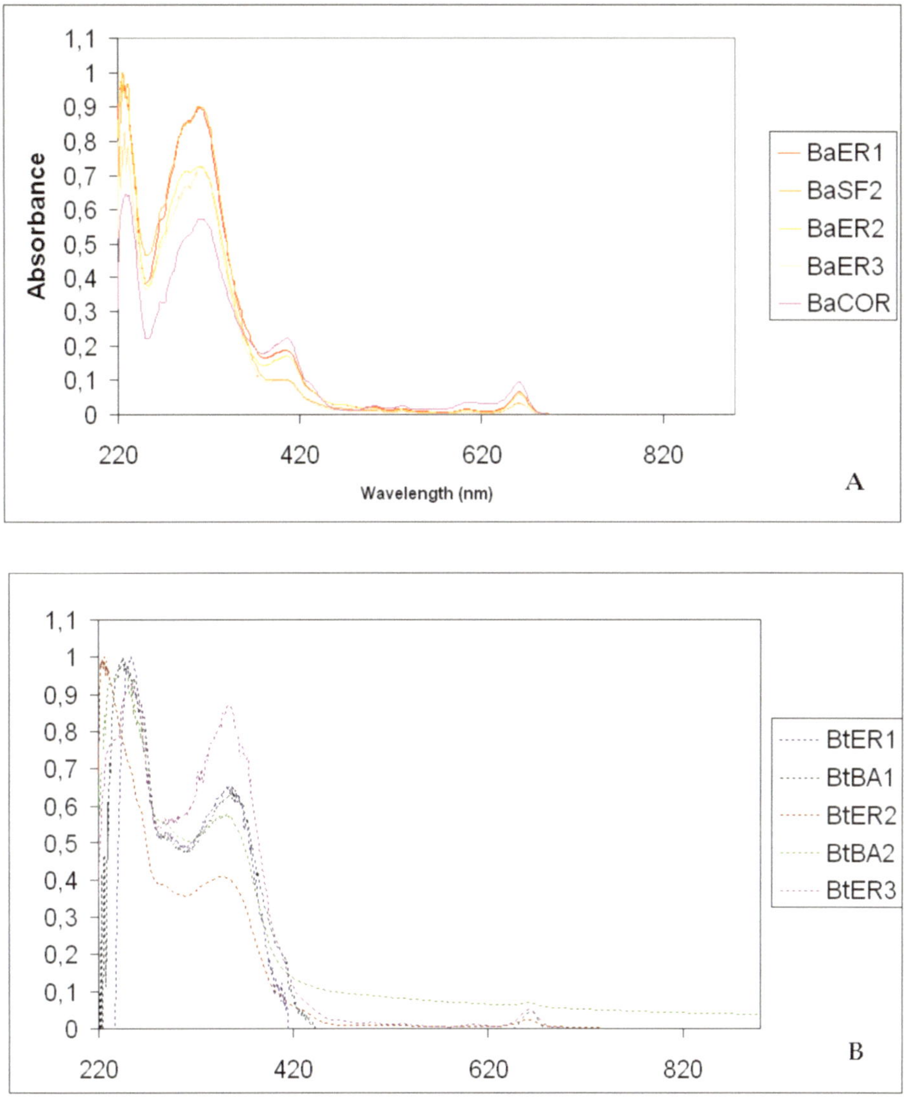

Fig. 2. Absorption UV/Visible spectra. **A**, *Baccharis articulata* (Ba); **B**, *Baccharis trimera* (Bt) populations. BA, Buenos Aires; COR, Corrientes; ER, Entre Ríos; SF, Santa Fe. Numbers indicate when there is more than one population of the same species

2. An example of differentiation of *Baccharis* L. species belonging to sect – Caulopterae DC. (Asteraceae) using UV/Visible spectrophotometry data and multivariate analysis

Baccharis L. is an exclusive American genus comprising approximately 500 species. Its distribution area covers the whole of South America and continues northwards up to the south of USA including the Atlantic coast up to Massachusetts, although its presence is greater in the intertropical and subtropical zones (Cuatrecasas, 1967; Müller, 2006). Several

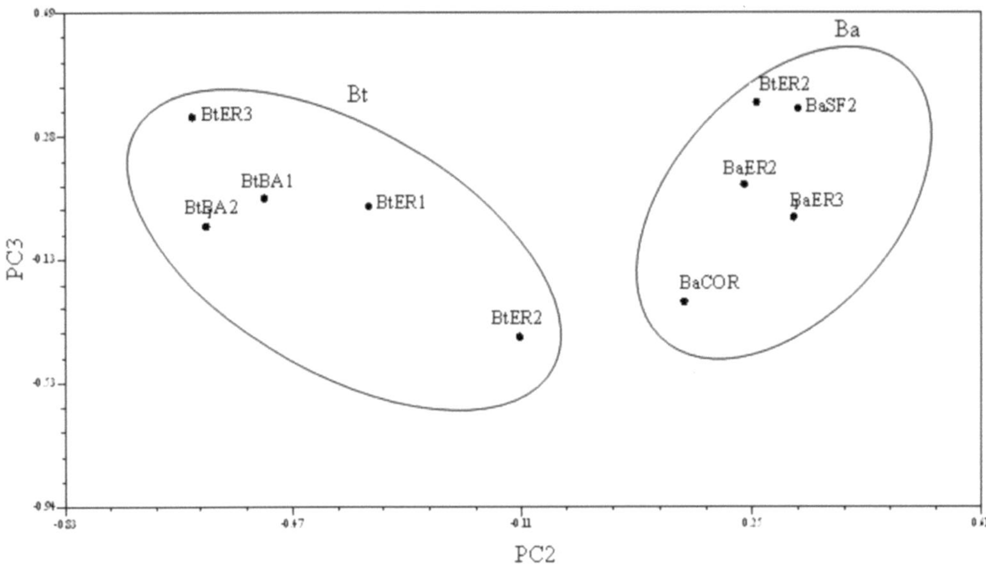

Fig. 3. Two-dimensional model derived from PCA of 700 quantitative variables of 10 *Baccharis* populations. Ba, *B. articulata*; Bt, *B. trimera*. BA, Buenos Aires; COR, Corrientes; ER, Entre Rios; SF, Santa Fe populations. Numbers indicate when there is more than one population of the same species.

authors have contributed to the infrageneric classification of *Baccharis* in general and regional floras (Ariza Espinar, 1973; Baker, 1882-1884; Barroso, 1976; Cuatrecasas, 1967; De Candolle, 1836; Giuliano, 2001; Heering, 1904; Lessing, 1831; Weddell, 1855-1856) and it was De Candolle (1836) who was the first to subdivide the genus in eight sections, mainly based on leaf morphology. More recently, Giuliano (2001) grouped 96 Argentine *Baccharis* species into 15 sections, among which the sect. Caulopterae DC. is characterized by the presence of species with alate stems. Two of nine Argentine species from this sect., i.e., *Baccharis articulata* (Lam.) Pers. and *Baccharis crispa* Spreng. are included in the National Argentine Pharmacopeia Ed. VI (1978), and a third one, *Baccharis trimera* (Less.) DC. in the Brazilian Pharmacopeia Ed. IV (2002) with the common name of "Carquejas". The nine species of this sect. are traditionally used in infusions or decoctions, as hepatic, colagogue, diuretic, ulcer healing and external antiseptics. They are also used in herbal remedies and phytotherapy and in the preparation of spirits and soft drinks (Correa, 1985; Gupta, 1995; Hieronymus, 1882; Martínez Crovetto, 1981; Sorarú & Bandoni, 1978; Toursarkissian, 1980). Beneficial effects of these species can be attributed in part to their high content of flavonoids. The chemistry of the flavonoids is predictive of their free radical scavenging activity, which confers the antioxidant activity (Harborne & Williams, 1992; Rice-Evans et al., 1995).

Currently, the morphoanatomical studies of these species in sect. Caulopterae only provide incomplete information which makes it difficult to differentiate each one properly in the non-flowering condition when the size of capitulum of each one varies (Giuliano, 2000; Müller, 2006). This fact has lead to the misuse of the same common name for botanically diverse species, which surely have different chemical compositions and therefore different pharmacological properties (Abdel-Malek et al., 1996; Desmarchelier et al., 1997; De Oliveira et al., 2003; Gene et al., 1992, 1996; Lapa et al., 1992; Palacios et al., 1983; Stoicke & Leng-Peschlow, 1987). Also the high frequency of errors committed during the collection of these species for medicinal purposes is understandable due to the coexistence of these entities in certain habitats. Hence, considering that numerical methods have been shown to be useful for multi-component metabolic classification studies, we decided to carry out population studies including 53 samples of these nine sect. Caulopterae species, combining their spectrophotometric profiles with multivariate analysis. These nine species belonging to sect. Caulopterae are: *Baccharis articulata* (Lam.) Pers., *B. crispa* Spreng., *B. gaudichaudiana* DC., *B. microcephala* (Less.) DC., *B. phyteumoides* (Less.) DC., *B. penningtonii* Heering., *B. sagittalis* (Less.) DC., *B. triangularis* Hauman and *Baccharis trimera* (Less) DC. Of these nine species, only three are official and its use is permitted but all are indiscriminately collected for medicinal purposes due to their similar morphotypes as they have alate stem.

2.1 Material and methods

Fifty three samples of nine *Baccharis* species were collected from wild materials in different locations in Argentina. All samples were botanically identified by our group and voucher specimens were deposited at the herbarium of the National University of Rosario, Argentina (Table 1).

The aerial parts of the dried plants (5 g) were macerated (24 h, 3x) with absolute ethyl alcohol. The ethanolic extract was filtered and concentrated in a rotary evaporator at a temperature lower than 100 °C. Thirty mg of dry extract were mix in 3 ml dichloromethane (DCM) and left for 1 h and then filtered with common filter paper. A dilution of 50 µl of this solution in 950 µl of methanol was prepared and filtered twice with a 0.45 µm Millipore filter (Lonni et al., 2003, 2005).

Spectrophotometric analyses were carried out using a Biochrom Model Libra S12 UV/Visible Spectrophotometer, equipped with tungsten halogen and deuterium arc light sources with a single solid state silicon photodiode detector, and operating software.

TLC analyses were carried out using silica gel 60 F254, Merck; mobile phase, DCM: Hexane:

MeOH (4:2:1). Chromatograms were evaluated under UV light at 254 and 365 nm to detect the presence of flavonoids. TLC was additionally sprayed with a diphenylborinic acid ethanolamine/polyethylene glycol reagent. Apigenin, chlorogenic acid, genkwanin, luteolin, quercetin and rutin were used as markers (purchased from Extrasynthèse, France).

HPLC analyses were carried out using a Spectra Physics Model SP8800 ternary pump chromatograph with Spectra 100 UV/Visible detector, having as chromatographic conditions, methanol eluent, 1 ml min^{-1} flow, Luna C18 phenomenex (250 x 4.6 mm, 5 µm particle size). The injection volume was 100 µl and elution was monitored al 254 nm. Apigenin, genkwanin and luteolin were used as markers (purchased from Extrasynthèse, France).

TLC and HPLC analysis were applied in order to complement the studies carried out by PCA of the spectrophotometric data and to find potential markers of the species that could not be characterized by the previous method.

Sample	Voucher	Date	Sample	Voucher	Date
BaCO1	1570	Apr-05	BmFO	1656	Feb-06
BaCO2	1607	Feb-06	BmMI	1657	Feb-06
BaCO3	1617	Mar-06	BphySF1	1888	Feb-07
BaCO4	1618	Mar-06	BphySF2	1939	Mar-08
BaCO5	1619	Mar-06	BpER	1594	Jan-06
BaCO6	1620	Mar-06	BpSF1	1887	Feb-07
BaCO7	1621	Mar-06	BpSF2	1938	Mar-08
BaCOR	1903	Mar-07	BsRN1	1906	Mar-07
BaER1	1927	Apr-07	BsRN2	1953	Nov-08
BaER2	1928	Aug-07	BtrBA[a]	SI Burkart[b]	Nov-1972
BaER3	1929	Apr-07	BtrCHU[a]	SI Dacnik[b]	Feb-1969
BaSF1	1930	Aug-07	BtrLP[a]	3262	Dec-1975
BaSF2	1916	Jan-07	BtrSL	1907	Mar-07
BaSF3	1917	Jan-07	BtBA1	1955	Aug-05
BcCO1	1543	Mar-05	BtBA2	1956	Mar-05
BcCO2	1623	Mar-06	BtBA3	1668	May-06
BcCO3	1624	Mar-06	BtCOR1	1539	Mar-05
BcME1	1590	Jan-06	BtCOR2	1553	Mar-05
BcME2	1591	Jan-06	BtCOR3	1574	Jul-05
BcME3	1909	Jan-07	BtCOR4	1954	Feb-06
BcME4	1910	Jan-07	BtER1	1535	Mar-05
BgMI1	1564	Oct-05	BtER2	1537	Mar-05
BgMI2	1566	Oct-05	BtER 3	1542	Dec-04
BgMI3	1655	Feb-06	BtER4	1583	Feb-06
BgMI4	1569	Mar-05	BtER5	1645	Mar-06
BgMI5	1654	Feb-06	BtER6	1926	Apr-07
BmCOR	1572	Oct-05			

Table 1. Collection data of analysed samples of *Baccharis* species. The abbreviation mean: species: Ba, *B. articulata*; Bc, *B. crispa*; Bg, *B. gaudichaudiana*; Bm, *B. microcephala*; Bp, *B. penningtonii*; Bphy, *B. phyteumoides*; Bs, *B. sagittalis*; Btr, *B. triangularis*; Bt, *B. trimera*; provinces: BA, Buenos Aires; CHU, Chubut; CO, Córdoba; COR, Corrientes; ER, Entre Ríos; FO, Formosa; LP, La Pampa; MI, Misiones; RN, Río Negro; SF, Santa Fe; SL, San Luis. Numbers indicate when there is more than one population of the same species. [a] Material extracted from Herbarium; [b] Names of collectors, numbers not provided by the Herbarium.

2.1.1 Statistical analysis

Principal components analysis (PCA) was applied to the population study. The analysis was performed using the NTSYS-pc 2.11w (Numerical Taxonomy and Multivariate Analysis System) designed by Rohlf (1998).

Basic data matrix was prepared considering 700 absorbance values (quantitative variable) of each analysed sample (53 extracts in total).

Before PCA the data were pre-processed with normalization to unit area technique (Beebe et al., 1998).

2.2 Results and discussion

2.2.1 Spectrophotometric analysis

Seven hundred absorbance values were utilised as quantitative variables for population analysis. Samples were collected in different provinces and seasons, mainly taking into account the quantitative variability of secondary metabolites during the year (Table 1). The principal component analysis showed that the first nine components explain almost 98.81% of the total variability. The second (PC2) and the third (PC3) principal components gathered relevant information for classifying species. Figure 4, shows a two dimension plot of PC2 vs. PC3 using all the variables. Samples could be classified in five large groups containing the species *B. crispa*, *B. microcephala*, *B. phyteumoides*, *B. triangularis* and *B. trimera*. PC2 clearly separates *B. microcephala* and *B. trimera* populations from *B. crispa* and *B. phyteumoides* populations. Moreover, samples corresponding to *B. triangularis* species were separated from those belonging to *B. microcephala*, *B. trimera*, *B. crispa* and *B. phyteumoides* populations by PC2. While PC3 separates *B. microcephala* samples from the rest of the species, it also separates *B. crispa* samples from *B. phyteumoides*, *B. trimera* and *B. triangularis* and between samples of the two latter species. However, PC3 did not completely distinguish *B. phyteumoides* from *B. trimera* and *B. triangularis* samples.

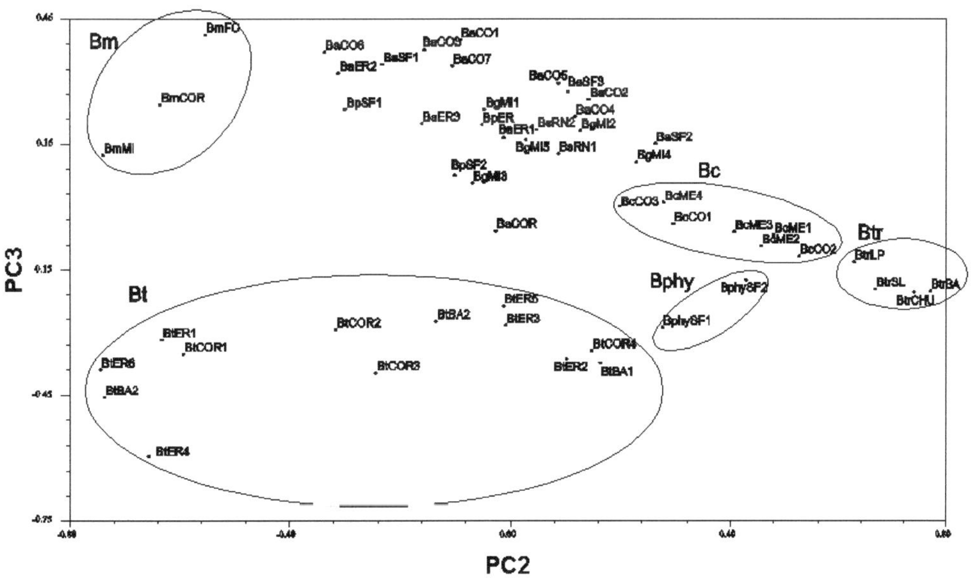

Fig. 4. Two dimensional model of PC2 *vs.* PC3 (15.63 and 7.80 %, respectively) derived from PCA of 700 quantitative variables of 53 *Baccharis* populations. Ba, *B. articulata*; Bc, *B. crispa*; Bg, *B. gaudichaudiana*; Bm, *B. microcephala*; Bp, *B. penningtonii*; Bphy, *B. phyteumoides*; Bs, *B. sagittalis*; Btr, *B. triangularis*; Bt, *B. trimera*; BA, Buenos Aires; CHU, Chubut; CO, Córdoba; COR, Corrientes; ER, Entre Ríos; FO, Formosa; LP, La Pampa; MI, Misiones; RN, Río Negro; SF, Santa Fe; SL, San Luis.

Figure 5, shows a two dimensional plot of PC7 vs. PC9 using all the variables. Here, the seventh (PC7) and ninth (PC9) principal components enable the *B. penningtonii* samples to be separated from the rest of the species.

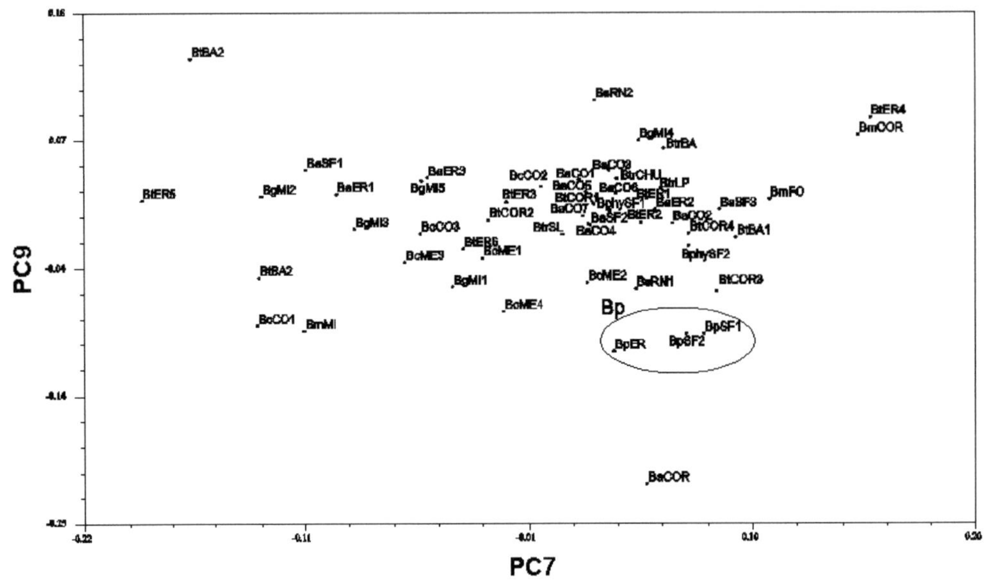

Fig. 5. Two dimensional model of PC7 *vs.* PC9 (0.62 and 0.36 %, respectively) derived from PCA of 700 quantitative variables of 53 *Baccharis* populations. Ba, *B. articulata;* Bc, *B. crispa;* Bg, *B. gaudichaudiana;* Bm, *B. microcephala;* Bp, *B. penningtonii;* Bphy, *B. phyteumoides;* Bs, *B. sagittalis;* Btr, *B. triangularis;* Bt, *B. trimera;* BA, Buenos Aires; CHU, Chubut; CO, Córdoba; COR, Corrientes; ER, Entre Ríos; FO, Formosa; LP, La Pampa; MI, Misiones; RN, Río Negro; SF, Santa Fe; SL, San Luis.

Spectrophotometric PCA data allowed the distinction of six out of the nine species examined in this study. However the three species *B. articulata, B. gaudichaudiana* and *B. sagittalis* could not be separated which is shown in the figure 6, where it is seen that the average spectra of the species are very similar.

As can be observed in Figure 4, the *B. microcephala* and *B. trimera* samples have PC2 scores with opposite signs to those of the *B. crispa, B. phyteumoides* and *B. triangularis* samples. This contrast can be explained with the help of Figure 7 A; C. Figure 7 A shows a graph of the loading values (eigenvalues) on PC2 vs. λ. Positive values are situated in a region between 200 and 230 nm. It is possible to verify from Figure 7 C that the analytical signals for *B. crispa, B. phyteumoides* and *B. triangularis* in this region are more intense than those of *B. microcephala* and *B. trimera,* and that the *B. crispa, B. phyteumoides* and *B. triangularis* samples have positive scores.

Figure 7 A also shows that three regions present negative values between: 225 and 275, 325 and 375 and 375 and 450 nm. In Figure 7 C, one can verify that in these intervals the most intense analytical signals belong to samples of the *B. microcephala* and *B. trimera* species. For this reason the *B. microcephala* and *B. trimera* species samples have negative PC2 scores.

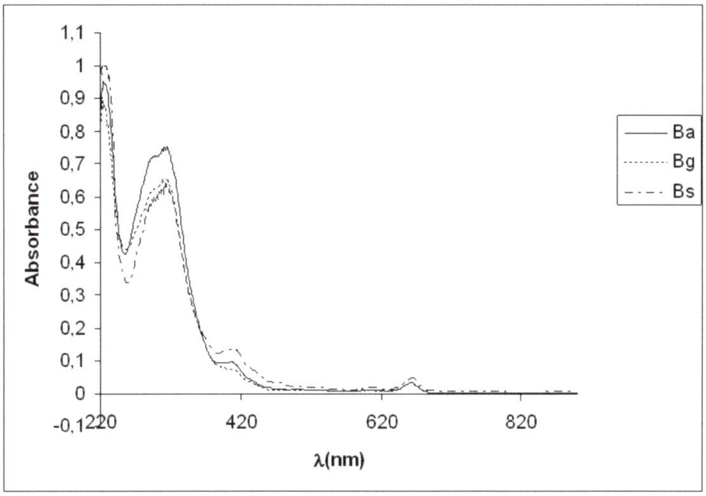

Fig. 6. Representative UV-Visible spectra of *B. articulata* (Ba), *B. gaudichaudiana* (Bg), *B. sagittalis* (Bs).

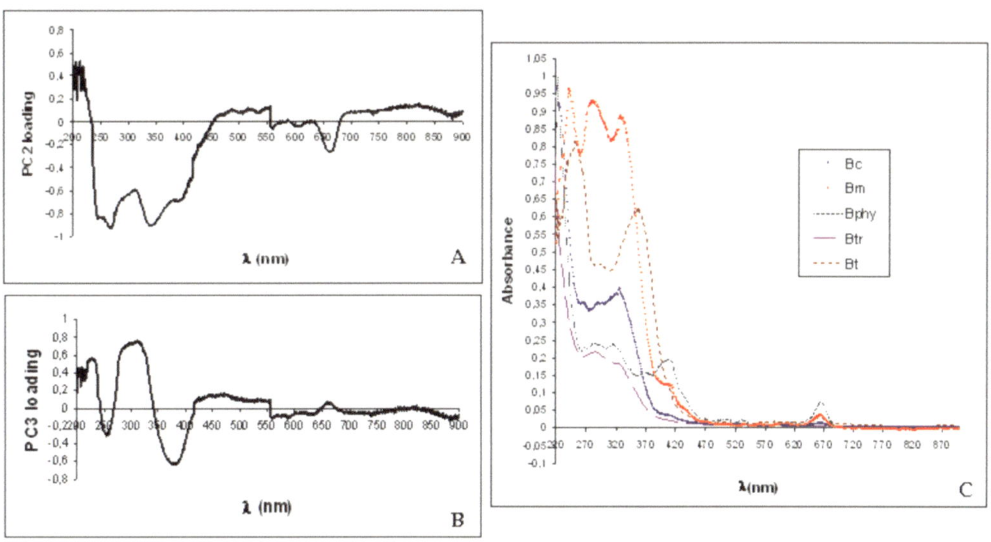

Fig. 7. A-B, PC loadings graph. **A**, PC2 (explains 15.63 % of the total data variance); **B**, PC3 (explains 7.80 % of the total data variance). C, Representative UV-Visible spectra of *B. crispa (Bc)*, *B. microcephala (Bm)*, *B. phyteumoides (Bphy)*, *B. triangularis (Btr)*, *B. trimera (Bt)*.

Figure 7 B (PC3 eigenvalues) shows two regions, one with positive values (270 to 350 nm) and one with negative values (350-420 nm). From Figure 7 C it is possible to verify that in the first region the analytical signals for *B. microcephala* are more intense than those of *B. trimera*, *B. crispa*, *B. phyteumoides* and *B. triangularis*, and that the *B. microcephala* samples have positive PC3 scores. The rest of the species have negative PC3 scores. In Figure 7 B, the region with negative values (270 to 350 nm) matches the more intense analytical signals for *B. crispa*, *B. phyteumoides* and *B. triangularis* in Figure 7 C.

B. penningtonii species samples have negative PC9 scores; PC9 separates *B. penningtonii* samples from the rest as seen in Figure 5. Figure 8 A shows a graph of the loading values (eigenvalues) on PC9 vs. λ; it also shows that two regions present negative values between: 200 and 250 and 320 and 370 nm and it can be observed that the most intense analytical signals for *B. penningtonii* species samples are in these regions (Figure 8 B).

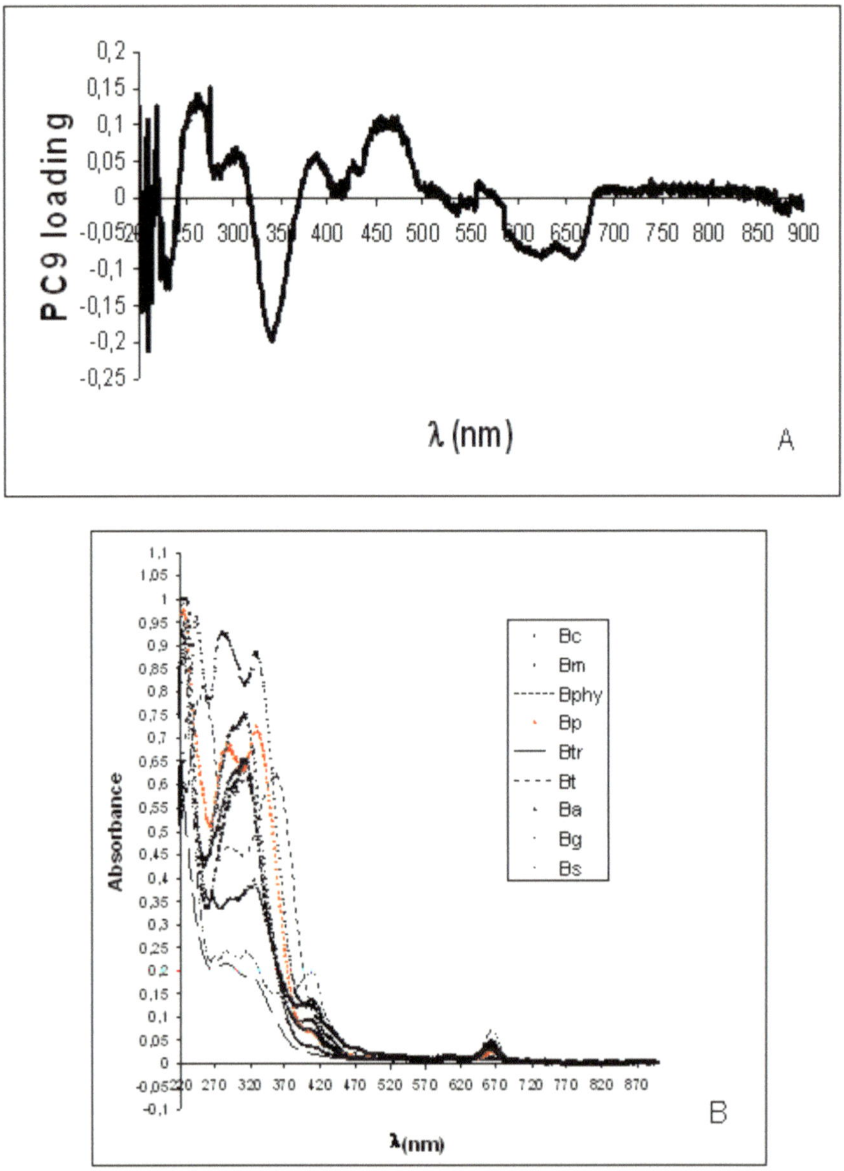

Fig. 8. A, PC loadings graph. PC9 (explains 0.36 % of the total data variance);
B, Representative UV-Visible spectra of *B. articulata* (Ba), *B. crispa* (Bc), *B. gaudichaudiana* (Bg), *B. microcephala* (Bm), *B. phyteumoides* (Bphy), *B. penningtonii* (Bp), *B. sagittalis* (Bs), *B. triangularis* (Btr), *B. trimera* (Bt).

These results suggest that the substances responsible for the discrimination between species are those which have peaks around 225, 250, 300, 325, 350, 375 and 425 nm. Considering that the ethanol extracts used in this study might possess mainly flavonoids and that the region studied is where they present their main absorption bands (Greenham et al., 2003), it is likely that the components responsible for the spectra in this work would be flavonoids. It is worth taking into account that previous studies on the phytochemistry of *Baccharis* spp. have shown the presence of flavonoids, mainly flavanones and flavones (Coelho et al., 2004; Gene et al., 1996; Torres et al., 2000). From species of this genus 298 flavonoids were isolated. Among them 24 were flavanones and 85 were flavones and among them 48% are oxygenated in the C-3 (Gonzaga Verdi et al., 2005). They were described as good chemotaxonomic markers for the lower hierarchical levels of the Asteraceae family (Emerciano et al., 2001). In contrast to other characterization studies in which it is necessary to isolate and identify chemical substances, this study uses the complete ultraviolet-visible spectra measured between 200 nm to 900 nm, avoiding the isolation, purification and characterization of chemical compounds (Lonni et al., 2005). This methodology does not distinguish the spectra of individual components in the extract. Instead, the full spectrum ranges are analyzed as a whole. Absorption spectra of figure 7 C show that the qualitative composition of flavonoids is different in the different species.

It has previously been observed that there is variability in the contents of several compounds in the different *Baccharis* species (Dresch et al., 2006; Gonzaga Verdi et al., 2005). With regards to the seasons, Borella et al. (2001) observed variations in the content of the total flavonoids, the largest number being found in a drug in the summer, which is an expected result due to the large number of functions attributed to them (Harborne & Williams, 2000). Our results are consistent with previous ones, showing different heights of peaks in the spectra in Figure 6, 7C and 8 B, even though these spectra were standardized. In a fertilization trial of *B. trimera,* in which the nutrient content of the soil was varied, no variation was observed in the contents of flavonoids (Borella et al., 2001). In our study we observed some changes in the content of flavonoids between collecting regions (Figure 4 and 5) but these did not prevent the grouping of various populations of the same species.

A routine step in multivariate data analysis is ordinarily to obtain a low-dimensional representation of the data. If two or three main components gives an accurate representation, a bi-or three-dimensional graph could be realized which mere observation is instructive. Clusters are usually easy to detect. After analysis of the eigenvalues (PC loading), more discriminant original variables are obtained. Then an ANOVA must be performing of each of these original variables between the OTUS (here species). The higher the eigenvalues, regardless of the sign will be more efficient in discriminating the OTUS. Variables that have negative eigenvalues (-) means that they are characterizing in the opposite direction in relation to the variables that have positive eigenvalues (+) and vice versa.

Thus, absorbance values of the wavelengths (225, 250, 300, 325, 350, 375 and 425 nm) obtained from the analysis of the eigenvalues were submitted to ANOVA and we have established the wavelengths that differ between pairs of species (Table 2). Thus we confirm that there are differences in the UV / Visible spectra not observable to the naked eye but are expressed as different clusters after a principal component analysis is applied.

	Ba	Bc	Bg	Bm	Bp	Bphy	Bs	Btr	Bt
Ba		300 325		250 325 350 375	250 350	300 325 350 425		225, 250, 300, 325 350	225, 250, 300, 325, 350, 375
Bc			300 325	250, 300, 325, 350 375	250 300 325 350	325 425	300 425	225 300 325 350	250 350 375
Bg				250 300 325 350	350	300 325 425		225, 250 300, 325 350	250 300 350 375
Bm					250 300	250 300 325 350	250 300 325 350	225, 250 300, 325 350, 375	300 325 375
Bp						250 300 325 350	250 350	225 250 300 325 350	250 300 325 375
Bphy							250 300 325	225 425	250, 300 325,350 375
Bs								225 300 325 425	250 300 350 375
Btr									225, 250 325, 350 375, 425
Bt									

Table 2. Wavelengths (nm) with statistically significant differences ($p < 0.05$) among the species. *B. articulata* (Ba), *B. crispa* (Bc), *B. gaudichaudiana* (Bg), *B. microcephala* (Bm), *B. penningtonii* (Bp), *B. phyteumoides* (Bphy), *B. sagittalis* (Bs), *B. triangularis* (Btr) and *B. trimera* (Bt)

2.2.2 TLC analysis

A TLC analysis was initiated of the dichloromethane extracts of *B. articulata, B. gaudichaudiana* and *B. sagittalis* used to perform UV-Visible spectra using the mobile phase DCM: Hexane: MeOH (4:2:1) in order to complement the studies carried out by PCA of the spectrophotometric data and to find potential markers of the species that could not be characterized by the previous method. NP-PEG reagent under UV 365 nm and UV 254 nm were used to detect the polyphenolic compounds present in the extracts (Wagner & Bladt, 1996), in accordance with that published for the *Baccharis* genus on account of the high occurrence of these compounds in the genus (Bohm & Stuessy, 2001; Gonzaga Verdi et al., 2005). Different color bands will be detected with the NP-PEG reagent under UV 365 nm or quenching bands will be detected under UV 254 nm if these compounds are present in the extract. On the other hand, given that one of our objectives was the identification of the species in the state of a crude drug, it was very important to select appropriate components of easy access for chemical quality control. So apigenin, genkwanin and luteolin, which are compounds present in several *Baccharis* species, were selected as markers. Genkwanin has been reported in *B. articulata* (Gianello & Giordano, 1984) and apigenin in *B. gaudichaudiana* (Fullas et al., 1994). Luteolin has not yet been reported in any of the three species analyzed by TLC, but it has been found in the following species from the same section: *B. microcephala* (Bohlmann et al., 1985), *B. trimera* (Soicke & Leng-Peschlow, 1987) and *B. triangularis* (Pettenati et al., 2007). Our results are shown in Table 3. The TLC chromatograms showed differences for the three species studied (Figures 9A and 9B). The flavonoids apigenin and genkwanin were found in all three species, although the band that corresponds to genkwanin somewhat overlaps in *B. gaudichaudiana* and *B. sagittalis.* In the case of the other marker, we observed that the band corresponding to luteolin appeared in *B. gaudichaudiana* and *B. sagittalis*. There are at least two more bands at Rf 0,83 and 0,75 for *B. articulata* and there is another band for *B. gaudichaudiana* at Rf 0,58 and three more bands for *B. sagittalis* at Rf 0,33, 0,5 and 0,58 (Figure 9 A).

	Apigenin	Genkwanin	Luteolin	Band 0,83	Band 0,75	Band 0,58	Band 0,50	Band 0,33
B. articulata	X	X	-	X	X	-	-	-
B. gaudichaudiana	X	X	X	-	-	X	-	-
B. sagittalis	X	X	X	-	-	X	X	X

Table 3. Summary of the bands obtained by TLC for *B. articulata, B. gaudichaudiana* and *B. sagittalis.* Mobile phase: DCM: Hexane: MeOH (4:2:1). x indicates presence of the band, - indicates absence of the band.

2.2.3 HPLC analysis

HPLC analysis was carried out on the same extracts as used for the studies with UV-Visible spectrophotometry and for TLC in *B. articulata, B. gaudichaudiana* and *B. sagittalis.* The chromatophotographic profiles showed the main peaks with the following retention times for *B. articulata*: 2.32, 3.00, 3.22 and 3.30 min; for *B. gaudichaudiana*: 2.26, 3.00, 3.12 and 3.30 min and for *B. sagittalis*: 2.26, 2.40, 3.00, 3.12 and 3.30 min. The retention time for the apigenin marker was 3.00 min and for genkwanin was 3.30 min. These peaks appear in all three species studied. In *B. gaudichaudiana* and *B. sagittalis* there is also a peak at 3.12 min, and this retention time corresponds to the luteolin marker (Figure 10).

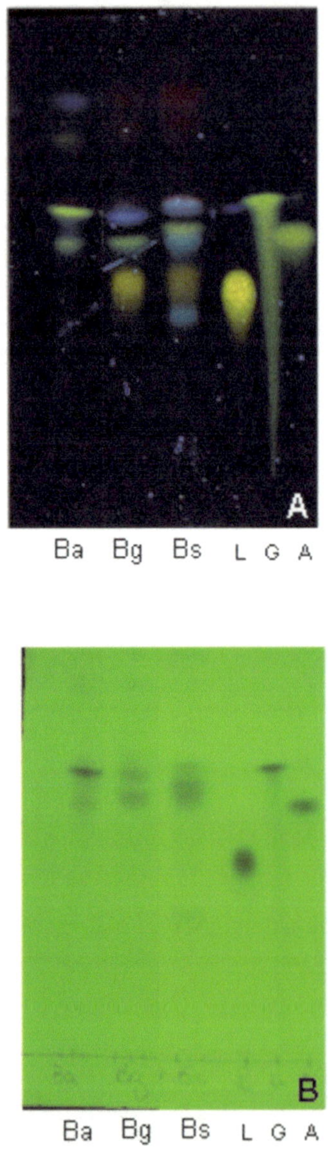

Fig. 9. TLC of *B. articulata (Ba)*, *B. gaudichaudiana (Bg)* and *B. sagittalis (Bs)*. Chromatograms were sprayed with NP-PEG and observed under UV 365 (A) or under UV 254 without chemical treatment (B). A, apigenin; G genkwanin; L: luteolin. Mobile phase: DCM: Hexane:MeOH (4:2:1)

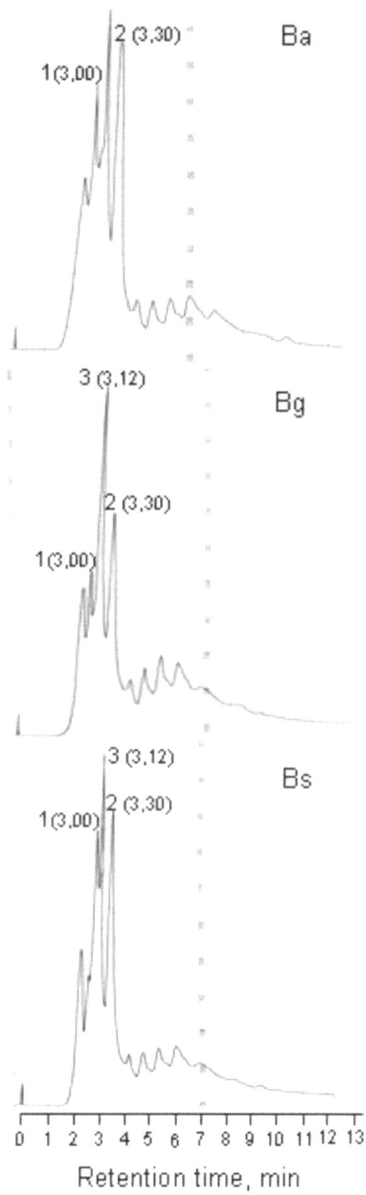

Fig. 10. Original chromatograms obtained by HPLC of *B. articulat*a (Ba), *B. gaudichaudiana* (Bg) and *B. sagittalis* (Bs),. Compounds: 1, apigenin; 2, genkwanin; 3, luteolin.

3. Conclusion

The growing tendency to use high quality, standardized plant extracts, with a guarantee of security and efficacy, is going to continue. Therefore, all efforts should be directed at researching the chemical-pharmalogical profiles of the extracts and combinations and in rationalizing their therapeutic applications.

In the present example, sect. Caulopterae species are very similar between them and only *B. articulata, B. crispa* and *B. trimera* are official in pharmacopeia. UV/Visible spectrophotometry coupled to PCA grouped populations of these three species in different areas in a two dimensional graph. Three additional species were grouped in different areas in the same graph too (*B. microcephala, B. phyteumoides* and *B. penningtonii*). Populations of species that fall outside the areas of official species are unfit for medicinal purposes. It should be noted that in the case of the official species *B. articulata,* additional techniques, such as TLC should be applied to distinguish between *B. articulata, B. gaudichaudiana* and *B. sagittalis*. According to this, the combination of these techniques could to be used for routine quality control of official herbal medicines.

4. Acknowledgment

The present work was supported by a grant from the Agencia Nacional de Promoción Científica y Tecnológica de Argentina, PICT 1494. Dr. M V Rodriguez and Dr. Marcos Derita are research workers of CONICET.

5. References

Abdel-Malek, S., Bastien, J.W., Mahler, W.F., Jia, Q., Reinecke, M.G., Robinson, W.E., Shu, Y.H. & Zalles-Asin, J. (1996). Drug leads from the kallawaya herbalists of Bolivia. 1. Background, rationale, protocol and anti-HIV activity. *Journal of Ethnopharmacology,* Vol 50, No. 4 (March 1996), pp. 157-166, ISSN 0378-8741.

Akaike, S., Sumino, M., Sekine, T., Seo, S., Kimura, N. & Ikegami, F. (2003). A New *ent*-Clerodane Diterpene from the Aerial Parts of *Baccharis gaudichaudiana. Chemical & Pharmaceutical Bulletin,* Vol 51, No. 2 (February 2003), pp.197-199, ISSN 0009-2363.

Ariza Espinar, L. (1973). Las especies de *Baccharis* (Compositae) de Argentina Central. *Boletín de la Academia Nacional de Ciencias,* Vol 50 (Septiembre 1973), pp.175-305, ISSN 0325-2051.

Baker, J.G. (1882-1884). Compositae III. Asteroideae, Inuloideae. IV. Heliathoideae, Anthemideae, Senecionideae, Cynaroideae, Ligulatae, Mutisiaceae, In: *Flora brasiliensis: enumeratio plantarum 6 (3),* C.F. Martius, A.W. Eichler, (Eds.), 1-100, F. Fleischer, Leipzig, Munich.

Barroso, G.M. (1976). Compositae, Subtribo Baccharidinae Hoffman. Estudo das espécies ocorrentes no Brasil. *Rodriguésia,* Vol 28, pp.3-273.

Borella, J.C., Fontoura, A., Menezes, Jr.A. & França, S.C. (2001). Influência da adubação mineral (N-P-K) e sazonalidade no rendimento e teor de flavonóides em indivíduos masculinos de *Baccharis trimera* Less. (Asteraceae) - Carqueja. *Revista Brasileira de Plantas Medicinais,*Vol 4, No.1 (Outubro 2001), 101-104, ISSN 1516-0572.

Bohlmann, F., Banerjee, S., Jalupovic, J., Grenz, M., Misra, L.N., Hirschmann, G.S., King, R.M. & Robinson, H. (1985). Clerodane and labdane diterpenoids from *Baccharis* species. *Phytochemistry,* Vol 24, No. 3 (October 1985), pp. 511-515, ISSN 0031-9422.

Bohm, B.A. & Stuessy, T.F. (2001) *Flavonoids of the Sunflower Family (Asteraceae).* Springer, Wien, New York.

Blumenthal, M. (1998). *The complete german commission E monographs. Herbal guide to herbal medicines*, American Botanical council, Austin.

Coelho, M.G.P., Reis, P.A., Gava, V.B., Marques, P.R., Gayer, C.R., Laranja, G.A.T., Felzenswalb, I. & Sabino, K.C.C. (2004). Antiarthritic effect and subacute toxicological evaluation of *Baccharis genistelloides* aqueous extract. *Toxicology Letters*, Vol 154, No. 1-2 (December 2004), pp. 69-80, ISSN 0378-4274.

Correa, M.P. (1985). *Dicionário das plantas úteis do Brasil e das exóticas cultivadas*, IBDF, Río de Janeiro, Brasil.

Cuatrecasas, J. (1967). Revisión de las especies colombianas del género *Baccharis*. *Revista de la Academia Colombiana de Ciencias Exactas*, Vol 13, pp. 5-102.

De Candolle, A.P. (1836). Compositae: *Baccharis*. *Prodromus Systematis Naturalis Regni Vegetabilis*, Vol 5, pp. 398-429.

De Oliveira, S.Q., Dal-Pizzol, F., Gosmann, G., Guillaume, D. Moreira, J.C. & Schenckel, F.P. (2003). Antioxidant activity of *Baccharis articulata* extracts: isolation of a new compound with antioxidant activity. *Free Radical Research*, Vol 37, No. 5 (January 2003), pp.555-559, ISSN 1071-5762.

Desmarchelier, G., Bermúdez, M.J.N., Coussio, J., Ciccia, G. & Boveris, A. (1997). Antioxidant and prooxidant activities in aqueous extracts of argentine plants. *International Journal of Pharmacognosy*, Vol 35, No.2 (January 1997), 116-120, ISSN 1388-0209.

Dresch, A.P., Montanha, J.A., Matzenba, N.E. & Mentz, C.A. (2006). Controle De Qualidade De Espécies Do Gênero *Baccharis* L.(Asteraceae) Por CCD A Partir De Extratos Rápidos. *Infarma*, Vol 18, No. 11-12, pp. 37-40, ISSN 0104-0219.

Emerciano, V.P., Militão, J.S.L., Campos, C.C., Romoff, P., Kaplan, M.A.C., Zambon, M. & Brant, A.J.C. (2001). Flavonoids as chemotaxonomic markers for Asteraceae. Biochem. *Biochemical Systematic and Ecology*, Vol 29, No. 9 (October 2001), pp.947-957.

ESCOP (European Scientific Cooperative on Phytotherapy) (2003). *ESCOP Monographs. The scientific foundation for herbal medicinal products*, ESCOP, Exeter; Georg Thieme Verlag, Sttutgart; Thieme NewYork, New York.

Farmacopea Nacional Argentina VI ed. (1978). *Codex Medecamentarius Argentino*. Buenos Aires, Argentina.

Farmacopéia Brasileira IV Edição. (2002). Oficializada Governo Federal. Atheneu Editora S. A., Sao Paulo LTDA.

Fullas, F., Hussain, R.A., Chai, H., Pezzuto, J.M., Soejarto, D.D. & Kinghorn, A.D. (1994). Cytotoxic constituents of *Baccharis gaudichaudiana*. *Journal of Natural Products*, Vol 57, No. 6 (June 1996), pp.801-807, ISSN 0163-3864.

Gené, R.M., Marín, E. & Adzet, T. (1992). Anti-Inflammatory Effect of Aqueous Extracts of Three Species of the Genus *Baccharis*. *Planta Medica*, Vol 58 (January 1992), pp.565-566, ISSN 0032-0943.

Gené, R.M., Cartañá, C., Adzet, T., Marín, E., Parella, T. & Cañigueral, S. (1996). Anti-Inflammatory and Analgesic Activity of *Baccharis trimera*: Identification of its Active Constituents. *Planta Medica*, Vol 62, pp. 232-235, ISSN 0032-0943.

Gianello, J.C. & Giordano, O.S. (1984). Exámen químico en seis especies del género *Baccharis*. *Revista Latinamericana de Quimica*, Vol 15, pp.84-86, ISSN 0370-5943.

Giuliano, D.A. (2001). Clasificación Infragenérica de las especies argentinas de *Baccharis* (Asteraceae, Astereae). *Darwiniana*, Vol 39, No. 1-2, pp. 138-154, ISSN 0370-5943.

Gonzaga Verdi, L., Costa Brighente, I.M. & Pizzolatti, M.G. (2005). Gênero Baccharis (Asteraceae): Aspectos Químicos, Econômicos e Biológicos. Química Nova, Vol 28, No.1, pp. 85-94, ISSN 0100-4042.

Gupta, M.P. (1995). *270 Plantas medicinales iberoamericanas*. Programa Iberoamericano de Ciencia y Tecnología para el Desarrollo, CYTED. Convenio Andrés Bello Santafé de Bogotá, Colombia.

Greenham, J., Harborne, J.B. & Williams, C.A. (2003). Identification of Lipophilic Flavones and Flavonols by Comparative HPLC, TLC and UV Spectral Analysis. *Phytochemical Analysis*, Vol 14, No. 2 (March/April 2003) pp. 100-118, ISSN 1099-1565.

Harborne, J.B. & Williams, C.A. (2000). Advances in flavonoid research since 1992. *Phytochemistry*, Vol 55, No. 6 (November 2000), pp. 481-504, ISSN 0031-9422.

Heering, W.C. (1904). Die *Baccharis*-Arten des Hamburger Herbars. *Jahrbuch der Hamburgischen Wissenschaftlichen Anstalten*, Vol 21, pp. 1-46.

Hieronymus, J. (1882). Plantae Diaforicae, Florae Argentinae. Tomo VI. *Boletín de la Academia Nacional de Ciencias*, Vol 4, pp. 159-160, ISSN 0325-2051.

IFPMA (International Federation of Pharmaceutical Manufacturers Associations) (1997). Major steps towards global drugs regulations. Brussels.

Keller, K. (1991). Legal requirements for the use of phytopharmaceutical drugs in the Federal Republic of Germany. *Journal of Ethnopharmacology*, Vol 32, pp. 225-229, ISSN 0378-8741.

Lapa, A.J., Fischman, L.A. & Gamberini, M.T. (1992). Inhibitors of gastric secretion from Brazilian folk medicinal plants, In: *Natural Drugs and the Digestive Tract*, F. Capasso, N. Mascolo (Eds.), 63-68, EMSI, Roma.

Lessing, C.F. (1831). Synanthereae: Molins-Alatae. *Linnaea*, Vol 6, pp. 83-170.

Liang, Y., Peishan, X. & Chan, K. (2004) Quality control of herbal medicines. *Journal of Chromatography B*, Vol 812 (September 2004), pp. 53-70, ISSN 1570-0232.

Lonni, A.A.S.G., Scarminio, I.S., Silva, L.M.C. & Ferreira, D.T. (2003). Differentiation of Species of the Baccharis Genus by HPLC and Chemometric Methods. *Analytical Sciences*, Vol 19, pp. 1013-1017, ISSN 0910-6340.

Lonni, A.A.S.G., Scarminio, I.S., Silva, L.M.C. & Ferreira, D.T. (2005). Numerical Taxonomy Characterization of *Baccharis* Genus Species by Ultraviolet-Visible Spectrophotometry. *Analytical Sciences*, Vol 21, pp. 235-239, ISSN 0910-6340.

López, J.A. & Hidalgo, M.D. (1994) Análisis de Componentes Principales y Análisis Factorial, En: *Fundamentos de Estadística con Systat*, M. Ato, J.J. López, (Eds.), 457-503, Addison Wesley Iberoamericana.

Martínez Crovetto, R. (1981). *Plantas utilizadas en medicina en el NO de Corrientes*. Fund. Miguel Lillo, Miscelán, 69, Tucumán.

Müller, J. (2006). *Systematics of Baccharis (Compositae, Astereae) in Bolivia, including an overview of the genus.* (Systematics Botany Monographs v. 76). The American Society of Plant Taxonomists, ISBN 0912861762, Michigan.

Palacios, P., Gutkind, G., Randina, R.V.D., De Torres, R. & Coussio, J.D. (1983). Antimicrobial activity of *B.crispa* and *B. notosergila*. Genus Baccharis II. *Planta Medica*, Vol 49, pp. 128, ISSN 0032-0943.

Petenatti, E.M., Petenatti, M.E., Cifuente, D.A., Gianello, J.C., Giordano, O.S., Tonn, C.E. & Del Vitto, L.A. (2007). Medicamentos Herbarios en el Centro-Oeste Argentino. VI. Caracterización y Control de Calidad de dos Especies de "Carquejas": *Baccharis sagittalis* y *B. triangularis* (Asteraceae). *Latin American Journal of Pharmacy*, Vol 26 No. 2, pp. 201-208, ISSN 0326-2383.

Pla, L.E. (1986) *Análisis Multivariado: Método de Componentes Principales.* Secretaría de la Organización de Estados Americanos (OEA), Washington, D.C.

Rice- Evans, C.A., Miller, N.J. & Paganga, G. (1996). Structure-Antioxidant Activity Relationships of Flavonoids and Fenolic Acids. *Free Radical Biology and Medicine*, Vol 20, pp. 933-956, ISSN 0891-5849.

Rodriguez M.V. (2010a). Caracterización y Normalización de un recurso vegetal autóctono medicinal: especies del género *Baccharis* L. Facultad de Ciencias Bioquímicas y Farmacéuticas, Universidad Nacional de Rosario, Argentina.

Rodriguez M.V., Martínez M. L., Cortadi A. A., Bandoni A., Giuliano D. A., Gattuso S. J. & Gattuso M. A. (2010b). Characterization of three sect. Caulopterae species (*Baccharis*-Asteraceae) inferred from morphoanatomy, polypeptide profiles and spectrophotometry data. *Plant Systematic and Evolution*, Vol 286, No. 3-4 (June 2010), pp. 175-190, ISSN 0378-2697.

Rohlf, F.J. (1998). On applications of geometric morphometrics to studies of ontogeny and phylogeny. *Systematic Biology*, Vol 47, pp. 147-158, ISSN 1063-5157.

Sorarú, S.B. & Bandoni, A.L. (1978). *Plantas de la Medicina Popular Argentina*. Albatros, Buenos Aires.

Stoicke, H. & Leng-Peschlow, E. (1987). Characterization of flavonoids from *Baccharis trimera* and their antihepatotoxic properties. *Planta Medica,* Vol 53, pp. 37-39, ISSN 0032-0943.

Torres, L.M.B., Gamberini, M.T., Roque, N.F., Lima-Landman, M.T., Souccar, C. & Lapa, A.J. (2000). Diterpene from *Baccharis trimera* with a Relaxant Effect on Rat Vascular Smooth Muscle. *Phytochemistry*, Vol 55, No. 6 (November 2000), pp. 617-619, ISSN 0031-9422.

Toursarkissian, M. (1980). *Plantas Medicinales de la Argentina*. Hemisferio Sur S. A., Buenos Aires.

Tyler, V.E. (1999). Phytomedicines: Back to the future. *Journal of Natural Products*, Vol 62, pp. 1589-1592, ISSN 0163-3864.

Wagner, H. & Bladt, S. (1996) *Plant Drug Analysis. A Thin Chromatography Atlas*. Springer-Verlag, Berlin.

Weddell, H.A. (1855-1856). *Chloris andina*, vol. 1. P. Bertrand, Paris.

WHO (World Health Organization) (1991). Guidelines for the Assessment of herbal medicines. WHO/TRM/91.4, Ginebra.

WHO (World Health Organization) (1999). WHO monographs on selected medicinal plants. Vol 1, Ginebra.
WHO (World Health Organization) (2001). Legal status of traditional medicine and complementary alternative medicine: A world wide review. WHO/EDM/TRM/2001.2, Geneva.
WHO (World Health Organization) (2002). WHO monographs on selected medicinal plants. Vol 2, Ginebra.

20

Flow-Injection Spectrophotometric Analysis of Iron (II), Iron (III) and Total Iron

Ibrahim Isildak
Yildiz Technical University, Faculty of Chemical and Metallurgical Engineering, Bioengineering Department, İstanbul, Turkey

1. Introduction

Determination of iron in analytical chemistry has become a routine procedure because of its importance in our life. Various chemical forms of iron can be found in natural waters depending on geological area and chemical components present in the environment. The main source of iron in natural waters is from the weathering and leaching of rocks and soils (Dojlido & Best, 1993). Also, metallic iron and its compounds are used in various industrial processes and may enter natural waters through the discharge of wastes. Iron(II) is normally less present in river water (Sangi et al., 2004) and iron (III) can precipitate rapidly by the formation of hydrous iron oxide and hydroxides, which they can absorb other trace metals. Thus, iron ion controls the mobility, bioavailability and toxicity of other trace metals in the natural water system (Wirat, 2008; Lunvongsa et al., 2006). Amounts of iron are widely present in tap, pond, well and underground water, and this metallic ion is essential for biological systems (Ohno et al., 2004; Kawakubo et al., 2004).

As iron is one of the most frequently determined analyte in environmental (water, soil and sediment) samples, many spectrophotometric and/or flow-injection spectrophotometric methods have been developed for iron determination. When trace levels of the iron are concerned, the detection methods applicable are reduced (Tarafder et al., 2005; Weeks & Bruland, 2002; Giokas et al., 2002; Themelis et al., 2001; Bagheri et al., 2000; Pascual-Reguera et al., 1997; Teshima et al., 1996; Tesfaldet et al., 2004; Udnan et al., 2004; Pojanagaroon et al., 2002; van Staden & Kluever, 1998; Asan et al., 2003, 2008; Andac at al., 2009). Flow-injection analysis, as a rapid and precise technique, has found wide application in the determination of iron in several sample matrices (Bowie A.R., et al. 1998; Hirata S., et al. 1999; Qin W., et al. 1998; Kass M., et al. 2002; Saitoh K., et al. 1998; Weeks D.A., et al. 2002; Giokas D.L., et al. 2002; Themelis D.G., et al. 2001; Bagheri H., et al. 2000; Molina-Diaz A., et al. 1998; Teshima N., et al. 1996).

Highly sensitive, selective and rapid flow-injection spectrophotometric analysis (FIA) methods for the determination of iron (II), iron (III) and total iron will be defined under proposed chapter of the book. The methods were based on the reactions of iron (II) and iron (III) with different complexing agents in different carrier solutions in FIA (Asan A. et al., 2010; Andac M. et al., 2009; Asan A. et al., 2008). Several parameters acting on the determination of iron (II) and iron (III) were examined. The developed methods have been successfully applied to the determination of iron (II), iron (III) and total iron in water and ore samples. The methods were also verified by applying certified reference materials.

2. A very sensitive flow-injection spectrophotometric determination method for iron(II) and total iron using 2', 3, 4', 5, 7-pentahydroxyflavone

Spectrophotometric detection based on the measurement of the absorbance at a characteristic wavelength of complex formed between a chelating agent and iron has been mainly applied (Kass M. and Ivaska A. 2002; Saitoh K., et al. 1998; Weeks D.A. and Bruland K.W. 2002; Giokas D.L., et al. 2002; Themelis D.G., et al. 2001; Bagheri H., et al. 2000; Molina-Diaz A., et al. 1998; Teshima N., et al. 1996; Tesfaldet Z.O., et al. 200); Udnan Y., et al. 2004; Morelli B., et al. 1983; Pojanagaroon T., et al. 2002; van Staden J.F. and Kluever L.G. 1998). A number of other chelating agents that have been reported for the spectrophotometric and/or flow-injection spectrophotometric determination of iron (III) and total iron include 2-thiobarbituric acid (Morelli B., et al. 1983), norfloxacin (Pojanagaroon T., et al. 2002), tiron (Mulaudzi L.V., et al. 2002; Van Staden J.F. and Kluever L.G. 1998), tetracycline (Ahmed M.J. and Roy U.K. 2009) and chlortetracycline (Sultan S.M. and Suliman F. 1992). Flow-injection spectrophotometric methods based on above chelating agents are not either selective, or a masking agent should be used (Wirat R., 2008). However, highly selective, simple and economical methods are still required for the routine determination of iron (II) in different sample matrices. An ultra-sensitive and highly selective, rapid flow-injection spectrophotometric method for the determination of iron (II) and total iron has been proposed. The method was based on the reaction between iron (II) and 2', 3, 4', 5, 7-pentahydroxyflavone (Morin) in slightly acidic solution (pH:4.50) with a strong absorption at 415 nm. The chemical structure of Morin is shown Fig. 1. The reagent itself is sparingly soluble in water and does not absorb in the visible region of the spectrum, therefore, might be well suited for flow-injection analysis of iron (II) and total iron. The method has been successfully applied to the determination of iron (II) and total iron in water samples and ore samples.

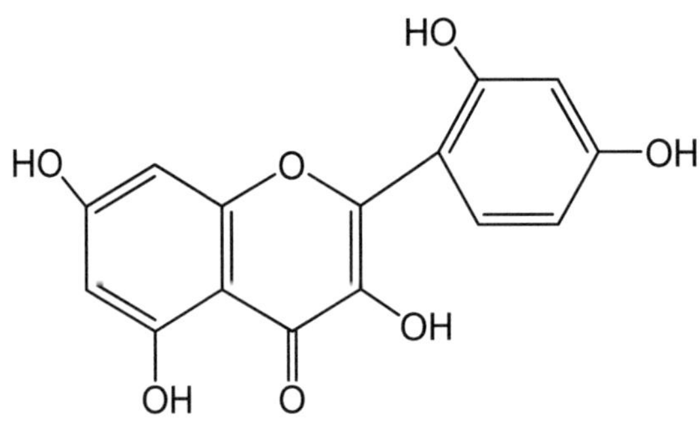

Fig. 1. The chemical structure of 2', 3', 4', 5', 7-pentahydroxyflavone (Morin)

2.1 Experimental

2.1.1 Reagent and standards

All chemicals used were of analytical reagent grade or the highest purity available. Doubly distilled deionized water was used throughout the study. Glass vessels were cleaned by soaking in acidified solutions of $KMnO_4$ or $K_2Cr_2O_7$ followed by washing with concentrated HNO_3, and were rinsed several times with high-purity deionized water. Stock solutions and environmental water samples (1000 mL each) were kept in polypropylene bottles containing 1 mL of concentrated hydrochloric acid. Standard iron (II) and iron (III) stock solutions were prepared by solving 278.02 mg of iron (II) and 489.96 mg of iron (III) sulphate (Merck) in 0.01 M 100 mL hydrochloric acid to give 0.01 M stock solution of iron (II) and iron (III). Iron (II) and iron (III) working standard solutions were prepared daily by suitable dilution of stock solutions with double deionized water. Standard reference material consisting of 0.085 % Fe (Zn/Al/Cu 43XZ3F) was provided from MBH Analytical Ltd. (UK). Hydrogen peroxide solution 30 % (v/v) used was from Merck.

A stock solution of Morin ($5x10^{-3}$M) was prepared by dissolving requisite amount of Morin (BDH Chemicals) in 100 mL of ethanol:water (4:96 v/v) because of it's low solubility in water only. For spectrophotometric study, morin complex solutions of various metals were prepared by mixing 1 mL of $1x10^{-4}$ M standard solution of each metal in double deionised water with a suitable volume of $1x10^{-4}$ M Morin solution. All stock solutions were stored in polyethylene containers. All polyethylene containers and glassware used for aqueous solutions containing metallic cations were cleaned with (1+1) nitric acid while the rest were cleaned with 3 % Decon 90, all were rinsed with deionized water before use. The working standard solutions were prepared by appropriate dilution immediately before use. All solutions were degassed before use using a sonicator (LC 30). Reagent carrier solution was composed of Morin in 0.1 M HAc/Ac^- buffer (pH:4.50) solution consisting of metanol 4 %.

2.1.2 Apparatus

UV-Visible spectra of metal-AcSHA complexes were taken with a Unicam spectrophotometer (GBC Cintra 20, Australia). A Jenway 3040 Model digital pH-meter was used for the pH measurements.

In the FIA system, peristaltic pump (ISMATEC; IPC, Switzerland) 0.50 mm i.d. PFTE tubing was used to propel the samples and reagent solutions. Samples were injected into the carrier stream by a 7125 model stainless steel high pressure Rheodyne injection valve provided with a 20 µL loop. The absorbance of the coloured complex formed (λ_{max} 415 nm) was measured with a UV-Visible spectrophotometer equipped with a flow-through micro cell (Spectra SYSTEM UV 3000 HR, Thermo Separation Products, USA), and connected to a computer incorporated with a PC1000 software programme.

A UNICAM 929 model (Shimadzu AA-68006) flame atomic absorption spectrophotometer with deuterium-lamp background correction was used for the determination of iron in reference to the FIA method. The measuring conditions were as follows: UNICAM hollow cathode lamp, 10 cm 1-slot burner, air-acetylene flame (fuel gas flow-rate 1.50 L/min), 0.2 nm spectral bandwidth, and 7 mm burner height. The wavelength and the lamp current of Fe was respectively 248 nm and 5 mA.

2.1.3 General procedure

The FIA system used was simple as shown schematically in Fig.2. The sample solution was introduced into the reagent carrier solution by the Rhodyne injection valve. The complex (λ_{max}=415 nm) was formed on passage of the reagent and iron (II) ion solution through the mixing coil. A PTFE tubing (50 cm long) was attached before the flow-through detection cell as a mixing coil. The absorbance of the coloured complex was selectively monitored in the flow-through spectrophotometric cell at 415 nm. The transient signal was recorded as a peak, the height of which was proportional to the iron (II) concentration in the sample, and was used for all measurements. Five replicate injections per sample were made.

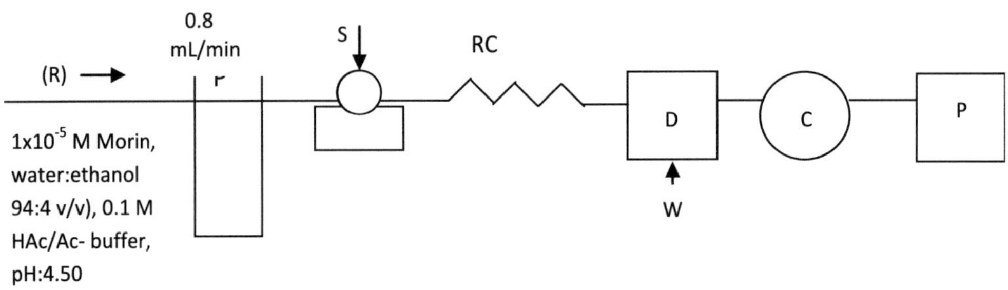

Fig. 2. Flow diagram of the FIA system used. R; reagent carrier solution (1x10^{-5} M Morin in ethanol: water (4:96 v/v) in 0.1 M HAc/Ac^- buffer (pH:4.50)), P, Peristaltic pump, S; Rheodyne sample injection valve, RC; reaction coil (50 cm long, 0.5 mm i.d), D; spectrophotometric detector (λ_{max} = 415 nm), W; waste, C; computer, P; printer.

2.1.4 Sample preparation procedures

Sea, river and industrial water samples collected in Nalgene plastics were acidified by adding 1 mL of hydrochloric acid (0.1 M) per 100 mL of sample solution behind filtration over 0.45 μm Millipore Filter (Millford, MA). After filtration, 20 μL of water samples were injected directly into the FIA system for the determination of iron (II). Total iron was determined by reducing of all forms of iron to iron (II) in the procedure described (van Staden J.F. and Kluever L.G. 1998; Asan A., et al. 2003).

A 0.10 g sample of the certified metal alloy (Zn/Al/Cu 43XZ3F) was dissolved in 12 mL of concentrated HCl+HNO_3 (3:1 v/v) in 100 mL beaker. The mixture was heated on a hot plate nearly to dryness; 5 mL HNO_3 was added to complete dissolution and diluted to 100 mL with deionized water. The solution was filtered and transferred quantitatively to 1000 mL volumetric flask and made up to volume with deionized water. 9 mL of this solution was treated with 1 mL of sodium azide (2.5 % w/v) for iron (III) reduction. After the reduction step, 20 μL of this solution was used for the determination of total iron (van Staden J.F. and Kluever L.G. 1998).

Metal ore samples (0.10 g) were powdered (≥ 500 mesh) and prepared as in the procedure described above. All analyses were performed with the least possible delay.

2.2 Results and discussion

2.2.1 Spectrophotometric studies of the Morin-metal complexes

The reaction mechanism of the present method was as reported earlier (Busev A.I., et al. 1981). Job's method of continuous variation and the molar ratio method were applied to ascertain the stoichiometric composition of the complex (MacCarthy P. and Zachary D.H.,

1986). A Fe(II)-Morin (1:2) complex was indicated by both methods. The reaction was very fast. Metal ions react with Morin in aqueous medium in the range pH: 2.0-7.0 forming coloured complexes with different stoichiometry. Absorption spectra's those correspond to solutions of $5x10^{-5}$ M of iron (II)-Morin complex was measured against a reagent blank and the average molar absorption coefficient of 6.82 x 10^4 L mol^{-1} cm^{-1} are shown in Fig. 3.

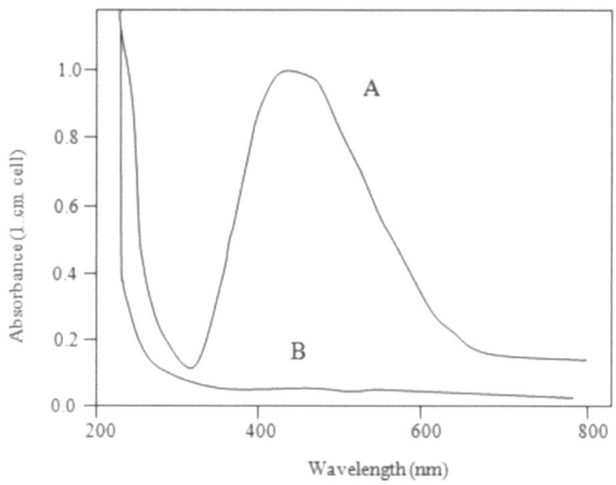

Fig. 3. Absorption spectras of iron (II)-Morin complex and Morin itself. (A) absorption spectra of iron (II)-Morin complex (5x 10^{-5} M) and (B) absorption spectra of the Morin in aqueous solution.

As can be seen from the Fig. 3, the iron (II) Morin complex that has an absorbance maxima at 415 nm. At this wavelength, the Morin itself has no absorption while Morin complexes of all of the tested metal ions and the anions (not shown) exhibited a negligible absorption.

In order to develop an FIA method based on the above phenomenon, the FIA setup shown in Fig.1 was used. In the FIA system, a complex was formed with an absorption spectrum that showed a maximum at 415 nm, which was in agreement with the value obtained in the spectrophotometric study.

2.2.2 Optimisation of chemical variables and the FIA manifold

Various variables closely related to the iron determination were examined using the simple flow-injection analysis system with a fixed iron (II) concentration of 5 μg L^{-1}. The Morin concentration was varied from $1x10^{-6}$ M to $1x10^{-2}$ M. The peak height was found to increase with increasing Morin concentration up to $1x10^{-5}$ M and no noticeable increase was found at higher concentrations. Therefore, $1x10^{-5}$ M Morin was decided as colour developing component of the carrier solution.

With the concentration of the Morin fixed $1x10^{-5}$ M, the pH of the carrier solution was varied from 2.0 to 7.0. The interference effect of the iron (III) were found to increase with increasing pH up to 4.5 and remain constant at higher pH. Also, the peak heights were found to increase with increasing pH up to 4.0, remain constant to 4.5 and decreased slightly above that.

The pH of the reagent carrier was however adjusted to 4.5 to obtain maximum peak height and minimum iron (III) interference in the analysis. In order to proceed with the final system design, the effect of sample volume, mixing coil length and flow-rate were studied using Morin at fixed concentration of $2.5x10^{-4}$ M and pH 4.5.

The sample volume was varied from 5-50 μL. The peak height was decreased by decreasing sample size, and the peaks were broadened with increasing sample size due to sample zone dispersion. A sample injection volume of 20 μL was selected as a compromise between sensitivity and sample throughput rate.

The mixing coil (RC) was examined by using PTFE tubing's (0,5 mm i.d.) at different lengths ranging between 10 and 150 cm. The peak height was increased with increasing mixing coil length from 10-50 cm. The peak height was decreased for lower concentrations and broadened for higher concentrations at longer coil lengths. A mixing coil length 50 cm was decided convenient for better peak height and shape.

The flow-rate was varied from 0.2 to 2 mL min^{-1}. The peak height decreased by increasing flow-rate, probably the extent of reaction decreased. A flow-rate of 0.8 mL min^{-1} was selected as a compromise between sample throughput rate and sensitivity.

2.2.3 Calibration, accuracy and precision

The developed analytical method was validated by evaluating the linear dynamic range, precision, accurate, limit of detection (LOD) and limit of quantification (LOQ) as well as by applying the standard addition technique. Under the optimized experimental conditions, a linear calibration graph was obtained for 0.01-120 mg L^{-1} iron (II) under the optimum conditions with a regression coefficient of 0.9914. The relative standard deviation for the determination of 5 μg L^{-1} iron (II). was 0.85 % for 10 replicate injections. The limit of detection (blank signal plus three times the standard deviation of the blank) was 0.4 μg L^{-1}. The sample throughput of the proposed method was almost 60 sample h^{-1}.

2.2.4 Interference studies

The interference effects of many cations and anions on the determination of 5 μg L^{-1} iron (II) were examined. The results summarized in Table 1.

Tolerance limit (μg L^{-1})	Foreign ion
Over 50000	Cr(III), Al(III), Cd(II), Mn(II), K(I), Na(I), Ag(I), Ca(II), Mg(II), Ba(II), Hg(II), CN^-, NO_3^-, NO_2^-, SO_4^{2-}, CO_3^{2-}, Cl^-, Br^-, PO_4^{3-}, NH_4^+, SCN^-, tartrate, oxalate, citrate, thio-urea
Over 200	Fe (III)

Table 1. Effect of foreign ions on the determination of 5 μg L^{-1} of iron (II) in solution

In the table, the tolerable concentration of each diverse ion was taken as a highest concentration causing an error of ± 3 %. Most of the ions examined did not interfere with the determination of iron (II). The major interference was iron (III) at the amounts of 200 μg L^{-1}.

It is apparent from the Table 1 that the proposed method can tolerate all of the interfering species tested in satisfactory amounts and it is therefore adequately selective for the determination of Fe (II) and total iron.

2.2.5 Applications

The FIA method was applied to the determination of iron (II) and total iron in water samples and ore samples. In order to evaluate the accuracy of the proposed method, the determination of total iron in a standard reference material (Zn/Al/Cu 43XZ3F) and metal alloy sample was carried out. The analytical results obtained by the proposed method are in good agreement with the certified values as is shown in Table 2.

Sample	Total Fe(1) (%)	Certified Fe (%)
Alloy (1)	8.23(0.12)	8.58
Alloy (2)	16.15(0.16)	16.62
Std Zn/Al/Cu 43XZ3 F	0.083(0.02)	0.085

Values in parenthesis are the relative standard deviations for n=5 with confidence level of 95 %.

Table 2. Total iron content of iron alloys and standard reference material

For the application of the proposed FIA method to river and sea water samples collected from different sources were analyzed by using both calibration curve and standard addition methods. The values obtained from the calibration curve and the standard addition methods are in good agreement with each other as shown in Table 3.

Samples(1)	Iron (II)(2) (μg L-1)		Total iron (2) (μg L-1)		
	Found(3)	Found(4)	Found(3)	Found(4)	AAS
Kurtun river water	38.33(0.24)	38.55(0.12)	42.33(0.02)	42.91(0.18)	43.65(0.17)
Seaport sea water	68.84(0.32)	68.65(0.24)	85.13(0.12)	85.75(0.06)	86.12(0.12)
Baruthane sea water	47.51(0.18)	47.62(0.14)	57.24(0.04)	57.65(0.15)	58.97(0.24)
Organized industry water	78.84(0.22)	78.65(0.18)	78.13(0.14)	98.75(0.07)	99.12(0.10)

(1) Samples were collected at Samsun, Turkey.
(2) Values in parenthesis are the relative standard deviations for n=5 with confidence level of 95 %.
(3) Calibration curve method.
(4) Standard addition method.

Table 3. Determination of iron (II) and total iron in river and sea water samples

Atomic absorption measurements taken in water samples were also given for comparison in Table 3. The analytical value of total iron in water is slightly in good agreement with that obtained by the AAS method. The results obtained show that the proposed method can be applied in the determination of iron (II) and total iron content in the water samples without a pre-concentration process.

3. Flow injection spectrophotometric determination of iron (III) using diphenylamine-4-sulfonic acid sodium salt (Reproduced with permission from the paper of Asan Adem et al., 2008. Copyright of Institute of Chemistry, Slovak Academy of Sciences)

In recent years, low cost automatic and userfriendly analytical methods have become attractive for the determination of trace levels of iron in many kinds of samples (Chen et al., 2006; Pons et al., 2005a; Lunvongsa et al., 2006b). Among these, flow-injection analysis (FIA) is a well accepted technique owing to its high sample throughput, cost effective performance, versatility, flexibility, and ease of operation. Also, FIA is compatible with a wide range of detection systems (Guo & Baasner, 1993; Ensafi et al., 2004). Up to date, FIA for the determination of iron(III) has been generally combined with optical detectors (Pulido-Tofino et al., 2000; Saitoh et al., 1998). The spectrophotometric detector based on measuring the absorbance of colored complexes formed with various chromogenic reagents is one of the most frequently used detectors for the determination of iron in many kinds of samples (Yegorov et al., 1993; Yamamura and Sikes, 1966; Ampan et al., 2002; Bruno et al., 2002; Tesfaldet et al., 2004; Van Staden and Kluever, 1998; Mulaudzi et al., 2002; Reguera et al., 1997; Pojanagaron et al., 2002; Araujo et al., 1997; Asan et al., 2003; Udnan et al., 2004; Alonso et al., 1989; M¨uller et al., 1990; Themelis et al., 2001; Kass and Ivaska, 2002; Weeks and Bruland, 2002). A large number of flow-injection spectrophotometric methods have been developed for the determination of iron using desferal (Yegorov et al., 1993), 1,10-phenantroline (Yamamura and Sikes, 1966; Ampan et al., 2002; Bruno et al., 2002; Tesfaldet et al., 2004), tiron (Van Staden and Kluever, 1998; Mulaudzi et al., 2002), ferrozine (Reguera et al., 1997), norfloxaxin (Pojanagaron et al., 2002), thiocyanate (Araujo et al., 1997), DMF (Asan et al., 2003), and salicylate (Udnan et al., 2004) as chromogenic reagents. However, many of the proposed methods have a high limit of detection (Alonso et al., 1989; M¨uller et al., 1990), suffer from many interfering metal ions, such as Zn and Co (Guo and Baasner, 1993), have a short linear dynamic range (Themelis et al., 2001; Kass and Ivaska, 2002), tedious procedures (Pons et al., 2005b), or low sampling rates (Teixeira and Rocha, 2007; Lunvongsa et al., 2006a).

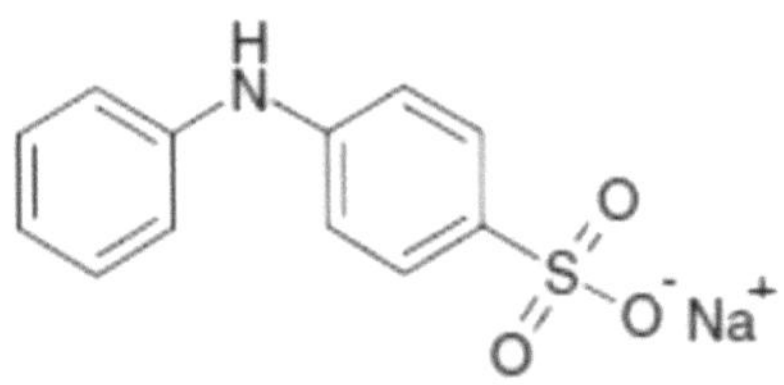

Fig. 4. Structure of the diphenylamine-4-sulfonic acid sodiumSalt

In this study, a highly sensitive and very simple spectrophotometric flow-injection analysis (FIA) method for the determination of iron (III) at low concentration levels is presented. The method is based on the measurement of absorbance intensity of the red complex at 410 nm formed by iron (III) and diphenylamine-4-sulfonic acid sodium salt (DPA-4-SA). It is a simple, highly sensitive, fast and low cost alternative method using the color developing reagent DPA-4-SA in acetate buffer at pH 5.50 and the flow-rate of 1 mL min^{-1} with the sample throughput of 60 h^{-1}. The accuracy of the method was evaluated using the standard addition method and checked by the analysis of the certified material Std Zn/Al/Cu 43 XZ3F.

3.1 Experimental

3.1.1 Reagents, chemicals, equipment

All reagents used were of analytical reagent grade and the solutions were prepared with double distilled and deionized water. The reagent diphenylamine-4- sulfonic acid sodium salt (DPA-4-SA) was provided by Merck. The chemical formula of the DPA-4-SA is shown in Fig. 4. Standard iron(III) (1 mg mL^{-1}) and iron(II) (5 mg mL^{-1}) solutions were prepared by dissolving $FeCl_3$ $6H_2O$ and $FeCl_2$ $4H_2O$ in 0.05 M nitric acid and standardized by titration with EDTA. The stock solution of DPA-4-SA ($1\times10_{-2}$ M) was prepared by dissolving the diphenylamine-4-sulfonic acid sodium salt in deionized water. All stock solutions were stored in polyethylene containers. All polyethylene containers and glassware used for aqueous solutions containing metallic cations were cleaned with (1+1) nitric acid while the rest were cleaned with 3 % Decon 90, all were rinsed with deionized water before use. The working standard solutions were prepared by appropriate dilution immediately before use. Interference studies were carried out using chloride or nitrate salts of the metal cations, and sodium or potassium salts of anions. All solutions were degassed before use using a sonicator (LC 30). A certified metal alloy sample consisting of 0.085 % Fe (Zn/Al/Cu 43XZ3F) was provided by MBH Analytical Ltd. (UK).

The pH measurements were carried out using a Jenway 3040 Model digital pH-meter consisting of a contained glass pH electrode. UV-Visible spectra of the DPA-4-SA reagent and metal-DPA-4-SA complexes were taken with a Unicam spectrophotometer (GBC Cintra 20, Australia). A peristaltic pump (ISMATEC; IPC, Switzerland) was used to propel the samples and reagent solutions. Samples were injected into the carrier stream by a 7125 model stainless steel high-pressure Rheodyne injection valve provided with a 20 µL injection loop. Absorbance of the colored complex formed in the flow system was measured using a UV-visible spectrophotometer equipped with a flowthrough micro cell (Spectra SYSTEM UV 3000 HR, Thermo Separation Products, USA), and connected to a computer (IPX Spectra SYSTEM SN 4000) incorporated with a PC 1000 software program. The reaction coil was made of PTFE tubing (1 m, 0.5 mm, i.d.). A UNICAM 929 model (Shimadzu AA-68006) flame atomic absorption spectrophotometer with deuterium- lamp background correction was used for the determination of iron in reference to the FIA method. The measuring conditions were as follows: UNICAM hollow cathode lamp, 10 cm 1-slot burner, airacetylene flame (fuel gas flow-rate 1.50 L min^{-1}), 0.2 nm spectral bandwidth, and 7 mm burner height. The wavelength and the lamp current of iron were 248 nm and 5 mA, respectively.

The manifold of the flow-injection system was similar to that proposed in our previous study (Asan et al., 2003). The peristaltic pump was used for propelling the reagent carrier solution at a flow-rate of 1 mL min^{-1}. Samples were injected into the reagent carrier solution, soon load the reaction coil. The reaction zone containing the complex was moving towards the flow-through spectrophotometric detector cell in which the presence of iron(III)-DPA-4-SA complex was selectively monitored, and the absorbance of the complex at 410 nm was continuously recorded.

3.1.2 Preparation of water samples and certified metal alloy solution

Sea and river water samples collected in Nalgene plastics were acidified by adding 1 mL of nitric acid (0.1 M) per 100 mL of sample solution after filtration over a 0.45 µm Millipore

Filter (Millford, MA). After the filtration and pre-treatment, water samples were injected directly into the FIA system for the determination of iron(III).

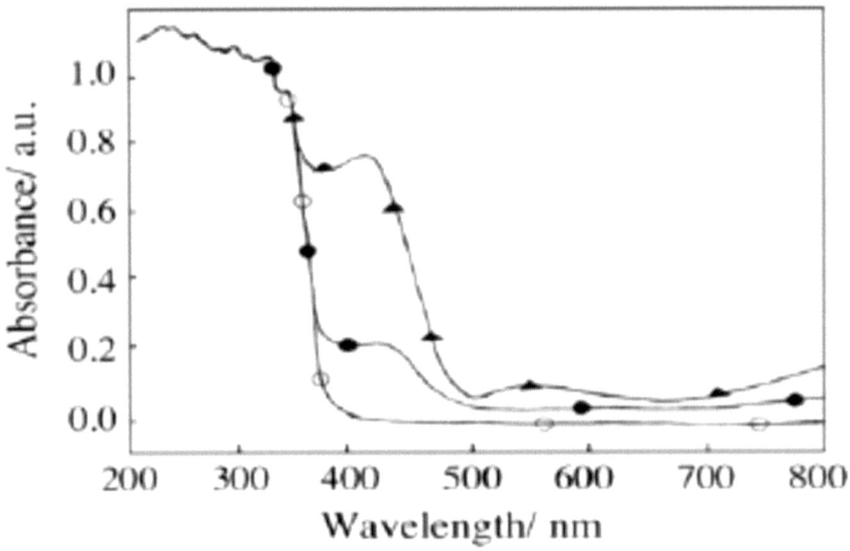

Fig. 5. Absorption spectra for DPA-4-SA and metal-DPA-4- SA complexes: (○) 5 × 10^{-5} M DPA-4-SA and 5× 10^{-5} M of each Co(II), Cu(II), Cr(III), Al(III), Cd(II), Ni(II), Mn(II), Ba(II), Ca(II), Ag(I), K(I), Na(I), Hg(II), Zn(II), and Mg(II); (•) 5 × 10^{-5} M of Fe(II)-DPA-4-SA; and (▲) 5 × 10^{-5} M of Fe(III)-DPA-4-SA.

Oxidizing iron(II) to iron(III) was used to determine the total iron amount. Hydrogen peroxide was chosen as the oxidizing agent for the determination of total iron. A concentration of 0.25 mol L^{-1} of H_2O_2 ensured total oxidation of iron(II) to iron(III) (Pons et al., 2005). Before the determination, H_2O_2 (10 mass %) was added to the water sample solution for complete oxidation of iron(II) to iron(III). Then, 20 µL of this solution were injected into the system, as in the procedure described above. Analyses were performed with the least possible delay.

A 0.10 g sample of the certified metal alloy (Zn/Al/Cu 43XZ3F) was dissolved in 12 mL of concentrated HCl + HNO3 (3 : 1) in a 100 mL beaker. The mixture was heated on a hot plate nearly to dryness; 5 mL of HNO3 were added to complete the dissolution and were diluted to 100 mL with deionized water. The solution was filtered and transferred quantitatively to a 1000 mL volumetric flask and filled up to the volume with deionized water. The volume of 10 mL of this solution was treated with H_2O_2 (10 mass %) for iron(II) oxidation. After the oxidation step, the solution was diluted 100 fold, and then, 20 µL of this solution were used for the determination of total iron.

3.2 Results and discussion

According to the spectrophotometric studies, iron (II) and iron(III) react with DPA-4-SA in aqueous medium to form complexes. As shown in Fig. 5, the absorption spectra corresponding to solutions of 5×10^{-5} M of each metal complex in water demonstrate strong absorption for the Fe(III)-DPA-4-SA complex. Iron(III) reacts with DPA-4-SA in the pH range of 2.0–6.0 forming a complex with absorption maxima at 410 nm and molar absorptivity of 1.60×10^{4} L mol^{-1} cm^{-1}. The Fe(II)-DPA-4-SA complex presents only a slight absorption at this wavelength. Seventeen different metals do hardly react with DPA-4-SA in aqueous medium to from complexes. This can be an important advantage when developing a simplified FIA method for iron(III). Therefore, the specific absorbance maximum of the Fe(III)-DPA-4-SA complex at this wavelength can be applied in the selective determination of iron(III) in the flow-injection system.

The optimum experimental conditions were determined using a standard iron(III) solution. The concentration of DPA-4-SA, pH, and the flow-rate were the main variables influencing the intensity of the signal in the FIA system. Optimization of the FIA system was therefore performed by changing these variables one by one while applying 10 µg L^{-1} and 90 µg L^{-1} of iron(III) standard solutions in order to obtain the highest signal and better reproducibility at different concentration levels.

Influence of the DPA-4-SA concentration in the carrier solution on the peak height was examined by changing the DPA-4-SA concentration in the range of 1×10^{-2} M to 5×10^{-4} M in an acetate buffer solution (pH = 5.5), at the flow rate of 1 mL min^{-1}. Maximum peak heights were found using 1×10^{-3} M of the DPA-4-SA solution for both, 10 µg L^{-1} and 90 µg L^{-1}, iron(III) levels. Therefore, 1×10^{-3} M of DPA-4-SA was chosen as the color-developing component of the carrier solution.

The effect of flow-rate on the peak height of 10 µg L^{-1} and 90 µg L^{-1} iron(III) was examined by varying flow-rates from 0.2 mL min^{-1} to 2.0 mL min^{-1}. Peak heights decreased at flow-rates above 1.2 mL min^{-1} and below 0.7 mL min^{-1}. Flow-rates below 0.7 mL min^{-1} peaks also broadened. In the flow-rates range of 0.7–1.2 mL min^{-1}, there were slight differences in the peak heights. However, taking into consideration the stability of the pump, peak shape, and sampling time, the flow-rate of the reagent carrier solution was adjusted to 1 mL min^{-1}. This provided a sampling frequency of 60 h^{-1}.

pH of the carrier solution consisting of 1×10^{-3} M of DPA-4-SA was adjusted by adding simple acids and bases into the buffer to obtain the pH range of 3.30– 6.10. The peak shape and height were found maximum at pH 5.5. Therefore, 1×10^{-2} M of the acetate buffer solution at pH 5.5 was used throughout the study.

Reaction coil was used for the interaction of iron(III) and DPA-4-SA in the flow-injection system. The effect of the reaction coil (RC) length was examined by changing the coil length from 10 cm to 150 cm. The peak height decreased with the increase of length due to fast kinetics of the color forming reaction. The 10 cm length reaction coil was chosen since it produced the best peak height together with a good reproducibility.

The calibration graph for the determination of iron(III) was maintained under the optimized conditions as described above. A good linear relationship was observed for iron(III) ranging from 5 µg L^{-1} to 200 µg L^{-1}. The calibration curve equation was $A = 0.4018C + 2.0196$; $r2 = 0.9958$; $n = 6$, where A represents the absorbance measured as peak height and C the iron concentration in µg L^{-1}. The confidence limits of the intercept and the slope were calculated at the 95 % confidence level. The same calibration graph can be used for the determination of total iron. The detection limit estimated (S/N = 3) was 1 µg L−1 of iron(III).

The limit of quantification(LOQ) was calculated as recommended (Currie, 1995); based on a ten fold of the standard deviation of 10 consecutive injections of the blank, the value of 1.65 $\mu g\ L^{-1}$ was obtained. The reproducibility of the method calculated as the relative standard deviation (RSD) of peak heights obtained from 5 injections of 10 $\mu g\ L^{-1}$ iron(III) was 3.5 %.

Possible interferences in the determination of iron(III) were examined under the optimum experimental conditions. The effect of potential interfering ions on the determination of iron was investigated at the 5 % interference level. To carry out this study, 20 µL of a 20 $\mu g\ L^{-1}$ iron(III) standard were injected. Table 4 summarizes the tolerance limits of the interfering ions. Most of the ions examined did not interfere with the iron(III) determination up to at least a 50000 fold excesses. The only interfering ion was iron(II), even 2 $mg\ L^{-1}$ of iron(II) gave a positive interference.

Tolerance limit ($mg\ L^{-1}$)	Foreign ion
Over 1000	Co(II), Cr(III), Al(III), Cu(II), Cd(II), Ni(II), Pb(II), Sn(II), Mn(II), Zn(II), K(I), Na(I), Ag(I), Ca(II), Mg(II), Ba(II), Hg(II), CN^-, NO_3^-, NO_2^-, SO_4^{2-}, CO_3^{2-}, Cl^-, Br^-, PO_4^{3-}, NH_4^+
Over 2	Fe(II)

Table 4. Effect of foreign ions on the determination of 20 $\mu g\ L^{-1}$ of iron(III) in solution

The proposed method was applied in the determination of total iron in river and seawater samples. Iron(III) and total iron were determined according to the FIA procedure as described in the experimental section. The results obtained by both, standard addition and calibration curve, methods were in good agreement with each other. Atomic absorption measurements taken in water samples 1 and 2 are also given for comparison (Table 5).

Sample	Fe(III)[2] ($\mu g\ L^{-1}$)		Total iron[2] ($\mu g\ L^{-1}$)		Total iron[2] ($\mu g\ L^{-1}$)
	Found[3]	Found[4]	Found[3]	Found[4]	AAS
Seaport (Sea water)	45.16 (0.06)	45.92 (0.21)	53.46 (0.19)	53.78 (0.27)	54.93(0.24)
Industry (Sea water)	56.28 (0.18)	56.11 (0.14)	76.45 (0.27)	76.13 (0.15)	78.19(0.16)
Atakum (River water)	21.45 (0.05)	21.18 (0.12)	32.69 (0.08)	31.85 (0.24)	34.47(0.36)
Mert (River water)	38.17 (0.11)	38.12 (0.19)	1.18 (0.04)	41.27 (0.16)	43.76(0.32)

1. Samples were collected at Samsun, Turkey.
2. Values in parantheses are the relative standard deviations for n =5 with confidence level of 95 %.
3. Calibration curve method.
4. Standard addition method.

Table 5. Analytical results of iron(III) and total iron in natural water samples[1]

The analytical value of total iron in water is in good agreement with that obtained by the AAS method. The accuracy of the proposed method was tested by the analysis of a certified metal alloy solution (MBH Zn/Al/Cu 43XZ3F). Three replicates of the solution using the sampling volume of 20 µL were analyzed. The certified and the obtained values were 0.085 % and (0.084 ± 0.006) of iron, respectively. An excellent agreement between the found and the certified values has been obtained for the certified metal alloy solution. The results obtained show that the proposed method can be applied in the determination of iron(III) and total iron content in water samples without a preconcentration process.

4. Flow injection spectrofluorimetric determination of iron (III) in water using salicylic acid (Reproduced with permission from the paper of Asan Adem et al., 2010. Copyright of Institute of Chemistry, Slovak Academy of Sciences)

In general terms, sensitivity of the spectrofluorimetric method is much higher than that of the spectrophotometric method. However, fluorescence reagents and methods suitable for the determination of iron are scarce and they suffer from serious interference of some metal cations such as aluminium, copper, and tin or they require a matrix separation step. Also, the reagents used for the determination of iron have a risk of toxicity (Tamm & Kalb, 1993; Yan et al., 1992; Cha et al., 1996; Ragos et al., 1998). Therefore, it is still important to develop simple and economical procedures that could be directly applied to real samples without the matrix separation step and with minimized reagent consumption.

In literature (Cha et al., 1998), salicylic acid has been used as a fluorescence reagent for the spectrofluorimetric determination of iron(III) in batch conditions. Experimentally it was found to be a very sensitive emission reagent for the spectrofluorimetric determination of iron(III) in the absence of iron (II). A very strong emission peak of salicylic acid in aqueous solution, which decreased linearly with the addition of iron(III), occurred at 409 nm with excitation at 299 nm. Also, salicylic acid is a commercially available reagent and it does not have a risk of serious toxicity when compared to the reagents used previously.

A simple and fast flow injection fluorescence quenching method for the determination of low levels of iron(III) in water has been developed. For this purpose, a preconcentration minicolumn consisting of cation-exchange resin was coupled to the FIA system. The use of mini-column in the system provided an improvement in sensitivity and the developed FIA method was successfully applied to the on-line determination of low levels of iron in real samples without the pre-concentration process. Fluorimetric determination was based on the measurement of the quenching effect of iron on salicylic acid fluorescence. An emission peak of salicylic acid in aqueous solution occurs at 409 nm with excitation at 299 nm. The effect of interferences from various metals and anions commonly present in water was also studied. The method was successfully applied to the determination of low levels of iron in real samples (river, sea, and spring waters).

4.1 Experimental

Analytical reagent grade chemicals were employed for the preparation of the standard, and the solutions were prepared using double distilled water. Standard iron(III) and iron (II) stock solutions (5×10^{-3} mol L^{-1} Fe(III) and Fe(II)) were prepared by dissolving $FeNH_4(SO_4)_2 \cdot 12H_2O$ and $Fe(NH_4)2(SO_4)_2 \cdot 6H_2O$ in water and were standardized by titration with EDTA. Iron(II) and iron(III) working standard solutions were prepared by appropriate dilution of the stock solutions with water immediately before use. Hydrogen peroxide solution, 30 mass %, was purchased from Merck (Darmstadt, Germany). Standard solutions of other metal ions (all of them from Merck (Darmstadt, Germany)) at different concentrations were prepared with doubly distilled water.

Buffer solution, 0.1 mol L^{-1} NH_4^+ /NH_3 at pH: 8.5, was used to produce analytical signal in the FIA system. Salicylic acid was provided from Merck (Darmstadt, Germany). Standard salicylic acid solutions were prepared daily by dissolving the appropriate amount of salicylic acid in an ethanol:water mixture (30 : 70). The reagent carrier solution was composed of 2×10^{-6} mol L^{-1} salicylic acid and 0.1 mol L^{-1} NH_4^+ /NH_3 buffer solution (90:10) at pH 8.5.

Fluorescence measurements for the batch experiments were performed with an SPF-500 model spectrofluorometer (American Instrument Co, Jessup, USA) using 1 cm quartz cells. Instrument excitation and emission slits were fixed at 10 nm. The light source was a 150 W Xenon lamp (American Instrument Co, Jessup, USA). Excitation and emission wavelengths were set at 299 nm and 409 nm, respectively. An eight-channel ISMATEC IPC peristaltic pump (Z¨urich, Switzerland), 0.75 mm i.d. PFTE tubing, was used to propel the samples and reagent solutions. Samples were injected into the carrier stream by a Rheodyne injection valve provided with a 20 µL loop. A Varian 2070 spectrofluorometer (Tokyo, Japan) using a 15 µL flow cell was used for the on-line measurements of analytical signals. Instrument excitation and emission slits were set at 20 nm. The light source was an ozoneless 75 W Xenon lamp (Tokyo, Japan). A strip chart recorder was attached to the instrument. Cation-exchange resin, sodium form of A650 W (100–200 mesh), was provided by the BioRad Labs (Hercules, CA, USA). The cation-exchange resin minicolumn (6 cm long, 2 mm i.d) was prepared in our laboratory.

pH measurements were carried out using a Jenway digital pH-meter model 3040 (Essex, England). An ATI UNICAM 929 model AAS (Cambridge, UK) flame atomic absorption spectrophotometer with a deuterium-lamp background correction was used for the determination of iron in reference to the FIA method. The measuring conditions were as follows: UNICAM hollow cathode lamp, 10 cm 1-slot burner, air–acetylene flame (fuel gas flow-rate of 1.50 L min^{-1}), 0.2 nm spectral bandwidth, and 7 mm burner height. The wavelength and the lamp current of iron were 248 nm and 5 mA, respectively. The flow injection manifold was similar to that proposed in our previous study (Isildak et al., 1999). Peristaltic pump was used to transport the reagent carrier solution through the system. The sample was injected using an injection loop (20 µL). The reagent carrier solution and the sample were allowed to mix in the flow stream and in the mini-column. The decrease in the fluorescence intensity of the salicylic acid as a function of Fe(III) concentration was measured in the flow cell using 299 nm for excitation and 409 nm for emission. Water samples were obtained from different places of the river, sea and thermal spring in Samsun, Turkey. They were filtered through a 0.45 µm Millipore Filter (Millford, MA, USA). Water samples were split into two portions: one part was directly injected into the FIA system for the determination of iron(III). Before the analysis of the other part, 1 mL of H_2O_2 (10 mass %) was added to a 9 mL sample solution for complete oxidation of iron(II) to iron(III). Then, 20µL of this solution were injected into the system for the determination of total iron, as in the procedure described above.

A 0.10 g sample of the certified metal alloy (Zn/Al/Cu 43XZ3F) was dissolved in 12 mL of concentrated HCl + HNO_3 (3 : 1) in a 100 mL beaker. The mixture was heated on a hot plate nearly to dryness; 5 mL of HNO_3 were added to complete the dissolution, and the solution was diluted to 100 mL with deionized water, filtered and transferred quantitatively to a 1000 mL volumetric flask and filled up to the volume with deionized water. The volume of 10 mL of this solution was treated with H2O2 (10 mass %) for iron(II) oxidation. After the oxidation step, the solution was diluted 100 fold, and then, 20 μL of this solution were used for the determination of total iron.

4.2 Results and discussion

Fig. 6 shows the fluorescence emission spectra of 5×10^{-5} mol L^{-1} salicylic acid in a buffer solution at pH 8.5 before and after the reaction with 1×10^{-5} mol L^{-1} iron(II) and iron(III), respectively, in batch experiments. As can be seen, the intensity of salicylic acid fluorescence decreased significantly in the presence of iron(III). From these spectra, the emission wavelength chosen for the FIA measurement was 409 nm, using 299 nm for the fluorescence excitation.

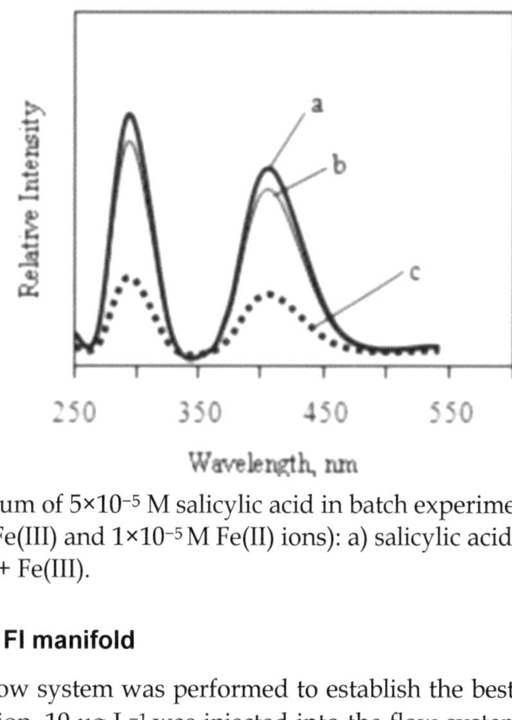

Fig. 6. Emission spectrum of 5×10^{-5} M salicylic acid in batch experiment (in the absence and presence of 1×10^{-5} M Fe(III) and 1×10^{-5} M Fe(II) ions): a) salicylic acid, b) salicylic acid + Fe(II), c) salicylic acid + Fe(III).

4.2.1 Optimization of FI manifold

Optimization of the flow system was performed to establish the best FIA variables. A fixed standard Fe (III) solution, 10 μg L^{-1} was injected into the flow system for the determination of optimum experimental conditions. The main variables influencing the intensity of the signal were: flow-rate, pH, and the concentration of salicylic acid. Therefore, optimization of the FIA system was carried out by changing these variables one by one.

The effect of salicylic acid in the carrier solution on the peak height was examined by changing the amount of salicylic acid in the range of 5×10^{-7}–5×10^{-5} mol L^{-1} in buffer solution at pH 8.5, at the flow rate of 1.0 mL min^{-1}. Peak heights were found maximum using a 2×10^{-6} mol L^{-1} salicylic acid solution for 10 μg L^{-1} iron(III) levels. Therefore, 2×10^{-6} mol L^{-1} salicylic acid was chosen as the fluorescence reagent in the carrier solution.

The effect of flow-rate on the peak height of iron(III) was examined by varying the flow-rate from 0.5 mL min^{-1} to 1.5 mL min^{-1}. Peak heights decreased at flow-rates above 1.2 mL min^{-1}

and below 0.8 mL min^{-1}. Below 0.8 mL min^{-1} the peaks also broadened. Between the flow-rates of 0.8–1.2 mL min^{-1}, there were slight differences in the peak heights. Considering the stability of the pump, peak height, and sampling time, the flow-rate of the reagent carrier solution was adjusted to 1.0 mL min^{-1}. This provided the sampling frequency of 60 h^{-1}. pH of the carrier solution consisting of 2×10^{-6} mol L^{-1} salicylic acid was adjusted by an NH_4^+ /NH_3 buffer solution to obtain the pH range of 8.0–10.0. The peak heights were found maximum at pH 8.5. Therefore, a 0.1 mol L^{-1} NH_4^+ /NH_3 buffer solution (90 : 10) at pH 8.5 was used throughout the study.

The use of a mini-column in the flow-injection system provided an improvement in the sensitivity and selectivity due to on-line pre-concentration and fast interaction of metal ions with reagent molecules in the carrier solution (Isildak et al., 1999). A mini-column packed with strong cation-exchange resin was selected because metal ions are strongly bound by the resin so that low amounts of the resin can be used. Higher amounts of the resin minimized the use of higher flowrates due to an increase in the hydrodynamic pressure. Sampling time in the FIA system depends on the retention time in the cation exchange mini-column and the residence time in the tubing in the flow-path. The effect of the column length was examined by changing the column length between 2 cm and 10 cm. From the results obtained, 6 cm column length brought the best results for the peak shape and sensitivity for iron for all concentration levels studied.

Also a mixing coil and a mini-column packet with silica and glass beads were inserted into the analytical path instead of the cation-exchange resin minicolumn. However, the observed peak height and sensitivity for iron(III) were lower and poorer, for all concentration levels studied. This result can originate from the short remaining time of iron(III) in each column, which means a narrow interacting zone of the sample. Finally, a mini-column packed with strong cation-exchange resin was used throughout the study for the determination of iron(III). Indeed, a significant improvement of the selectivity and sensitivity was observed.

4.2.2 Analytical performance characteristics

Analytical performance characteristics of the method were evaluated under optimum conditions. Fig. 7 shows typical flow signals for iron(III) obtained by the proposed method. The reaction of iron(III) with salicylic acid resulted in negative peaks due to the fluorescence quenching of salicylic acid. Under the optimum working conditions, calibration graphs were prepared from the results of triplicate measurements of iron(III) standard solutions of increasing concentration. The calibration graph showed a good linearity from 5–100 µg L^{-1} iron(III) with the linear regression equation: $Y = 0.0353X + 0.0909$, where Y is the peak height (cm) and X is the concentration of iron(III) in µg L^{-1}. The correlation coefficient was $r^2 = 0.9963$ and the relative Standard deviation (RSD) of the method based on five replicate measurements of 10 µg L^{-1} iron(III) was 1.25 % for a 20 µL injection volume. The limit of detection (determined as three times the standard deviation of the blank) was 0.3 µg L^{-1} and the sampling rate was 60 h^{-1}. The limit of quantification (LOQ) was calculated as recommended (Currie, 1995); based on a ten fold standard deviation of ten consecutive injections of the blank, the value of 1.12 µg L^{-1} was obtained.

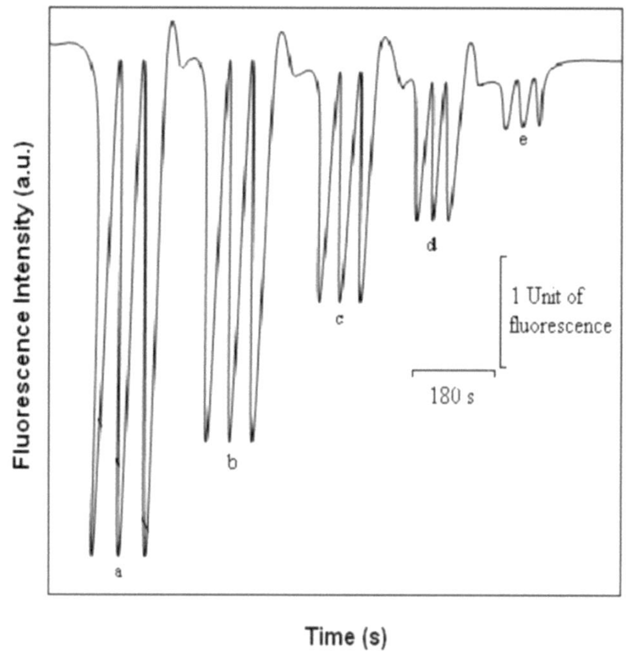

Fig. 7. Flow signal for iron(III) standard solutions by fluorescence quenching-FIA a) 100 µg L−1, b) 75 µg L−1, c) 50 µg L−1, d) 25 µg L−1, and e) 5 µg L−1 when using the optimized FIA system.

4.2.3 Interference study

The effect of diverse ions on the detection of iron by the present system were examined using a solution containing 10 µg L^{-1} iron(III) and one of the other ions. The tolerable concentration of each diverse ion was taken as the highest concentration causing the error of ± 5 %. The results are summarized in Table 6.

Tolerance limit (mg L^{-1})	Foreign ion
No interfere	CO_3^{2-}, SCN^-, Br^-, SO_4^{2-}, Ca^{2+}, Zn^{2+}
Over 50 000	Co(II), Cr(III), Al(III), Cu(II), Cd(II), Ni(II), Pb(II), Mn(II), K(I), Na(I), Ag(I), Mg(II), Ba(II), Hg(II), CN^-, NO_3^-, NO_2^-, Cl^-, PO_4^{3-}, NH_4^+
Over 100	Fe(II)

Table 6. Effect of foreign ions on the determination of 10 µg L^{-1} of iron(III) in solution

4.2.4 Analysis of water samples

The proposed method was applied to the determination of iron in river, sea, and thermal spring water samples to evaluate its applicability. Iron(III) and total iron were determined according to the FIA procedure as described in the experimental section. Table 7 shows the analytical results of iron(III) and total iron. Atomic absorption measurements taken were also given for comparison. The results obtained with the standard addition and the calibration curve methods, and the AAS measurements were in good agreement with each other.

Sample	Fe(III)[2] (µg L^{-1})		Total iron [2] (µg L^{-1})			
	Found[3]	Found[4]	Found[3]	Found[4]	AAS	*Ec* (%)
Seaport (Sea water)	52.16 (0.12)	52.92 (0.21)	67.25 (0.10)	67.52 (0.12)	66.98 (0.05)	0.60
Atakum(River water)	25.41 (0.10)	26.01 (0.19)	37.41 (0.14)	38.01 (0.17)	38.15 (0.07)	1.16
Kurtun river	32.84 (0.24)	33.57 (0.28)	48.14 (0.19)	48.57 (0.27)	49.12 (0.09)	1.55
Spring water (1)	10.95 (0.15)	11.25 (0.27)	16.75 (0.32)	16.20 (0.28)	16.62 (0.18)	0.88
Spring water (2)	12.65 (0.09)	13.18 (0.12)	21.83 (0.08)	21.32 (0.24)	21.75 (0.14)	0.81
Spring water (3)	38.17 (0.11)	38.12 (0.19)	52.54(0.04)	52.73 (0.16)	52.95 (0.12)	0.60

1. Samples were collected at Samsun, Turkey.
2. Values in parantheses are the relative standard deviations for *n* =5 with confidence level of 95 %.
3. Calibration curve method.
4. Standard addition method.

Table 7. Determination of total iron in water samples[1]

Accuracy of the proposed method was also tested by analyzing a certified metal alloy solution (MBH Zn/Al/Cu 43XZ3F). Three replicates of the solution using the sampling volume of 20 µL were analyzed. The certified and the obtained values were 0.085 % and (0.084 ± 0.006) % of iron, respectively. An excellent agreement between the found and the certified values was obtained for the certified metal alloy solution. The obtained results show that the proposed method can be applied to the determination of iron(III) and total iron content in water samples without a pre-concentration process.

5. A simple flow injection spectrophotometric determination method for iron (III) based on O-acetylsalicylhydroxamic acid complexation (Reproduced with permission from the paper of Andac Muberra et al., 2009. Copyright of Institute of Chemistry, Slovak Academy of Sciences)

1,10-phenanthroline and salicylic acid are the most reported chelating agents applied for the determination of iron(III) and total iron after oxidation to iron(III) (Tesfaldet et al., 2004; Udnan et al., 2004). A number of other chelating agents that have been reported for the spectrophotometric and/or flow-injection spectrophotometric determination of iron(III) and total iron include 2-thiobarbituric acid (Morelli, 1983), norfloxacin (Pojanagaron et al., 2002) tiron (van Staden & Kluever, 2002) DMF (Asan et al., 2003), tetracycline (Sultan et al., 1992) and chlortetracycline (Wirat, 2008). Flow-injection spectrophotometric methods based on the above chelating agents are either not selective, or a masking agent has to be used. However, highly selective, simple and economical methods for routine determination of iron(III) in different sample matrices are still required. In the present study, a simple and rapid flow-injection spectrophotometric method for the determination of iron (III) and total iron is proposed. The method is based on the reaction between iron (III) and *O*-acetylsalicylhydroxamic acid (AcSHA) in a 2 % methanol solution resulting in an intense violet complex with strong absorption at 475 nm. The reagent itself is sparingly soluble in water and did not absorb in the visible region of the spectrum, therefore, it might be well suited for flow-injection analysis of iron(III) and total iron. An addition of copper sulphate (1×10^{-4} mol L^{-1}) into the reagent carrier solution resulted in baseline absorbance, and possible interfering ions were eliminated without a significant decrease in the sensitivity of the method. The method was successfully applied in the determination of iron (III) and total iron in water and ore samples. The method was verified by analysing a certified reference material Zn/Al/Cu 43XZ3F and also by the AAS method.

5.1 Experimental

All chemicals used were of analytical reagent grade, and solutions were prepared from double deionised water. Standard iron(II) and iron(III) stock solutions were prepared by dissolving 278.02 mg of iron(II) and 489.96 mg of iron(III) sulphate (Merck; Darmstadt, Germany) in 100 mL of 0.01 mol L^{-1} hydrochloric acid to give 0.01 mol L^{-1} stock solution of iron(II) and iron(III). Iron(II) and iron(III) working standard solutions were prepared daily by suitable dilution of the stock solutions with double deionised water. Standard reference material consisting of 0.085 % Fe (Zn/Al/Cu 43XZ3F) was provided from MBH Analytical Ltd. (UK). Hydrogen peroxide solution of 30 vol. % was obtained from Merck. AcSHA was synthesised according to the procedure described previously (Asan et al., 2003). A stock solution of AcSHA (0.01 mol L−1) was prepared by dissolving 0.095 g of AcSHA in 100 mL of aqueous methanol (2 vol. %). For the spectrophotometric study, AcSHA complex solutions of various metals were prepared by mixing 1 mL of 1×10−4 mol L−1standard solution of each metal in double deionised water with the suitable volume of 1×10−4 mol L−1 AcSHA stock solution. Reagent carrier solution was composed of AcSHA in a 2 % methanol solution and 1×10−4 mol L−1 CuSO4 in 0.001 mol L−1 HCl 98 % (pH 2.85). UV-VIS spectra of metal-AcSHA complexes were taken with a Unicam spectrophotometer (GBC Cintra 20, Australia). A Jenway 3040 Model digital pH-meter was used for the pH measurements. In the FIA system, a peristaltic pump (ISMATEC; IPC, Switzerland) 0.50 mm i.d. PTFE tubing was used to propel the samples and reagent solutions. Samples were injected into the carrier stream by a 7125 model stainless steel high pressure Rheodyne injection valve provided with a 20 µL loop. Absorbance of the coloured complex formed was measured with a UV-VIS spectrophotometer equipped with a flowthrough micro cell (Spectra SYSTEM UV 3000 HR,Thermo Separation Products, USA), and connected to a computer incorporated with a PC1000 software programme. A UNICAM 929 model (Shimadzu AA-68006) flame atomic absorption spectrophotometer with a deuterium-lamp background correction was used for the determination of iron in reference to the FIA method. The measuring conditions were as follows: UNICAM hollow cathode lamp, 10 cm 1-slot burner, air-acetylene flame (fuel gas flow-rate 1.50 L min−1), 0.2 nm spectral bandwidth, and 7 mm burner height. The wavelength and the lamp current of iron were 248 nm and 5 mA, respectively.

The FIA system used, similar to that proposed in our previous works (Asan et al., 2003), is quite simple. The sample solution was introduced into the reagent carrier solution by the Rhodyne injection valve. A water-soluble complex (λmax = 475 nm) was then formed on the passage of the reagent carrier solution in the mixing coil. As a mixing coil, PTFE tubing (50 cm long) was attached before the flow-through detection cell. The absorbance of the coloured complex was selectively monitored in the cell at 475 nm. The transient signal was recorded as a peak, the height of which was proportional to the iron(III) concentration in the sample, and it was used in all measurements. Five replicate injections per sample were made.

Sea and river water samples collected in Nalgene plastics were acidified by adding 1 mL of nitric acid (0.1 mol L^{-1}) per 100 mL of sample solution after filtration over a 0.45 µm Millipore Filter (Millford, MA). After the filtration, water samples were injected directly into the FIA system for the determination of iron(III).

Total iron was determined by oxidising iron(II) to iron(III). Hydrogen peroxide was chosen as the oxidising agent for the determination of total iron. A 0.25 mol L^{-1} H_2O_2 concentration ensured total oxidation of iron(II) into iron(III) (Pons, et al., 2005). Before the determination of total iron, H_2O_2 (10 mass %) was added to the water sample solution for complete oxidation of iron(II) to iron(III). Then, 20 µL of this solution were injected into the system, as in the procedure described above. A 0.10 g sample of the certified metal alloy (Zn/Al/Cu 43XZ3F) was dissolved in 12 mL of concentrated HCl and HNO_3 (3 : 1) in a 100 mL beaker. The mixture was heated on a hot plate nearly to dryness; 5 mL of HNO_3 were added to complete the dissolution and the solution was diluted to 100 mL with deionised water. The solution was filtered and transferred quantitatively to a 1000 mL volumetric flask and filled up to volume with deionised water. 9 mL of this solution were treated with 1 mL of H_2O_2 (10 mass %) for iron(II) oxidation. After the oxidation step, 20 µL of this solution were used in the determination of total iron. Metal ore samples (0.10 g) were powdered (≥ 500 mesh) and prepared as in the procedure described above. All analyses were performed with the least possible delay.

5.2 Results and discussion

5.2.1 Spectrophotometric studies of AcSHA-metal complexes

Metal ions react with AcSHA in aqueous media in the range of pH 2.0–10.0 forming coloured complexes with different stoichiometry. These complexes are fairly soluble in aqueous media (O'Brien et al., 1997). Their absorption spectra corresponding to solutions of 5×10^{-5} mol L^{-1} metal complexes measured against a reagent blank are shown in Fig. 8.

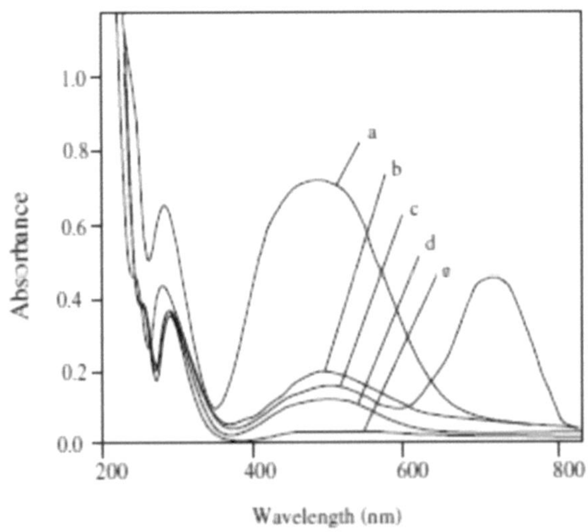

Fig. 8. Absorption spectras of 5x10^{-5} M AcSHA and M-$(AcSHA)_n$ complexes. a) Fe(III)-$(AcSHA)_n$; b) Fe(II)-$(AcSHA)_n$; c) Cu-$(AcSHA)_n$; d) M-$(AcSHA)_n$; (M: Ni, Co, Zn, Pb); e) AcSHA only.

As can be seen from Fig. 8, only AcSHA reacted efficiently with iron to form iron-(AcSHA)*n* complexes with the absorbance maxima at 475 nm. At this wavelength, AcSHA itself has no absorption while Ac- SHA complexes of copper(II), nickel(II), cobalt(II), and zinc(II), among all metal ions with the anions tested, show a negligible absorption. The FIA setup shown in Fig. 9. was used in order to develop an FIA method based on the above phenomenon.

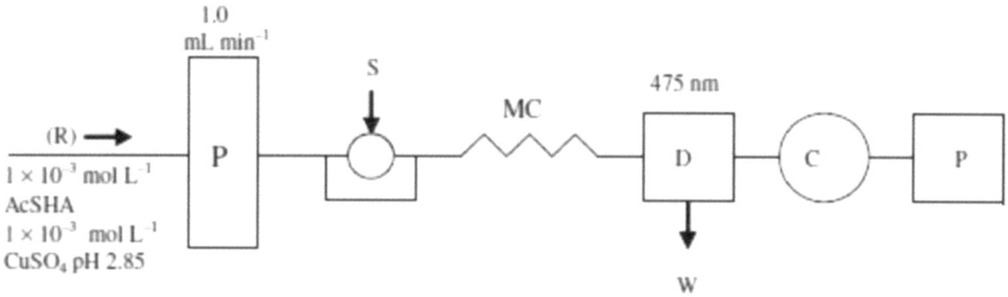

Fig. 9. Flow diagram of the flow-injection analysis system used for the determination of iron (III) and total iron, R; reagent carrier solution (1x10^{-4} M AcSHA, 1x10^{-4} M $CuSO_4$, pH: 2.85), P, Peristaltic pump, S; Rheodyne sample injection valve, MC; mixing coil (50 cm long, 0.5 mm i.d), D; spectrophotometric detector (λ_{max} = 475 nm), W; waste, C; computer, P; printer.

5.2.2 Optimisation of chemical variables and FIA manifold

Various variables closely related to iron determination were examined using a simple flow-injection analysis system with a fixed iron(III) concentration of 5 µg L^{-1}. The AcSHA concentration was varied from 1×10^{-5} mol L^{-1} to 1×10^{-2} mol L^{-1}. The peak height was found to increase with the AcSHA concentration increasing up to 1×10^{-4} mol L^{-1}, no noticeable increase was found at higher concentrations. Therefore, 1×10^{-4} mol L^{-1} AcSHA was used as the colour developing component of the carrier solution. With the concentration of AcSHA fixed at 1×10^{-4} mol L^{-1}, pH of the carrier solution was varied from 1.5 to 5.5. The interference effect of iron(II) was found to increase with pH increasing up to 3.5 and to remain constant at higher pH. Also, the peak heights were found to increase with pH increasing up to 3.0, to remain constant up to 4.0, and to decrease slightly above this value. pH of the reagent carrier was, however, adjusted to 2.85 to obtain the maximum peak height and minimum iron(II) interference in the analysis. To obtain a reasonable background of absorption and a smooth baseline, $CuSO_4$ was added into the carrier solution. The $CuSO_4$ concentration was varied from 1×10^{-5} mol L^{-1} to 1×10^{-2} mol L^{-1}. When the concentration of $CuSO_4$ was 1×10^{-4} mol L^{-1}, the baseline was stable and the interference effects of nickel(II), cobalt(II), and zinc(II) were found minimum. Over the $CuSO_4$ concentration of 1×10^{-4} mol L^{-1}, the sensitivity of the method decreased.

In order to proceed with the final system design, the effects of sample volume, mixing coil length and flow-rate were studied at the optimal pH (2.85), and fixed concentrations of AcSHA (1×10^{-4} mol L^{-1}) and $CuSO_4$ (1×10^{-4} mol L^{-1}). The sample volume was varied from 5-50 µL. The peak height was decreased by decreasing the sample size, and the peaks were broadened with the increasing sample size due to the sample zone dispersion. The sample injection volume of 20 µL was selected as a compromise between the sensitivity and sample throughput rate. The mixing coil (MC) was examined using PTFE tubing (0.5 mm i.d.) of different lengths ranging between 10 cm and 150 cm. The peak height increased with the increasing mixing coil length from 10–50 cm, decreased at lower concentrations and broadened at higher concentrations and longer coil lengths. The mixing coil length of 50 cm was chosen since it resulted in the best peak height and good reproducibility.

The flow-rate was varied from 0.2 mL min^{-1} to 2 mL min^{-1}. The peak height decreased with the increasing flow-rate, probably due to the extent of the reaction decrease. The flow-rate of 0.8 mL min^{-1} was selected as a compromise between the sample throughput rate and sensitivity. A linear calibration graph for 4–150 µg L^{-1} iron(III), with the regression coefficient of 0.9914, was obtained under optimum conditions. The relative standard deviation for the determination of 5 µg L^{-1} iron(III) was 0.85 % (10 replicate injections), RSD of the data was below 3 %. The limit of detection (blank signal plus three times the standard deviation of the blank) was 0.5 µg L^{-1}. The sample throughput of the proposed method was almost 60 h^{-1}.

Tolerance limit (µg L^{-1})	Foreign ion
Over 50000	Cr(III), Al(III), Cd(II), Mn(II), K(I), Na(I), Ag(I), Ca(II), Mg(II), Ba(II), Hg(II), CN^-, NO_3^-, NO_2^-, SO_4^{2-}, CO_3^{2-}, Cl^-, Br^-, PO_4^{3-}, NH_4^+
Over 100	Fe (II)

Table 8. Effect of foreign ions on the determination of 5 µg L^{-1} of iron (III) in solution

The interference effects of many cations and anions on the determination of 5 µg L^{-1} iron(III) were examined. The results summarised in Table 8 represent tolerable concentrations of each diverse ion taken as the highest concentration causing an error of 3 %. Most of the ions examined did not interfere with the determination of iron(III). The major interference was caused by iron(II) at the amount of 100 µg L^{-1}. It is known that zinc and cobalt are the main interference metal ions in the determination of iron (Ensafi et al., 2004). In this study, the interference of these ions was completely eliminated by an addition of copper sulphate (1×10^{-4} mol L^{-1}) to the reagent carrier solution. Background absorbance of copper(II) maintained in the reagent carrier solution eliminated possible interfering ions and improved the determination of iron(III). It is apparent from Table 1 that the proposed method tolerates all interfering species tested in satisfactory amounts, and it is therefore adequately selective for the determination of iron(III) and total iron.

5.2.3 Applications

The FIA method was applied in the determination of iron(III) and total iron in water and ore samples. In order to evaluate the accuracy of the proposed method, the determination of total iron in a standard reference material (Zn/Al/Cu 43XZ3F) and in a metal alloy sample was carried out. The analytical results obtained by the proposed method are in good agreement with the certified values as shown in Table 9.

For the application of the proposed FIA method to water samples; river and sea water samples collected from different sources were analysed using both the calibration curve and the standard addition methods. The values obtained from the calibration curve and the standard addition methods are in good agreement as shown in Table 10. Atomic absorption measurements taken in water samples are also given for comparison (Table 10). The analytical value of total iron in water is in good agreement with that obtained by the AAS method.

Sample	Total Fe(1) (%)	Certified Fe (%)
Alloy (1)	8.23(0.24)	8.58
Alloy (2)	16.15(0.17)	16.62
Std Zn/Al/Cu 43XZ3 F	0.083(0.022)	0.085

(1) Values in parenthesis are the relative standard deviations for n=5 with confidence level of 95 %.

Table 9. Total iron content of iron alloys and standard reference material

Samples(1)	Iron (III)(2) (μg L-1)		Total iron(2) (μg L-1)		
	Found(3)	Found(4)	Found(3)	Found(4)	AAS
Kurtun river	38.33(0.24)	38.55(0.12)	42.33(0.02)	42.91(0.18)	43.65(0.17)
Seaport	78.84(0.32)	78.65(0.24)	95.13(0.12)	95.75(0.06)	97.12(0.12)
Baruthane sea water	47.51(0.18)	47.62(0.14)	57.24(0.04)	57.65(0.15)	58.97(0.24)

(1) Samples were collected at Samsun, Turkey.
(2) Values in parenthesis are the relative standard deviations for n=5 with confidence level of 95 %.
(3) Calibration curve method.
(4) Standard addition method.

Table 10. Determination of iron (III) and total iron in river and sea water samples

The results obtained show that the proposed method can be applied in the determination of iron(III) and total iron content in water samples without a preconcentration process.

6. Conclusions

A number of highly sensitive, selective and rapid flow-injection spectrophotometric and spectrofluorimetric analysis methods for the determination of iron (II), iron (III) and total iron in a wide concentration range, without employing any further treatment, have been described. The methods were based on the reactions of iron (II) and iron (III) with different complexing agents in different carrier solutions in FIA. In addition to the simplicity and low reagent consumption of the methods, the complexing agents used are commercially available and may not have a risk of serious toxicity, thus enhancing the potential applicability of the methods for iron analysis in real samples. Several parameters affecting to the determination of iron (II) and iron (III) were examined. The methods developed have been successfully applied to the determination of iron (II), iron (III) and total iron different types of water samples including river, sea, industry and spring water samples. The methods were also verified by applying certified reference materials.

7. References

Ahmed M. J. and Roy U. K. (2009) A simple spectrophotometric method for the determination of iron(II) aqueous solutions. Turkish journal of chemistry. 33, 709-726

Alonso J., Bartroli J., Valle M. D. and Barber R. (1989) Sandwich techniques in flow-injection analysis Part 2. Simultaneous determination of iron(II) and total iron. Analytica Chimica Acta, 219, 345–350.

Ampan P., Lapanantnoppakhum S., Sooksamiti P., Jakmunee J., Hartwell S. K., Jayasvati S., Christian G. D. and Grudpan K. (2002) Determination of trace iron in beer using flow-injection systems with in-valve column and bead injection. Talanta, 58, 1327–1334.

Andac M., Asan A. and Isildak I. (2001) Flow-injection spectrophotometric determination of cobalt/II) at low μg l^{-1} levels with 4-benzylpiperidinedithiocarbamate. Analytica Chimica Acta, 434, 143–147.

Andac M., Asan A. and Isildak I. (2009) A simple flow injection spectrophotometric determination method for iron(III) based on O-acetylsalicylhydroxamic acid complexation. Chemical Papers, 63, 268–273

Araujo A. N., Gracia J., Lima J. L. F. C., Poch M., Luica M. and Saraiva M. F. S. (1997) Colorimetric determination of iron in infant fortified formulas by sequential injection analysis. Fresenius' Journal of Analytical Chemistry, 357, 1153–1156.

Asan A., Andac M. and Isildak I. (2010) Flow injection spectrofluorimetric determination of iron(III) in water using salicylic acid. Chemical Papers, 64(4) 424-428.

Asan A., Andac M., Isildak I. and Tinkilic N. (2008) Flow Injection Spectrophotometric Determination of Iron(III) Using Diphenylamine-4-sulfonic Acid Sodium Salt. Chemical Papers, 62 (4), 345-349.

Asan A., Andac M. and Isildak I. (2003) Flow-injection spectrophotometric determination of nanogram levels of iron(III) with N,N-dimethylformamide. Anal. Sci., 19 (7), 1033-1036.

Asan A., Isildak I., Andac M. and Yilmaz F. (2003) A simple and selective flow-injection spectrophotometric determination of copper(II) by using acetylsalicylhydroxamic acid. Talanta, 60, 861–866.

Bagheri H., Gholami A. and Najafi A. (2000) Simultaneous preconcentration and speciation of iron(II) and iron(III) in water samples by 2-mercaptobenzimidazole-silica gel sorbent and flow injection analysis system. Analytica Chimica Acta, 424, 233–242.

Bruno H. A., Andrade F. J., Luna P. C. and Tudino M. B. (2002) Kinetic control of reagent dissolution for the flow injection determination of iron at trace levels. Analyst, 127, 990–994.

Busev A. I. , Tiptsova V. G. and Ivanov V. M. (1981) Anal. Chem. of rare elements, 385.

Cha K. W. and Park C. I. (1996) Spectrofluorimetric determination of iron(III) with 2-pyridinecarbaldhyde-5-nitropyridylhydrazone in the presence of hexadecyltrimethylammonium bromide surfactant. Talanta, 43, 1335–1341.

Cha K.. W. and Park K.. W. (1998) Determination of iron(III) with salicylic acid by the fluorescence quenching method. Talanta, 46, 1567–1571.

Chen J. Q., Gao W. and Song J. F. (2006) Flow-injection determination of iron(III) in soil by biamperometry using two independent redox couples. Sensors and Actuators B, 113,194–200.

Bowie A. R., Achterberg E. P., Mantoura R. F. C. and Worsfold, P. J. (1998) Determination of sub-nanomolar levels of iron in seawater using flow injection with chemiluminescence detection. Analytica Chimica Acta, 36, 189-200.

Currie L. A. (1995) Nomenclature ın evaluation of analytical methods including detection and quantiication capabilities. Pure and Applied Chemistry, 67, 1699-1723.

Dojlido J. R. and Best G. A. (1993) Chemistry of water and water pollution. Chichester: Ellis Horwood series in water and wastewater technology. Prentice Hall Inc. Englewood Cliffs, 21: 251.

Ensafi A. A., Chamjangali M. A. and Mansour H. R. (2004) Sequential determination of iron(II) and iron(III) in pharmaceutical by flow-injection analysis with spectrophotometric detection. Analytical Sciences, 20, 645-650.

Giokas D. L., Paleologos E. K.. and Karayannis M. I. (2002) Speciation of Fe(II) and Fe(III) by the modified ferrozine method, FIA-spectrophotometry, and flame AAS after cloudpoint extraction. Analytical and Bioanalytical Chemistry, 373, 237-243.

Guo T. and Baasner J. (1993) Determination of mercury in urine by flow-injection cold vapour atomic absorption spectrometry. Analytica Chimica Acta, 278, 189-196.

Hirata S., Yoshihara H. and Aihara M. (1999) Determination of iron(II) and total iron in environmental water samples by flow injection analysis with column preconcentration of chelating resin functionalized with Nhydroxyethylethylenediamine ligands and chemiluminescence detection. Talanta, 49, 1059-1067.

Isildak I., Asan A. and Andac M. (1999) Spectrophotometric determination of copper(II) at low μg l−1 levels using cationexchange microcolumn in flow-injection. Talanta, 48, 219- 224.

Kawakubo S., Natio A., Fujihara A. and Iwatsuki M. (2004) Field determination of trace iron in fresh water samples by visual and spectrophotometric methods. Analytical Sciences, 20, 1159-1163.

Kass M. and Ivaska A. (2002) Spectrophotometric determination of iron(III) and total iron by sequential injection analysis technique. Talanta, 58, 1131-1137.

Lunvongsa S., Takayanagi T., Oshima M. and Motomizu S. (2006) Novel catalytic oxidative coupling reaction of N,N-dimethyl-p-phenylenediamine with 1,3-phenylenediamine and its applications to the determination of copper and iron at trace levels by flow injection technique. Analytica Chimica Acta, 576, 261-269.

Lunvongsa S., Oshima M. and Motomizu S. (2006) Determination of total and dissolved amount of iron in water samples using catalytic spectrophotometric flow injection analysis. Talanta, 68, 969-973.

MacCarthy P. and Zachary D. H. (1986) Journal of Chemical Education 63 (3): 162-167.

Molina-Diaz A., Ortega-Carmona I. and Pascual-Reguera M.I. (1998) Indirect spectrophotometric determination of ascorbic acid with ferrozine by flow-injection analysis. Talanta, 47:531-536

Mohammad R. S., Deepika J., Jonathan P. K. and Keith A. H. (2004) Determination of labile Cu^{2+} in fresh waters by chemiluminescence: interference by iron and other cations. Talanta, 62: 924-930

Morelli B. (1983) Determination of iron(III) and copper(II) by zeroth, first and second derivative spectrophotometry with 2-thiobarbituric acid (4,6-dihydroxy-2-mercaptopyrimidine) as reagent. Analyst, 108, 870–879.

Mulaudzi L. V., van Standen J. F. and Stefan R. J. (2002) On-line determination of iron(II) and iron(III) using a spectrophotometric sequential injection system. Analytica Chimica Acta, 467, 35–49.

Muller H., Muller V. and Hansen E. H. (1990) Simultaneous differential rate determination of iron(II) and iron(III) by flow-injection analysis. Analytica Chimica Acta, 230, 113–123.

O'Brien E. C., Roy S. L., Levaillain J., Fitzgerald D. J. and Nolan K. B. (1997) Metal complexes of salicylhydroxamic acid and O-acetylsalicylhydroxamic acid. Inorganica Chimica Acta, 266, 117–120.

Ohno S., Tanaka M., Teshima N. and Sakai T. (2004) Successive determination of copper and iron by a flow injectioncatalytic photometric method using a serial flow cell. Analytical Sciences, 20, 171–175.

Qin W., Zhang Z. J. and Wang F. C. (1998) Chemiluminescence flow system for the determination of Fe(II) and Fe(III) in water. Fresenius' Journal of Analytical Chemistry, 360, 130–132.

Pascual-Reguera M. I., Ortega-Carmona I. and Molina-Diaz A. (1997) Spectrophotometric determination of iron with ferrozine by flow-injection analysis. Talanta, 44, 1793–1801.

Pojanagaron T., Watanesk S., Rattanaphani V. and Liawruangrath S. (2002) Reverse flow-injection spectrophotometric determination of iron(III) using norfloxacin. Talanta, 58, 1293-1300

Pons C., Forteza R. and Cerd`a V. (2005) Multi-pumping flow-system for the determination, solid-phase extraction and speciation analysis of iron. Analytica Chimica Acta, 550, 33–39.

Pons C., Forteza R. and Cerd`a V. (2005) The use of anion exchange disks in an optrode coupled to a multi-syringe flowinjection system for the determination and speciation analysis of iron in natural water samples. Talanta, 66, 210–217.

Pulido-Tofino P., Barrero-Moreno J. M. and Perez-Conde M. C. (2000) A flow-through fluorescent sensor to determine Fe(III) and total inorganic iron. Talanta, 51, 537–545.

Ragos G. C., Demertzis M. A. and Issopoulos P. B. (1998) A high-sensitive spectrofluorimetric method for the determination of micromolar concentrations of iron(III) in bovine liver with 4-hydroxyquinoline. II Farmaco, 53, 611–616.

Reguera M. I. P., Carmona I. O. and Diaz A. M. (1997) Spectrophotometric determination of iron with ferrozine by flow-injection analysis. Talanta, 44, 1793–1801.

Saitoh K.., Hasebe T., Teshima N., Kurihara M. and Kawashima T. (1998) Simultaneous flow-injection determination of iron(II) and total iron by micelle enhanced luminol chemiluminescence. Analytica Chimica Acta, 376, 247–254.

Sangi M. R., Jayatissa D., Kim J. P. and Hunter K. A. (2004) Determination of labile Cu2+ in fresh waters by chemiluminescence: interference by iron and other cations. Talanta, 62, 924–930.

Sultan S. M., Suliman F. O., Duffuaa S. O. and Abu-Abdoun I. I. (1992) Simplex-optimized and flow injection spectrophotometric assay of tetracycline antibiotics in drug formulations. Analyst, 117, 1179–1183.

Sultan S.M. and Suliman F.O. (1992) Flow-Injection spectrophotometric determination of the antibiotic ciprofloxacin in drug formulations. Analyst. 117: 1523-1526

Tamm L. K.. and Kalb E. (1993) Microspectrofluorometry on supported planar membranes. In S. G. Schulman (Ed.), Molecular luminescence spectroscopy. Methods and applications: Part 3 (pp. 303). New York, NY, USA: Wiley.

Tarafder P. K. and Thakur R. (2005) Surfactant-mediated extraction of iron and its spectrophotometric determination in rocks, minerals, soils, stream sediments and water samples. Microchemical Journal, 80, 39–43.

Teshima N., Ayukawa K.. and Kawashima T. (1996). Simultaneous flow injection determination of iron(III) and vanadium(V) and of iron(III) and chromium(VI) based on redox reactions. Talanta, 43, 1755–1760.

Teixeira L. S. G. and Rocha F. R. P. (2007) A green analytical procedure for sensitive and selective determination of iron in water samples by flow-injection solid phase spectrophotometry. Talanta, 71, 1507–1511.

Tesfaldet Z. O., van Standen J. F. and Stefan R. J. (2004) Sequential injection spectrophotometric determination of iron as Fe(II) in multi-vitamin preparations using 1,10-phenanthroline as complexing agent. Talanta, 64, 1189–1195.

Themelis D. G., Tzanavaras P. D., Kika F. S. and Sofoniou M. C. (2001) Flow-injection manifold for the simultaneous spectrophotometric determination of Fe(II) and Fe(III) using 2,2′-dipyridyl-2-pyridylhydrazone and a single-line double injection approach. Fresenius' Journal of Analytical Chemistry, 371, 364–368.

Udnan Y., Jakmunee J., Jayasavati S., Christian G. S., Synovec R. E. and Grudpan K.. (2004). Cost-effective flow injection spectrophotometric assay of iron content in pharmaceutical preparations using salicylate reagent. Talanta, 64, 1237–1240.

van Staden J. F. and Kluever L. G. (1998) Determination of total iron in ground waters and multivitamin tablets using a solid-phase reactor with tiron immobilised on amberlite ion-exchange resin in a flow injection system. Fresenius' Journal of Analytical Chemistry, 362, 319–323.

Weeks D. A. and Bruland K. W. (2002) Improved method for shipboard determination of iron in seawater by flow injection analysis. Analytica Chimica Acta, 453, 21–32.

Wirat R. (2008) Reverse flow injection spectrophotometric determination of iron(III) using chlortetracycline reagent. Talanta, 74, 1236–1241.

Yamamura S. S. and Sikes J. H. (1966) Use of citrate-EDTA masking for selective determination of iron with 1,10-phenanthroline. Analytical Chemistry, 38, 793–795.

Yan G. F., Shi G. R. and Liu Y. M. (1992) Fluorimetric determination of iron with 5-(4-methylphenylazo)-8- aminoquinoline in the presence of surfactants. Analytica Chimica Acta, 264, 121–124.

Yegorov D. Y., Kozlov A. V., Azizova O. A. and Vladimirov Y. A. (1993) Simultaneous determination of Fe(III) and Fe(II) in water solutions and tissue homogenates using desferal and 1,10-phenanthroline. Free Radical Biology & Medicine, 15, 565–574.

Yilmaz V. T. and Yilmaz F. (1999) Acetylsalicylhydroxamic acid and its cobalt(II), nickel(II), copper(II) and zinc(II) complexes. Transition Metal Chem., 24, 726-729.